THE WILEY PROJECT
ENGINEER'S DESK REFERENCE

THE WILEY PROJECT ENGINEER'S DESK REFERENCE
Project Engineering, Operations, and Management

SANFORD I. HEISLER, P.E.

A Wiley-Interscience Publication
JOHN WILEY & SONS, INC.
New York / Chichester / Brisbane / Toronto / Singapore

Library of Congress Cataloging in Publication Data:
Heisler, Sanford I.
     The Wiley project engineer's desk reference: project engineering,
     operations, and management / by Sanford I. Heisler.
          p.   cm.
     Includes index.
     ISBN 0-471-54677-1 (alk. paper)
     1. Engineering—Handbooks, manuals, etc. 2. Engineering—
Management—Handbooks, manuals, etc. I. Title.
TA151.H425   1993
     658.4'04—dc20   93-14694

Printed in the United States of America

10 9 8 7 6 5 4 3 2 1

To the Strongest and Quickest Mind It is
Far Easier to Learn than to Invent.
—*Samuel Johnson*

CONTENTS

PREFACE

Most of the time of the working engineer, as well as that of the engineering supervisor or manager, is spent dealing with operational engineering matters which differ significantly from the purely technical demands of projects. Engineers need to routinely deal with organizational, procurement, project and general management, contract, cost, estimating, financial, marketing, sales, and schedule matters. These activities represent the method of implementation for the engineer's technical work. While much operational work is not defined by fixed methods or formulas, and appears to be qualitative, it is at least partially quantifiable in terms of time or money.

This volume is an attempt to describe and to define these operational activities and to provide techniques and data to assist such work. Its purpose is to provide a reference that can be used to guide the engineer in areas that may not be totally familiar to him or her. I have tried to include enough data to also assist the more experienced engineers, supervisors, and managers. Some of the data will be a refresher to material they may not have worked with recently, and some will reflect more current practices in the workplace.

Numerous examples are included to illustrate the elements or factors that enter into and provide control of the work process or product. As one would expect, they vary in importance and sophistication from company to company and from project to project. In actual application, the documents would be tailored to the particular use intended. Samples of manually prepared documents for small or low-technology projects are shown. These manual documents represent the building blocks from which larger, more complex systems are developed, and they provide a convenient way to study system elements and expand them for more involved applications, often using computer data bases. For the reader who wishes to dig deeper, a list of references has been included.

Clearly, with the interrelated nature of project work, it is not possible to cleanly categorize aspects of it as only a schedule, cost, or procurement activity. The chapter titles used as a convenience in organizing should not be considered as absolute divisions for project operations. Similarily, with integrated systems for project control becoming more widespread, several functions formerly performed separately now are integrated into one form or subsystem. To facilitate their explanation, they are often described as if they were independent.

In deciding how best to organize the material in such a volume, I considered two approaches: One was to arrange the material in the order in which a typical project unfolds, and another to present the material in groupings based on the functional requirements of the activities. Each has its problems. The sequential approach is awkward, since activities overlap, are often performed in different sequences, and frequently parallel each other. The functional approach represents a somewhat restricted grouping, but real world organizations, particularly large ones, are often set up on a functional basis. Clearly the functional approach seemed the better way to arrange the material. However, in making this distinction, the operational dependency and interrelationship of the material could not always be precise. Consequently in some chapters additional related material on earlier topics is found under different headings. For this reason readers are encouraged to use the index to locate sections where supplementary information on a particular topic can be found. To facilitate presentation, the volume is divided into four principal areas: The Project, The Engineering Process, Project Operations, and Professional Practice. Within these major divisions the material is further subdivided into functional topics.

Because of the ever-increasing scope of technical knowledge, our colleges and universities are unable to include much operational information in their curricula. The typical engineer must acquire this knowledge in a fragmented, and often limited, way as a part of a continuing learning process extending over his or her professional career. In addition, given the fact that one rarely remains in the same field of work for an entire career, the engineer will continually be working in new or unique areas where a source of operational information can be useful. For some engineers responsibilities may broaden to include work in nontechnical managerial areas. Having a ready reference, such as this book on the standard practices of project engineering, will permit the work of the engineer to proceed in a more assured manner.

This volume represents a continuation and expansion of my earlier work *The Wiley Engineer's Desk Reference* © 1984 by John Wiley & Sons. Together these works cover the major areas of both the technology of engineering and its operational methodology.

With the broad and comprehensive scope of engineering knowledge there are few absolutes in the practice of engineering. In many respects, when a technical field joins an operational one, the implementation of the work is as much an art as a science. A successful project engineer or manager understands this and maintains a sense of balance between the conflicting demands of a project. While this balance can be learned by experience and exposure, some broad principles do apply. One purpose of this volume is to identify and highlight these principles so that they are not overlooked, particularly by the young practitioner.

While in practice there are differences in the details of techniques, the principles of application are the same, and the relevance of this material will be apparent. To reflect actual conditions found in operations, much stress is laid on the questions of schedule and schedule maintenance. In practice the work is often driven and controlled by schedule considerations, and many of the decisions made are a result of these factors.

This work would not have been possible without the assistance and cooperation of many people who gave freely of their time and knowledge to provide commentary and advice. But for them, this would still be merely an idea. Particular thanks are due to the personnel of ICF Kaiser Engineers, including Ray List, Shashi Bubna, Rich Nunes, Jr., and Chip Golde, as well as Bill Friend and Bernie Meyers of Bechtel Corporation, and Rose Gionta of Ebasco Services Inc. These companies were most generous in permitting use of their data including forms and standards. In addition the assistance and commentary of Mike Chessman and Linda Hook of Varian Associates, Bruce Thiel of Torchiana Mostrov & Associates, Wayne Edwards and Ahmet Taskpinar of Pacific Gas and Electric Co., Quade Hansen of Hughes Aircraft, Bob Templeton of M. W. Kellogg, Professors Bill Ibbs and Vic Cole of the University of California at Berkeley, Dan Ono of AT&T, Sanford Blank CPA, Ronald Winters CPA, Richard Radd Esq., Clyde Mitchell and Joe Velasco my old colleagues, and the many friends who patiently endured during preparation of this work has been most helpful. My editors Thurman Poston and later Frank Cerra of John Wiley and Sons, the publisher, have provided encouragement and great assistance throughout the sometimes agonizing process of writing a manuscript. The Wiley production team headed by Rose Ann Campise has also been a major source of help in the actual production of this volume. But perhaps most of all, without the active encouragement and assistance of my wife Lois, this work would not have been possible. I owe her a great debt.

Hopefully this volume will assist engineers in avoiding common pitfalls. To make this and future editions more useful, suggestions for improvement or expansion of this book are welcomed.

SANFORD I. HEISLER

Foster City, California

PART I

THE PROJECT

CHAPTER 1

THE PROJECT

The subject of **project engineering** covers a wide range of activities, at the core consisting of engineering but supplemented by extensive operational work. This includes activities such as schedule development and control, estimating, material acquisition, contractual and legal considerations, and human relations work. When put in the context of a project, the tasks become more extensive and varied and often involve the function of project management. All of these activities are undertaken to implement the goals of a project and the focus of the work of the project team members is directed to this end.

1.1 SCOPE

The range, scope, complexity, emphasis, and importance of these factors will vary widely between the public and private sectors, from industry to industry, and within an industry from project to project. Further, because projects flow from one phase to another without interruption, the activities may be continual or intermittent, occur at different times, or be repeated during the life of the project. As a result during the actual operations of a project there is a blending of the various activities and a blurring of the clean distinctions between both the phases of the project and the classifications of activities.

While many engineers and managers are engaged in work that is relatively open-ended, their activities are often also of a project nature. One definition of a **project** is that it has discrete boundaries, is a specific planned undertaking, and has associated with it considerations of time and cost. Projects are often of considerable size and exceed the ability of a single person, and thus for successful completion require groups of people, with different training and backgrounds. Projects can be large or small, but they all include some elements of data development, analysis of findings, resolution of uncertainties, and development and detailing of recommendations and results. Table 1.1 lists the major Project Functional Activities.

The different operational activities of a project are often categorized on a functional basis as follows: **Contractural** activities establish the basic legal, and commercial relationships between the parties and will in most cases also define the project and its important attributes. **Project definition** begins at or even before the project is established and continues over the life of the work. The process of definition loses its unique nature as the work proceeds and the project evolves into an actuality with all of its developed elements.

Project execution is the work performed to meet the project goals. Depending on the type of project, it may involve design, fabrication, manufacturing, test, training, development of software and special management or operational systems, or other specific activities. Often project execution requirements will include start-up of the facility or system, and testing and performance demonstrations. These activities may or may not be specifically called out in the contract.

Scheduling establishes key milestones and provides the method by which project performance is measured against time. It may also in some incentive-type contracts be integrated with performance payments.

Cost control provides the means by which the project's cost is **estimated** and converted into **budgets.** The budgets later, usually in conjunction with schedules, are used to measure and, in some cases, pay for actual performance. Often the cost control group is called upon to support financing activities, including the preparation of special economic studies.

Material acquisition covers all activities that deal with outside furnished materials. This includes the soliciting and obtaining of bids, their analysis, award of contracts and purchase orders for materials and equipment, their expediting and inspection, and shipping and receiving.

TABLE 1.1 Project Functional Activities
Contractural
Project definition
Project execution
Scheduling
Cost control
Material acquisition
Project control
External activities

Project control involves all activities that are put in place to control both the internal activities of the project team and the external activities of suppliers, subcontractors, and other parties. Among such items are the project procedures manual, project quality assurance manual, and engineering or other departmental procedures. Cost and schedule data also are two of the tools used in exercising control over the project.

External activities include all activities performed off project such as the obtaining of permits and licenses, training of client personnel, development and implementation of a public relations campaign, establishment of favorable community relations and similar activities.

Projects assume different forms and have differing emphasis placed upon design, drafting, procurement, and similar functions depending on their industry. Thus projects of a developmental nature such as those in the high-technology fields will have a larger portion of their effort spent on design and engineering analysis than, say, a process plant project where the technology is proven and where the challange is to minimize cost and maximize output and product quality. In the **defense sector** the project may often consist of a total design effort, including qualification of the design, construction of prototypes, and testing. In some cases the design team completes its project when its design is qualified by testing. This may be followed by what is considered a second project, the actual production of the item, frequently consisting of assembly of buyouts produced to the specifications developed by the design team.

In the **communications** and **electronics** industries, the projects tend to resemble those of the defense sector with a very large portion of the effort spent upon design itself and only a minor amount on other functions.

One common factor found in virtually all projects is the concern over minimizing schedule time. As a result there is a major emphasis to shorten or collapse the schedule through **concurrent engineering.** This occurs where the several engineering disciplines or engineering and production proceed with their work in parallel, rather than sequentially and thus complete their activities significantly sooner. As compared to totally sequential work, often savings of half of the schedule time can be achieved.

1.2   ROLES OF THE PROJECT ENGINEER AND PROJECT MANAGER

Projects are normally headed by an individual who often holds the title **Project Engineer** (PE), if a project largely involves engineering. The title **Project Manager** (PM) is commonly used if the task includes responsibility for activities besides those usually recognized as engineering, such as legal, accounting, and material acquisition. Where projects are headed by a PE, he or she is in responsible charge of the work undertaken and is usually the most senior person on the project. For large projects the PE or PM often works on the project full time and has that as his or her sole responsibility. For smaller projects the PE or PM may carry responsibility for a number of projects at the same time.

Where projects involve significant effort in other areas such as fabrication, construction, or extensive external activities, in addition to engineering itself, a PM in overall charge of the work is often named. Normally the PE is in responsible charge of the engineering work, with the PM handling administrative duties and other activities like scheduling, cost control, legal, and business operations.

As the leader of the team the PE is responsible for the direction of the work of the group. To this end it is necessary that the work be properly organized and scheduled to ensure that the design elements are performed in the proper sequence, that the design is completed on time, and that the necessary resources of personnel, including specialists, are employed to support the design work. A proper **work breakdown** is necessary to ensure that this occurs, with this same breakdown being used to monitor the progress of the group. A **schedule** is the usual tool for this, with various types of schedules used for controlling different parts of the project, certain schedules being **resource loaded** to ensure the requirements for personnel or materials are established.

Large projects often extend over a long time, and a significant number of personnel may be required to perform the engineering design, project commissioning, and testing, demonstration, and closeout activities.

1.3 PROJECT PHASES

The **phases** of projects have been defined in many ways but perhaps the most convenient is to separate a project into the **conceptual** and **definition,** the **execution,** and the **closeout** or **final** phases. With the exception of the initiation of engineering and the final acceptance of the project, the phases flow into one another, since later activities are dependent upon earlier step-by-step developments and decisions. Thus often different disciplines on the project find themselves in different phases of the project at the same time. Figure 1.1 illustrates the various phases both of a typical design and of a construction or fabrication project.

As the name implies, the conceptual phase covers the development of the concept for the project, including the basic decision of selecting a concept that will be used for the execution of the work.

For example, a project to develop a new aircraft engine might look at several engine types such as air cooled radial, air cooled in line, fluid cooled in line, and gas turbine prior to making a decision as to which type to develop further. The process of elimination to determine the final choice in some cases can be done by an inspection of the general performance levels and characteristics of the engines. In other cases more detailed economic and

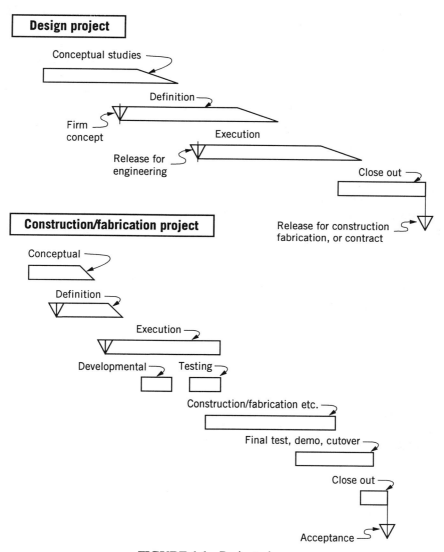

FIGURE 1.1   Project phases.

manufacturing studies may be required. Once the basic decision is made, the project passes from the conceptual phase to the definition phase where the design effort is directed to refining the concept in sufficient detail and firming up the approach to ensure the project is sound. In this way no major unknowns will reverse or significantly change the project's basic direction at a later date. In effect the definition phase reduces the risk of having formulated a wrong conceptual approach.

The normal development of the design has the effect of firming up **project** (or in the case of a manufactured product **life cycle**) **costs** relatively early in the development of the design. For this reason it is essential that the early decisions reached are sound both technically and economically. This is depicted in Fig. 1.2.

On a large or complex project there may be several points of the design that are subject to conceptual and definition phases. When these occur sequentially, a smooth evolution of the final design results. As mentioned earlier, however, the project schedule often does not permit a sequential development. Where parts of a design need to be developed simultaneously, separate teams may be established to work at the same time. This work may involve different aspects of the project or alternate approaches to specific areas. This method has the advantage of permitting a reduction in the elapsed schedule time, thus collapsing the schedule. It has the disadvantage of being somewhat more costly and possibly ending up with design studies that take somewhat different approaches or utilize different evaluation factors, making comparison difficult. While normally not a major problem, the parallel work effort should utilize uniform criteria or equivalent approaches to facilitate comparison of the alternates and the final decision(s).

The execution phase involves the actual design, fabrication, or construction. During this phase the detailed requirements are established and factors such as **"form, fit,** and **function"** are firmed up. The execution phase involves the issuance of firm information in the form of drawings and specifications

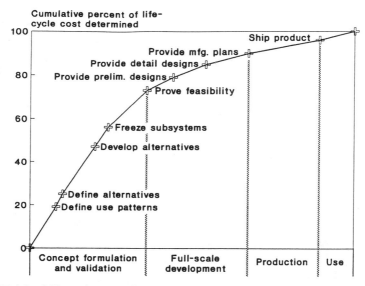

FIGURE 1.2   Life cycle costs. From *Concurrent Design of Products and Processes* by James L. Nevins and Daniel E. Whitney, copyright 1989. Reproduced with permission of McGraw-Hill, Inc.

to others outside the project, such as suppliers and subcontractors. It usually involves **commitments** to outside parties to purchase materials, perform product design, fabricate incorporated items, arrange for staffing and for numerous other activities. At this point significant changes have major effects, usually adverse, upon both schedule and cost because the design has already been determined. During the execution phase the project will normally be at its maximum size, and the space and support requirements peak. Often the execution phase will involve testing to evaluate the adequacy and performance of the design solution. This may be an iterative process with several stages of development.

During the closeout or final phase, the design is completed, fabrication or other activities are concluded, and **testing** and **demonstration** of the completed project, product, facility, or system takes place. The completion of these activities requires strict adherence to prior decisions and, if not earlier instituted, rigorous systems to ensure **control of** proposed **changes** are normally put in place. Among other things, **training** of operational personnel may be a requirement as well as demonstrations of performance capability. If not a factor earlier, extensive involvement with the client may now occur as preparations are made for the client to take over and operate the facility, system, or equipment. Wrap-up of technical loose ends normally also occurs, together with final contractural activities such as release of retentions, certification of capability, and formal contract closeout.

1.4 CONCEPTUAL OR DEFINITION PHASE

The conceptual or definition phase includes the activities earlier noted, with many more specific ones being required. The usual first task in undertaking a project is to name a Project Engineer or Project Manager. Following that, the PE or PM develop preliminary **staffing plans** that establish the first estimate or personnel required for the work. As an adjunct, preliminary estimates of facilities are drawn up for housing the project team and requirements are developed for tooling, fabrication, special modeling, and testing facilities.

The overall project parameters are defined, including the performance required, with preliminary flow sheets, logic, or block diagrams developed to identify significant functional or performance requirements. Then detailed preliminary design parameters are developed, including the space envelope required, power and utility requirements, overall criteria—such as operational modes, limitations on weight, forces, ambient conditions—identification of applicable codes, and overall approaches to automation. These considerations normally take the form of a **scope book** which documents the early definition of the project and is modified and expanded as the project proceeds. Also the scope book identifies significant areas that need additional work and any extensive definition. Thus, for example, if a radio telescope were under design, it would be essential to develop sufficient data on the structural system supporting the dish to ensure its integrity over a wide range of weather conditions. As a part of this concern it is necessary to define the maximum wind forces it can withstand, as well as maximum and minimum temperatures; lightning protection required; what probable loads will be imposed during erection and maintenance, and so forth. The list is likely to be long, but in general the longer the list, the firmer the design basis. The time taken initially to set down the criteria will be saved many times over during the project. With this information gathered in the scope book agreement on specific design requirements will be expedited and the design can proceed with less concern over errors or omissions.

If not already completed, the development and execution of a **contract** for the work must be undertaken at the earliest possible date. Where a PE or PM has been named, they can often greatly assist the contract development by providing input and commenting on contract provisions, particularly those that relate to feasibility of the project—performance, schedule, and similar matters. In some cases they may be called upon to participate in the contract negotiations.

Development of a preliminary **project schedule** is a priority item for this phase of the project. The schedule can be a simple 20-line bar chart schedule with only a relatively few **milestones** shown. While the number of milestones depends on the project data available, as few as 15 or 20 provide sufficient definition to permit this schedule to be used for the first one-quarter to one-third of the project life. The milestones should identify critical dates or activities that restrain other activities and establish the release of information or authorization to outside parties to take actions, such as major equipment commitments, test dates, and acquisition of permits.

Concurrent with the establishment of a project schedule is the development of an **order-of-magnitude** cost estimate. The estimate need not be in extreme detail, nor is extensive backup required. Rather historical data and rough estimates can be used to establish likely overall costs for the project and many of its components. The cost estimate may be of the maximum/minimum type. As with the schedule, the estimate will be further developed and detailed as project definition occurs.

From the project scope book, the schedule, and the order-of-magnitude estimate, data can be developed to support **project-financing** requirements, and firm commitments arranged. In many cases projects are funded from cash flows of ongoing operations, using the project schedule as the basis for developing a cash flow estimate. Where overall project funding is established initially, there is less pressure for cash flow projections at the beginning of the project, but fairly early in the life of the project, they will be required to permit effective financial management.

With their increasing numbers and often long lead times, early identification and scheduling of the **licenses** and **permits** required for the project are becoming increasingly important. As a consequence the project schedule should include them as one or more line items, with issuance of the most significant permits indicated as milestone events.

To control project activities and establish standard methods for the various project functions, early development of the **project procedures manual** should be undertaken. For projects where a formal quality assurance program is required, preparation of a project **quality assurance manual** should also begin at the earliest date. This will have the effect of minimizing the impact of retrofitting to demonstrate compliance with program requirements. Similarly, where extensive public involvement is anticipated, a public relations and/ or advertising program should be prepared.

Where several groups are involved, in addition to the normal engineering disciplines, a division of responsibility document should be prepared to ensure that each significant "deliverable" (i.e., completed piece of work) is assigned to a specific group. Where there are complex interfaces, the division of responsibility should identify the various activities: for example, concept formulation, review, input, outside review, costing, schedule evaluation, final approval and implementation.

As soon as the major equipment or buyouts are identified, and the preliminary schedule prepared, project components involving **long lead times** should be given priority in the preparation of specifications and bidding documents. This may require some out-of-step detailed early engineering. Since such components are the pacing ones in the schedule and require expedited work, this procedure is not particularly troublesome. To support these activities standard bidding forms and procedures are required, and their development has to proceed in parallel. Thus evaluation methods, suitable economic or operational factors, procedures including approvals, and similar activities need to be put in place to permit the orderly placement of early purchase orders, contracts, or subcontracts.

In parallel with many of the above-mentioned activities is the **mobilization** of the project team. Also there may need to be created a timetable for occupying project facilities for installing special communication or reproduction equipment, and for procuring and installing tooling or production equipment to support the project schedule. There are many other secondary items, but these are among those commonly considered as having primary impact, and hence priority, in the early phases of a project.

1.5   EXECUTION PHASE

Broadly the **execution phase** covers the development and detailing of the work begun in the conceptual/definition phase. Basically all of the activities are carried forward, though with different emphasis. For example, while contract development work may be concluded, maintenance and administration of project activities in accordance with the contract will be required.

More importantly, the execution phase involves the validation and selection of experimental or developmental features of the design. Depending on the type of project, these may be the pacing items on some projects. On others with sufficient schedule float, it may be possible to work around them by defining the likely **space envelope** (or other key parameters) to accommodate the final selection. Even if workarounds are possible, the alternates always seem to significantly affect secondary elements of the design, such as service utilities, and thus it is important to not overly delay their selection.

The **execution phase** of the project covers the continuation of the previous activities with particular effort on the work commonly thought of as typical engineering. There is sometimes a milestone date appearing on the schedules for the start of the execution phase—typically called "Release for Engineering." In this phase designs are developed, equipment data obtained or developed, drawings and specifications prepared, and the usual work of engineering pursued. This phase somewhat parallels that of the conceptual and definition phases in that it proceeds from the general to the specific.

Overall **system design** begins during this phase with emphasis on fulfilling these six steps:

1. Establishing criteria and standards.
2. Developing key documents.
3. Preparing preliminary drawing and specification lists.
4. Identifying the studies likely to be required.
5. Developing standard details for the various disciplines.
6. Establishing lists of drawings and specifications and other data for independent off-project review.

As work proceeds during this phase these documents will be maintained and updated to reflect the evolution of the project.

Detailed drawings prepared during this phase show the Area, Piping, Structural, Machine or Parts Design, Electrical Control Schematic, Wiring, Hydraulic, and Electronic aspects. All of these represent the development and detailing of the controlling drawings previously listed. These drawings typically number several times the quantity of drawings prepared during the initial phases of the project and are precisely developed to be suitable for manufacture, construction, release for purchase, and so forth. One general arrangement drawing can yield a family of drawings in each of the design disciplines—mechanical, piping, electrical, civil, instrumentation, and so forth—to define explicitly a configuration on the general arrangement drawing. The detailed drawings evolve from and are supported by quantities of calculations, supplier information on outside purchased materials, and equipment and development of design interfaces such as fit, finish, or arrangement. The supporting documentation represents refinement and evolution of the basic concepts. Obviously, then, the later in the design cycle changes are introduced, the greater their impact is on the design and the detailed drawings, and the more changes must be made. When changes must be made, there is always the possibility that changes required for other affected documents will be overlooked, and this greatly increases the likelihood of conflict, omission, error, and rework. For this reason it is critical that design decisions be reached early and only changed for compelling reasons. One way to control design changes is to establish a **design freeze** that places strict controls on the manner in which changes can be introduced. Figure 1.3 depicts the interrelationship between drawings and the development during various phases of the project. Note that not only do the number of drawings increase dramatically, but their firmness and detail increases as well.

Design, manufacture, or fabrication takes place during the execution phase when technical or other skills are mobilized to complete the work. As work progresses, detailed design, fabrication, and so forth, will naturally flow together, though on a phased basis, with essential early details being completed first.

Clearly schedule compliance is important and staffing the project with the correct number of appropriate personnel and equipment is critical. The staffing plan drawn up in the initial phase is updated and used as a guide, but with the understanding that, regardless of the plan, performance is the basic goal and adjustments may be necessary.

The preliminary schedule must be firmed up and established as a basic project schedule. Following that, most projects prepare a more detailed schedule usually of the **precedence** type with carefully chosen logic and **restraints** that depict the relationships and **dependencies** among activities. Suitable milestones are established, and activities having lengthy **durations** are subdivided into several smaller ones so that intermediate progress can be assessed. From these data a **critical path** is established, and with adjustment to sequences and workarounds to minimize the elapsed time, a final schedule is adopted and used to control project activities.

Similarly, and using in part, data from the project schedule, a **preliminary** and later a **definitive** project **estimate** is developed. Other inputs include quotations and estimates of equipment cost, developmental activities, fabrication, construction, and testing, activities. Like the project schedule, the project estimate will become the controlling budget for expenditures. To provide data for improved financial management, time-dependent **cash flow** estimates are developed to forecast cash requirements, their purpose, and timing. As the work progresses, billings for work are tendered.

The initial **procurement** activities during this period are the bidding, evaluation, and award of purchase orders, contracts, and subcontracts for long lead items. These are followed by orders for secondary items or activities, as they are developed by design activities. Finally, procurement of **bulk** items, such as nonfabricated piping, electrical conduit and cable, and miscellaneous instruments, takes place. At this point the review and processing of vendor drawings and data and administration of the technical aspects of material acquisition assume increasing importance. As a part of the design process, dimensional, performance and other data on suppliers equipment are reviewed their scope and content commented upon and resolved, and

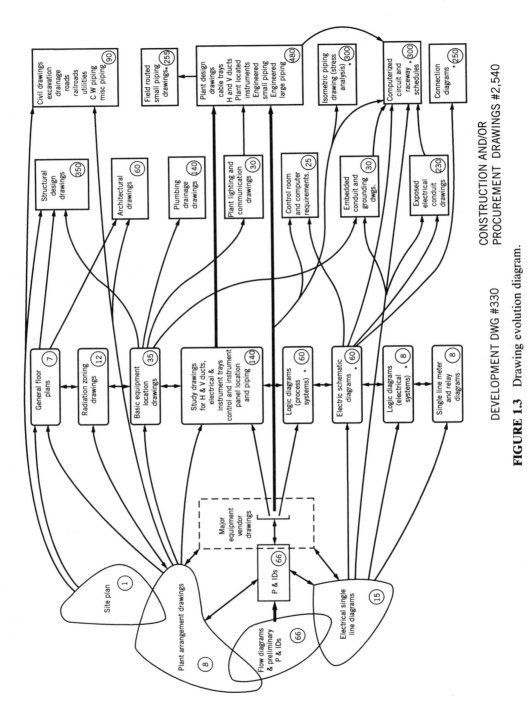

FIGURE 1.3 Drawing evolution diagram.

CONSTRUCTION AND/OR
PROCUREMENT DRAWINGS #2,540

DEVELOPMENT DWG #330

14

finally approved for fabrication and supply. The engineer's comments are incorporated into the detailed design, and any impacts on and adjustments to pricing and delivery of materials and equipment are administered on an ongoing basis.

The **quality assurance program** established earlier is now detailed by the preparation of specific procedures covering the activities in the program. The procedures are normally developed to cover the activities of the quality assurance group, which deals with external confirmation and assurance that the activities of the operational groups are following their approved methods. In addition **quality control procedures,** the internal controls on their own activities, are established by operational groups such as engineering, fabrication, and procurement. **Audit programs,** both programatic and technical, may be initiated during this phase to demonstrate conformance.

Most complex projects today require a significant training effort for the operational personnel. It is usual for selection and training of these personnel to begin during the execution phase of the project. Often these personnel are integrated into the design team, and this practice allows them to develop a fuller understanding of the systems, their features, trade-offs, and limitations than if they are trained at a later date.

1.6 CLOSEOUT OR FINAL PHASE

As the term implies, the **closeout** or final phase of the project concludes the major work and involves wrapping up all activities. In this phase all outstanding questions, decisions, and actions are resolved. The final phases of the engineering work will take different form depending on the activities involved. If, for example, the engineering work involves no procurement or follow-on work, such as construction or fabrication, the final phases will consist of assembling the engineering product, primarily drawings, and specifications into a suitable package or packages that may then be used to solicit bids or to direct the work of others.

If follow-on work is required, and the work is done by the same company as the design, schedules may be collapsed or overlapped, and the completion of the execution phase would not necessarily occur at a single date but rather consist of a series of releases for manufacture, purchase, or construction. In this instance the design engineering group will have a continuing responsibility and involvement with the other department(s). On large or complex construction projects there will usually be a clearly identifiable **testing** and **start-up** phase that follows the completion of construction. It is advantageous for design engineers to be assigned to this follow-on work, be it construction or testing and start-up work, because they have intimate knowledge of the design, can judge whether the performance of the components and systems is satisfactory, and can rapidly resolve fabrication or construction questions.

In the operational areas this phase involves finishing the activities and, more important, completing and assembling documentation for each activity. During this phase, when the stress is on finalization, loose ends are not permitted. To facilitate this, **punch** or completion **lists** may be prepared so that there can be an orderly closeout of open items. Pickup items are completed, and final corrections and adjustments performed so that the work can be turned over to the client. If not done earlier, plant or process **data books** are prepared, as well as **operating manuals,** if necessary. Operational personnel are trained during this period, and initial start-up of the project is performed. **Performance demonstrations** are held during this phase, with results used to repair or upgrade the plant or system to meet operational requirements. When all of these activities are complete, sale of surplus materials and equipment often occurs to permit some recovery of the values of these assets. Finally, from a contractural point of view, it is necessary to obtain some form of **release** indicating completion of the contract and with that release of retentions and final payment under the contract. For some firms it is also usual to prepare project historical data, with technical information, cost, and other data to assist future estimating and proposal work.

PART II

THE ENGINEERING PROCESS

CHAPTER 2

THE ENGINEERING PROCESS

The engineering process can be the most creative as well as demanding aspect of an engineers work. Understanding the impact of the laws of physics and their application to a particular design is the ultimate test of the skill of an engineer. While few engineers have difficulty applying their training in the design process, many feel the necessary documentation and administrative work to be distractions and impediments. Still the proper pursuit of operational activities is essential, and they give the engineer an opportunity to broaden professional development. The demand for absolute precision in this work as well as the technical side of engineering further compels the engineer to be more disciplined and rigorous in conducting project work.

2.1 DESIGN EXECUTION

The process of **design execution** involves myriad decisions that play off against each other. Besides detailed knowledge of the interaction of the design elements and the effect of the decisions made on each of them, the design process involves organizational decisions; judgments on purely technical questions; the preparation of calculations, drawings and specifications, reports, schedules, and cost estimates; participation in drafting, negotiating contracts, and the material acquisition; preparation of proposals; many routine tasks of personnel administration; and involvement in the day-by-day decisions of business operations. A significant effort also goes into the areas of the establishment of goals and the **leadership** of the project design team. In the manufacturing sector the way in which these factors are resolved may determine the success or failure of a product, and sometimes a company in the marketplace.

Many of the design considerations mentioned here are present in every project, and some projects may even involve all of them, depending on the scope of the work, size of the engineering organization and its charter. Being in responsible charge of the project, the project engineer (PE) carries the **overall responsibility** for the execution of the work, including motivation and direction of the project team, and is responsible for their assigned activities. While the question of responsibility is a large and serious one, the rewards, both tangible and intangible, from a successful project can be truly significant.

The PE must of course recognize that the primary responsibility to protect the **health** and **safety** of the public can not be compromised, and this consideration must be constantly borne in mind. State licensing laws are designed to ensure this performance, but many engineers are not licensed, and some states do not require that drawings or designs be sealed or even that the work be done under the direction of a licensed engineer. None of these factors, if present, relieves the engineer of the basic responsibility that the design not present a hazard to the public.

2.2 RESPONSIBILITY

It may sound elementary to say that a clear and unequivocal understanding of the responsibility for the execution of the engineering design is the cornerstone of a well-executed effort, but it is absolutely true. Many projects of vast scale and cost have foundered because of unclear or divided responsibility for performance of the design function. At the outset of the work it is essential to establish who is responsible for the design, or in the case of several technologies, who is responsible for what portion of the work, and in addition who is responsible for the overall direction of the work. Often this responsibility is reduced to writing in a project manual, which will include other matters as well.

Responsibility for the design includes several areas which inherently conflict with each other. The **trade-offs** include the effect of scope or complexity of the project, against the requirement for as short a project schedule as possible, with the concern to complete a project at minimum cost—all developed by a design effort that also has a minimum cost. **Scope** and **complexity** will often have the primary effect upon the project schedule and cost and, if the technology for the project is not developed, may involve extensive developmental work. The requirement for a short schedule can alone significantly increase costs particularly where the schedule is expedited by double shifting, parallel operations, and comparable techniques.

Maintenance of a minimum cost approach can affect the scope and technology employed as well as the degree of refinement developed in the design. Funding limitations on project costs often extend the schedule for multiyear projects. This can greatly increase the ultimate cost of the project, often by a factor of two or more. Finally, while a minor item, the cost of the design effort itself needs to be considered. A less economic design can, however, increase the project's overall cost. Inherent in the design effort are also concerns for technical adequacy, development of the design in the minimum time, improved reliability, and establishment of the required life cycle, as well as low maintenance requirements, and minimizing weight and utility requirements. All these attributes are played off against each other. As a result any design is a **compromise** and attempts to strike a reasonable balance between the design objectives. Figure 2.1 indicates the major trade-offs of complexity, cost, schedule, and design cost for project development activities.

An excellent checklist of design input considerations has been promulgated by the American National Standards Institute (ANSI) and the American Society of Mechanical Engineers (ASME) in standards N45.2.11 and NQA 1. Although developed primarily from a quality assurance point of view, because of their value and applicability, these standards have been widely adopted for general use. Figure 2.2 abstracts this material.

Of all the design objectives the most important is **technical adequacy.** If the design is not technically correct and cannot meet its performance requirements, the design is unsuccessful. For this reason at the beginning of the work the technical or performance criteria must be clearly established. These **criteria** need to be stated in specific quantified terms with ranges of accuracy given where appropriate.

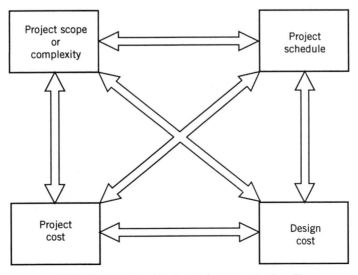

FIGURE 2.1 Project development trade-offs.

DESIGN INPUT CONSIDERATIONS
(Excerpt from ANSI N45.2.11 and ANSI/ASME NQA-1)

1. Basic functions of each structure, system and component.

2. Performance requirements such as capacity, rating and system output.

3. Codes, standards and regulatory requirements including the applicable issue and/or addenda.

4. Design conditions such as pressure, temperature, fluid chemistry and voltage.

5. Loads such as seismic, wind, thermal and dynamic.

6. Environmental conditions anticipated during storage, construction, and operation such as pressure, temperature, humidity, corrosiveness, site elevation, wind direction, nuclear radiation, electromagnetic radiation and duration of exposure.

7. Interface requirements including definition of the functional and physical interfaces and the effects of cumulative tolerances involving structures, systems and components.

8. Material requirements including such items as compatibility, electrical insulation properties, protective coating and corrosion resistance.

9. Mechanical requirements such as vibration, stress, shock and reaction forces.

10. Structural requirements covering such items as equipment foundations and pipe supports.

11. Hydraulic requirements such as pump net positive suction heads (NPSH), allowable pressure drops and allowable fluid velocities.

12. Chemistry requirements such as provisions for sampling and limitations on water chemistry.

13. Electrical requirements such as source of power, voltage, raceway requirements, electrical insulation and motor requirements.

14. Layout and arrangement requirements.

15. Operational requirements under various conditions such as plant startup, normal plant operation, plant shutdown, plant emergency operation, special or infrequent operation and system abnormal or emergency operation.

16. Instrumentation and control requirements including indicating instruments, controls and alarms required for operation, testing and maintenance. Other requirements such as the type of instrument, installed spares, range of measurement and location of indication are included.

17. Access and administrative control requirements for plant security.

FIGURE 2.2 Design input considerations. Reprinted by permission of American Society of Mechanical Engineers.

For example, a pump design project might begin with the criteria that it should have a capacity to pump a minimum of 125 U.S. gal. per minute of clear water, ranging from 40° to 120°F at a discharge pressure of 80 ft with a maximum suction lift of 10 ft. The pump shall be self-priming, powered by a gasoline engine with the entire unit mounted on a portable skid, able to be moved by two workers. It should have sufficient fuel capacity to run for one hour without refueling and must be able to operate outdoors in weather ranging from 20° to 140°F in direct sun as well as in heavy rain.

The specification above, while brief, provides the design team with a good bit of detail and answers many of the questions that may arise during the

18. Redundancy, diversity and separation requirements of structures, systems and components.

19. Failure effects requirements of structures, systems and components including a definition of those events and accidents which they must be designed to withstand.

20. Test requirements including pre-operational and subsequent periodic in-plant tests and the conditions under which they will be performed.

21. Accessibility, maintenance, repair and inservice inspection requirements for the plant including the conditions under which these will be performed.

22. Personnel requirements and limitations including the qualifications and number of personnel available for plant operation, maintenance, testing and inspection, and radiation exposures to the public and plant personnel.

23. Transportability requirements such as size and shipping weight, limitation and I.C.C. regulations.

24. Fire protection or resistance requirements.

25. Handling, storage, cleaning and shipping requirements.

26. Other requirements to prevent undue risk to the health and safety of the public.

27. Materials, processes, parts and equipment suitable for application.

28. Safety requirements for preventing personnel injury including such items as radiation safety, criticality safety, restricting the use of dangerous materials, escape provisions from enclosures and grounding of electrical systems.

29. Quality and quality assurance requirements.

30. Reliability requirements of structures, systems and components including their interaction, which may impair functions important to safety.

31. Interface requirements between plant equipment and operation and maintenance personnel.

32. Requirements for criticality control and accountability of nuclear materials.

FIGURE 2.2 (continued).   Design input considerations.

course of design. While a statement of the criteria cannot and should not take the place of proper design decisions, it can answer many of the questions that arise early in the effort and thus avoid false starts and misdirection of the work. It offers a further advantage, since the important performance requirements are clearly stated at the beginning of the work, that decisions will not be taken up piecemeal during the work causing dislocations. Further since the design process is often very hectic, if important criteria are left to evolve during the design work, it is difficult, if not impossible, to later determine when they were introduced, by whom, and the rationale behind them. Also important criteria, on many large or complex projects are subject to review and approval by **boards of review,** or similar bodies, and the value of these review steps is lost if they occur in a developmental fashion rather than given the consideration they merit.

Another major responsibility of the PE is to ensure that the design is **economical.** While every aspect of a design may not be optimized, it is important to develop a design that is viable from the point of view of cost. A design is merely a device for implementation by others, and as such, it must be responsive to its economic environment. Where the design is one of a kind such as a hydroelectric project, the economic decisions have large dollar value but are limited to a few items. For projects of this type there will be fewer but higher-value economic studies undertaken. For designs that affect **large volume** repetitive items, such as a manufactured product like a clothes dryer, the dollar value of the individual parts is smaller and the economic studies will lead to decisions based upon smaller unit cost differences. In this case many economic studies would be made to optimize costs for virtually all components. This is crucial because such items are sold in a commercial market where there may be severe price competition.

To some extent each project is unique and includes developmental aspects. This, together with the basic inclination of engineers to make improvements to designs, creates a continual temptation to refine the design rather than to freeze it, and thus permit its implementation. As a result there can be a significant impact upon the cost of the design as well as potential delay in the completion of the design and its release for implementation. Of course, if a market window is missed, and the product comes on the market after other competing products, a major adverse impact will be felt. Thus another important responsibility of the PE is to keep the project design development **on schedule** and to prevent design iterations that are of marginal value to the end product.

One concern familiar to all practicing engineers is the absolutely fundamental requirement of **attention to detail.** There is no area that will create difficulty more quickly than a design where the details have not been properly thought out or developed. Regardless of the intellectual prowess of the engineer who developed the concept or formulated the basic design solution, if the details are not well executed, the design will fail, and the entire project team will suffer, with the major blame being directed at the PE. Some engineers like to think of themselves as conceptual engineers who leave the detail work to others, they may make valuable contributions to the work of the project team, but the PE must be certain that there are sufficient resources on the team to properly execute these concepts and that details are properly and completely developed.

To permit the PE to effectively direct the work of the group, it is necessary that the work be carefully organized and scheduled. The design elements must be performed in the correct sequence, the design must be completed on time, and that the necessary personnel, including specialists, must be available to support the design work. A suitable work breakdown is necessary to ensure the planning of activities, with this same breakdown used to monitor the progress of the work. A **schedule** is the usual tool for this, with various types of schedules used for scheduling different parts of the project. Certain schedules are **resource loaded** to establish the requirements for personnel and material. More information on scheduling is found in Chapter 5.

2.3 DESIGN CHECKLISTS

The checklists that follow indicate the general types of tasks performed during the different phases of the project. While intended for a process plant design and not totally inclusive, they provide general guidance on placing design work into categories for other projects as well. Because of the large number of activities during the execution phase of the work, the checklists are subdivided into initial and follow-on engineering.

TABLE 2.1   Design Task Checklists

Definition Phase Design Checklist

1. Firm up detailed scope of work.
2. Complete special studies.
3. What are the plant base conditions?
4. Establish process flow sheet or cycle and/or alternates.
5. What is plant rating?
6. Is overload or "stretch" capacity required?
7. What are local, state, or governmental codes that affect
    Site and land usage
    Height of structures
    Stream or air pollution
    Thermal/chemical/biological/particulate/noise, or aesthetic limitations
    Civil/structural/seismic design
    Mechanical/electrical/nuclear design
    Safety or other special design codes
8. Water access or clearances required?
9. Establish ultimate development of site and project impact.
10. Are there overall funding limitations, or by function or on an expenditure/time basis?
11. Establish level of design detail required.
12. Are additional studies required to verify major design features?
13. What are client standards for
    Design/reliability/redundancy
    Control and communications
    Operational aspects/access/manning/habitability
    Interconnections
14. Are there unique design features such as computerized control?
15. Determine fuels/delivery methods/storage requirements.
16. Are data books or operating manuals included?
17. Is operator training included?
18. Which utilities or services are available on site?
19. Obtain site data, maps/borings, etc.
20. Is client design liasion required?
21. Are either static or dynamic models required?
22. What are site access requirements?

Design Task Checklists (continued)

---

23. Are there legal requirements regarding the practice of engineering, signing of drawings, etc.?
24. Are there special information security requirements?
25. Are special briefings or presentations to the client required?
26. What are the number, type, and timing of estimates?
27. Are cash flow forecasts required?
28. Are there special financial arrangements such as letters of credit, which may require technical support?

Execution Phase Design Checklist
General

1. Establish basic economic evaluation factors.
2. Set overall project schedule.
3. Firm up critical path schedule diagram for project.
4. Establish and schedule additional studies.
5. Reach understanding on use of outside consultants.
6. Establish drawing sizes and title blocks.
7. Set up equipment list and procurement control.
8. Set up control registers, drawing, foreign print, requisition, etc.
9. Identify client counterpart personnel.
10. Set frequency and type of client meetings.
11. Establish interim procurement procedure and bid list to let work or obtain major equipment.
12. Determine if client will require a resident(s) in design office.
13. Are permitting hearings required, and if so, which and when?
14. Establish type of construction:
    All subcontract
    Large contract packages
    Single prime contractor
15. Establish special communications required.
16. Establish client/project cost code.
17. Will client require a set of engineering, purchase order, or requisition files at end of job?
18. Hold client kickoff meeting.
19. Establish start-up requirements.
20. Begin cost trend work.
21. Determine extent of computerization:
    Control
    Data logging
    Entire plant or selected systems
22. Has client purchased or committed any major equipment.
23. Soilicit bids for major equipment.
24. Determine buyer of spare parts.
25. Develop preliminary plant arrangement—obtain client approval if necessary.
26. Establish general architectural concept for the plant.
27. Establish general routing of major piping, tray, etc.
28. Are special operator access or maintenance provisions required?
29. Develop handling or transportation methods for large or heavy loads or lifts.

Design Task Checklists (continued)

30. Establish level of design detail required by client if beyond normal standard.
31. Begin microfilming selected calculations and drawings

Mechanical

1. Begin development of P&IDs.
2. Establish material balance and basic heat balance.
3. Begin piping specification.
4. Size major equipment, prepare specifications, and obtain bids.
5. Establish basic heating, ventilating, and air-conditioning (HV&AC) system design.
6. Determine source and quality of makeup water.
7. Develop basic water treatment approach.
8. Establish extent of field fabrication of piping.

Civil

1. Perform site reconnaissance.
2. Initiate soils investigations.
3. Establish site access requirements and arrangements.
4. Begin foundation design and excavation drawings.
5. Begin superstructure design.
6. Establish structural framing system, column spacing, floor elevations, etc.
7. Firm up requirements for service building(s).
8. Prepare artists sketch of plant.
9. Solicit bids for piling and structural steel.[a]
10. Place bulk rebar order.[a]
11. Place miscellaneous iron order.[a]
12. Validate major process water systems design.
13. Begin design of major concrete structures.
14. Establish if concrete is to be supplied from an on-site batch plant or by transit mix.[a]

Electrical

1. Begin single line drawings.
2. Begin load study.
3. Size major electrical equipment, prepare specifications, and obtain bids.
4. Begin grounding arrangement.
5. Begin cathodic protection design.
6. Establish emergency power requirements and services.
7. Establish plant start-up power requirements.
8. Establish temporary power provisions.[a]

Administrative

1. Prepare project procedure manual.
2. Establish quality assurance program.
3. Prepare quality control procedures.

Design Task Checklists (continued)

4. Establish approved bidders list.
5. Set up calculation binders.
6. Establish procurement procedures.
7. Prepare preliminary cost estimate.
8. Begin monthly progress report.
9. Establish vendor print approval stamp and begin vendor drawing review.

Final Phase Design Checklist
General

1. Continue design for all engineering disciplines.
2. Freeze design and issue drawings for construction.
3. Set station services, cranes, elevators, tankage, etc.
4. Begin issuing advanced copies of drawings to field.[a]
5. Prepare plant data books.

Mechanical

1. Perform stress analysis and final design of critical piping.
2. Final design of all primary components.
3. Complete design of secondary equipment and auxiliaries.
4. Complete P&IDs.
5. Complete design of HV&AC systems, and procure equipment.
6. Continue design of piping systems with particular emphasis on vents, drains, and pipe hangers.

Electrical

1. Complete logic diagrams.
2. Begin electrical schematics.
3. Complete secondary electrical design, including low voltage circuits, lighting, etc.
4. Purchase secondary electrical equipment, small motors, lighting panels, small transformers, MCCs, etc.
5. Make the bulk purchase of electrical materials, cable, conduit, etc.[a]

Administrative

1. Continue vendor print approval.
2. Begin expediting and inspection report screening.
3. Prepare final cost estimate.
4. Terminate routine issuance of administrative control documents.
5. Obtain or issue completion letter to client.
6. Retire records, break down into categories, and destroy as procedures dictate.

---

[a] Turnkey work—where the design and construction organizations are part of the same company.

2.4 CALCULATIONS

It is normal practice that the basis for the design be clearly stated at the beginning of the calculation. Typically the calculation includes statements for items that are **given,** items that are **to be found** (or calculated), and the necessary **assumptions** used in the calculations. It is important that all calculations be **signed** and **dated** by the person performing them. Calculations must be legible and arranged in a logical sequence so that someone unfamiliar with them, performing an independent check, can readily follow them and verify their accuracy. Further the calculations should be so arranged that someone of experience equal to the originator can readily check them and determine their accuracy. Thus they should call out codes, standards, references used, and so on, so that both the originator and the checker can return to them at a later date, if necessary, and reestablish both their applicability and their accuracy.

All calculations should be checked. The checking may include any of these methods:

A rough rule-of-thumb check.

An independent separate check using different but equivalent formulas.

An estimate by a highly experienced person.

A detailed, meticulous check of each of the steps in the calculation.

Calculations can have different levels of accuracy depending on their end use. A calculation for a precision machine part might have a very high level of accuracy, while a calculation used in a gross analysis of earth to be moved for a large construction project might be suitable at a level of rough approximation only. It is important to state the level of accuracy desired in the calculations so that the usage is clear and the purpose of the calculation is clearly understood.

Calculations are often prepared on quadrille paper, usually of the vellum variety. This provides ease of reproduction by the ozalid as well as the photocopy processes. The quadrille ruling makes it convenient to prepare those sketches that are a part of the calculation.

With the development and widespread use of **personal computers** the vast majority of all the calculations performed in today's engineering offices are computerized. Three basic principles must be followed:

1. The calculation should be appropriate to the complexity and risk of the design.
2. Input data must be accurate and data entry must be error free.
3. The program used must be accurate and of proven applicability.

It matters little whether the calculation is performed on a large mainframe or on a PC because the major concern is the **suitability** of the program used to perform the calculation. Since the steps of the calculation are internal to the program and hence hidden from the view of the engineer, it is imperative to provide some form of control and overview of the programs used.

The normal program concerns include **validation** of the accuracy of the program, obtaining or establishing **documentation** on the program, and controlling the **distribution** and **use** of the program, including revisions or changes to it. To provide for this, larger firms usually establish a sponsor who controls and approves the program, approves all modifications to it, including options, and maintains its documentation, including establishing limits for usage. Since these programs are developed on a design discipline basis, it is usual to place this responsibility on the chief engineers of each of the design disciplines.

A key concern is obtaining or establishing the documentation for the specific program. This usually consists of three elements, the **user's manual** which instructs the user on how to enter data, establish limits, run the program, and interpret results; a **theoretical manual** which establishes the theoretical basis for the program; and a **validation report** which describes the testing of the program, its validation and any limits which may be necessary.

To ensure control, all program issuances, changes, modifications and any other activities concerned with the program, its accuracy, and applicability are under the direct control and guidance of the **program sponsor.** To provide assurance that the proper controls are being exercised both by the sponsor and user, a system of notices and receipts is often used for both the program issuance and any subsequent changes. Where formal quality assurance programs are in place, the system for documentation of these activities is mandatory and is rigorously followed. Normally only programs approved by the sponsor may be used.

Despite the rigorous controls described above, errors may occur or be detected either in the programs or in their use. To correct this problem, a system of computer error reporting is usually established. The system provides for the classification of errors and their correction. **Error classification** normally uses three levels:

1. Errors that have no effect on the validity of the results.
2. Errors that give obvious meaningless or absurd results.
3. Errors that can be interpreted as being valid.

When an error is detected, it is, until proved otherwise, classified as being in the third category. The program sponsor is immediately notified and normally requires that users suspend use of the program pending resolution of the problem. Where the program is shown to be incorrect, the program is corrected, and copies are distributed to the users. In addition, and most important, all **previous** calculations performed using the program are reviewed to determine the error's impact and fixed where necessary. Because of the far-reaching effect of an error, it is imperative that the program sponsor exercise extreme care when authorizing programs for use.

As mentioned earlier, all calculations should be independently checked and the date of the check and the checkers initials added to the calculation sheet. Where the checker has used a method different from the originator, it is helpful to briefly indicate the method used to verify that the calculation is correct.

Since calculations are performed for the benefit of and paid for by the owner, requests to provide them, or copies, to the owner or other outside parties should be complied with. Where calculations are provided to owners during the design process, frequent questions may arise which necessitate considerable discussion to resolve. This of course has an adverse impact upon the time of the senior project personnel and as a result may delay completion of the design. The best course of action is to provide the calculations at the conclusion of the project as a part of the close out activities.

Calculations should be oriented to using standard sizes of materials and equipment wherever possible. In general, the cost to produce a special off-standard size or type of material is prohibitive and, even if justified initially, often creates maintenance problems later.

2.5 DRAWINGS

While traditional hand drafting continues to be available the continuing expansion of **Computer Aided Design (CAD)** equipment capability has relegated it to either small projects or quick sketches. As a result many drawings today are prepared on CAD equipment.

A major advantage of the CAD drawings is their uniformity, legibility, and the ease of making additions and changes. CAD drawings do not suffer from handling or staining and do not become torn, dog-eared, or faded. Erasures are clean, and there is no problem with repeated handling. Where engineering is pursued concurrently with manufacturing or other external activities the ease of revision of drawings and preparation of bills of material greatly facilitate the work. In addition these drawings can be routinely transmitted electronically, thus speeding up the comment and distribution processes. CAD drawings can be manipulated electronically to produce bills of material and in some cases have **extraction** capability that permits the automatic preparation of specialized discipline drawings. Some CAD programs have a **reference file** feature that ensures that the different workstations using the system are using the same data base and that the drawing information is current. This avoids the problem of two or more engineering disciplines revising the same drawing and ending up with two different versions of it.

Experience to date indicates that for 2D (conventional drafting) CAD systems using a common data base with two or more workstations and with dedicated CAD operators, when the cost to amortize the computer equipment and programs are added to the actual cost of the time of the CAD operator, it is about 6 to 10% cheaper in cost then traditionally prepared hand-drafted drawings.

The initial investment in a central data-base-type 2D CAD workstation, having capacity to handle say 500 E size drawings, is significant on the order of $25,000 to $30,000 per workstation, including the CAD equipment, X-Y plotters, and software. While the cost of the workstation hardware is constantly declining, the cost of the software to drive them is tending to increase somewhat due to continuing expansion of capability. Today this increase is still less than the equipment decrease, resulting in a net reduction in cost. Formerly these systems were often operated in a two-shift mode: 16 hours per day and sometimes six or seven days a week to reduce the unit cost that must be charged to the drawing(s) produced. However, with the continuing reduction in overall cost and the increased productivity from advanced software, this is the exception rather than the rule.

To overcome the high cost associated with these type of systems, **stand-alone workstations** are available that range in price from $7,000 to over $100,000 depending on CPU, processing speed, memory, disk drive capacity, size and resolution of the monitor, number of monitors, and so forth. Centralized CPU-based systems are little used, and the current approach is to connect combinations of stand-alone workstations through a Local Area Network (LAN) by a file server. The file server normally has additional disk drives as well as plotters attached to it. This approach is useful on smaller installations, as small as two or three workstations where one of the workstations acts as the **file server.** The state of the art today is to utilize stand-alone workstations, with development pointed toward higher-capacity workstations in the future.

While most CAD work is performed as a replacement for conventional drafting, so-called 2D CAD, a new system of using **3D (three-dimensional) CAD** is becoming more common. The 3D CAD system is essentially an evolution of the 2D CAD systems, and much of the basic CAD technology and hardware is used. However, instead of developing drawings directly, a three-dimensional **mathematical model** of the item, or facility, is developed, and drawings are prepared by the computer based upon data resident in the mathematical model.

Both 2D and 3D systems begin with intelligent flow diagrams and P&IDs prepared using equipment symbol cells. A common data base that includes all information related to the specific piece of equipment is used, or developed. This information is used on the P&IDs, flow diagrams, and electrical schematics.

In 3D CAD many of the drawings are prepared by cutting sections through the three-dimensional model, the sections becoming the drawings. In some cases, for example, structural drawings, **automatic drawing extraction** is available that prepares structural steel plans and elevations from the data base, without showing other features of the design, such as piping and equipment. In addition to the preparation of orthographic and extracted drawings, 3D CAD permits the preparation of perspective and walk-thru-type drawings as well.

To construct the model requires that dimensional and other data on components, piping, structural systems, and similar elements be entered into the computer. In the case of equipment and many major components, a **library** of them is established and drawn upon for principal characteristics such as the space envelope, weight, connection size, and location, with the engineer, designer, or drafter providing orientation and centerline location data. Some equipment vendors are beginning to provide equipment data in **electronic format** that can be input directly into the electronic library for use by these 3D systems. For piping and wiring components, entering centerline location, routing, system identifiers, locations of valves, junction and pull boxes, and similar data permit the computer to establish all the significant parameters and incorporate their configuration into the design. The same approach is used with the structural systems and other elements of the design and creates the three-dimensional model in the computer. The preparation of **orthographic** design drawings from this model involves the cutting of sections thru the model and then printing them on **high resolution** plotters along with necessary dimensions and notes. The systems can produce ghost backgrounds to permit the highlighting of a particular type of work, such as piping. In addition it is possible to omit certain types of data such as structural systems from drawings to permit easier reading.

While the system only requires from 50 to 60% of the time of conventional drafting, there is a considerable investment in establishing the data base or library of standard equipment, piping, electrical equipment, structural members, and so on. The investment for a 3D CAD system that can produce designs for facilities in the $50,000,000 range is on the order of $150,000, to $200,000, which includes the cost for the equipment, operator training, and plotters. Additionally setting up the basic electronic library with equipment, piping, structural, and similar data can run as much as $100,000 or more. As a result the hourly rate for personnel, including overheads to recover the investment, would be somewhat higher than for conventional 2D CAD work but, with the larger reduction in hours, will usually cost on the order of 20 to 25% less for the design work overall. Initially the cost to utilize the 3D system is on the order of twice as much as conventional or 2D CAD work, but with successive projects the cost falls rapidly and after the third or fourth project achieves the values noted above. In some cases, to encourage the use of the 3D system, it is advantageous to keep the charges to the project for use of the CAD systems the same for both 2D and 3D equipment. One major advantage to 3D CAD is the reduction not only in the physical drafting effort but also the reduction in **interferences** in the field. The combined saving in these two items often equals the total cost of the engineering effort.

Figure 2.3 gives a cost comparison for different methods of drawing preparation and is reasonably representative of the economic factors present. Figure 2.4 shows the typical configuration of a multi-workstation 2D CAD system. To convert the system to 3D two of the PS2-90 workstations would be replaced with high-performance stand-alone workstations, and the PS2-

EXAMPLE PROJECT DOCUMENT PRODUCTION			
	MANUAL NO CAE	BASIC CAE 2D DRAFTING	FULL CAE 3D PDS
NUMBER OF DRAWINGS (1)	909	909	875
TOTAL MANHOURS (1)	101,324	88,861	65,073
HOURS/DRAWING	111	98	74
MANUAL DRAFTING HOURS (1)	24,925		
CAE HOURS (1)		12,463	21,546
TOTAL LABOR	$5,319,491	$4,976,771	$3,954,983
$/HOUR	$52.50	$56.01	$60.78
MULTIPLIER (2)	2.10	2.24	2.43
DIRECT LABOR $/HOUR	$25.00	$25.00	$25.00
SAVINGS FROM MANUAL		$342,720	$1,364,509
% SAVINGS FROM MANUAL		6.44%	25.65%
TOTAL LABOR (A)	$5,319,491	$4,976,771	$3,954,983
EQUIPMENT COST	$45,500,000	$45,500,000	$45,500,000
CONSTRUCTION COST (C)	$84,500,000	$84,500,000	$84,500,000
TOTAL PROJECT COST (B)	$130,000,000	$130,000,000	$130,000,000
% A OF B	4.09%	3.83%	3.04%
% A OF C	6.30%	5.89%	4.68%
SAVINGS FROM MANUAL DURING ENGINEERING DURING CONSTRUCTION		$342,720	$1,364,509 $1,690,000
TOTAL SAVINGS		$342,720	$3,054,509
% SAVINGS OF B	0.00%	0.26%	2.35%
% SAVINGS OF C	0.00%	0.41%	3.61%

NOTES:
1. The preparation of the following drawings have been excluded from the number of drawings, total manhours, manual drafting hours and CAE hours, above:
   o 470 Piping Isometrics
   o 130 Instrument Loop Diagrams
   o   20 Perspectives (3D General Arrangements) in FULL CAE 3D PDS, only
2. The cost of CAE has been included in the multiplier. The multiplier on labor, only, is 2.10. The cost of CAE is $25 per hour.

FIGURE 2.3 Drawing preparation cost comparison. Reprinted by permission of ICF Kaiser Engineers.

90 file server would be replaced with a high-capacity file server. In addition more disk drive capability would be required.

With either form of CAD, automated **checking** and **interference** identification are built into the system, and bills of material can be automatically produced. This is more extensive and accurate in the 3D than the 2D systems. In 2D CAD there is no physical tie between plans and sections or plans and elevations. Plans are only 2D, and no third dimension exists. As a result the quantity of material in the third dimension is an approximate allowance or has to be manually input. Similarly 2D CAD provides interference detection by overlay detection but, with different elements appearing at different elevations, may not be entirely accurate. As a result computerized data cannot be totally relied upon, and some manual checking is required. With either

Configuration

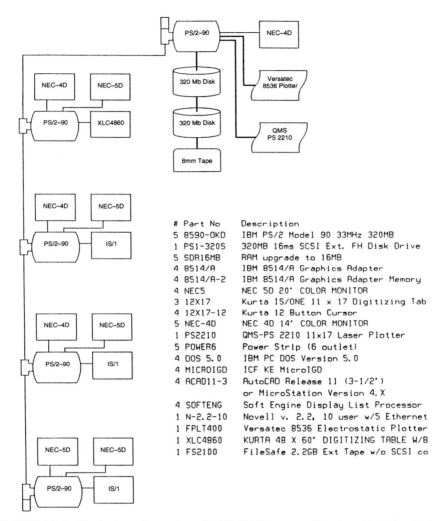

#	Part No	Description
5	8590-0KD	IBM PS/2 Model 90 33MHz 320MB
1	PS1-320S	320MB 16ms SCSI Ext. FH Disk Drive
5	SDR16MB	RAM upgrade to 16MB
4	8514/A	IBM 8514/A Graphics Adapter
4	8514/A-2	IBM 8514/A Graphics Adapter Memory
4	NEC5	NEC 5D 20" COLOR MONITOR
3	12X17	Kurta IS/ONE 11 x 17 Digitizing Tab
4	12X17-12	Kurta 12 Button Cursor
5	NEC-4D	NEC 4D 14" COLOR MONITOR
1	PS2210	QMS-PS 2210 11x17 Laser Plotter
5	POWER6	Power Strip (6 outlet)
4	DOS 5.0	IBM PC DOS Version 5.0
4	MICROIGD	ICF KE MicroIGD
4	ACAD11-3	AutoCAD Release 11 (3-1/2")
		or MicroStation Version 4.X
4	SOFTENG	Soft Engine Display List Processor
1	N-2.2-10	Novell v. 2.2, 10 user w/5 Ethernet
1	FPLT400	Versatec 8536 Electrostatic Plotter
1	XLC4860	KURTA 48 X 60" DIGITIZING TABLE W/B
1	FS2100	FileSafe 2.2GB Ext Tape w/o SCSI co

FIGURE 2.4   Typical configuration of a 2D CAD system. Reprinted by permission of ICF Kaiser Engineers.

form of CAD, **bills of material** can be automatically produced and then electronically input to requisitions to begin the purchasing process. Because of the greater memory and processing capacity, there is an exponential increase in this capability in the 3D systems. Phased drawings can be produced to indicate both sequence of operations and completion with time, enabling the preparation of drawings and diagrams to assist in scheduling and estimating. **Walk thrus,** which simulate the appearance of the facility, can be produced to permit review of clearances, access, and similar considerations. Architectural and other type presentation drawings can be prepared, and they are often of great value in dealing with licensing or permitting bodies.

One method of estimating the **overall design effort,** including design and drafting, assigns hours to drawings broken down by type and difficulty. In this method a detailed drawing list is prepared, and this forms the basis of the estimate. An indication of the general level of effort to produce these drawings by different types of drafting systems is shown in Chapter 5.

The requirement for checking of calculations stated earlier applies to drawings as well. It is important that the check truly be an independent one and that the checkers initials and date appear in the title or revision block of the drawing. A major advantage of CAD-prepared drawings, as described earlier, is their accuracy and identification of interferences.

The **sealing** of drawings by a licensed professional engineer is a common requirement of many jurisdictions and where drawings are presented for specific licensing and code actions. Sealing of the drawing should be performed only after all reviews and checks have been completed. The drawing is then presented to the engineer in charge who, if the drawing is satisfactory, will apply his or her seal (or more usually stamp). Increasingly outside reviews of project drawings are required. These range from general arrangement and presentation type drawings to those containing specific details of the design. The reviews may be mandated by governmental groups such as planning commissions, licensing and permitting agencies, clients who wish greater involvement in the project design, or groups concerned with the health and safety of the public. While it is useful to bring these parties into the design process early enough to permit their requirements to be integrated into the design development, it is not wise to bring them into the process until the design concepts and requirements have been reasonably well established and some preliminary engineering, sizing, and so forth, completed. Where the parties are brought into the process too soon, insufficient information is available, and often misdirection will result. If brought in too late, it is possible that some completed work will have to be redone. It is a delicate balance, but in general a bit earlier is better than a bit later.

On all but the smallest projects, during the course of design, **coordination** prints are required to provide a way to include comments/changes, and to update or exchange information from one part of the design group to another. These prints normally are shown with a check and a date in the revision block. Comments on these coordination prints should be dated and carry the name or initials of the commentor so that the originator can review the comments to ensure a clear understanding of them. Where coordination prints are reviewed by several persons, it is common practice for each reviewer to comment in different color, permitting ready identification. In some cases use of a formal internal routing stamp may be appropriate.

Regardless of the method of handling it is important that coordination prints be given high priority to ensure that change or corrections are rapidly provided to the originator and that the design process is not delayed.

When any drawing is issued to persons outside the design organization or a new revision is prepared, one hard copy should be marked a **"record print"** and retained in the project archives.

Graphic standards are an important part of the drafting requirements for any office. Standard sizes for drawings are $8\frac{1}{2} \times 11$ inches (outside sheet dimensions) or multiples thereof:

$$8\frac{1}{2} \times 11 \text{ inches} \qquad 11 \times 17 \text{ inches}$$
$$17 \times 22 \text{ inches} \qquad 22 \times 34 \text{ inches}$$
$$34 \times 44 \text{ inches}$$

This arrangement of sizes permits folding into letter size ($8\frac{1}{2} \times 11$ inches) for convenient mailing or filing in standard office files.

Drawing sheets using preprinted borders are preferred to hand-drawn borders and are typically available from engineering and drafting supply firms in the sizes noted; in many cases sheets with preprinted title blocks bearing the company name, standard numbering systems, and so forth, are stocked.

Scale requirements vary widely depending on the type and purpose of the drawing. Scales of $\frac{3}{32}$ inch and $\frac{3}{16}$ inch equal one foot should be avoided where possible. Since most drawings are reproduced, transparent drawing sheet material is typically used. Vellum has been the standard material used very satisfactorily for several decades. More recently Mylar drawing sheets have come into wide use. Mylar has the advantage of permitting more repeated erasures without damaging the transparent characteristics of the sheets, as well as cleaner erasures without "ghosts," and has greater resistance to tearing or creasing than vellum. The disadvantages of Mylar are its higher cost and the requirements to use special leads and erasers.

Some drafting practices that are economic and minimize error or confusion include:

Draw repetitive details only once and reference on other drawings.

Eliminate unnecessary views.

Make full use of industry standard details.

Use templates for drawing symbols and common shapes.

Cross-referencing must be accurate.

Independent checking is essential and should be indicated in the sign-off block.

Details of backgrounds should be omitted unless essential to a clear understanding and use of the drawing.

2.6 DRAWING TYPES

Drawings prepared during the conceptual or developmental phases of a project establish the overall controlling arrangement and configuration of the project or system and as such set the framework for the development of detailed drawings during the execution phase. Figure 1.3 depicted this evolution and the interrelationship of their development. These drawings are typically of the following types: Flow Diagrams, Block Diagrams, Material and Energy Balances, Piping and Instrumentation Diagrams, Logic Diagrams, and Single-Line Electrical Diagrams. Examples of Piping and Instrumentation, Logic and Electrical Single-Line Diagrams are shown, together with their typical legends, as Figs. 2.5 through 2.10.

2.7 OVERALL CONTROL

Overall control of the design process is achieved by the same methods used to control entire projects. This includes review of progress against schedule, cost expended against plan, and personnel available versus requirements. In addition the evaluation should include an overview of the status of the work, incorporating all of the foregoing plus an overall judgment of the general status and outlook. This should include factors not reflected in the formal control documents but which have the potential to significantly affect the progress of the work. Typical of these may be the likelihood of cash flow limitations on design expenditures or the possibility of changes in licensing requirements. The actual methods of analyzing the work and exercising control are described in more detail in Chapter 3.

Progress reviews are often held with clients or licensing bodies to provide information on design progress and in some cases to reach agreement on alternate approaches. Such meetings should be carefully structured and detailed minutes kept and distributed to the parties to ensure a common understanding of decisions reached. This is particularly important because contractual issues are often involved and a clear record is essential to proper contract administration.

2.8 CLIENT INVOLVEMENT

Client involvement is for most design work a highly desired relationship. It permits incorporation of the wishes and experiences of the client as the design evolves and, where required, can facilitate their approval and agreement.

FIGURE 2.5 Selected P & ID legend.

TYPICAL INSTRUMENT LEGEND

Symbol	Measured or Initiating Variable	Primary Element	Transmitter Blind	Transmitter Indicating	Recorder	Indicator	Alarm Low	Alarm High	Alarm High & Low (Note 5)	Controller Blind	Controller Indicating	Controller Recording	Control Valve (Note 6)	Self Actuated Valve	Switch (Note 3)
A	Analysis (Note 2)	AE	AT	AIT	AR	AI	AAL	AAH	AAHL	AC	AIC	ARC	AV		AS
B	Burner Fire & Flame	BE						BAH							
C	Conductivity	CE	CT	CIT	CR	CI	CAL	CAH	CAHL		CIC	CRC	CV		CS
D	Density	DE	DT	DIT	DR	DI	DAL	DAH	DAHL	DC	DIC	DRC	DV		DS
E	Voltage (EMF)	EE	ET	EIT	ER	EI	EAL	EAH	EAHL	EC	EIC	ERC			ES
F	Flow	FE	FT	FIT	FR	FI	FAL	FAH	FAHL	FC	FIC	FRC	FV	FCV	FS
G	Gaging	GE	GT	GIT	GR	GI	GAL	GAH	GAHL	GC	GIC	GRC			GS
H	Hand (Manual)									HC	HIC	HRC	HV	HCV	HS (Note 9)
I	Current	IE	IT	IIT	IR	II	IAL	IAH	IAHL	IC	IIC	IRC	IV	ICV	IS
J	Power	JE	JT	JIT	JR	JI	JAL	JAH	JAHL	JC	JIC	JRC			JS
K	Time	KE	KT	KIT	KR	KI	KAL	KAH	KAHL	KC	KIC	KRC	KV		KS
L	Level	LE	LT	LIT	LR	LI	LAL	LAH	LAHL	LC	LIC	LRC	LV	LCV	LS
M	Moisture	ME	MT	MIT	MR	MI	MAL	MAH	MAHL	MC	MIC	MRC	MV		MS
N	Unclassified (Note 4)														
O	Torque	OE	OT	OIT	OR	OI	OAL	OAH	OAHL	OC	OIC	ORC			OS
P	Pressure	PE	PT	PIT	PR	PI	PAL	PAH	PAHL	PC	PIC	PRC	PV	PSV (Note 7)	PS
PD	Pressure Differential		PDT	PDIT	PDR	PDI	PDAL	PDAH	PDAHL	PDC	PDIC	PDRC	PDV	PDCV	PDS
Q	Quantity or Event		QT	QIT	QR	QI	QAL	QAH	QAHL	QC	QIC	QRC	QV		QS
R	Radioactivity	RE	RT	RIT	RR	RI	RAL	RAH	RAHL	RC	RIC	RRC	RV		RS
S	Speed or Frequency		ST	SIT	SR	SI	SAL	SAH	SAHL	SC	SIC	SRC			SS
T	Temperature	TE	TT	TIT	TR	TI	TAL	TAH	TAHL	TC	TIC	TRC		TCV	TS
TV	Television	TVE	TVT	TVIT	TVR	TVI				TVC	TVIC	TVRC			TVS
U	Multivariable		UT	UIT	UR	UI	UAL	UAH	UAHL	UC	UIC	URC	UV		US
V	Viscosity	VE	VT	VIT	VR	VI	VAL	VAH	VAHL	VC	VIC	VRC	VV		VS
W	Weight (Note 10)	WE	WT	WIT	WR	WI	WAL	WAH	WAHL	WC	WIC	WRC	WV		WS
X	Unclassified (Note 4)														
Y	Object or Motion Sensor	YE	YT	YIT	YR	YI	YAL	YAH	YAHL	YC	YIC	YRC	YV		YS
Z	Position	ZE (Note 11)	ZT	ZIT	ZR	ZI	ZAL	ZAH	ZAHL	ZC	ZIC	ZRC			ZS

NOTE:

1. THE INSTRUMENT LEGEND IS BASED ON ISA STANDARD S5.1 1975.

2. THE LETTER "A" IS USED FOR ALL ANALYSIS VARIABLES. TERMS ARE PLACED OUTSIDE THE INSTRUMENT CIRCLE OF A LOOP TO DENOTE THE SPECIFIC VARIABLES. SOME EXAMPLES ARE

   CO — CARBON MONOXIDE DO — DISSOLVED OXYGEN
   COMB — COMBUSTIBLES O_2 — GASEOUS OXYGEN
   H — DISSOLVED HYDROGEN pH — PERCENT HYDROGEN
   H_2 — GASEOUS HYDROGEN Cl_2 — CHLORINE
   Na — SODIUM SMOKE — SMOKE DENSITY
   M — METHANE SO_2 — SULPHUR DIOXIDE
   NO_x — NITROGEN OXIDES TURB — TURBIDITY
   TSP — TOTAL SUSPENDED PARTICULATE

3. A DEVICE THAT CONNECTS, DISCONNECTS, MODIFIES OR TRANSFERS ONE OR MORE CIRCUITS MAY BE EITHER A SWITCH, A RELAY, OR AN ON-OFF CONTROLLER, DEPENDING ON THE APPLICATION.
   - A SWITCH IF IT IS ACTUATED BY HAND OR THE DEVICE IS USED FOR ALARM, PILOT LIGHT, SELECTION, INTERLOCK, OR SAFETY.
   - A CONTROLLER IF THE DEVICE IS USED FOR NORMAL ON-OFF OPERATING CONTROL, SUCH AS A SIMPLE HEATING THERMOSTAT.
   THE LETTERS H AND L ARE ADDED TO THE MEASURED VARIABLES FOR HIGH AND LOW RESPECTIVELY. LETTER AS FOR ALARMS (LSH, PSL, ETC)
   - A CONTROL OR SENSING DEVICE HAVING A DISPLAY FUNCTION SHOULD HAVE THE APPROPRIATE DISPLAY LETTERS ADDED AFTER THE MEASURED VARIABLE DESIGNATION. E.G. A/C DESIGNATES ANALYSIS INDICATING CONTROL STATION.

4. AN UNCLASSIFIED LETTER MAY BE USED FOR UNLISTED MEANINGS THAT WILL BE USED REPETITIVELY ON A PARTICULAR PROJECT. THE MEANING WILL BE DEFINED ONLY ONCE FOR THAT PROJECT AND HAVE ONE MEANING AS THE FIRST LETTER AND ANOTHER SINGLE MEANING AS THE SUCCEEDING LETTER.

5. HIGH-HIGH AND LOW-LOW ALARMS HAVE ( ) AHH AND LOW AND LOW ALARMS HAVE ( ) ALL IN INSTRUMENT CIRCLE E.G. LAHH — DESIGNATES HIGH-HIGH LEVEL ALARM. LALL — DESIGNATES LOW-LOW LEVEL ALARM.

6. VALVES
   - IF A DEVICE MANIPULATES A FLUID PROCESS STREAM AND IS NOT A MANUALLY ACTUATED ON-OFF BLOCK VALVE, IT SHALL BE DESIGNATED AS A CONTROL VALVE.
   - A HAND CONTROL VALVE HCV IS A MANUALLY ACTUATED VALVE THAT EITHER MODULATES (THROTTLES) A PROCESS STREAM OR IS USED AS AN INSTRUMENT DEVICE.
   - MOTORIZED VALVES ARE DESIGNATED THE SAME AS OTHER CONTROL VALVES. E.G. FV, PV, HCV, HV ETC
   - AN ON-OFF VALVE REMOTELY CONTROLLED BY A HAND SWITCH IS DESIGNATED AS A HAND VALVE HV.

7. THE DESIGNATION PSV APPLIES TO ALL VALVES INTENDED TO PROTECT AGAINST EMERGENCY PRESSURE CONDITIONS. RUPTURE DISCS SHALL BE DESIGNATED PSV.

8. USE OF MODIFYING TERMS HIGH, LOW, AND MIDDLE OR INTERMEDIATE SHALL CORRESPOND TO VALUES OF THE MEASURED VARIABLE, NOT OF THE SIGNAL, UNLESS OTHERWISE NOTED.

9. SPECIAL SWITCH CONTROLS SHALL BE DENOTED IN POSITION C OF SWITCH SYMBOLS
   JOG — JOGGLE CONTROLS
   KEY — KEY LOCKED SWITCHES

10. LOAD AND PRESSURE CELLS ARE USUALLY DENOTED "WE"

11. DENOTES STRAIN GAUGE

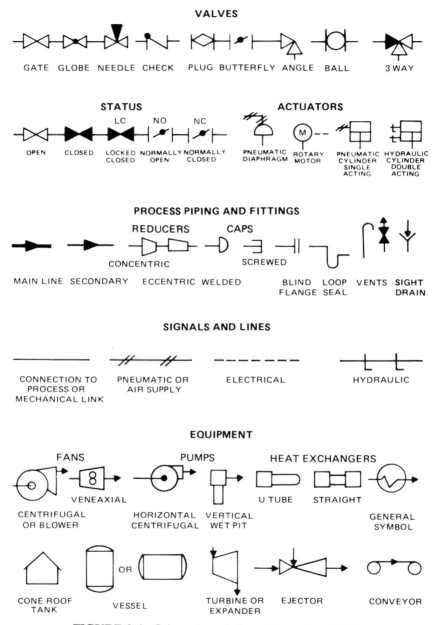

FIGURE 2.6 Selected symbols and legend on P&IDs.

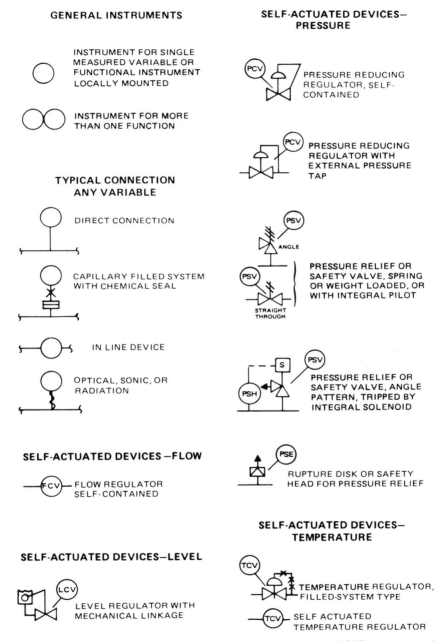

FIGURE 2.6 (continued). Selected symbols and legend on P&IDs (*continues*).

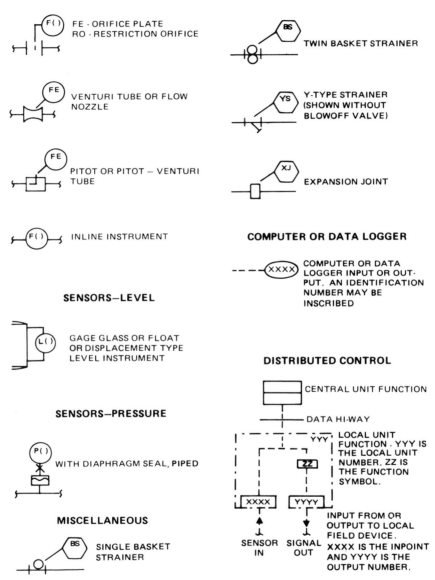

FIGURE 2.6 (continued).   Selected symbols and legend on P&IDs.

There are some potential pitfalls on the extent and type of client involvement which should be constantly borne in mind. Perhaps the most basic is
whether at the outset a proper scope description of the work exists. While
it should not be a drawn-out process, the preparation of a written **scope**

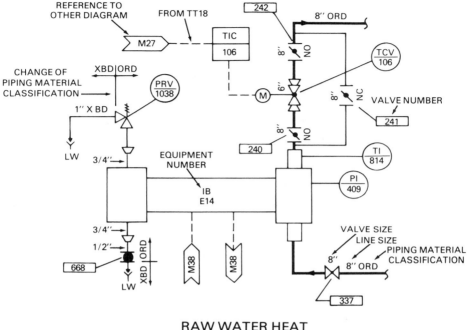

RAW WATER HEAT
EXCHANGER

FIGURE 2.7   Typical P&ID.

document should be the first project work item. On major projects client representatives will often be located in the design office and may wish to become involved in the scope development. This is hazardous as scope developed by client representatives is often fragmented, not balanced, and may even be contrary to what the parent (client) organization really wants. If the client wishes their representatives to take this role, there should be a clear understanding of the authority and responsibility that their representatives are given. This arrangement should be put in writing to avoid later misunderstandings. Review and approval of the scope document, including its later revisions, should be arranged between the PE and his or her client counterpart, be relatively formal, and be documented to avoid later misunderstandings.

  If the design work is on a tight schedule, as they always seem to be, client representatives who are often technical specialists may appear to be indifferent to the schedule requirements and delay the design work by asking for more studies and analyses than contemplated. Further, where the client has an approval role, delays often ensue, adversely affecting the design schedule. Typically this occurs because the client personnel have little or no responsibility for overall schedule performance and thus do not share the same set of priorities as the project team.

GENERAL NOTES

1. Logic symbols represent system functions and do not necessarily duplicate circuit arrangement or devices. System control logic diagrams do not inherently imply energized, de-energized, or other circuit operation states.

2. Process equipment will change state when a change is initiated, and will remain in that state until a change to another state is initiated.

3. Process equipment will remain in, or return to, the original state after a loss and restoration of power, unless otherwise noted.

4. Inherent equipment interlocks such as circuit breaker trip free and reversing starter cross interlocks are not shown.

5. Some protection actions are shown also as start permissives. Trip free design prevents equipment operation whenever a protection action exists, even if a start permissive is not provided.

6. Final instrument set points are shown elsewhere. Set points shown on system control logic diagrams are approximate.

7. See electrical drawings for details of equipment electrical overcurrent, short circuit, and differential protection and space heaters.

8. The memory, reset, and start permissive logic associated with the operation of electrical protection devices is not shown. Electrical auxiliary system breakers are reset by operation of the control room switch to trip. Mechanical auxiliary system circuits are reset by operation of a switch at the switchgear or motor control center.

9. The test control switches at the switchgear which function only when a circuit breaker is in the test position are not shown.

10. All circuit controls, except interlocks with other equipment, function when a circuit breaker is in the test position to allow circuit testing.

11. The logic to show that valve and damper position lights are both on when the equipment is in an intermediate position is not shown.

12. Limit and torque switches to stop valve and damper motor actuators at the end of travel are not shown in the logic. The valve type and required actions will be noted on the diagram when available.

ABBREVIATIONS

C01	– Unit control panel
C02	– Auxiliary control panel
C03	– Hot shutdown panel
L	– Local to controlled equipment
MCC	– Motor control center
SWGR	– Switchgear

Function	Symbol	Definition
MANUAL INPUT		Momentary hand switch input to logic
		Maintained hand switch input to logic
AND		Output exists only when all inputs are present.
OR		Output exists only when one or more inputs are present.
NOT		Output exists only when input is not present.
ON DELAY		Output exists only when input has been continuously present for a preset time and remains present.
OFF DELAY (TIMED MEMORY)		Output exists only when input is present and for a preset time after the input is not present.
MEMORY	S / R	Set output exists when set input is present and continues until the reset input is present. Reset output exists only when set output is not present.
COINCIDENCE MATRIX	A/B	Output exists only when at least A out of B inputs are present.
LOW BISTABLE	S.P.	Digital output exists only when analog input is lower than set point.
HIGH BISTABLE	S.P.	Digital output exists only when analog input is higher than set point.
ISOLATION	ISO	Output is electrically isolated from input.
TEST DEVICE	T · R TEST	Test signal can be inserted manually in place of normal signal.
LIGHT		RED–Operating, flowing, or increasing GREEN–Not operating, not flowing, or decreasing AMBER–Automatic, standby, or intermediate WHITE–Manual or protective trip BLUE–
ANNUNCIATOR		Input to annunciator
COMPUTER		Input to computer
CONTINUATION		Logic continuation

FIGURE 2.8 Logic diagram legend.

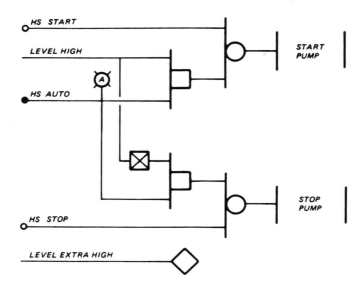

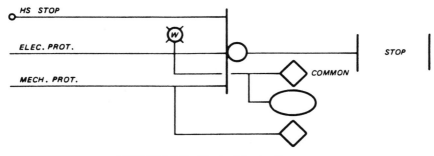

FIGURE 2.9 Typical logic diagram.

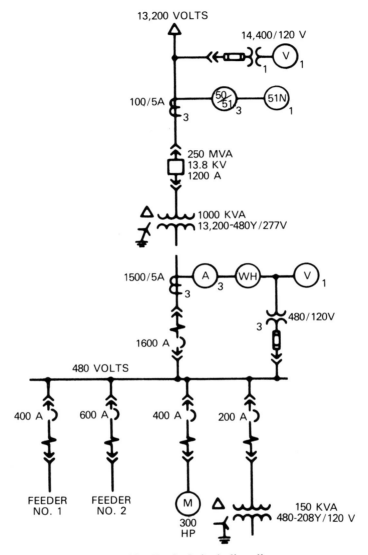

FIGURE 2.10   Typical single line diagram.

Where the design work is being performed on a lump sum or **fixed price** basis, the design organization can take the position that, since the work is being performed for a fixed price, the total responsibility for execution lies with the design organization and the client has no role in the design effort apart from providing design criteria and standards as required in the contract. While true, this approach is generally unacceptable to a client, and it is reasonable to expect some involvement by them even in this type work.

The best solution, in either instance, is to reach early agreement with the client as to what degree of involvement will exist, the responsibility and authority of client personnel, and, most important, the specific approvals required, their timing, and the time (duration) the client requires for review, from submittal until approval. This will tend to introduce more discipline into client activities and will highlight the responsibility they assume by virtue of their review and approval role. This process can be given additional visibility by incorporating these activities in the project schedule which, if also signed by the client, additionally validates their understanding of their role and responsibility.

It should also be remembered that the work is being undertaken for the benefit of the client and thus that the two organizations have the same goal. This commonality can often be used effectively to reduce some of the polarization that might develop otherwise.

The client will often furnish both **criteria** and **basic process** information to the project. This may be in the form of Flow Diagrams. Material and Energy Balances, or other documents—in some cases well developed. This information should be made a part of the basic scope book for the project and clearly identified as client furnished. These data may also be proprietary, creating a requirement for added security to prevent its disclosure. The need for security of these data should be established during the initial phases of the work.

2.9 GOVERNMENTAL INVOLVEMENT

Governmental involvement takes two forms, involvement in defining and establishing standards for the project and setting requirements for the conduct of the work. More and more planning requirements are mandated; these typically deal with environmental matters such as discharges from a facility, impact upon traffic, and visual appearance. Often environmental requirements are only broadly defined and the permitting process by which they are enforced involves negotiation with licensing bodies. In some cases citizen groups having great influence, but not necessarily any legal standing, are involved. Often the problem is less one of reaching agreement on a solution than of reaching agreement on some kind of reasonable schedule. Indeed some groups will use schedule delays as a way to kill a project, without necessarily taking a direct position in opposition to it. Little guidance can be given in this case except to point out that the persons putting the project forward are not helpless. Considerable influence on public opinion and hence support for the project can be obtained from labor unions whose members would benefit from the project, chambers of commerce, employers, local government officials, the media, and other interested parties. Where appropriate, they can be brought together to sponsor the work and provide balance to the claims and concerns of those who oppose it.

Governmental influence is felt also in setting requirements for the work in the area of applicable **codes and standards.** Frequently local jurisdictions will utilize codes and standards no longer appropriate. It is useful to review the codes and request updating of those that are out of date or inappropriate. As with other areas this is often highly political and very sensitive, and every effort should be made to present this material in a way that will minimize its psychological impact. Where useful, the aid of the groups mentioned above should be employed.

Governmental requirements for the performance of the work is another area with considerable impact. Wage requirements on certain governmental work, workplace and safety requirements, affirmative action and minority, or disadvantaged group, contract requirements are all constraints that affect the way in which the work is executed. No simple set of rules can be put forth except that all of these requirements are a matter of public law and as such should be known to senior management and legal counsel. If there is a significant concern about which laws apply to a particular block of work, or a project, legal counsel should be obtained.

As a matter of general policy it is prudent, in addition to complying with the letter of the law, to always make a good faith effort to comply with its spirit. This is of particular importance where the law appears to be evolving and a good faith effort will usually put the firm ahead of current practice and hence in a less vulnerable position.

2.10   VENDOR DATA AND RESPONSIBILITIES OF VENDORS

Rarely can a design proceed without information from **vendors.** Not only is information usually required on the configuration of the vendor-supplied equipment and components, but performance data and foundation, utility, and special maintenance requirements will require provisions in the design. In most cases vendor equipment is pre-engineered, and much of the required data are available in the form of catalog tear sheets or standard vendor drawings. These data can normally be obtained without any commitment to the vendor, and vendors are usually happy to work with the engineer to provide the drawings and data required. When dealing with the vendors to obtain these data, they should be told that while receipt of the data is appreciated and helpful, there is likely to (or will) be competitive bidding at a later date. This means also that the designer should ensure that the design (data) being used does not close out the possibility of other vendors supplying the material or equipment if they are successful in the competitive bidding. As a result the designer should be alert to special provisions of the vendor-proposed equipment that is unique and would create this complication.

Where several vendors can supply the same equipment but where the equipment is significantly different in requirements, it is good practice to design such that the largest types, the greatest utility requirements, or the heaviest foundations can be incorporated. If it is possible to utilize this method, it avoids being locked into one vendor's offering, and the margins resulting when a vendor is finally chosen are normally helpful.

Where a design requires special or **custom-designed** equipment, it may be necessary to obtain preliminary quotations and to make a tentative or conditional award to a vendor to provide the design information required. This can be done with a minimum of cost if the vendor is required to quote the cost for this design work as a breakout price in the quotation. Generally this includes the stipulation that the vendor prepare the design only, and is not authorized to order materials or commence fabrication without the prior express written permission of the purchaser (design organization). This arrangement commits the designer to certain costs that should be considered when preparing the project cash flow requirements. These advance payments are not charged against the design work but are usually considered a prepaid portion of the cost of equipment.

Security of both design and vendor information is essential during the design phase. Vendors are naturally curious and interested in the progress of the design and the likelihood of gaining a commercial advantage by the incorporation of their equipment in the design. To maintain competition, it is important to avoid unintended disclosure of bid information, design information, or design status to vendors. For this reason it is poor policy to allow vendors access to the design area. A separate small conference room is extremely useful for meetings with vendors and avoids their presence in the area where the engineers and drafters are working and where sensitive information may be available. Similarly care should be taken in telephone calls and correspondence to avoid disclosure of this type information to unauthorized persons.

The need for security should not prevent the designer from maintaining cordial relations with equipment vendors. The vendors are the experts on the characteristics and operation of their equipment and can provide valuable information to the designers. This can be particularly helpful where the designers view the vendor-supplied components as "black boxes," with significant interfaces with the design. Often vendors or their companies will have had experience incorporating or utilizing their equipment in the type of application being considered, and they can cite application problems or advantages not foreseen by the design personnel. If vendors are treated fairly and professionally, they will be willing to provide large amounts of data that can be very useful. If a competitive bid situation is to be maintained, vendors should be told of this as early as possible to avoid misunderstandings and later difficulties.

Vendors will often attempt to improve their business relationship with the designers by providing gifts, meals, entertainment, and, on occasion, trips. Many firms have specific policies dealing with such circumstances. Where such policies do not exist, and as long as they are not excessive or compromising to the designer, they can be accepted with good grace. One rule of thumb to determine if they are appropriate is, "Can they be returned in kind by the recipient?" If the answer is yes, there is usually no problem with accepting the entertainment or bottle of wine, and so forth. Where the answer is no, some thought needs to be given to the response. None of this is to forestall the natural development of friendships between design personnel and vendors, but some caution must be exercised to ensure that the friendship does not impinge upon fairness to all vendors, particularly where a competitive bid situation is involved.

2.11   PREPARATION OF SPECIFICATIONS

Specifications together with drawings form the principal method by which the design is described and transmitted to others. Specifications are of several types: those necessary for procurement activities, those used for construction or subcontracting work, and those used for test and acceptance work. For replacement parts, or for simple or inexpensive items, a formal specification may not be necessary; merely a statement setting forth the model number or the basic performance requirements may be adequate. Where, however, a formal specification is necessary, certain principles are common to the preparation of them regardless of the purpose. These include clarity, simplicity, precision, quantification, use of tolerances, and brevity.

Clarity is certainly a fundamental requirement. The specification needs to be written in a direct way using simple unambiguous language to define the requirements. Often specifications are used by personnel who have little or no formal technical training, and the writer must keep this in mind. Technical terms should be used where necessary, but the specification should never be used as an exercise to demonstrate the deep technical knowledge of the writer. Rather it is a way to convey meaning, intent, and requirements and as such should, if possible, be written at the educational level of those who will use it.

Where there is any question of **clarity,** that portion of the specification should be rewritten to ensure improved understanding. Often having someone unfamiliar with the work read it and ask questions will highlight areas where rewriting is necessary. Precision of language tends to improve the clarity and understanding also. Where there is a choice of words, the most definitive and precise ones should be used. Certain phrases in common use add little to a specification and merely become a basis for argument. Examples of these are "of the highest commercial quality," or "suitable for the service intended." Although desirable goals, such highly subjective language does little to help the vendor or constructor. It is far better to include **quantifiable** values such as specifying a "B-10 bearing life of 20,000 hours," requiring "that the minimum safety factor for rotating parts be 3.5," and so on. This approach removes the subjectiveness in determining whether the criteria is met and provides an objective measurable standard against which to judge **specification compliance.**

Related to clarity is the question of avoidance of **conflict** in the specification documents. The only real way to guard against this is to state the requirement only once. If the requirement is stated in more than one place, it can almost be guaranteed that in the course of design evolution, the value will change, and it will only be corrected in one place thus leading to uncertainty and conflict.

Standard specifications are extremely useful if developed early in the project. They can than be included as attachments in specifications where the component is incorporated. For example, standard specifications for motor starters and finish painting could be included in a procurement specification for a conveyor system, where motor starters and finish painting are included. This has the advantage of permitting the designer of the system—in this case, a mechanical engineer—to concentrate on the area of his or her expertise and incorporate well-thought-out standards, developed by others, that apply to the entire project. There is an added advantage in that standard specifications can identify particular brands and models of components. There is a great advantage to simplifying the stock of spare parts and the later maintenance of the equipment or facility.

National codes and **standards** as well as government specifications, very often the Military (MIL) series, provide a convenient source for standard specifications. They are available for all the disciplines and most common products. In addition technical professional societies and manufacturer's groups have prepared standard specifications that can be a great help. It is not necessary to incorporate all portions of these standard specifications; to do so would in most cases be excessive. They can, however, be effectively used to generate a specification and often serve as a checklist to ensure no major features or requirements are overlooked. **Manufacturer's specifications** can be the starting point, but particular care should be taken that they do not result in a specification that limits competition, for they will sometimes include clauses and requirements that are proprietary or met only by their particular size or features.

An important consideration that needs to be decided upon early is the format of the specifications to be used. It will probably be different from the guide specs used and should be followed for all specifications prepared, although not all sections are necessary in all specifications. One commonly used format has the following sections; the example shown is the technical specification for procurement of a centrifugal pump:*

---

* Reprinted from I. J. Karassik and William C. Krutzch, *The Pump Handbook,* by permission of McGraw-Hill, Inc. Copyright © 1986 by McGraw-Hill, Inc.

The technical specification should consist of a series of carefully defined and distinct sections. The more complete and specific the specification, the more competitive will be the bid prices. A typical specification might contain the following:

1. *Scope of work.* Pump, baseplate, driver (if included), interconnecting piping, lubricating oil pump and piping, spare parts, instrumentation (pump-mounted), erection supervision.

2. *Work not included.* Foundations, installation labor, anchor bolts, external piping, external wiring, motor starter.

3. *Rating and service conditions.* Fluid pumped, chemical composition, temperature, flow, head, speed range preference, load conditions, overpressure, runout, off-standard operating requirements, transients.

4. *Design and construction.* (Care should be taken to provide latitude in this section, as this borders on dictating construction requirements.) Codes, standards, materials, type of casing, stage arrangement, balancing, nozzle orientation, special requirements for nozzle forces and moments (if known), weld-end standards, supports, vents and drains, bearing type, shaft seals, baseplates, interconnecting piping, resistance temperature detectors, instruments, insulation, appearance jacket.

5. *Lubricating oil system (if applicable).* System type, components, piping, mode of operation, interlocks, instrumentation.

6. *Driver.* Motor voltage standards, power supply and regulation, local panel requirements, wiring standards, terminal boxes, electric devices, for internal combustion drivers, fuel type preferred (or required), number of cylinders, cooling system, speed governing, self-starting or manual, couplings or clutches, exhaust muffler.

7. *Cleaning.* Cleaning, painting, preparation for shipment, allowable primers and finish coats, flange and nozzle protection, integral piping protection, storage requirements.

8. *Performance testing.* Satisfactory for the service, smooth-running, free of cavitation and vibration, shop tests (Hydraulic Institute standards) for pump and spare rotating elements, overspeed tests, hydrostatic tests, test curves, field testing.

9. *Drawing and data.* Drawings and data to be furnished, outline, speed versus torque curves, WK^2 data, instruction manuals, completed data sheets, recommended spare parts.

10. *Tools.* One set of any special tools, including wheeled carriage for rotor if needed for servicing and maintenance.

11. *Evaluation basis.* Power, efficiency, proven design.

Supplementing these may be technical specifications relating to other requirements of the order, such as specifications for the electric motor, steam turbine, or other type of driver; a specification on marking for shipment; a specification on painting; and requirements for any supplementary quality control testing (particularly important for pumps for nuclear services where quality control and quality assurance are critical requirements of the Nuclear Regulatory Commission).

In addition it is important that any unusual requirements be listed in the technical specification so that the manufacturers are aware of them. Examples of these are special requirements for repair of defects in pump castings, a sketch of the intake arrangement for wet-pit applications, and special requirements regarding unique testing, for example, metallurgical testing that may be required during manufacture apart from performance testing.

It is helpful to the pump supplier to provide system-head curves, sketches of the piping system (dimensioned, if this is significant), listings of piping and accessories required, etc.

Pump data sheets are extremely useful in providing a summary of information to the bidder and also in allowing the ready comparison of bids by various manufacturers. Some of the items on these sheets are filled in by the purchaser and the balance by the bidder to provide a complete summary of the characteristics of the pump, the materials to be furnished, accessories, weight, etc. The data sheets should be included with the technical specification.

One common way to develop a specification is to rework previous specifications. The **"cut** and **paste"** method has the advantage of speed and in general ensures that the specification format and structure fit the standard previously used. Its main weakness is that less-experienced personnel will often include items not required, and the specification may become much more complicated and far-reaching than needed. Perhaps more important, items needed may be omitted. Often when this method is used, old requirements are carried forward, and many instances exist where specifications contain requirements or references that had ceased to exist many years earlier. This has the effect not only of increasing the price for the item but forgoes the benefit of technological development. It further often has the effect of excluding bids that may be favorable in both technology and price. When using this method then, it is important that the writer think critically about each clause and requirement in the specification to ensure that it is truly necessary. Besides improving precision, this approach has the further advantage of making the specification shorter and more understandable. Where there is truly a doubt whether or not to include a requirement, and where advice is not readily available, it is usually satisfactory to omit the requirement, since the bidders will normally have their standards that cover such requirements.

To summarize, there are a number of sources for standard or guide specifications that can be used as a starting point for specification preparation:

Previous specifications
Industry standard specifications
Government specifications

National codes and standards

Manufacturer's "guide" specifications

Specifications fall into two distinct categories **performance** or **design. Performance specifications** state input and output parameters but do not normally restrict or establish the way in which they are met. They may also in a nondetailed way establish the features required. They have the advantage of permitting the supplier to offer a standard unit (or alternative units) most nearly matching the overall required performance and place no responsibility on the purchaser. They are most applicable to standard items for which some design, manufacturing, and operating experience is available.

Design specifications normally state not only input and output parameters but also specific design requirements, materials of construction, and so on. They establish not only what must be achieved, but to some extent how to do it as well. As a consequence standard designs rarely apply, alternates are limited, and the purchaser may bear some degree of responsibility if performance is not met. They are most appropriate for innovative designs or developmental work, particularly where new technologies are involved. Frequently the number of sources of supply is limited, and prequalification of bidders may be necessary to ensure technical competence.

When preparing specifications, preciseness is essential. Terms such as "fitness for use," and "good commercial quality" are not sufficient to define requirements for complex, costly, or high-impact items. For these cases specific **acceptance criteria** should be stated, and, where codes are involved, they should be referenced or preferably included in the body of the specification.

Often specifications will take a **"black box"** approach where there is little or no interest in the internals of a supplied package and the only parameters spelled out in the specification are the inputs and outputs with perhaps limitations of space or weight if they are important. This can be a convenient way to avoid getting embroiled in a supplier's offering, but some care must be exercised to ensure that there are no hidden problems which arise because of integration requirements.

Each industry or trade has standards of **tolerance.** It is good practice to list or refer to these standards in the specification. Where the standards do not exist, or where the writer wishes something tighter, the tolerances required should be stated. This can be extremely important to the cost of the item specified because tightened tolerancing can increase costs dramatically, particularly if only a few items are to be produced or purchased. The caution here, is to never impose tolerances tighter than actually needed for the item.

Allowing for **alternates** in procurement specifications usually helps the purchaser. Equipment is generally produced to pre-engineered designs in standard sizes, and bidders are more familiar with their equipment capability, features, and size breaks than are purchasers. As a result it is often possible for a supplier to offer equipment that may differ slightly from some details of the specification but that complies with the performance requirements and may have significant advantages in terms of cost or other considerations.

An important part of specifications often overlooked is the requirement for the supplier to provide **drawings and data** to assist the designers with their work. A "Drawings and Data Requirements" form can be included to define which documents are to be furnished and when they are needed. An example of an engineering documents requirements form is shown as Fig. 2.11.

2.12   QUALITY AND QUALITY ASSURANCE

Quality is necessarily a fundamental requirement in the execution of a project. Quality as used in this sense and contrary to a popular concept is not the finest, the most expensive, or the most reliable. Quality as used by nonquality professionals can best be described by the term **"fitness for use"**; that is, it is a level of quality in the design or product that suits the service intended with adequate, but not excessive, margins and that provides an economic answer to the design or fabrication problem. Quality is also defined by those in the field in a more precise way as **conformance with specifications,** which deals with the question of predictability of output.

Much has been written upon the subject, but a brief overview may be helpful. From the producer's point of view quality systems exist to ensure conformance to specifications and to provide both **predictability** and **repeatability** of operations. The objective is to achieve this through processes that limit nonconformances in output. From the customers point of view quality systems provide a level of assurance that the product is acceptable and that it will be suitable for the use intended. High-quality systems need not be more expensive than lower-quality ones, and in fact organizations with higher-quality systems tend to produce more efficiently and cheaply. Poor-quality systems may lose market share while the opposite is true for high-quality systems, which, besides increasing market share, often permit premium pricing of the product.

In general, quality results from doing the right things **(effectiveness)** correctly **(efficiency)** the first time. Quality involves and depends on the motivation and commitment of the work force. To assist in evaluating quality, it is important to measure the results of the work and to develop standards to which actual performance can be compared **(measurement).** Quality also involves an approach where one learns from previous performance and seeks to continually improve upon it **(feedback).** Attention to detail is a major factor used to achieve these goals. Above all, quality requires a commonsense approach to implementation. Considerable caution should be exercised to ensure that the quality programs and practices do not become ends in themselves; they are merely tools to help produce a superior product.

ENGINEERING DOCUMENT REQUIREMENTS

DOCUMENT CATEGORY NUMBER	2. SPECIFICATION PARAGRAPH REFERENCE	3. DOCUMENT DESCRIPTION	4. PERMISSION TO PROCEED REQUIRED		5. SUBMITTAL SCHEDULE *	6. QUANTITY REQUIRED		7. KIND OF COPIES	8. REMARKS
			YES	NO		INITIAL	FINAL		
1.1	1.4.1	Workpoints, elevations, key dimensions	X		4	1	1	R	
1.1	1.4.1	Outline dim., Services, Foundation/MTG. Details	X		4	1	1	R	
3.0	1.4.1	Completed Data Sheets	X		6	1	1	R	
1.1	1.4.1b	General Arrangement Dwgs.	X		4	1	1	R	With Bill of Materials
1.6	1.4.1	Flow Diag/P&IDs and Input/Output List to DCS	X		6	1	1	R	
2.0	1.4.1K	Parts List and Cost		X	20	1	1	R	
1.2, 1.3	1.4.1C	Assembly Drawings			8	1	1	R	
8.0	1.4.1J	Design Calculations	X		4	1	1	R	
8.0	1.4.1L	Calculations for Structural Steel**	X		8	1	1	R	
1.5	1.4.1g	Control Logic Diags, Func. Block Diags, Descriptions	X		8	1	1	R	
1.4	1.4.1g	Single Line Power Dist, Elem. & Schematic Diags.	X		8	1	1	R	
1.4	1.4.1g	Wiring & Interconnection Diagrams	X		8	1	1	R	
5.0	1.4.1i	Schedule Engineering & Fabrication/Erection	X		2	1 1		R R	With Proposal
4.4	3.2.2	Site Storage & Handling		X	22	1		R	
4.1	1.4.1h	Erection/Installation		X	22	1	10	O	
4.2	1.4.1h	Operating Instructions			22	1	10	O	
4.3	1.4.1h	Maintenance Instructions		X	22	1	10	O	

9. FORWARD COPIES TO

SPECIAL INSTRUCTIONS   * Weeks after award

     ** Final structural drawings & related calculations to be stamped and signed by a structural or civil engineer registered in

10.

11. JOB NO.

12. SPEC NUMBER

ENGINEERING DOCUMENT REQUIREMENTS

SHEET 1 OF 3   REV 0

FIGURE 2.11 Engineering documents requirements.

ENGINEERING
DOCUMENT CATEGORY DEFINITIONS

(E) Engineering Documents. This term comprises procedures, drawings, specifications, QA plans, prototype qualification test reports, and other similar documents that require   permission to proceed prior to fabrication, or prior to use of the document on the design, fabrication, installation, or other work programs. The term is also applied to price lists, and instructions for erection/installation, operation, maintenance, and site storage and handling.

A. DEFINITION OF TERMS

(Note: Standard abbreviated titles follow the category definitions).

Supplier – This is a comprehensive term and includes seller, vendor, contractor, subcontractor, subsupplier, etc.

Original – The initial document of which copies are made, i.e., handwritten copy, typed copy, printed matter, tracings or drawings and photographs.

Reproducible - A master copy which can be legibly duplicated by either microreproduction, diazo or electrostatic process. Diazo sepias may be submitted, only if the meet and satisfy   microfilming requirements.

Microfilm - Film containing an image reduced in size from the original and capable of being enlarged to a clear reproduction of the original.

Permission to Proceed Required -   review required prior to use of documents in the design, fabrication, installation, or other work processes.

Initial - The first submittal of a document in accordance with the schedule mutually agreed to by   and the supplier.

Final - The submittal that reflects the required resolution of review comments or the complete submittal required. Drawings submitted as final shall show   job title, job number, procurement document number, line, equipment, tag or code number and the manufacturer's serial number(s).

B. SUBMITTAL

In column 5,   to place the following codes where applicable:

F - Before Fabrication   I - Before Installation   W - With Shipment
S - Before Shipment   P - Before Final Payment   D - Before Design

or

Expressed in calendar days after notice of award.

In column 7,   engineering to place the following letter as applicable:

M - Microfilm
R - Reproducible
O - Original

In column 8, supplier to indicate its schedule if different than shown, and agreed with by

C. DISTRIBUTION

Items and/or documents required to be provided by the G-321-E shall be forwarded to the   designated under entry No. 9, "Forward Copies To:"

D. DOCUMENT CATEGORY NUMBERS & ABBREVIATED DESCRIPTIONS

Engineering Documents are identified and defined as follows:

1.0 DRAWINGS (DWG)

1.1 Outline Dimensions, Services, Foundations and Mounting Details (OUTLINE DIM, SERVICE & FDN/MTG DETS) - Drawings providing external envelope, including lugs, centerline(s), location and size for electrical cable, conduit, fluid, and other service connections. Isometrics and details related to foundations and mountings.
1.2 Assembly Drawings (ASSEMBLY DWGS) - Detailed drawings indicating sufficient information to facilitate assembly of the component parts of an equipment item.
1.3 Shop Detail Drawings (SHOP DET DWGS) - Drawings which provide sufficient detail to facilitate fabrication, manufacture, or installation. This includes pipe spool drawings, internal piping and wiring details, cross-section details and structural and architectural details.
1.4 Wiring Diagrams (WIRING DIAGS) - Drawings which show schematic diagrams, equipment internal wiring diagrams, and interconnection wiring diagrams for electrical items.
1.5 Control Logic Diagrams (CONT LOGIC DIAGS) - Drawings which show paths which input signals must follow to accomplish the required responses.
1.6 Piping and Instrumentation Diagrams (P&IDs) - Drawings which show piping system scheme and control elements.

2.0 PARTS LIST AND COST - Sectional view with identified parts and recommended spare parts for one year's operation or specified with unit cost.

3.0 COMPLETED   DATA SHEETS (COMP DATA SHT) - Information provided by a supplier on data sheets furnished by

4.0 INSTRUCTIONS

4.1 Erection/Installation (EREC/INSTL) - Detailed written procedures, instructions and drawings required to erect or install material or equipment.
4.2 Operating - Detailed written instructions describing how an item or system should be operated.
4.3 Maintenance - Detailed written instructions required to disassemble, reassemble and maintain items or systems in an operating condition.

FIGURE 2.11 (continued). Engineering documents requirements.

4.4 Site Storage and Handling - (SITE STOR & HDLG) - Detailed written instructions which define the requirements and time period for lubrication, rotation, heating, lifting or other handling requirements to prevent damage or deterioration during storage and handling at jobsite. This includes return shipping instructions.

5.0 SCHEDULES: ENGINEERING AND FABRICATION/ERECTION (SCHED)(ENGRG & FAB EREC) - Bar charts or critical path method diagrams which detail the chronological sequence of activities.

6.0 QUALITY ASSURANCE MANUAL/PROCEDURES (QA MNL/PROC) - The document(s) which describe(s) the planned and systematic measures that are used to ensure that structures, systems, and components will meet the requirements of the procurement documents.

7.0 SEISMIC DATA REPORT - The analytical or test data which provides data and demonstrates suitability of material, component or system in relation to the conditions imposed by the stated seismic criteria.

8.0 ANALYSIS AND DESIGN REPORT (ANAL & DSGN RPRT) - The analytical data (stress, electrical loading, fluid dynamics, etc.) which demonstrates that an item satisfies specified requirements.

9.0 ACOUSTIC DATA REPORT (ACST DATA RPRT) - The noise, sound and other acoustic vibration data required by the procurement document.

10.0 SAMPLES
  10.1 Typical Quality Verification Documents (TYP QUAL VERIF DOC) - A representative data package which will be submitted for the items furnished as required in the procurement documents.
  10.2 Typical Material Used (TYP MAT USED) - A representative example of the material to be used.

11.0 MATERIAL DESCRIPTION (MAT DESCRT) - The technical data describing a material which a supplier proposes to use. This usually applies to architectural items, e.g., metal siding, decking, doors, paints, coatings.

12.0 WELDING PROCEDURES AND QUALIFICATIONS (WLDG PROC & QUALF) - The welding procedure, specification and supporting qualification records required for welding, hard facing, overlay, brazing and soldering.

13.0 MATERIAL CONTROL PROCEDURES (MATERIAL CONT PROC) - The procedures for controlling issuance, handling, storage and traceability of materials such as weld rod.

14.0 REPAIR PROCEDURES (REPAIR PROC) - The procedures for controlling material removal and replacement by welding, brazing, etc., subsequent thermal treatments, and final acceptance inspection.

15.0 CLEANING AND COATING PROCEDURES (CLGN & CTG PROC) - The procedures for removal of dirt, grease or other surface contamination and preparation and application of protective coatings.

16.0 HEAT TREATMENT PROCEDURES (HEAT TR PROC) - The procedures for controlling temperature and time at temperature as a function of thickness, furnace atmosphere, cooling rate and method, etc.

19.0 UT - ULTRASONIC EXAMINATION PROCEDURES (UT PROC) - Procedures for detection of presence and certain characteristics of discontinuities and inclusions in materials by the use of high frequency acoustic energy.

20.0 RT - RADIOGRAPHIC EXAMINATION PROCEDURES (RT PROC) - Procedures for detection of presence and certain characteristics of discontinuities and inclusions in materials by x-ray or gamma ray exposure of photographic film.

21.0 MT - MAGNETIC PARTICLE EXAMINATION PROCEDURES (MT PROC) - Procedures for detection of surface (or near surface) discontinuities in magnetic materials by distortion of an applied magnetic field.

22.0 PT - LIQUID PENETRANT EXAMINATION PROCEDURES (PT PROC) - Procedures for detection of surface discontinuities in materials by application of a penetrating liquid in conjunction with suitable developing techniques.

23.0 EDDY CURRENT EXAMINATION PROCEDURES (EDDY CUR EXAM PROC) - Procedures for detection of discontinuities in material by distortion of an applied electromagnetic field.

24.0 PRESSURE TEST - HYDRO, AIR, LEAK, BUBBLE, OR VACUUM TEST PROCEDURE (PRESS TEST - HYDRO, AIR, BUBBLE - VAC TEST PROC) - Procedures for performing hydrostatic or pneumatic structural integrity and leakage tests.

25.0 INSPECTION PROCEDURE (INSPECTION PROC) - Organized process followed for the purpose of determining that specified requirements (dimensions, properties, performance results, etc.) are met.

26.0 PERFORMANCE TEST PROCEDURES (PFRM TEST PROC) - Tests performed to demonstrate that functional design and operational parameters are met.
  26.1 Mechanical Tests (MECH TEST) - e.g., pump performance data, valve stroking, load, temperature rise, calibration, environmental, etc.
  26.2 Electrical Tests (ELEC TEST) - e.g., impulse, overload, voltage, temperature rise, calibration, saturation, loss, etc.

27.0 PROTOTYPE TEST REPORT (PROTO TYP TEST REPORT) - Report of a test which is performed on a standard or typical example of equipment or item and is not required for each item produced in order to substantiate the acceptability of equal items. This may include tests which result in damage to the item(s) tested.

28.0 PERSONNEL QUALIFICATION PROCEDURES (PERSONL QUAL PROC) - Procedures for qualifying welders, inspectors and other special process personnel.

29.0 SUPPLIER SHIPPING PREPARATION PROCEDURE (SPLR SHPNG PREP PROC) - The procedure use by a supplier to prepare finished materials or equipment for shipment from its facility to the jobsite.

30.0 (OPEN)
31.0 (OPEN)
32.0 (OPEN)

FIGURE 2.11 (continued). Engineering documents requirements.

The villain in quality is **variability**—that is, the normal variations in materials and fabrication, tooling, wear, human error, and other similar factors that contribute to unpredictable changes in the output of the process. Since the main function of quality systems is to remove the variability from the production processes, quality systems measure these variations and put into place methods that avoid or identify nonconformance before they produce out of specification work. When nonconformance occur, these systems identify the deviation and set into action both **remedial** (cause) and **corrective** (output) measures.

Quality systems require not only a statement of the managerial goals, including the importance of quality and the overall approach which will be taken, but also the more detailed methods by which quality will be achieved and ensured. There are in effect two levels at which the subject is addressed: the **program** level and the **procedural** one.

Typically quality programs focus on the managerial aspects of the programs with the details of implementation found in departmental procedural documents. The principles set forth in the programs are sufficiently broad to be applicable to most situations and thus can be applied with a high degree of confidence. As would be expected, the implementation (procedural) aspects will vary widely from company to company, product to product, and industry to industry.

W. Edwards Deming, widely credited with introducing modern quality methods to Japan and more recently to the United States, has developed a series of principles on the managerial approach to the subject of quality. His approach has been adopted by some major firms. While more far-reaching than most quality standards, it is worth reviewing them as they establish a basic philosophical and managerial approach to the subject (see Fig. 2.12).

There are several **quality programs** used in the United States and abroad that provide overall direction to the content and requirements and are suitable for a variety of industries and products.

The ANSI/ASQC Q90-1987 series of quality standards have been widely adopted for use in the United States and include generally the standards used in other countries. (In particular these standards are considered to be the equivalent of International Standards Organization's ISO 9000, described further below). Q90 consists of an **overall** standard that is supplemented by a series of **daughter** standards which are to be applied to the type and scope of work undertaken. This family of standards is as follows:

ANSI/ASQC Q90-1987 Quality Management and Quality Assurance Standards—Guidelines for Selection and Use.

ANSI/ASQC Q91-1987 Quality Systems—Model for Quality Assurance in Design Development, Production, and Installation.

ANSI/ASQC Q92-1987 Quality Systems—Model for Quality Assurance in Production and Installation.

ANSI/ASQC Q93-1987 Quality Systems—Model for Quality Assurance in Final Inspection and Test.

ANSI/ASQC Q94-1987 Quality Management and Quality Systems Elements—Guidelines.

Another quality systems standard that has been widely adopted and is considered essential for selling in the European market is ISO (International

DR. DEMING'S 14 POINTS

1. Create and publish to all employees a statement of the aims and purposes of the company or other organization. The management must demonstrate constantly their commitment to this statement.
2. Learn the new philosophy, top management and everybody.
3. Understand the purpose of inspection, for improvement of processes and reduction of cost.
4. End the practice of awarding business on the basis of price tag alone.
5. Improve constantly and forever the system of production and service.
6. Institute training.
7. Teach and institute leadership.
8. Drive out fear. Create trust. Create a climate for innovation.
9. Optimize toward the aims and purposes of the company the efforts of teams, groups, staff areas.
10. Eliminate exhortations for the work force.
11.a. Eliminate numerical quotas for production. Instead, learn and institute methods for improvement.
   b. Eliminate M.B.O. (Management By Objectives) Instead, learn the capabilities of processes, and how to improve them.
12. Remove barriers that rob people of pride of workmanship.
13. Encourage education and self-improvement for everyone.
14. Take action to accomplish the transformation.

FIGURE 2.12   Dr. Deming's 14 points. Reprinted from "Out of the Crisis" (Revised) by W. Edwards Deming by permission of the Massachusetts Institute of Technology and W. Edwards Deming. Published by the MIT Center for Advanced Engineering Study, Cambridge, MA 02139. Copyright 1986 by W. Edwards Deming. Revised January 1990.

Standards Organization) 9000-1987. This standard is identical to BS (British Standard) 5750. ISO 9000 again uses an overall standard that establishes concepts and applications and a series of daughter standards as follows:

ISO 9000, Quality systems—Principal concepts and applications.

ISO 9001, Quality systems—Model for quality assurance in design/development, production, installation, and servicing.

ISO 9002, Quality systems—Model for quality assurance in production and installation.

ISO 9003, Quality systems—Model for quality assurance in final inspection and test.

ISO 9004, Quality management and quality system elements—Guidelines.

With these American and foreign standards, the intent is to apply a particular standard to the activities performed, and not all portions of a standard need to be implemented. Thus a design organization would utilize those portions of 9001 that deal with design activities, and requirements for production control, installation activities, and such would not be utilized.

At the operational level numerous organizations have established more detailed and specific requirements for the scope and content of quality programs. Usually these quality programs make a distinction between **quality assurance** and **quality control.** Here quality assurance is considered the overall program and the activities (usually of a quality assurance organization) that ensure the quality of the product. The term "quality control" is generally applied to production activities, where the interest is in controlling accuracy of output. Production as used in this sense can apply to an engineering design operation. Further distinctions are usually made between quality assurance and reliability. Thus quality assurance practices are based on organization and control of activities whereas reliability deals with predictability of results.

For example, U.S. federal regulations set forth 18 criteria to which nuclear projects must conform (see 10 Code of Federal Regulations, Part 50, Appendix B). These regulations are typical of the philosophy and scope of formal quality assurance programs applied throughout industry, although all elements may not be appropriate for a particular activity or component. In summary form, these elements are stated as follows:

  I. An organization with sufficient independence shall be established to control and verify the performance of functions affecting quality.
 II. A documented program defining scope and responsibilities shall be established.
III. Measures to assure proper control of design activities including verification or testing shall be applied.
 IV. Documents which control procurements shall include the applicable Quality Assurance program requirements.
  V. Activities affecting quality shall be documented, accomplished in accordance with these documents (procedures) and include acceptance criteria.
 VI. The review, issuance and change of documents affecting quality is controlled.
VII. Conformance to procurement documents of purchased materials, equipment and services shall be established, including both control and documentation.
VIII. Identification and control of material, parts and components shall be established.
 IX. Welding, heat treating, nondestructive testing and other special processes are to be controlled.
  X. A program for the inspections of activities affecting quality shall be established.

    XI. A controlled testing program shall be established and its results shall be documented.

    XII. Measuring and test equipment shall be controlled and calibrated.

    XIII. Measures to control handling, shipping and storage to prevent damage or deterioration shall be established.

    XIV. Measures shall be established to control and indicate inspection, test and operating status.

    XV. Nonconforming materials shall be identified and controlled to prevent their installation or inadvertent use.

    XVI. Conditions adverse to quality shall be identified, corrected and for significant items reported to management.

    XVII. Sufficient records shall be maintained to furnish evidence of activities affecting quality.

    XVIII. A system of planned, periodic audits shall be established to verify compliance with and the effectiveness of the Quality Assurance program.

To avoid excessive cost and waste, it is important that the requirements for quality assurance and reliability be selectively applied only to necessary components and/or characteristics.

In the defense sector similar quality assurance programs are used but with somewhat more specific requirements. Typical of this are MIL-Q-9858 A and MIL-STD-1535 A (USAF).

Because of the criticality of **inspection** additional standards have been developed and are widely used. Typical of these are MIL-I-45208 A, Inspection System Requirements and MIL-STD-105D, Sampling Procedures and Tables for Inspection by Attributes. Certain industries also have developed quality assurance programs such as API Spec. Q1, for work in the petroleum industries and ANSI/ASME NQA-1 for work in the nuclear field. In addition some standards organizations have developed **generic** standards such as ANSI/ASQC Z-1.15. For the most part these standards are similar to the general standards described earlier, and overall compliance involves the same type of organization and approach. Because there are some differences, it is important to establish early in the work which quality program will be used. The standard should then be compared carefully with the program in place, and reviewed in terms of the appropriate departmental procedures and corrections or additions made to ensure compliance.

Another management level approach, and one that is finding increasing acceptance, is the concept of **Total Quality Management** (TQM). This methodology is applicable to all types of activities regardless of whether product or service oriented. TQM includes such different industries as hospitals and electronic manufacturing firms. Basically the TQM programs focus on control of all of the processes of the company. This includes the orientation and motivation of the work force, starting with and including the senior management, analysis of current operating practices and the development of corrective and remedial approaches, and finally implementation of the improvements identified. The concept of TQM is to involve the entire organization at all levels in a continuing process of improvement in the activities of the business. A proper TQM program will involve such diverse groups as sales, legal, and accounting, as well as the technical and production personnel. It may in some cases also include customers both within and outside the organization. The methodology also may include techniques such as **partnering** with suppliers, the use of **"just-in-time"** practices to reduce in-process inventories in manufacturing firms, and an overall approach that stresses "do it right, once, the first time." The TQM programs are tailored to each firm and are highly individualized—they cannot merely be adapted from an existing program. As with other quality programs, when properly introduced and supported by top management, dramatic improvements have often been seen within a very short time. Some clients now require that bidders submit their TQM programs for evaluation before being permitted to offer proposals.

As stated earlier, the quality program for a typical firm establishes requirements that cover the procedures and practices used to **control** the work, as distinct from the work itself. The actual control of the work product is left in the hands of the operational personnel and appears in their procedures. In an engineering office, for example, the design personnel would use some form of quality control (QC) system to control their day-to-day work. The quality assurance group will normally concern itself with checking the procedures and methods and their **compliance** and will not normally perform reviews of the technical adequacy of the work itself. In many cases, because of overriding concern for accuracy, **technical audits** are conducted on the work product itself to ensure that it is correct and in accordance with the required criteria.

To perform its function, the quality assurance (QA) group will normally review and approve quality-related procedures prepared by the design group and thereafter periodically, perhaps semiannually, audit the design group for compliance. Noncompliances are noted and graded on a scale, with **observations** being the lowest category, **deviations** the next, and **violations** the most severe of the categories.

Where a firm or project does not use a board of review or where the chief engineer's personnel are extremely busy, one or more **technical** audits in specific limited areas can be useful. A technical audit, for example, reviewing the details of a specific design area, is usually much more in-depth and often more valuable. Such an audit might compare the final design to each of the initial criteria established at the beginning of the project. Included would be confirmation of the reasonableness of assumptions, checks of calculations, reviews of drawings, and all of the other elements that entered into the design to ensure they are not only correct but nearly optimum. This involves independent engineering judgment and requires that the audit team be as technically qualified as the group that originated the design. Since the audit will involve judgment of the adequacy of the original work, the audit must be conducted with considerable tact and understanding, and the audit team should be made up of senior personnel whose judgment is sound and who will be careful to only report on important items. Technical audits are costly and can be disruptive, since they can take a large amount of the time of the design personnel and thus are often limited in scope. These audits should be used only where there is concern that a specific area may not be properly executed. Normally the design groups initially resist such audits, but in the long run both the firm and the personnel will benefit from them if they are properly managed and, more important, used constructively.

The most important portion of the audit concerns the establishment of corrective and remedial action. Corrective action deals with the correction of the particular noncompliance, while remedial action deals with the correction of the methodology that permitted the nonconformance to occur and to remain unobserved.

For many smaller engineering offices such far-reaching and formal programs are not necessary, and there is no need to establish detailed systems such as those used in nuclear or very high-tech work. For such cases common sense should be the rule with a few principles for overall guidance:

- Quality starts at the top.
- Keep it simple.
- Check it periodically—independently.
- The doers are responsible and create quality—not outsiders.
- Establish and maintain a desk book of how a job is performed. Make sure it is both accurate and realistic, and revise it when practices change.
- Get input from the doers, and permit them to make improvements.
- Fit the practices to the goals and risks of the design.
- If there's an error, fix it and the cause as well.
- Quality is a continual process and improvements in methods, tooling, activities, and systems must be continual over time.

2.13   STATISTICS

Statistics provide a methodology by which characteristics may be evaluated and predicted. **Central tendency,** as its name implies, evaluates the clustering of values. The two most useful measures of central tendency are the **arithmetic mean** and the **median.**

The **arithmetic mean** is

$$\bar{x} = \Sigma \frac{x}{n}$$

where

$\bar{x}$ = arithmetic mean,
x = individual observed values,
n = number of values observed.

The **median** is the individual observed value that has an equal number of observed values above and below it; that is, it is the middle observed value in a series. In small or skewed statistical populations it may vary widely from the arithmetic mean, whereas in large nonskewed populations it will closely approach the arithmetic mean. **Population** is the term applied to the entire group whose characteristics will be determined.

Observations will tend to cluster around the arithmetic mean with the degree of their dispersion defined as the **standard deviation:**

$$s = \sqrt{\frac{\Sigma(x - \bar{x})^2}{n - 1}},$$

where

s = standard deviation.

For normal (nonskewed) distributions a predictable proportion of the values will fall within a range of standard deviations (see Fig. 2.13):

Range	Percent of Values Included
$-1s$ to $+1s$	68.26
$-1.96s$ to $+1.96s$	95.0
$-2s$ to $+2s$	95.46
$-3s$ to $+3s$	99.73

The **coeffecient of variation** is a useful measure to compare dispersion of data from different sources or bases. This can be particularly useful when establishing correlations:

$$V = \frac{s}{\bar{x}}$$

where

V = coefficient of variation.

Sampling provides a method by which the characteristics of a population can be predicted by examination of only a small portion of the population.

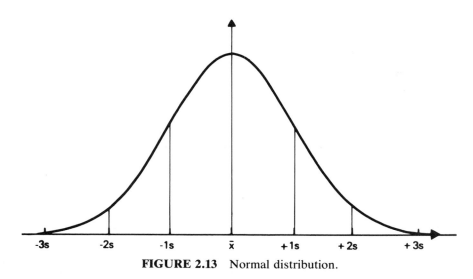

FIGURE 2.13 Normal distribution.

To achieve this, it is necessary that the sample be representative (usually achieved by random sampling). Where all items in the population are examined, the sample is 100%. This is sometimes also referred to as **exhaustive sampling.**

Sampling frequently uses a single sample whose size varies depending on the size of the population (or lot) and the **confidence level** desired. The confidence level desired will have the larger effect on sample size required. Predetermined acceptance levels are compared to the sample characteristics to establish acceptability or rejectability of the lot.

Where a **single sample** indicates noncompliance with acceptable criteria, **double (or multiple) sampling** is frequently employed. This requires larger progressive samples with a reduced proportion of nonacceptable items permitted.

After determining the **acceptable quality level (AQL)** (the level of defects acceptable) standard published tables of acceptance sampling plans can be referred to for lot sizes, sample sizes, acceptance and reject quantities for each, and so on. Among the more widely used are the following:

MIL-STD-105 (Military Standard 105)
Dodge-Romig Sampling Plans
DOD Interim Handbooks H106, H107, and H108

Probability establishes the mathematical likelihood of an event occurring. The probability of a single event is expressed by a number between 1.0

(certainty) and 0 (impossibility). The probability of a series of events occurring is the product of their individual probabilities:

$$P_s = P_a \cdot P_b \cdot P_c \cdots P_n,$$

where

P_s = probability of the series of events,

P_a, P_b, P_c, P_n = probability of the individual events.

The probability of a series of events can be greatly improved by identifying and improving items having low probability. Alternatively, the probability or performance of a system of components can be improved by establishing parallel circuits:

$$P_s = 1 - (1 - P_a)(1 - P_b) \cdots (1 - P_n),$$

where

P_s = probability of the system.

Thus for a **series system** (Fig. 2.14) with a low-probability (reliability) component,

$$P_a = 0.98,$$
$$P_b = 0.88,$$
$$P_c = 0.65,$$
$$P_d = 0.98,$$
$$P_s = (0.98)(0.88)(0.65)(0.98) = 0.549.$$

Duplication the low-reliability component in **parallel** (Fig. 2.15) yields

$$P_s = (0.98)(0.88)[1 - (1 - 0.65)(1 - 0.65)](0.98)$$
$$= [1 - (0.35)(0.35)]$$
$$= [1 - 0.123],$$

$$(0.98)(0.88)(0.88)(0.98) = 0.744$$

A **frequency diagram** is a powerful diagnostic tool used to establish causes of events. Where events are grouped by cause (or type) and the number of occurrences of each are plotted as a simple bar graph, the diagram will clearly establish those items that occur most frequently and that deserve first attention.

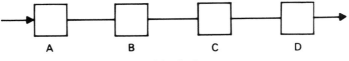

FIGURE 2.14 Series system.

As is apparent in Fig. 2.16, event types E, B, and H are the major repetitive ones and should be given first attention. Of the 80 events, 46 are of only three types while the remaining 34 comprise 10 different event types. The **Pareto principle,** which describes this, states that only a few causes account for the majority of the events. When testing data, a rule of thumb often used is "20% of the causes generate 80% of the events"—**the 80–20 rule.**

2.14 ECONOMIC CHOICE

Economic choice, if reduced to its basic approaches, presents two ways in which to evaluate alternatives: by the **time** required to recover the investment or by the **rate of return** that an investment yields. Even these two are somewhat mirror images of each other since from one, the other can be derived, but for convenience they are commonly considered independently. Each approach involves a variety of analytical methods that typically includes consideration of the worth or value of time. Some of the methods are simple and direct, while others are relatively complex.

An often overlooked way of analyzing economic choice is in terms of **risk** or **sensitivity.** Both are secondary aspects of the two approaches above, and a properly performed analysis will take them into consideration where appropriate. The risks to be evaluated should only be those that are truly credible; it may mean in some cases even setting a threshold of 0.10 for them. Running the analysis for several alternative values of the principle parameters, both larger and smaller than those in the basic study, will provide a basis for judging sensitivity to changed conditions.

The simplest approach is the **payout analysis** approach with no consideration of the discounting of future values. One rule of thumb widely used for

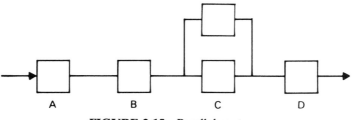

FIGURE 2.15 Parallel system.

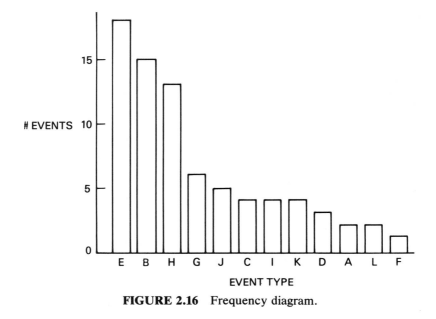

FIGURE 2.16   Frequency diagram.

industrial projects and facilities requires that the payout period not exceed five to six years. This equates to earnings in the range of 16 to 20% per year when future earnings are not discounted. Although the result is approximate, payout is easy to calculate, surprisingly useful, and is widely used for projects that have technological lives of at least five to seven years. The main reason for its wide acceptance, apart from its simplicity, is the likelihood of the effect of outside factors exceeding the improved accuracy of more precise estimates, and this tends to compensate for the risks and uncertainties.

For facilities that have a long service life and do not face early technological obsolescence, payout periods of 10 to 15 years are used. For infrastructure projects, permanent utility systems, and similar projects where there is no alternative once the investment is made, the payout period can be even longer, in some cases approaching the economic life of a project measured in decades. For projects at the cutting edge of technology where change occurs rapidly, very short payout periods may be necessary.

When payout periods exceed six to eight years, it is wise to perform a **discounted evaluation,** such as **present worth** analysis to reflect the effect time has on the value of future revenues or payments. Even at discount rates of a few percent, with a time horizon of more than a few years, the discounting of future revenues can be significant. The basic principle to follow is that the payout period must not significantly exceed its likely competitive economic life. In many cases the industry uses preestablished standards for maximum payout time, which the analyses must meet to demonstrate economic acceptability. (More information on financial ratios often used for analytical work is found in Chapter 13.)

Rate-of-return analyses tend to be more complex than the payout period methodology because they establish a criteria that the alternate must meet. The requirement stated simply is that the return on the investment, or funds used, must meet or exceed a predetermined value based on either internal or external criteria. The internal criteria are normally the alternate earnings (usually in annualized percent) that would be possible if the investment were made in the existing product lines of the company. This is often called the **internal** or **pool rate** of earnings, and if the alternate investment cannot generate that rate, it is not made.

The external approach is to compare the rate of return to that possible were the investment to be made in outside financial instruments such as high-grade bonds or stocks. This has the advantage of comparing the investment to the ability to earn on the open market with little or no risk and effort of a project development. The rates considered acceptable will vary widely depending upon whether the industry is capital intensive, subject to rapid technological change, and so forth. Often the rate is based upon the prime lending rate plus a specific adder such as 4 or 7%.

The actual calculation of the rate of return will follow the standard methods found in engineering economics references. In virtually all cases the value of payments, or receipts at future dates, is discounted by the present worth method, or some variation on it, to better determine the likely financial performance of the investment.

Factors considered in these methods of evaluation should include:

Capital cost
Depreciation
Salvage value
Discount factor
Cost of money
Annualized operating cost(s)
Capacity factors, efficiency, quality of output, and similar factors
Personnel costs
Risks to the investment
Sensitivity of the investment to outside factors

2.15   DESIGNING FOR MANUFACTURE

The trade-offs for the design process include considerations of product performance, product cost, development program expense, and development speed. While each must be considered the primary concerns must be for product **performance** and product **cost.** Even though early entry into the market is often critical to capturing market share, development speed and expense must be considered secondary to the basic functional requirement for product performance and quality and the general goal of optimum product cost. Great care should be taken to ensure that in the rush to bring a product to market, deficiencies and defects are not introduced.

Designing for manufacture requires above all a commonsense approach to the goal of minimum product cost, without sacrificing product quality or performance. To effectively achieve this, it is necessary to establish specific, measurable product requirements and to limit expansion or enhancement. The design team has several "customers" apart from the actual purchaser. They include manufacturing, installation, service, test, and maintenance. In a sense, even marketing and sales are customers because they will sell the product. Of course the restraints and requirements of the manufacturing organization must be considered and incorporated by the design organization, since these are fundamental to being able to produce the product.

The principles that should be utilized are well known and follow what experience normally dictates. The total number of parts should be minimized. In that way not only does it avoid the cost of producing unnecessary parts, but also it cuts down on the time and effort for the design, administration, material handling, and manufacturing effort. Where one part can be designed to include the functions of two or more parts, the savings may be significant, even though the cost to produce the single more complex part may be higher. Since the potential for interaction between components roughly doubles with the addition of each new element, the effort to reduce the number of elements, and thus reduce the **interactions** and improve **reliability,** can be easily justified. Care needs to be taken to maintain an optimum design and not merely to reduce the number of parts for its own sake.

Figure 2.17 shows how these interactions increase with added elements, while Fig. 2.18 indicates the difficulty of maintaining a 99% reliability with increasing numbers of elements in the product.

In all systems some components have a significantly higher failure rate than others. One way to deal with this is to group developmental functions that have a potential for higher failure rates in a single component. This will greatly improve the reliability of the entire assembly. It also will permit isolation and remedy of problems more quickly. Thus the design will also be more easily maintained in service. **Modular designs** favor this risk concentration approach and should be used where possible. Figure 2.19 indicates how risk concentration and risk distribution among the design components affects the overall reliability of the design.

Standard components should be used wherever possible. Their reliability is known, and developmental problems can be avoided, they are readily available, often provided by several sources both at competitive cost and offer assurance against loss of supply. Use of standard components also frees up in-house resources for more creative work.

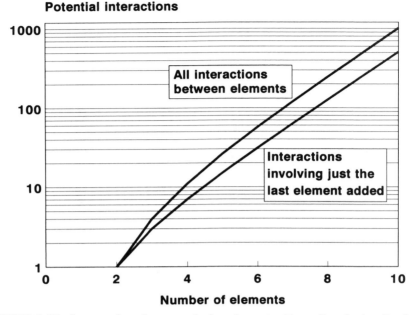

FIGURE 2.17 Interractions between design elements. From *Developing Products in Half the Time* by Preston G. Smith and Donald G. Reinertsen, Van Nostrand Reinhold, 1991.

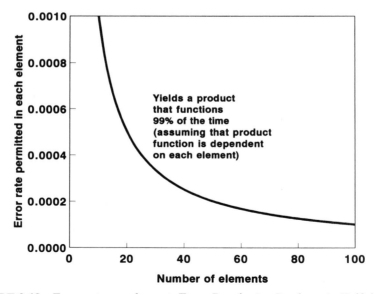

FIGURE 2.18 Error rate per element. From *Developing Products in Half the Time* by Preston G. Smith and Donald G. Reinertsen, Van Nostrand Reinhold, 1991.

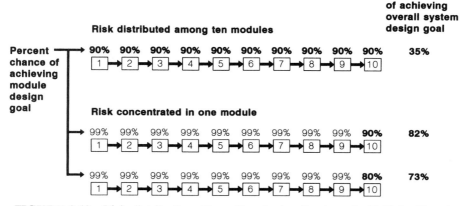

FIGURE 2.19 Risk distribution. From *Developing Products in Half the Time* by Preston G. Smith and Donald G. Reinertsen, Van Nostrand Reinhold, 1991.

Parts must be designed for ease of **assembly.** The use of suitable tolerances, proper radii, generous tapers, self-guiding features, and similar considerations will do much to prevent assembly difficulties and high cost.

The use of separate fasteners should be minimized, and integral clips or similar features should be provided. Where fasteners are necessary, they should be of the captive or self-guiding type designed to provide maximum flexibility and provision for ease of alignment during assembly.

Where possible parts should be designed to perform several functions. As mentioned above, a more expensive part may in the long run offer a significant advantage in enabling simplification of the production process. A corollary to this is for a part to be designed to have multiple uses. For example, a small shaft could also be a guide pin, a hinge, or a similar component on the same assembly, thus simplifying handling, storage, stocking, and, more important, production of the part. However, care must be taken to avoid the temptation to use custom-designed components where the functional gain is only modest, or the cost impact negative.

Tolerances established for the part should be based on rational **statistical** decisions. Often designers will decrease tolerances to ensure that there is sufficient margin to accept normal variations in the manufacturing or fabrication process. The manufacturing planners then may further tighten them to ensure a lower rejection rate after production. The result of this is a cutting of tolerances, often to $\frac{1}{4}$ or less of that necessary, with a resultant large increase in cost. It further classifies items that may be usable as out of tolerance. This requires that they be reviewed and dispositioned by a Material Review Board or similar technical group. Out of tolerance parts not only generate unnecessary paperwork and effort, but they undermine the integrity of the design, since many of the deviations then are dispositioned as **"use**

as is." The loss of confidence in the design can become a real problem and a point of controversy, particularly when tolerances are not maintained for critical items. By far the best and most economic policy is to determine what tolerances are necessary, state them, and most importantly, enforce them.

The design of individual parts should minimize manufacturing cost by finding the lowest-cost method of production considering production volume, available tooling, and so forth. Sometimes what may seem to be the cheapest manufacturing process may not be the lowest cost after all, as assembly, fastening, finish, and other processes add to the cost of the completed assembly.

Design should ideally provide for assembly from **one direction** only. This will avoid additional handling and positioning of either the assembly itself or positioning and assembling parts and fasteners in other planes and axes. For **robotic** assisted manufacturing, this can often avoid the need for robots with five or six directions of freedom of motion. As a corollary, all handling of the parts and assembly should be minimized. There should be as little repositioning as possible, and to the greatest extent handling should be linear in direction.

Manufacturing tolerances should be established statistically where possible, and if approaches such as the 6-sigma philosophy (described below) are used, inspection can be reduced and even avoided.

For large production runs, a methodology of 6-sigma quality has been adopted by some firms. This approach uses a plus or minus 6-sigma band (12 sigma) between the upper and lower specification limits and permits a 1.5-sigma drift of the mean in response to normal manufacturing variations, while limiting the defect rate to 3.4 parts per million. The use of sigma, the standard deviation, to define the quality band permits the linking of the degree of process variation to specification limits and allows the derivation of correlation and capability ratios that simplify application. Where less rigorous approaches are adopted, the possibility of nonconforming parts will increase and automatic or self-inspection will be required to detect and eliminate them.

Improvements in product design should be made in **small increments** and be continual. This permits a shorter time to market and reduces the risk of significant problems being discovered at a later date.

In some cases it may be prudent to consider the use of nonstandard sizes and components to foil attempts at counterfeiting or the unauthorized production of replacement parts. But this extreme step, though sometimes justified, should be carefully weighed before implementation.

In general, manufacturing systems should be designed to minimize production time without **value being added** to the product. The timing of operations involving order entry, queueing time, in-process storage, material movement, inspections, time awaiting management decisions, and Material Review Boards should be minimized, and eliminated, to cut down on manufacturing costs. Just-in-time inventory and production control measures might need to be instituted to reduce such costs.

The use of robotic equipment to assist in the manufacturing process has been long recognized. In general, robots can be separated into two types: those that transfer or move materials and those that manipulate a component and perform some manufacturing operation. The decision to design for and install robotic equipment involves both design and manufacturing considerations. The mere act of designing for robotics, but using people to perform the work actually yields about 85% of the benefit. To gain the 15% additional benefit yield, high-volume production is normally required. Although if elements of the work are performed on NC tooling where the programming is convenient, the break-even size of the lot can be significantly reduced.

The great advantage of using robotics is due not so much to production volume as to **accuracy.** With predictability of production rates and increased amortization costs, firmer economic data are available for pricing. Economic analyses of robotics manufacturing need to quantify all of these factors to properly reflect their impact.

The economic justification for robotics should additionally include criteria such as **return on investment (ROI).** Suppose that the ROI analysis for automatic test equipment does not indicate a clear cost advantage. Since robotic equipment can reduce the need for highly trained test technicians (where there is a potential for increasing production levels), the time needed for ramp up to a second or third shift would be reduced and thus its selection, while higher in cost may be the correct decision.

Yet another critical factor in the design process is the time it takes to complete the design, implement it, and produce and distribute the product to the market. Throughout the last quarter century the effective life of products has been continually decreasing. As a result the time required to develop and produce a new product, that is, **"time to market,"** has become increasingly important. Clearly the company reaching the market first can command higher prices, capture a significant share of the market, and establish name recognition and perhaps customer loyalty. In addition a firm that establishes a reputation for innovation ensures public acceptance of successive models and products. Later, as products of other firms enter the market, there is less differentiation between competing products, so profit margins are reduced, and customer (brand) loyalty becomes increasingly important. In other words, a delay in reaching market results in a loss of market share with a consequent loss of the early sales having the highest profit margins. These effects are shown graphically in Figs. 2.20 and 2.21.

One example of the concern given to the time to market concept is shown in Fig. 2.22. In the figure note the reduction in recent development cycle times for a variety of products.

"Break-even time" is a recent development that promises to be an effective alternate to the "time to market" approach. "Break-even time" refers to the time it takes to develop, market, and profit from a new product. This includes the time to recover not only the production costs but also the development costs. It affords the advantage of establishing the true cost of the product rather than permitting development costs to be spread across

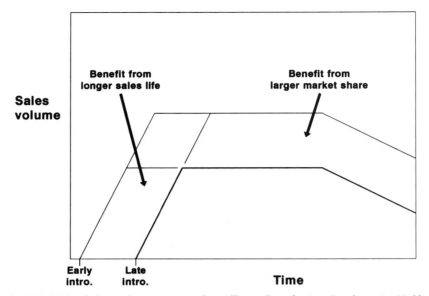

FIGURE 2.20 Sales volume versus time. From *Developing Products in Half the Time* by Preston G. Smith and Donald G. Reinertsen, Van Nostrand Reinhold, 1991.

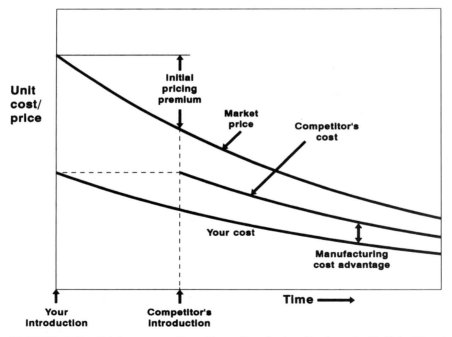

FIGURE 2.21 Pricing versus time. From *Developing Products in Half the Time* by Preston G. Smith and Donald G. Reinertsen, Van Nostrand Reinhold, 1991.

■ **This product** ▦ **Prior experience**

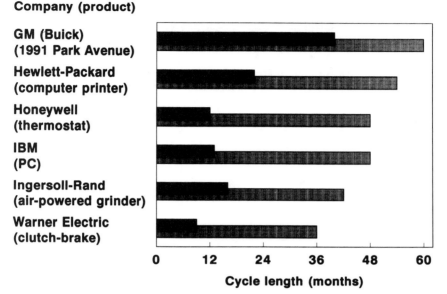

FIGURE 2.22   Development cycle time reductions. From *Developing Products in Half the Time* by Preston G. Smith and Donald G. Reinertsen, Van Nostrand Reinhold, 1991.

other product costs or carried as an overhead charge. This is based in part on the concept of charging individual production operations with all identifiable charges, such as the appropriate tooling and overhead, to ensure that all the costs attributable to a specific operation appear in the economic analyses.

Concurrent design development is widely used to achieve many of the reductions discussed above. In this approach design teams are established to pursue different components of the design. Each team is autonomous and has sufficient authority to obtain priority support from the organization. Each team can call upon groups as diverse as sales, maintenance, and shop personnel, in addition to the manufacturing engineering personnel. The product design is developed in parallel rather than sequentially, and this shortens the schedule. For example, a design including a power supply, electronic components, structural and mechanical systems, and plastic appearance housings might have each of these elements designed at the same time, relatively independently of each other. To minimize the risk of major problems, a preliminary space envelope, configuration, and performance characteristics of each component is agreed upon very early in the work and as the design of each evolves, frequent coordination meetings take place to update the groups on the development of the design, particularly as it may affect their interfaces. If the group is housed in a common area, the coordination may be ongoing, with the need for formal meetings reduced. With this system some rework will be needed but it will normally be offset many times over by the saving in time to complete the design. As the design teams work on additional projects, their initial estimates become more accurate, so less rework would be required.

One of the important elements in the development of a design is liason with the **shop personnel** who will produce the product. The liason with the shop floor is essential if the design is to be readily producible. The design engineers must maintain a close and continuing relationship with the shop floor personnel who will actually produce the product to ensure that the design is practical and suited to the capability of both the personnel and the production facility. The attitude of the design personnel toward the shop personnel is also important. It is easy to forget that the shop work force that will produce the item have accumulated experience and skill that often spans decades. They should therefore be treated with respect and their views given proper weight. At times shop personnel may be less articulate and convincing than designers, but that does not lessen their contribution nor the importance of their views.

In some instances it might be useful to form an ad hoc committee of both design and shop personnel to work on development of a new or improved design. Such a group would meet periodically to review the development of the design and to suggest improvements and simplifications to facilitate production. The tendency to abdicate design responsibility to the shop personnel should be guarded against, but these people can make a valuable and very practical contribution to the design process.

Where new tooling is to be introduced, the shop personnel should be made a part of the decision process. They need not be given veto authority, but their views should be sought. This is particularly the case if automation, robotics, or other such improvements are contemplated. In these cases there is a significant impact on the work force, and it is better to deal with it early than to attempt to do it later with an unhappy work force. It is good practice to maintain the trust of the shop work force by letting them know early what is happening and what will be expected of them. Maintaining this trust goes a long way to easing the developmental problems and implementation difficulties associated with a changed design or a new product.

Another form of liason with the shop floor occurs where specially designed products are manufactured outside the organization. Wherever possible it is wise to consult with the outside vendors on the **manufacturability** of the item. Often design engineers will not be fully knowledgeable of manufacturing techniques and limitations, standard sizes of materials available, secondary effects of processes, and such, and will produce a design that, though functional, may be expensive to manufacture. It is not unusual for outside vendors to recommend nonfunctional design changes that can reduce cost and hence the purchase price by factors of 60 to 90% or more. An example might be a thin plate whose design calls for a series of small holes to be punched in it. A clever fabricator may wish to produce them by chemical milling at a fraction of the cost per hole, significantly cutting down on the cost of the part. Other examples could include part size selection that creates large quantities of scrap, surface finish, or appearance problems.

Economic lot size depends primarily upon the complexity of the product, the degree of precision needed in replicated parts, the tooling in use, and the cost target for the part produced. In most cases there is a trade-off between these variables. An individual case can only be analyzed by costing the various alternates. With labor costs becoming a smaller portion of the cost of the product, other production costs are becoming important. As a result, in estimating the costs of manufacturing parts, companies are increasingly turning to **activity-based financial management (ABM).** This method involves breaking down in some detail costs formerly charged to overhead or other general accounts and charging them to the part under consideration. Included are charges such as rent, utilities, depreciation, and interest on costs of special tooling, operator training, and so forth. This approach provides a more accurate and useful model upon which to base decisions.

For simple parts the economic lot size may be fairly small because replication is fairly easy. Similarly, for parts requiring low precision, relatively coarse tooling may be used, and it may be possible to produce the part satisfactorily with small lot sizes. Where high precision is required, the setup necessary to produce the part may require semi- or fully automatic machines, together with sophisticated inspection and gaging equipment. Where **numerically controlled (NC)** machines are not available, this requirement would involve a significant investment in either special machinery or jigs and fixtures, and thus a fairly large lot size to amortize the equipment and setup costs. With NC tools, once the programming is done, successive lots can be produced with virtually no special setup required. Often NC tools can lower the economic lot size to one unit. One example is the NC-directed laser cutting of light thickness parts.

Ultimately the **cost target** for the part will dictate the final production method employed. Very low cost targets will favor larger lot sizes and will often incur a higher degree of risk because of the difficulty of accurately predicting final costs, particularly with complex or new manufacturing systems. Further, where special tooling or processes are required, the machine tools utilized are often costly, and insufficient work may be in hand or confidently projected to amortize their cost. Faced with this circumstance, some manufacturers are contracting work out to other more specialized manufacturers. The advantages to this are lower market risk, simplified operations, and predictability of final costs (through the contract).

Steady improvements in the manufacturing processes will reduce the cost of successive units. The result is called the **learning curve,** and it applies to virtually any type of manufacture with more complex products having longer improvement curves. A rough rule of thumb that is used is that doubling the quantity produced causes a decline of 40 to 50% of the unit cost. This depends to a great extent upon the product, work force, duration of the work, and so on. More detailed information on predicting reductions in unit costs is found in Chapter 6.

2.16 SPECIALISTS

The use of **specialists** can be extremely helpful if the work is complex or unique. Before a specialist can be effectively employed, however, it is necessary to define the work product, the overall approach, and the schedule for the work, or at least its principal phases.

Specialists are available in virtually every engineering discipline. If they are not available in-house, personal references are an excellent way of obtaining them, because the quality of their performance will then be known. The telephone company business directories are also a good source. There should be an initial interview to determine qualifications, experience, references, availability, scope of work, fee schedule, and so on. After this, if the specialist is hired, a brief letter contract is usually sufficient to ensure agreement and understanding of the responsibilities, scope, fee structure, and other issues. A sample of a letter form contract is given in Fig. 2.23.

If the specialist is an organization to whom a work package is to be given, the entire interview and selection process will be much more rigorous and a formal contract will probably be used. The process of review and approval of the specialist's work will then be more frequent and intensive.

In using specialists, there is a tendency to keep them on a project longer than necessary. If the specialist is a part of the same company and is on loan to the project group, it may be easier to terminate his or her work on the project than if the specialist is a local consultant. If the specialist is based in another city and must travel to the design work location, there is a natural tendency to keep the person aboard even longer, since once the person returns to his or her location, the distance factor makes it less convenient to review further questions with them.

In general, when using someone outside the company, the person should be held responsible for the **correctness** of his or her work. A check or confirmation of the outside specialist's design is performed by the design group that hired that person. Despite this, if the specialists' work is inaccurate or in error, any changes or rework required should be performed by the specialist at no further charge.

As a part of its basic responsibility the design group should ensure that the work of the specialist is independently checked. It may in extreme cases even be necessary to hire a second specialist to perform an overall review of the results to ensure they are accurate. Errors in work of specialists do not usually occur in the detailed calculations or design but rather in the basic approach taken, or the assumptions made to facilitate the solution. The overall review mentioned above is generally sufficient to identify the problem areas.

2.17 PROCEDURES

Procedures to control and document the design process are essential. This does not mean that every design organization should have volumes of detailed step-by-step procedures to guide its every activity but rather that sufficient written direction be available. This assures that the design personnel will know the methodology and who has responsibility for specific actions and the correcting of them. In small organizations **desk books,** kept by the individual personnel can serve the purpose, and extensive detailed procedures are not needed. Naturally, the larger the organization, the more involved the practices become and the lengthier and more formalized the procedures. The important fact to bear in mind is that procedures are really an aid to a new employee or a reminder to an older one, and they should represent the actual way the work is conducted. It is important that the procedures be no more detailed than necessary. Remember, where a formal quality program is in effect, the quality organization will audit the design group for conformance to the procedures; any activities stated in the procedures but not followed will be written up as findings. A typical index of engineering department procedures for a large design firm doing complex work is shown in Chapter 3.

It is good practice and in some organizations a firm requirement that *all* documents originated by an individual be **initialed** and **dated.** These include notes of all types, calculations, memoranda, miscellaneous entries in files, and so on. As a result it is possible to determine not only who participated in the discussion/decision but the date of the activity—often critical in putting actions into perspective. Should there be any litigation, which these days occurs quite frequently, it serves the additional purpose of permitting a stronger case (or defense) to be put forward.

2.18 COMMUNICATIONS

Besides letter correspondence, the telephone and fax machine are the usual means used for external communication in today's engineering offices. Large offices will often have telex facilities as well. Still the telephone is far and away the most convenient method, and it has the advantage of immediate feedback from the other party, whose comments and tone of voice provide more information about the communication than both fax and teletype. Where telephone calls are used, there should be a **written record** of the call and its agreed action items—in particular, who has the next and subsequent

---

Consulting Agreement

This Agreement is made between ABC, Inc., a Massachusetts corporation (hereinafter referred to as "ABC") and John Smith (hereinafter referred to as "CONSULTANT"). ABC agrees to contract for the services of the CONSULTANT, and the CONSULTANT agrees to provide services under the terms and conditions in this Agreement.

I. STATEMENT OF WORK

The CONSULTANT shall provide consulting services on behalf of ABC in the area of thermal software development (see attached description) and any related matters ABC may request during the period of performance.

II. PAYMENT FOR SERVICES

In full consideration of the consulting services hereunder, ABC agrees to pay CONSULTANT at a rate of $75 per hour, plus reasonable expenses incurred at the request of ABC.

A monthly invoice describing services rendered and expenses incurred shall be submitted to ABC at the end of each month in which the services are rendered.

III. PERIOD OF PERFORMANCE

CONSULTANT shall be available for a maximum of 200 hours (25 days) for the period of performance beginning Jan. 1, 1985 and ending Sept. 30, 1985. This period of performance shall not be extended without written authorization by ABC.

IV. NOT-TO-EXCEED (N-T-E) LIMIT

Total payment under this contract shall not exceed $15,000, unless authorized in writing by ABC.

V. INDEPENDENT CONTRACTOR

It is understood and agreed that: CONSULTANT is an independent contractor in the performance of this Agreement, CONSULTANT is not an agent or employee of ABC, and CONSULTANT is not authorized to act on behalf of ABC.

CONSULTANT shall assume full responsibility for payment of all federal, state, and local taxes, and/or special levies required under unemployment insurance, social security, income tax, and/or other laws, with respect to performance of the CONSULTANT's obligations under this Agreement.

VI. RIGHT TO ACT AS CONSULTANT

CONSULTANT warrants to ABC that he is not subject to any obligations, contracts, or restrictions that would prevent him from entering into or carrying out the provisions of this Agreement.

VII. TERMINATION

This agreement may be terminated by either party at any time by giving written notice of such termination to the other party. Upon receipt of such written notice by either party, no further charges will be made under this Agreement. Termination shall not affect the CONSULTANT's obligations under articles IX and X.

VIII. HOLD HARMLESS

CONSULTANT shall indemnify and hold ABC harmless from any and all suits, claims, actions, damages, or losses whatever, resulting from any act or

---

FIGURE 2.23 Sample letter contract for a consultant.

omission of the CONSULTANT, its employees, agents, and sub-contractors in its performance hereunder.

IX. CONFIDENTIALITY

CONSULTANT acknowledges that information about the research, design, development, marketing, and manufacture of ABC's products, including findings, reports, and improvements made or conceived by the CONSULTANT under this Agreement, is confidential and of great value to ABC. Accordingly, CONSULTANT agrees not to disclose any such confidential information to any person not authorized by ABC to receive it. Upon completion of the work, CONSULTANT shall deliver to ABC all documents, drawings, specifications, and similar materials which were furnished by ABC to CONSULTANT or which were prepared by CONSULTANT in performance of services hereunder.

X. DISCOVERIES, INVENTIONS, AND COPYRIGHTS

CONSULTANT will promptly disclose to ABC all inventions, improvements, designs, and ideas made or conceived by CONSULTANT in the course of CONSULTANT's services under this Agreement. CONSULTANT assigns to ABC all right and title to such inventions, copyrights, and developments, and agrees to execute any and all such documents, including patent assignments, as ABC deems necessary to secure to it all right, title, and interest.

XI. AMENDMENT

This Agreement may be amended only by a written document, signed by both ABC and CONSULTANT.

XII. ASSIGNMENT

CONSULTANT may not assign this Agreement or any right hereunder. Any such attempted assignment shall be void.

XIII. GOVERNING LAW

This Agreement shall be governed by the laws of the Commonwealth of Massachusetts.

	Consultant		ABC
by	_____	by	_____
Name	_____	Name	_____
Title	_____	Title	_____
Date	_____	Date	_____

FIGURE 2.23 (continued).

actions and the scheduling of these actions. The telephone call record would be routed to all interested personnel for their information and reaction. It is also helpful to send a copy of the telephone notes to the other party to the call to assure a common understanding. Despite the relatively modest cost for individual telephone calls, in the aggregate they can be costly, particularly if long distance or international connections are involved. Therefore it is good practice to list the items to be covered in the call before making it to ensure complete coverage and brevity. This is especially useful when calling internationally or across several time zones, as the windows for communication are often reduced to only a few hours. In these situations it may be more convenient to employ fax or telex, since these messages can be received when an office is otherwise unattended.

The advantages of the telex are that both parties have time to reflect on their answers and a written record is produced. Further a two-way conversation can be held if both parties are at the telex machines in their respective offices.

For items of a more formal or contractual nature, including official transmittals of information, mail is still the most common method, although computers in separate offices can be linked to transmit data electronically. Regardless of the mode of communication, if the information is of an official nature, all correspondence should be **serially numbered** to permit easy reference and to ensure that each piece of the correspondence is received. One common system uses three letters followed by a unique sequential number. The first letter is the abbreviation of the sender, the last the receiver, and the middle letter, the type of document, for example, letter, memo, telex or fax. Thus a letter sent from Greene Engineers to the Ajax Resources Company could be numbered GLAxxx, with a telex sent the other way numbered ATGxxx. With this system both parties maintain and periodically exchange registers of the communications, thus ensuring that all communications are received. This system is usually only implemented between parties with which there is a continuing dialogue. The system can be used even with outside parties who may not wish to establish such a system for their own correspondence to at least ensure that all correspondence to them is controlled and accounted for.

2.19   REPORTS

Although many engineers consider **reports** the bane of their existence, all kinds of reports are essential for the orderly execution of a project. Table 2.2 lists the reports that may be produced on a large project and their frequency. For smaller projects the list would be scaled down in number and adjusted in frequency as appropriate. The burden of issuing these reports can be reduced if the control document that constitutes the report is kept up to date and merely reproduced when the report is required. Indeed all of the report documents must be kept current if they are to be of any value.

2.20   STUDIES

Every design project will involve the preparation of some **studies.** They may be as simple as half a page, outlining a design decision based upon an inspection of the requirements, or they may involve a full-blown design, cost and schedule analysis of the alternates considered. Regardless of its depth, each study must be approached with the same rigorousness to ensure that the results are correct.

TABLE 2.2 **Report Checklist**	
Item	Frequency
Progress report	Monthly
Equipment list	Monthly
Motor list	As required
Drawing control	Monthly
Microfilm selected calculations and drawings	Monthly
Engineering schedule update	Monthly
Requisition register	Monthly
Commitment register	Monthly
Purchase order register	Monthly
Cost trend report	Monthly
Conference notes, trip reports, telephone notes, etc.	Immediately after event

For large projects, or those where the work is performed on a firm price basis, it is useful to list, at least in a preliminary way, the studies anticipated. To provide the client with a complete evaluation of the project work, the studies contemplated, together with an estimate of the effort required, should be made available to them at an early stage of the work. Clients will often request additional studies after the work is undertaken. If the work is to be performed on a fixed price basis, an extra charge may be added for this work; in this circumstance it is important to recognize that not only is the cost to perform the work increased but additional time may be required in the schedule.

2.21 MODELS

Despite the extensive computational capabilities currently available to us, some engineering design problems can best be solved by **modeling.** Often these relate to **dynamic performance,** two examples being predicting the performance of breakwaters in severe storms and sedimentation behind dams. River flows and hydraulic mixing represent other similar examples.

A particularly valuable resource in modeling these systems and developing solutions are the laboratories at major universities. While the facilities to conduct the necessary model tests are found at numerous institutions, not all universities have them. The best course of action is to establish a consulting agreement with a professor who is an authority in the field and to utilize his advice in selecting the laboratory. Often this will be a laboratory at the same university or in the local area. The professor can assist not only in making the arrangements but in supervising the model tests and evaluating the test data. On occasion, an iterative approach is taken, with necessary adjustments after a series of progressive tests is conducted; in this circumstance the involvement of an expert can be especially helpful.

The use of three-dimensional scale models for architectural and presentation purposes continues to be important. The use of three-dimensional models to study design interferences, access, and similar factors is now greatly diminished with the improved capability of the currently available CAD programs. Some of these programs feature a "walk-thru" capability that permits the user to literally walk through the facility to establish clearances and general configuration as it would be viewed by someone passing through the interior of the actual facility. Other programs present external views from various heights, at different times of day and year, and so on.

2.22 DEGREE OF COMPLETION

The **degree of completion** of the design will depend upon the type of contract governing the design work and the responsibilities of firms performing follow-on work. Where the contract calls for a complete design, including working drawings, the design will be completed to its full extent. In other cases where the design is to merely provide general guidance, and will not be followed in detail, the design can be more general establishing overall dimensions, critical interface points and requirements, and essential performance data. Still the design should be specific on the codes and standards to be followed, as well as set forth any sizing criteria such as stress levels.

Regardless of which of these two levels of design is pursued, certain types of detailed drawings are not usually part of the design work and, if required, can be considered outside the scope of the work. Generally these drawings are fabrication drawings used to control shop activities, such as rebar bending drawings and schedules, pipe spool drawings, connection diagrams, piping and sheetmetal layout drawings.

Where the client has elected to shift part of the design responsibility to other parties, care should be taken to ensure the **contract** reflects the proper division of responsibility for the correctness of the design. Often, to ensure adequacy of the design, the client will want the original designer to review the design submittals of the downstream organization and in some cases to provide inspection and engineering representation at the point of fabrication/erection.

2.23 WRITING AND STYLE

As important as skill in technical engineering matters is the ability to effectively communicate ideas in **writing.** Unfortunately, most engineers have little or no formal training in technical writing or in writing in general. As a result their work is not well presented, and the value of their efforts may be underestimated. Clarity, brevity, and directness are important attributes of effective writing and should be kept in mind when preparing any written document. Two excellent guides to writing, style, composition, and English usage are recommended to every practitioner: *The Elements of Style* by William Strunk, Jr., and E. B. White (The Macmillan Company) and *Harbrace College Handbook* by John Hodges and Mary Whitten (Harcourt Brace Jovanovich).

In engineering work the purpose of writing is to communicate information and ideas, and often to recommend action. Even though the technical development of a solution is important, the communication itself must be effective. It is always useful to place oneself in the position of the reader and to view the document from that perspective.

When writing a document that is controversial or argumentative, it is good practice to allow a day or so to elapse between writing the document and issuing it. This is particularly the case with a document prepared with some emotion. Often what seems so obvious, upon reflection could benefit from some rewriting or addition.

Report writing is an essential skill of the engineer and must be performed with a view toward providing full information for the reader or user of the report. Typically reports are structured as shown in Fig. 2.24. A portion of the report (typically the introduction) would indicate who participated in the preparation of the report, when the report was prepared, and so forth. This provides a source for clarifications and also permits the reader to determine the validity of the report based on the credentials or reputation of the author(s) and contributors.

The **purpose** sets forth why the report was written, for instance, the factors that led to its writing. The **scope** describes the extent of the report, limitations on its coverage, and so on. This serves to put the background of the report in clear perspective so that the reader will understand fully not only the purpose of the report but any limitations imposed.

Conclusions and recommendations may be one section or separate sections, but should be placed toward the front of the report, so that a busy person can easily locate this section and readily determine the findings of the report. Each conclusion must be supported by material elsewhere in the report. Conclusions are rarely useful without recommendations. Therefore typically recommendations will be included and are essential to a full use of the report.

Sections on **discussion** and **appendixes** are self-evident and become a convenient way of dividing the material between that found in the course of the investigation (Discussion) and Appendixes, which typically are secondary or reference material of limited interest to most users of the report, although of possible interest to some.

For extensive or lengthy reports a separate **executive summary** is often prepared which, although placing major emphasis on the conclusions and recommendations, provides an overview of the entire report.

FORMAL STUDY REPORT FORMAT

1.0   INTRODUCTION

    1.1   Purpose
    1.2   Scope
    1.3   Objectives

2.0   SUMMARY OF STUDY RESULTS

3.0   CONCLUSIONS AND RECOMMENDATIONS

4.0   DESCRIPTION OF ALTERNATIVES/SOLUTIONS

    4.1   Study Bases

        4.1.1   Criteria
        4.1.2   Assumptions
        4.1.3   Methodology

    4.2   Alternatives

    [Include as appropriate]
    o   Title
    o   Description
    o   Advantages and Disadvantages
    o   Safety Considerations
    o   Environmental Considerations
    o   Cost and Schedule Estimates (Capital and Operating) and Uncertainties
    o   Other Decision Data

5.0   DISCUSSION OF RECOMMENDED ALTERNATIVE/SOLUTION

6.0   SOURCES (REFERENCES AND BIBLIOGRAPHY)

7.0   APPENDICES

Notes:

(1) The Introduction should be brief (1-2 pages). If any background information is needed, include it in Section 1.1 under Purpose.

(2) For Sections 4.1.1, 4.1.2, and 4.1.3, information should generally be listed in bullet format.

(3) Study alternative should be designated by letter (i.e., A,B,C); variations on a given alternative may be designated with a following number (e.g., B1, B2, B3).

(4) Key drawings, tables, etc. should be included in the body of the study. When needed, backup drawings and other supporting material should be included as appendices.

FIGURE 2.24   Formal study report format.

Reports should be written in clear, concise language, using short, direct sentences. Ponderous or obscure language should be avoided, and technical or complex material should be put into understandable language. Abstractions should be avoided, and descriptions made as concrete and specific as possible.

Graphical illustrations are preferred over tabular material, although tables can be placed in the discussion or appendixes if necessary.

2.24 DATA BOOKS AND OPERATING MANUALS

The preparation of **Data Books** or **Operating Manuals** may compromise budgets and project schedules by extending the engineering effort. There should be suitable provision in the budget and in the schedule for these manuals, with additional time allowed depending on the complexity of their preparation. Data Books and Operating Manuals are largely (75% or more) a collection of information provided by the vendors with some project flow diagrams, P&IDs, Electrical Single Line Drawings, and so on, to tie the data and system information together. If drawings and data required are not produced in the course of design, this additional effort must be negotiated as a part of the original contract for the work.

During the construction/fabrication phase, minor changes are frequently and routinely made in the configuration of the work. Often owners for the use and convenience of the operating personnel will request the preparation of **"as built"** drawings. These drawings show the actual configuration. While meritorious, this can impose significant additional work upon the design group and can produce information that often is of little practical value since most data can later be obtained from the parts and systems themselves. It is far better, if "as built" data are requested, to try to limit the drawing to **hidden** or embedded features, or underground lines and similar systems where it is not possible to readily determine the routing or configuration.

If possible, a sample data book should be shown to the client early in the negotiation process, and agreement obtained as to its type, depth, content, and arrangement. In addition the **number** of copies to be provided needs to be established before major procurements are underway to ensure that vendors provide sufficient copies of their operating instructions, spare parts lists, special drawings, and so forth, to be incorporated into these manuals. To facilitate the preparation of manuals, a holding file can be established where material can be accumulated until the later stages of the project when the manual is ready to be prepared. Calculations are almost never found in data books or operating manuals, and they should not be included.

2.25 RECORDS SECURITY AND RETENTION

To reduce the possibility of a major loss due to fire, flood, or similar disaster, some system must be in place for periodically duplicating important records. A **retention** program should be implemented where, for example, calculations and drawings are microfilmed monthly and computer calculations and drawings are backed up daily or weekly. Electronic media such as disk or tape storage are recommended over optical systems such as microfilm because they permit more data to be stored in less space. In such a program it is important to also duplicate other important project records such as confer-

TABLE 2.3   Record Retention Recommendations
Permanently Retained
All design record prints and selected vendor prints All calculations of permanent project design work Selected specifications and requisitions of major materials, equipment, and services Selected outline drawings, performance data, and instruction manuals for major equipment Plant data books Operating manuals Equipment lists and drawing controls Significant, selected client correspondence Reports, studies, and proposals Construction/fabrication photos
Five-Year Retention
Client correspondence Supporting data for special studies and reports
Two-Year Retention
Specifications, requisitions, and associated vendor prints and correspondence for equipment having one-year guarantees

ence notes, contracts, and major purchase orders. The backup data should be stored in a separate secure location to avoid a common disaster that might effect both locations. Temperature and humidity control may be necessary in the records storage area to prevent deterioration of the stored documents.

Appropriate retention of design records is important to the future position of the firm. An orderly system of record retention permits the **retrieval** of data for future projects. Such information is also useful in avoiding or defending against **litigation.** While legal advice should be obtained to develop a suitable standard, the sample shown in Table 2.3 can be used as a starting point. (Record retention for contractural documents is discussed in Chapter 11.)

CHAPTER 3

Design Control

Control is an important factor in the design process. Implemented properly, design control ensures that the engineering output will evolve in accordance with the agreed scope, that the design will meet functional requirements, that the design will be on schedule, and that costs will be properly controlled. Control is necessary to assure the work is to be of the proper quality and that the health and safety of the public is protected. Design control can be complex; it affects implementation and is essential to the economic and efficient discharge of project responsibilities. Project control procedures will vary from company to company and may be simple or intricate depending upon the type of work undertaken. Design control should be appropriate to the work and no more elaborate than necessary.

3.1 DESIGN CONTROL

Design control can be implemented a number of ways. Design control systems and documents can be made as simple or complex as necessary. They are tailored to each activity over the entire span of the project. For small or noncomplex work, documents prepared **manually** can be effectively used, but for large projects the computer is particularly helpful in updating and in producing timely status and exception reports. Computerized reports can become excessively detailed and burdensome if not thought out, modified, and controlled as experience indicates. In a large organization care must be taken to ensure that their preparation and maintenance does not become an end in itself.

In designing or utilizing these documents, the trick is to keep them **simple**—the simpler the better. It is better to err on the side of too little information than too much, particularly if the information is to be periodically updated. A remarks column on the control forms provides flexibility and allows the entry of important but infrequent data.

One person should be made responsible for the preparation and issuance of the documents. Much of the work of maintaining the documents is nontechnical. The person assigned the task does not need to be an engineer or a technician; the most important criteria are that the person be attentive to detail, thorough, and accurate. Where the project control system uses a computerized network, documents can be continually up dated by the design groups when routinely entering progress and forecast dates for their activities.

Large projects, with progress documents numbering in the several thousands, require the services of more than one full-time person. For large projects that have strict document control procedures, a **document control center** is established through which all documents pass to ensure that each document is controlled and tracked.

3.2   POLICIES AND PROCEDURES

Policies and **procedures** need to be well defined. The working level personnel must be aware of their responsibilities, so there must be a clear job description for each person. All personnel should have access to the policies and procedures documents. Flow diagrams are helpful in defining the responsibilities of the design groups, and these diagrams should show how the groups are expected to interact in the design process. The availability of these documents will reduce the orientation time of new employees.

Procedures can be standardized for a department or a design discipline, or they can be unique to a particular project. **Standard procedures** reflect not only the technical requirements of design execution but also the organization of the company. In addition they cover the normal methods used to coordinate between groups, conduct reviews, ensure accuracy, and so forth. **Project unique procedures** will typically evolve from the departmental standard procedures. Often they deal with project-specific nontechnical matters such as procedural steps and special reviews. **Economic evaluation factors** are the exception to this. They vary from project to project depending on the client and the economic conditions in the industry or geographic area. Usually these factors are part of the project criteria established either in the contract or during the initial phase of the design work. One example of a set of project procedures is given in Chapter 8.

Some procedures are used on all projects to ensure a common approach to the work and permit the same forms and methods to be used on all projects. This reduces the chance for error and also the time required to indoctrinate new employees.

The project unique procedures are prepared by the personnel performing the work and reviewed by the supervisor to ensure they represent the **actual** practices employed. Where quality-related requirements are stated, the quality (control or assurance) organization should review and approve them. One area where policies and procedures are particularly important is in the area of human resources; this is discussed in more detail in Chapter 12. One large engineering company working with the Department of Energy and the Nuclear Regulatory Commission, as well as numerous clients in the private sector, uses the engineering department procedures listed in Table 3.1.

3.3 TECHNICAL SCOPE DESCRIPTION

A technical **scope** description is fundamental to the design process. For projects that are complex, costly, subject to significant design evolution, or of long duration, the technical scope description will be lengthy. For projects

---

TABLE 3.1   Engineering Department Procedures

The EDP system	Material requisitions and purchase memorandums
Engineering planning and control	Technical services contracts
Design criteria	Supplier engineering and quality verification documents
Engineering standards and guides	
Configuration management for DOE projects[a]	Nonconformance reports[a]
Engineering studies	Field change requests/field change notices
Licensing documents[a]	
Design interface control[a]	Supplier deviation disposition requests[a]
Design verification[a]	
Identification of items/services subject to QA programs[a]	Information or material prepared for submittal to the NRC[a]
Off-project design review	Start-up field reports
Standard computer programs	Generic deficiency information processing
Design calculations	
Computer program error reporting	Protection of safeguards information[a]
Standard project document and component numbering	Indoctrination/orientation and training[a]
Project drawings	Bid evaluation
Engineering specifications	Specification of supplier quality assurance program requirements[a]
ASME III design specifications[a]	
Qualification of personnel authorized to perform ASME III code certifying activities[a]	Evaluation of supplier quality assurance programs[a]

[a] Denotes those procedures issued to support projects with regulatory quality program requirements. Use on other projects is optional.

that are simple, the technical description of the work may be only a few pages long.

If the work is being performed for another party (the owner) this method will define the work for the owner and may even be incorporated into the contract. The technical scope description establishes for the design team the general objectives of the work. The more detailed it is, the better it will guide the work and avoid unnecessary studies and false starts. A full definition of the extent (scope) of the work can prevent later misunderstandings between the designer and the owner, and it establishes a firm basis for estimating the cost and scheduling of the design. For complex projects the technical scope description is kept in a loose leaf binder, to which additional pages can be added. As design changes are approved, they can readily be incorporated thus providing a convenient up-to-date record of agreed-upon design configuration. Page revision and date control can be used to ensure accuracy.

The technical scope description begins with a definition of scope of the work, followed by sections that deal with the operational requirements of the facility or product on a design discipline basis. The organization of a scope book for an aircraft project is given in Table 3.2.

TABLE 3.2   Project Scope Book

Overall scope and performance requirements:
  Mission, general dimensions, cargo capacity, stability, service life, unique service requirements, overview of the aircraft
Airframe:
  Type of construction, stress levels, methods of analysis, structural and dynamic testing requirements, fabrication techniques, repair methods
Power plant:
  Type, fuel, fuel consumption, refueling features, tankage, maintenance requirements, mounting, engine changeover features
Control and guidance systems:
  Type, backups, use of boosters, avionics, navigation systems, instrument arrangement
Interior features:
  Number of flight personnel, seating, cockpit arrangement, access, lighting, communications, power, HV&AC, cargo stowage
Operational support requirements:
  Engine starting, power and ground utilities required, minor maintenance provisions, major overhauls, cargo handling
Spare parts:
  Requirements, commonality with other type aircraft, rate of consumption, stockage, replenishment, etc
Testing and performance demonstrations:
  Overall testing and performance demonstration program

3.4 DESIGN REVIEW

Design control involves ongoing progress reviews of deliverable drawings and specifications. Design review involves not only ascertaining that the design conforms to the agreed scope and technical content, but also that the **design elements** are properly integrated and that the design is progressing at a satisfactory rate.

To ensure a proper evaluation of design, ANSI N45.2.11 and ANSI/ASME NQ-1 have established design review criteria. These criteria are widely used and should be adopted for any program involving design review. They are excerpted and shown in Fig. 3.1.

In the manufacturing sector, design reviews must in addition include considerations of **manufacturability** and **serviceability.** These must be sufficiently broad to cover not only the suitability of the design but also design margins and special considerations such as noise immunity, vibration, adverse ambient conditions, shock, and acceleration. The specifics of the manufacturing process are vital and should also consider tooling, economics of special machinery and processes, and integration of the design with manufacturing to ensure overall optimization. For serviceability and warranty there are considerations of ease of access for and convenience of maintenance, safety to operators and maintenance personnel, and protection from damage due to inadvertant use.

When **concurrent engineering** is undertaken, the design review must include all of the potential inputs to the design process; in other words, sales, design, manufacturing, construction, maintenance, and post sales service participate in the developmental work. The design then becomes qualified and can be put in production. The guiding principle in this work is that there is not merely a single customer who must be satisfied with the product but a whole family of customers within and outside the company who have an interest in its final configuration.

To confirm proper scope and technical content, **reviews** are often conducted by the PE using the project scope description or the contract to ensure compliance. For projects where external licensing or permitting requirements exist, scope descriptions must be included in these reviews. For complex projects or those of long duration, the formal reviews should be conducted about twice a year. In some organizations a separate group is assigned the task of performing these reviews to ensure an unbiased report. In these cases the reviewers are technically competent and experienced in the design process.

A progress review of the drawing and specification controls can be used to provide an overall quantitive evaluation of the status of the work. Where a system of **earned** versus **expended** hours is in place, it can be very detailed.

One widely used informal method of evaluation is **"walking the boards."** This method consists of periodically reviewing the drawings, whether completed or in progress. Some time each day is set aside for the PE to walk

DESIGN REVIEW ELEMENTS
(Excerpt from ANSI N45.2.11 and ANSI/ASME NQA-1)

1. Were the inputs correctly selected and incorporated into design?

2. Are assumptions necessary to perform the design activity adequately described and reasonable? Where necessary, are the assumptions identified for subsequent re-verifications when the detailed design activities are completed?

3. Are the appropriate quality and quality assurance requirements specified?

4. Are the applicable codes, standards and regulatory requirements including issue and addenda properly identified, and are their requirements for design met?

5. Have applicable construction and operating experience been considered?

6. Have the design interface requirements been satisfied?

7. Was an appropriate design method used?

8. Is the output reasonable compared to inputs?

9. Are the specified parts, equipment, and processes suitable for the required application?

10. Are the specified materials compatible with each other and the design environmental conditions to which the material will be exposed?

11. Have adequate maintenance features and requirements been specified?

12. Are accessibility and other design provisions adequate for performance of needed maintenance and repair?

13. Has adequate accessibility been provided to perform the in-service inspection expected to be required during the plant life?

14. Has the design properly considered radiation exposure to the public and plant personnel?

15. Are the acceptance criteria incorporated in the design documents sufficient to allow verification that design requirements have been satisfactorily accomplished?

16. Have adequate pre-operational and subsequent periodic test requirements been appropriately specified?

17. Are adequate handling, storage, cleaning, and shipping requirements specified?

18. Are adequate identification requirements specified?

19. Are requirements for record preparation review, acceptance, retention, etc., adequately specified?

FIGURE 3.1   Design review elements. Reprinted by Permission of American Society of Mechanical Engineers.

through the drafting area and review the work in progress. This is not a formal review. The PE looks over the shoulder of the drafter and asks a few questions about the drawing currently under preparation. This may mean only a minute or two for a drawing or even a brief conference. The idea is that if someone's work appears to be departing from the agreed approach, it can be redirected while still at an early stage. The PE benefits by improving his or her knowledge of the status of the work and developing a good sense of the drafting work being done, which is often the final output of the design process.

Design control can also be achieved by means of **formal** review. Typically these reviews are called design reviews, constructability reviews, and—for more specialized jobs—operability, maintainability, safety, corrosion, and so forth.

Boards of Control, review boards, or consultants often perform these reviews. As its name implies, a Board of Control is a group that conducts reviews based on a system of periodic or milestone submittals of design work. Frequently the milestones are activities shown on the project schedule, and the approval of the board is necessary before further work can be undertaken. The work of the board is often structured, with formal minutes taken and a record maintained of the approvals, comments, or directives that the board issues.

Review boards are less formal. They are typically limited to making recommendations rather than issuing formal approvals. Review boards meet only when necessary, and not necessarily on schedule milestone dates. While seemingly having less authority than boards of control, their role can still be quite effective if supported by senior management.

The consultant frequently is one individual whose expertise is often limited to one topic or speciality. His or her effectiveness depends on the support senior management gives them and on the seriousness with which the findings and recommendations are taken.

On Department of Defense projects, where the output is a manufactured item, formal design reviews normally comprise one or more of the following four steps:

1. *Preliminary Design Review.* The functional design is reviewed to see if its concept and planned implementation is in accordance with the basic operational requirements.
2. *Critical Design Review.* The hardware and software details are reviewed to see if the preliminary design has been properly executed.
3. *Functional Readiness Review.* Proof-of-design hardware [first article(s)] tests are applied to see if the hardware is manufactured correctly. This is the last step prior to authorizing full-scale production of the item.
4. *Product Readiness Review.* Specifications and drawings are matched to the manufactured hardware to confirm that the controlling design documents will produce the required hardware.

3.5 PROJECT SCHEDULE

A project **schedule** is maintained to ensure the development and timely achievement of the design milestones. The schedule should be simple and limited to one page (usually 11 × 17 inches) about 20 to 40 lines long. The format is often a **bar chart** with four or five milestones (or restraints) shown for each line item. For a typical design-and-build project, about one-third of the items shown will affect the design, but it is essential to include the other items so that a schedule can be set for completion of blocks of work. For design-only projects the schedule can be very short; 20 lines will often adequately cover the main activities and their milestones.

If the project is large, complex, or lengthy, the schedule will be supported by additional, detailed schedules; as many as 10 or 20 schedules may be needed. Usually they are of the **precedence** type and are computerized to permit machine processing. The supplementary schedules are maintained by discipline group, each of which provides the details necessary for its portion of the schedule and updates the progress. Each schedule is coordinated with the master project schedule in terms of restraints and milestone dates. An example of a project schedule is shown in Fig. 5.4.

3.6   MATERIAL ASSIGNMENT SCHEDULE

Sometimes design work is performed by a construction or fabrication group. Examples are secondary utility systems, routing of 2 inch pipe and under, small foundations, field run conduits, and lighting. The **Material Assignment Schedule** records make-or-buy decisions along with other project responsibilities. Project responsibility must be well defined to avoid ommission or duplication of work. The Material Assignment Schedule is both a planning and control document, and it is developed early to prevent scheduling conflicts. An example of a Material Assignment Schedule is shown in Fig. 3.2.

3.7   DESIGN CONTROL CHECKLISTS

Design Control Check Lists define the key documents, design criteria, drawings, and specifications for review off project. These lists are established on a discipline basis at the outset. Once work is in place, they should only be expanded for compelling reasons. The chief engineers of the various design groups—or in a small design organization, other technically qualified personnel who are not associated with the project—act as the design reviewers. The design is compared to the criteria established in the Technical Scope Book, to determine if the methodology, calculations, and general approach are appropriate. Where a Technical Scope Book does not exist, the design parameters would be checked against the basic contract, applicable codes, standards, previous designs, and current industry practice. This review would include a check of important calculations only. The checking method need not involve a repetition of the designer's calculations but can utilize another method described in Chapter 2. It is important to allow enough time in the design schedule to permit the checking to be orderly and incorporate any important changes. It is prudent to allow modest durations in the schedule

MATERIALS ASSIGNMENT SCHEDULE

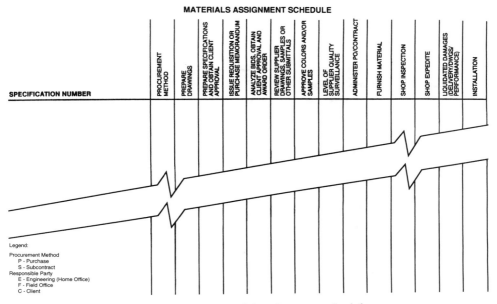

SPECIFICATION NUMBER

Column headers (left to right):
- PROCUREMENT METHOD
- PREPARE DRAWINGS
- PREPARE SPECIFICATIONS AND OBTAIN CLIENT APPROVAL
- ISSUE REQUISITION OR PURCHASE MEMORANDUM
- ANALYZE BIDS, OBTAIN CLIENT APPROVAL AND AWARD ORDER
- REVIEW SUPPLIER DRAWINGS, SAMPLES OR OTHER SUBMITTALS
- APPROVE COLORS AND/OR SAMPLES
- LEVEL OF SUPPLIER QUALITY SURVEILLANCE
- ADMINISTER PO/CONTRACT
- FURNISH MATERIAL
- SHOP INSPECTION
- SHOP EXPEDITE
- LIQUIDATED DAMAGES (DELIVERY/DWGS/ PERFORMANCE)
- INSTALLATION

Legend:
Procurement Method
  P - Purchase
  S - Subcontract
Responsible Party
  E - Engineering (Home Office)
  F - Field Office
  C - Client

FIGURE 3.2   Material assignment schedule.

for such changes. An example of a Design Control CheckList is shown as Fig. 3.3.

3.8 DRAWING CONTROL

The major work product of the design group will be drawings and specifications. A **Drawing Control** is used to control and monitor all the drawings planned, their scheduled and actual starts, and completion dates for both the initial versions and revisions. Care should be given to including even **conceptual** and **developmental** drawings. Among these are Flow Diagrams, Block Diagrams, Material and Energy Balances, Piping and Instrumentation Diagrams, Logic Diagrams, and Electrical Single-Line Drawings. These drawings will typically have several times as many revisions as the detailed drawings that evolve from them, and it is a good idea to allow extra space on the Drawing Control. The balance of the Drawing Control will note the drawings that result from this initial group.

For small or short-duration projects the progress of drawings can be represented by a bar graph that is filled in to indicate the estimated percentage of completion of the drawings. While not hard quantitative data, a bar graph can readily identify drawings that are significantly behind schedule.

A rational system of **numbering** drawings should be decided upon at the outset of the project. Some systems utilize area numbers, discipline letters,

DESIGN CONTROL CHECKLIST

DISCIPLINE   PAGE 1 OF

                                                    PROJECT:
                                                    JOB NUMBER:

Document Number	Item/Description		Remarks		
	Design Criteria, Design Specifications,				
	P&IDs, Flow Diagrams				
	Condesate Storage and Transfer System Makeup Water System		Shaded Documents have been added on this revision of DCCL		
	Reactor Component Cooling Water System Plant Service Water System Station Service Air System Fire Protection High-Pressure Nitrogen Gas Supply System Instrument Air System Auxiliary Steam Supply System				
	Reactor Building HVAC Systems Chilled Water System				
REV NO.	*DATE*	*REVISON*	*APPROVED GR SUPV*	*APPROVED CHIEF ENGR*	*APPROVED PROJ ENGR*

FIGURE 3.3   Design control checklist.

master part numbers, and so on. Thus a drawing number might be the project number followed by the area, the elevation, the sequence number, and the revision: for example, 736-14-572-24-Rev 3. A discipline letter system could have the project number followed by the discipline letter (E = electrical, M = mechanical, etc.), followed by a sequence, area, or part number, and then by a revision number: for example, 736-E-342-Rev 4. A straight part number system could use the product code, followed by the part or assembly number, then by a subpart number (if part of a larger assembly), and finally by a revision number: for example, V42-74917-07-Rev 4. In machine design work revisions are customarily not used; instead new part numbers are assigned to the modified design.

Drawings issued to outside groups for procurement, construction, or fabrication must be controlled more rigorously than those still being developed in-house. One practice is to use **lettered** revisions for in-house work and **numbered** revisions for drawings issued to others. This permits immediate recognition of the drawing's status by the type of revision indicated.

Each design group maintains a **stick file** with a copy of each current drawing as it is issued. The stick file is handy for both the designers and engineers, and they are usually located nearby. Markups of interim changes and corrections are made to the stick file copies, just prior to the next issue of a drawing, and are incorporated into the tracing or CAD file. When it is necessary to distribute minor changes or corrections to plan holders and rather than issue an entire drawing for a minor change, a **Drawing Change Notice** (DCN) is used to pass on the information. The DCN is usually an $8\frac{1}{2} \times 11$-inch, letter-size sketch indicating the change with suitable identification of the base drawing number, current revision, reason for change, and so forth. Copies are distributed, and one copy is attached to the stick file print. A memorandum register of DCNs is maintained to avoid duplication and to ensure that unique control numbers are used. After the accumulation of a specific number of DCNs (often six), the drawing is revised and reissued. When field changes are made, the field issues a **Field Change Notice** (FCN), which is handled in the same fashion as DCNs.

For design work performed by outside groups, it is useful to assign blocks of drawing numbers. The blocks should be generous enough to avoid broken sequences, should more drawings be needed than originally anticipated.

Certain types of **standard drawings** are used across company, project, or product. These standard drawings will bear a common drawing number, with each drawing within the set bearing a sheet number and a revision designation. It is good practice, even if these drawings are produced outside the project design group, to include these standards drawings in the project drawing control.

Once a number is assigned to a specific drawing, it cannot be reassigned. If canceled, or no longer needed, the number is withdrawn from further use. Doing so prevents old versions of superseded drawings from reappearing and creating confusion and error. Examples of a computerized Drawing Control and a manually prepared Drawing Control form are shown in Figs. 3.4 and 3.5.

3.9 SKETCH CONTROL

Sketches are always used in developing the design, but they do not require the same rigorous control as drawings. At some point, usually relatively early, they may be discarded. A sketch control register is an informal record listing the sketch title, a number, and a revision.

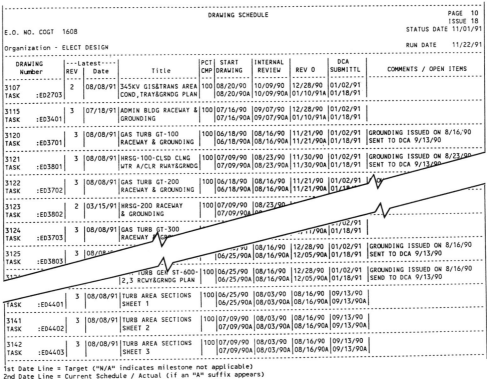

DRAWING Number	REV (Latest)	Date	Title	PCT CMP	START DRAWING	INTERNAL REVIEW	REV 0	DCA SUBMITTL	COMMENTS / OPEN ITEMS
3107 TASK :ED2703	2	08/08/91	345KV GIS&TRANS AREA COND,TRAY&GRNDG PLAN	100	08/20/90 / 08/20/90A	10/09/90 / 10/09/90A	12/28/90 / 01/10/91A	01/02/91 / 01/18/91	
3115 TASK :ED3401	3	07/18/91	ADMIN BLDG RACEWAY & GROUNDING	100	07/16/90 / 07/16/90A	09/07/90 / 09/07/90A	12/28/90 / 01/10/91A	01/02/91 / 01/18/91	
3120 TASK :ED3701	3	08/08/91	GAS TURB GT-100 RACEWAY & GROUNDING	100	06/18/90 / 06/18/90A	08/16/90 / 08/16/90A	11/21/90 / 11/21/90A	01/02/91 / 01/18/91	GROUNDING ISSUED ON 8/16/90 SENT TO DCA 9/13/90
3121 TASK :ED3801	3	08/08/91	HRSG-100-CLSD CLNG WTR A/CLR RWAY&GRNDG	100	07/09/90 / 07/09/90A	08/23/90 / 08/23/90A	11/30/90 / 11/30/90A	01/02/91 / 01/18/91	GROUNDING ISSUED ON 8/23/90 SENT TO DCA 9/13/90
3122 TASK :ED3702	3	08/08/91	GAS TURB GT-200 RACEWAY & GROUNDING	100	06/18/90 / 06/18/90A	08/16/90 / 08/16/90A	11/21/90 / 11/21/90A	01/02/91 / 01/18...	
3123 TASK :ED3802	2	03/15/91	HRSG-200 RACEWAY & GROUNDING	100	07/09/90 / 07/09/90A	08/23/90			
3124 TASK :ED3703	3	08/08/91	GAS TURB GT-300 RACEWAY &GR...				...11/90A	01/02/91 / 01/18/91	
3125 TASK :ED3803	3	08/0...			...90 / 06/25/90A	08/16/90 / 08/16/90A	12/28/90 / 12/05/90A	01/02/91 / 01/18/91	GROUNDING ISSUED ON 8/16/90 SENT TO DCA 9/13/90
312...			TURB GEN ST-600-2,3 RCWY&GRNDG PLAN	100	06/25/90 / 06/25/90A	08/16/90 / 08/16/90A	12/28/90 / 12/05/90A	01/02/91 / 01/18/91	GROUNDING ISSUED ON 8/16/90 SEND TO DCA 9/13/90
TASK :ED4401	3	08/08/91	TURB AREA SECTIONS SHEET 1	100	06/25/90 / 06/25/90A	08/03/90 / 08/03/90A	08/16/90 / 08/16/90A	09/13/90 / 09/13/90A	
3141 TASK :ED4402	3	08/08/91	TURB AREA SECTIONS SHEET 2	100	07/09/90 / 07/09/90A	08/03/90 / 08/03/90A	08/16/90 / 08/16/90A	09/13/90 / 09/13/90A	
3142 TASK :ED4403	3	08/08/91	TURB AREA SECTIONS SHEET 3	100	07/09/90 / 07/09/90A	08/03/90 / 08/03/90A	08/16/90 / 08/16/90A	09/13/90 / 09/13/90A	

DRAWING SCHEDULE — PAGE 10, ISSUE 18
E.O. NO. COGT 1608 — STATUS DATE 11/01/91
Organization - ELECT DESIGN — RUN DATE 11/22/91

1st Date Line = Target ("N/A" indicates milestone not applicable)
2nd Date Line = Current Schedule / Actual (if an "A" suffix appears)
START DRAWING = 10%, INTERNAL REVIEW = 50%, REV 0 = 20%,

FIGURE 3.4 Drawing control. Copyright © Ebasco Services, Inc., 1992.

3.10 VENDOR PRINT CONTROL

Drawings are also provided by **vendors** which furnish interface information or in some cases are incorporated as a reference into a design. A record of these drawings also needs to be maintained. Such drawings require a special project numbering system, since the numbers assigned to them by the vendor will not fit the project numbering system. A number of methods are possible. One popular system uses the purchase order number (under which the equipment or component is purchased), followed by a sequence received number, followed by the submittal number.

The special project number is added to the drawing(s) upon receipt, and this rather than the vendor's number is used for internal control. The purchaser has the right of approval of these drawings, and several submittals may occur. Submittals may be returned to the vendor for additions or corrections and subsequent resubmittal. An **approval stamp** is used, and the level or type of approval is noted. Since there is usually internal coordination and

LEGEND

- DRAWING STARTED (—)
- ISSUED FOR APPROVAL (◸)
- ISSUED FOR CONST. (▽)

PROGRESS LAST PERIOD ▬
PROGRESS THIS PERIOD ▬▥
ISSUED FOR A - APPROVAL, C - CONSTRUCTION, Q - QUOTATION, P - PURCHASE

DRAWING CONTROL

JOB NO. 30CC PERIOD ENDING 3-1-90
JOB SAMPLE
SHEET ____ OF ____ SHEETS

CODE	DRAWING NUMBER	TITLE	START DATE SCHED.	START DATE FOR	START DATE ACTUAL	COMP DATE SCHED.	COMP DATE FOR	COMP DATE ACTUAL	APP'VD DATE	DATE ISSUED REV. 0	REV. 1	REV. 2	REV. 3	REV. 4	REV. 5	REV. 6	REV. 7	REV. 8	REMARKS * DESIGNATES PRINCIPAL DRAWING
		EQUIPMENT LOCATION DRAWINGS																	
	M-1	Ground Floor Plan	4-15-89		4-13-89	8-1-89	9/1	9-15-89	9/17	A 9/15	C 9/20								
	M-2	Plan at Turbine Floor	5-3-89		5-1-89	8-1-89	10/1	9-26-89		A 9/27									
	M-3	Plan at Control Floor	7-20-89	8/15	8-10-89	2-1-90	3/4												
		F. O. HEATER AND PUMPS																	
	M-20	F. O. Heater Assembly	5-15-89		5-9-89	6-30-89		6-14-89	6/16	Q 6/9	A 6/14	P 6/23	C 6/25	C 7/13					
	M-21																		
		AREA NO. 2																	
	M-100	Plan Below Elev. 30'	8-1-89		7-21-	10-1-89		9-24-		A 9/25									
	M-101	Misc. Turbine Piping	2-1-90	3/10		3-1-90	4/10												

FIGURE 3.5 Drawing control form (manual).

Project
Steam_Turbine

* STATUS
A.DRAWING REVIEW
1.NO COMMENTS,PRINT TO VENDOR
2.COMMENTS AS NOTED,PRINT TO VENDOR
3.DRAWING NOT APPLICABLE,PRINT TO VENDOR
4.NO COMMENTS,NO PRINT TO VENDOR
5.FOR INFORMATION ONLY

B.FABRICATION
1.PROCEED NO FURTHER REPRODUCIBLE REQUIRED
2.PROCEED SUBMIT REVISED REPRODUCIBLE
3.DO NOT PROCEED NO FURTHER REPRODUCIBLE REQR'D
4.DO NOT PROCEED REVISED REPRODUCIBLE

==
1605.1002-Steam Turbine
General Electric

EBAS No.-Rev	Vendor No.-Rev	Drawing Title	Sheets	Date Rec'd	STATUS	Ret'd	Remarks
1605 0177 -1	165A483CX-C	ST TURB GEN-WOODWARD 501 DIGITAL CONTROL SYS(1-59)	3	08/15/91	5/1	09/19/91	
1605 0178 -1	B111OJ34-M	ANCHOR BOLT DATA - STEAM TURBINE GENERATOR	2	06/20/91	1/1	07/09/91	
1605 0179 -0	E195T22-1A-	ESTIMATED SATURATION & SYNCHRONOUS IMPEDANCE CURVE	2	04/10/91	1/1	06/13/91	STEAM TURBINE GENERATO
1605 0180 -0	E195T22-2-	ESTIMATED REACTIVE CAPABILITY CURVES S/T GENERATOR	2	04/10/91	1/1	06/13/91	
1605 0181 -0	E195T22-3A-	ESTIMATED EXCITATION V CURVES S/T GENERATOR	2	04/19/91	1/1	06/13/91	
1605 0182 -0	R&P-12-	ESTIMATED GENERATOR DATA - ELECT CONST S/T GEN.	2	04/10/91	1/1	06/13/91	
1605 0183 -0	RP-4C-	EXCITER CONSTANTS STEAM TURBINE GENERATOR	2	04/10/91	1/1	06/13/91	
1605 0184 -1	762A663002-D	WATER DETECTOR STEAM TURBINE GENERATOR	1	06/20/91	1/1	07/09/91	
1605 0185 -2	104E2821-A	MECHANICAL OUTLINE (SHEET 1 OF 3) S/T GENERATOR	1	07/26/91	5/1	09/19/91	
1605 0186 -2	104E2821-A	MECHANICAL OUTLINE (SHEET 2 OF 3) S/T GENERATOR	1	07/26/91	5/1	09/19/91	
1605 0187 -2	104E2821-A	MECHANICAL OUTLINE (SHEET 3 OF 3) S/T GENERATOR	1	07/26/91	5/1	09/19/91	
1605 0280 -0	B777A37E155326-0	ELECTRICAL OUTLINE-BLOCK DIAGRAM (SHT 1 OF 9) S/T	1	04/19/91	1/1	06/13/91	
1605 0281 -0	B777A37E155326-0	ELECTRICAL OUTLINE-TURBINE WIRING (SHT 2 OF 9) S/T	1	04/19/91	2/2	06/13/91	
1605 0282 -0	B777A37E155326-0	ELECTRICAL OUTLINE-TURBINE WIRING (SHT 3 OF 9) S/T	1	04/19/91	1/1	06/13/91	
1605 0283 -0	B777A37E155326-0	ELECTRICAL OUTLINE-TURBINE WIRING (SHT 4 OF 9) S/T	1	04/19/91	2/2	06/13/91	
1605 0284 -0	B777A37E155326-0	ELECT OUTLINE-TURBINE & TRIP WIRING(SHT 5 OF 9)S/T	1	04/19/91	1/1	06/13/91	
1605 0285 -0	B777A37E155326-0	ELECT OUTLINE-TURBINE & ACCESS WIRING (SHT 6 OF 9)	1	04/19/91	1/1	06/13/91	(S/T)
1605 0286 -0	B777A37E155326-0	ELECT OUTLINE-LUBE OIL SYS & ACCESS WIRING(7 OF 9)	1	04/19/91	2/2	06/13/91	(S/T)
1605 0287 -0	B777A37E155326-0	ELECT OUTLINE-LUBE OIL SYS & ACCESS WIRING(8 OF 9)	1	04/19/91	1/1	06/13/91	(S/T)
1605 0288 -0	B777A37E155326-0	ELECT OUTLINE-LUBE OIL SYS & ACCESS WIRING(9 OF 9)	1	04/10/91	1/1	06/13/91	(S/T)
1605 0459 -3	104E2760-C	PACKAGE WIRING STEAM TURBINE GENERATOR (SH 1 OF 6)	1	11/14/91	5/1	12/09/91	

circulation required for these vendor prints within the design group, it is normal to request that reproducibles be provided to simplify the reproduction, marking, and internal distribution. It is important that when a vendor print has received final approval, a copy is marked as the "record print" and retained in the project archives. This is analogous to the treatment of a design drawing prepared on project.

An example of a Vendor (or foreign) Print Register and a Vendor Approval Stamp are shown in Figs. 3.6 and 3.7.

Vendor drawings can be transmitted electronically, but the vendor and the designer must have compatible equipment. If not, some design organizations now input vendor drawings into their data base by reading a hard copy with an optical scanner. Comments to the drawing can be added, and subsequent versions of the drawing can be transmitted electronically, thus speeding up the entire approval process.

3.11 EQUIPMENT LIST

An **Equipment List** (or parts list) records all the pieces of equipment purchased or supplied, their project part number (often tag number), and some limited data on the size or performance of each item. The data are not exhaustive, merely enough to identify one item from another. Equipment lists are organized by type of items: for example, pumps, switchgear, heat exchangers, boilers, conveyers, and rectifiers. Specific categories sometimes require several pages. A portion of an Equipment List is shown in Fig. 3.8.

3.12 MOTOR LIST

For projects using large numbers of electric motors, it is common to prepare a **Motor List.** It is similar to the Equipment List mentioned above but is limited to motors. This list records the main characteristics of all motors. Motors may be grouped by type (voltage, etc.) for convenient reference. A portion of a Motor List is shown in Fig. 3.9.

3.13 CIRCUIT AND RACEWAY SCHEDULES

Circuit and **Raceway Schedules** are the products of a computerized design program that routes wire and cables through cable tray and conduit (raceways). The program determines the shortest routing available through the network. In doing so, it ensures that the percentage of fill of the raceway

---

FIGURE 3.6 Vendor drawing register. Copyright © Ebasco Services, Inc., 1992.

SUPPLIER DOCUMENT STATUS

STATUS 1

Information delineated on the document has been reviewed by   without comment and may be used by Supplier. The document requires no changes or additions. Matters remaining to be resolved by the Supplier will not require document changes and can be handled by correspondence. The document must meet project record retention requirements, including reproduction.

STATUS 2

Information delineated on the document is in basic accord with the specifications and purchase order. Minor deviations have been noted, some aspects of the subject matter are incompletely defined, or other minor technical changes are required to make the item usable.

STATUS 3

Information delineated on the document is not to be used by the Supplier because it:

o   does not conform to project criteria;

o   is of a design that is technically unusable without significant changes;

o   does not meet   requirements (for example, orientation of equipment, nozzles, conduit connections); or

o   does not conform to the procurement requirements.

STATUS 4

Information delineated on the document need not be reviewed prior to its use by the Supplier. This STATUS is given to those documents which depict supplier standard products, shop details not requiring   review, etc.

JOB NO.

SUPPLIER DOCUMENT STATUS

1. ☐   WORK MAY PROCEED

2. ☐   REVISE AND RESUBMIT. WORK MAY PROCEED SUBJECT TO RESOLUTION OF INDICATED COMMENTS

3. ☐   REVISE AND RESUBMIT. WORK MAY NOT PROCEED.

4. ☐   REVIEW NOT REQUIRED. WORK MAY PROCEED.

PERMISSION TO PROCEED DOES NOT CONSTITUTE ACCEPTANCE OR APPROVAL OF DESIGN DETAILS, CALCULATIONS, ANALYSES, TEST METHODS OR MATERIALS DEVELOPED OR SELECTED BY THE SUPPLIER AND DOES NOT RELIEVE SUPPLIER FROM FULL COMPLIANCE WITH CONTRACTURAL OBLIGATIONS.

REVIEWED	A	C	E	J	M	PD

G-321
DOCUMENT CATEGORY _____

RESPONSIBLE
ENGINEER _____ DATE _____

FIGURE 3.7 Supplier document status.

does not exceed the allowed percentage of cross section of the raceway, maintains the identity of cables, conductors, and similar information. The National Electrical Code is used to establish the percentages of fill permitted for various services, including ambient conditions and conductor and insulation type. Since the fill is limited by the heat dissipation characteristics of the mix of wire and cable, the program is important where projects are large or a complex mix is present. This program is used mainly for large or complex

EQUIPMENT LIST

Job Number: Client: Date:

Sorted By: Equipment Number   Project: Page:

Period Ending:

Equipment Number / Motor #	Description	Capacity / Power	Weight / Voltage	Volume / R.P.M.	Spec Numb / Encl	Supp Pack / Scope	Inst Pack	Cost Code	Remark	Sched. Code	On-Site Date	Installation Date / Resp
100-CP-101 / 100-M-101	Condensate Polisher Unit #1	3000 gpm / 15 kW	56 TN / 230.8	780 M3 / 5600	5001	5001	5001	100 5100	INSPECTION REQUIRED	H5000	* 1-APR-85	1 2-Jan-85 / ENG
100-DM-102 / E-455	Demin.-Cation & Anion #2 Unit	7000 gpm / 75 kW	23 TN / 215.0	450 M3 / 340	7004	5003 A	0500	100 5100		G6000	* 31-May-85	P 7-Feb-85 / ENG
100-E-101 / E-455	Current Limiting Reactor	550 cfm / 5 kW	56 TN / 355.9	100 M3 / 1500	5003	5001 A	0500	100 5100		H5000	* 1-Apr-85	I 7-Feb-85 / ENG
100-E-102 / E-333	2500 kVA 12.47/4.16 Transform	550 cfm / 5 kVA	123 TN / 120.0	1000 M3 / 2400	5005	5000 C	0500	100 5100		G5000	* 25-Jun-86	I 7-Feb-85 / ENG
100-E-103 / G-223	4.16 kV Med. Voltage Controller	7000 cfm / 75 kW	248 TN / 120.0	300 M3 / 3200	5002	5002 A	0500	100 5100	WAITING ON VENDOR DRAWING	R3000	* 28-Sep-85	D 7-Feb-85 / ENG
100-E-104 / A-100	480 V Switchgear	7000 cfm / 75 kW	23 TN / 215.0	450 M3 / 340	5004	5003 A	0500	100 5400		H8000	* 30-Jun-85	T 2-Jan-85 / ENG
100-E-105 / A-101	480 V Motor Control Center	3000 cfm / 15 kW	56 TN / 230.8	780 M3 / 5600	5006	5001 A	0500	100 5400		G6000	* 31-May-85	P 7-Feb-85 / ENG
100-EF-101 / E-333	Exhaust Fan	310 cfm / 40 hp	98 TN / 355.9	233 M3 / 2100	5002	5003 B	0500	100 5400		H5000	* 1-Apr-85	T 7-Feb-85 / ENG
100-EF-102 / Q-223	Exhaust Fan	310 cfm / 600 hp	65 TN / 120.0	341 M3 / 450	5002	5001 B	0500	100 5400	RECEIVED IN SEVERAL PARTS	C2000	* 21-Jan-85	P 2-Jan-85 / ENG
100-EF-103 / A-100	Exhaust Fan	310 cfm / 600 hp	342 TN / 120.0	222 M3 / 780	5004	5000 D	0500	100 5500		R5000	* 28-Sep-85	I 7-Feb-85 / ENG
100-EF-104 / A-101	Exhaust Fan	100 cfm / 5 hp	122 TN / 215.0	347 M3 / 900	5006	5002 A	0500	100 5500		R4000	* 31-May-85	D 7-Feb-85 / ENG

FIGURE 3.8 Equipment list. Reprinted by Permission of ICF Kaiser Engineers.

MOTOR LIST

| DRIVEN EQUIPMENT/ DESCRIPTION | EQUIP TOTAL QUANTITY | OPERATING QNTY | | MECH: BHP or KW: each | RATED MOTOR HP | POWER INPUT (KW) (Note 1) | | | | | VOLTG VOLT/ PHASE | NOMINAL EQUIPT SPD,RPM | REMARKS |
		NORMAL	PLNT-OFF			PLANT NORMAL OPER. SUMMER	WINTER	GUAR.	PLANT-OFF SUMMER	WINTER			
AIR COMPRESSOR AND DRYER:													
AIR COMPRESSOR	1	0	1	55	60			49.46	49.46	49.46	460/3/60		
FAN MOTOR	2	0	1	2.9	3			2.76	2.76	2.76			
DRYER CNTL PANELS	1	1	1	0.08		0.08	0.08	0.08	0.08	0.08	115/3/60		
						0.08	0.08	52.30	52.30	52.30			
AUXILIARY (STAND-BY) BOILERS:													STAND-BY UNITS 4-25%
AUX. BOILER BLOWER	4-25%	0	4		40			132.62	132.62	132.62			NOS. OF EQPT OPERATING MAY
AUX. BOILER FEED PUMP	2-50%	0	2		20			33.16	33.16	33.16			VARY FROM 1 TO 4 DEPENDING
CHEM FEED PUMPS	4-25%	0	4		0.333			1.66	1.66	1.66			ON AMT OF STEAM REQ'D.
CHEM FEED TANK AGITATOR	1	1	1		0.333	0.41	0.41	0.41	0.41	0.41			
						0.41	0.41	167.85	167.85	167.85			
AUX. COOLING WATER SYSTEM:													
AUX. COOLING WATER PUMP	2-100%	1	0	51	60	42.04	42.04	42.04					@ 2,150 GPM/ 70 FT TDH
						42.04	42.04	42.04					
CIRCULATING WATER SYSTEM:													
COOLING TOWER FAN	3-33.3%	3	0	293 *Total	150	331.08	237.79	220.72			460/3/60		*GUAR.- 2 FULL & 1 HALF SPEED.
		3	0	136	150						460/3/60	1800	SUMMER OPER. - MAX
		2	0	136	150						460/3/60	1800	WINTER OPER. - MIN
CIRC. WATER PUMP	3-33.3%	3	0	224	250	541.96	541.96	361.31					@ 10,900 GPM/ 75 FT TDH NORMAL
		3	0	224	250								SUMMER OPER. - MAX. & NORMAL
		2	0	224	250								WINTER OPER.
						873.04	779.75	582.02					

FIGURE 3.9 Motor list. Copyright © Ebasco Services, Inc., 1992.

projects where there may be thousands of different segments of raceway and hence a myriad of alternate paths for routing of the cable. The cable routing is done automatically by the computer; it provides a printout of the routing, and raceway fill, and pull cards for the use of the installers. The pull cards, have multiple parts that tell the installers which cable numbers and cable types are to be pulled in each section of the raceway (i.e., the routing) and which, where multiple cables are pulled at the same time, can be used to perform gang pulling. The program will compute maximum pull tension maintaining it within the cable tolerance. A second part of the two-part pull card, after pulling, is returned to the engineering office for entry into the program as pulled cable and thus permits current statusing of the installation work.

Since work at under 480 volts is less complex, a **Panelboard Schedule** combined with the standard electrical power and lighting plans and standard details provides sufficient definition, so a circuit and raceway program is not used. Figure 3.10 shows a portion of a typical panelboard schedule.

3.14 LINE DESIGNATION TABLES

Line Designation Tables are used for process or hydraulic work where there are numerous liquid lines to be sized and identified. These tables identify each length of line of a uniform size and indicate line specification, size, fluid velocity, and normal operating temperatures. They also may provide other variables such as insulation, paint specification, and test pressures. They are useful for verifying that lines are correctly sized, based on an inspection of the flow velocity and its pressure drop. The identification of each length of line is also frequently used to establish ''spools'' (individually fabricated piping assemblies). Selections from a typical Line Designation Table are shown as Fig. 3.11.

Similar to the line designation tables are **Valve Tables.** Valve tables list standard valves by service, type, size, and tag number. A mark number is assigned to each valve, indicating, general construction, pressure rating, material, trim, and other similar characteristics. This avoids the need to name a manufacturer and permits competitive bidding for the valve supply.

3.15 CONTROL VALVE AND INSTRUMENT LISTS

A large number of **control valves** and **instruments** are available. Some large projects may have hundreds of them thus lists and summaries are necessary for definition and control. Because of their variety and different characteristics, it is useful to group them by type on separate data sheets: for example,

PANEL	LOCATION		REF. DWG				120/208V, 3∅, 4W, 225A BUS, 175A MAIN CB			
C9-021LP	ELEC. ROOM #1		C-01-01-E				TOP FEED, SURFACE MOUNTED TYPE			

LEFT SIDE

CKT	SUPPLIES	LOADING (WATTS) A	B	C	BREAKER AMP
1	PNL. C9-021LP1 (EXT.LTG.)	4570			
3	PNL. C9-021LP1 (EXT.LTG.)		5110		100
5	PNL. C9-021LP1 (EXT.LTG.)			4520	
7	LTG. - EL. ROOM #1	955			20
9	REC. - EL. ROOM #1		600		20
11	REC. - EL. ROOM #1			600	20
13	TIME CLOCK CONTROL	100			20
15	SPARE				20
17	SPARE			1000	20
19	LTG. - TUNNEL	540			20
21	LTG. - TUNNEL		630		20
23	LTG. - TUNNEL			690	20
25	LTG. - DOME FLOODS	1000			20
27	LTG. - DOME FLOODS		1500		20
29	LTG. - DOME FLOODS			1000	20
31	LTG. - DOME WALL LIGHTS	1400			20
33	LTG. - DOME WALL LIGHTS		1400		20
35	LTG. - DOME WALL LIGHTS			1400	20
37	LTG. - GALLERY CONV. PLTF.	770			20
39	LTG. - GALLERY CONV. PLTF.		1180		20
41	LTG. - GALLERY CONV. PLTF.			680	20
	SUBTOTAL CONNECTED LOAD	9335	10420	8890	
	SUBTOTAL SPARES	-	1000	1000	
	SUBTOTAL (LEFT SIDE)	9335	11420	9890	
	TOTAL LOAD		44525		
	PANEL: C9-021LP				

RIGHT SIDE

BREAKER AMP	LOADING (WATTS) A	B	C	SUPPLIES	CKT
20	800			REC. -TOP OF RAW MAT. SILO	2
20		800		REC. -TOP OF RAW MAT. SILO	4
20			800	REC. -TOP OF RAW MAT. SILO	6
20	1000			REC. -TUNNEL & CONVEYOR	8
20		800		REC. -TUNNEL & CONVEYOR	10
20			1000	REC. -TUNNEL & CONVEYOR	12
20	800			REC. -DOME	14
20		800		REC. -DOME	16
20			600	REC. -DOME	18
20	280			REC. -DOME BATTERY UNITS	20
20		800		REC. -GALLERY CONVEYOR	22
20			600	REC. -GALLERY CONVEYOR	24
20	600			REC. -GALLERY CONVEYOR	26
20				SPARE	28
20			1000	SPARE	30
20	1000			SPARE	32
20		1000		SPARE	34
20			1000	SPARE	36
-	-			SPACE	38
-		-		SPACE	40
-			-	SPACE	42
	2680	3200	3000	CONNECTED LOAD (SUBTOTAL)	
	1000	2000	2000	SPARES (SUBTOTAL)	
	3680	5200	5000	SUBTOTAL (RIGHT SIDE)	

NOTE: NORMAL POWER

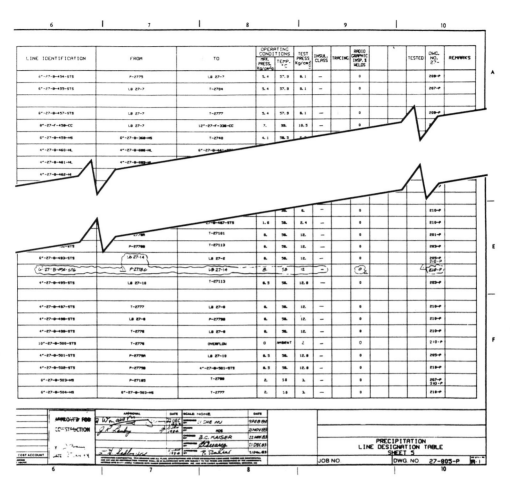

FIGURE 3.11 Portions of a line designation table. Reprinted by permission of ICF Kaiser Engineers.

pressure regulators, would be grouped together, and level switches similarly grouped. The summary sheets can group them by system, P&ID, Area Drawing, or another convenient design designation. The same practice is used for other specilities such as steam traps, and strainers. An example of a Control Valve Data Sheet is shown in Fig. 3.12.

FIGURE 3.10 Panelboard schedule. Reprinted by permission of ICF Kaiser Engineers.

3.16 REQUISITION REGISTER

A **Requisition Register** is required if materials are to be requisitioned for purchase. A requisition is basically a request upon the purchasing department to obtain bids for materials and equipment. The requisition has several parts: the requisition itself, which establishes the required delivery date(s); the cost code; whether inspection and/or expediting will be required; where the material shall be shipped; and other such general data. A list of required materials is included, and one or more technical specifications may be noted there. (More detailed information on requisitioning and purchasing procedures is found in Chapter 6.) The Requisition Register is a master list that establishes control on the development of the requisitioning process. For convenience the register is often arranged in chronological order, permitting by inspection a quick review of the status of the work. For smaller or less complex projects, often the **Requisition Register** and a **Specification Register,** which controls the preparation of the specifications in the same way, will be combined into a single document. The register lists the important dates in the evolution of the requisition, including its beginning, its issuance for internal review, and its finalization. Responsibility for preparation is included, as is other significant information such as material takeoffs that may be needed and any information to be furnished by persons outside the project.

3.17 DESIGN CALCULATIONS

Important to the proper performance of the work is maintenance of a rigorous system of control and custody of the **design calculations** as they are prepared. The project sets up a calculation binder in which all calculations prepared for the project are filed. For large projects this might comprise one or more binders for each of the engineering disciplines or interest areas. Hard copies of computerized calculations are included as well as manually prepared calculations. All calculations, including their current revisions, must be maintained in the binder(s) so that they are available for review or revision. There is nothing that will cause loss of credibility quicker than being unable to produce a calculation upon which a particular element of the design was based. To avoid this and the sheer frustration that occurs when the calculations are not readily available, it it important to designate one individual as responsible for maintenance of the binder(s). For convenience on medium- and large-size jobs, indexes of the calculations prepared should be maintained with the binders. Each of the design disciplines maintains the calculation binder for its own work. Preparation and checking of calculations is discussed in Chapter 2.

CONTROL VALVE DATA SHEET

SPEC NO: VO1
PAGE: 20
ORDER NO: 8668-1064 |R1
SUPPLIER: COPES VULCAN

PLANT:
SPECIFICATION FOR CONTROL VALVES WITH ACCESSORIES

REFERENCES:	ANSI B16.5	MSS SP-25	SSPC PA-1	ANSI B16.34 GEN REQ IC-1	REV 0 DATE: 8/4/89
	ANSI B16.10	MSS SP-55	SSPC SP-2	ANSI B16.104	REV 1 DATE: 10/09/89
	ANSI B16.11	MSS SP-61	SSPC SP-3	ANSI B31.1	REV 2 DATE: 12/21/89
	ANSI B16.25	MSS SP-67	SSPC SP-6	MSS SP-72	REV 3 DATE: 2/26/90
	ASME VIII DIV.1	MSS SP-84	ASTM A-216	ASTM A-351	

TAG NUMBER: BF-FCV-15A   SERVICE: HRSG #1 IP ECON. FW FLOW CONT.

SERVICE CONDITIONS

Fluid: HEATED FEEDWATER

	Units	Max Flow	Norm Flow	Min Flow	Shut-Off
Flow Rate	KPPH	93.2	79	34	-
Inlet Pressure	PSIA	362	368.4	409.3	
Outlet Pressure	PSIA	166.7	151.8	121	
Inlet Temperature	DEG F	259	244	259	
Spec Wt/Spec Grav/Mol Wt					-
Viscosity/Spec Heats Ratio					-
Vapor Pressure Pv					-
*Required Cv		13.75	11.03	4.13	-
*Travel	%				0
Allowable/*Predicted SPL	dBA	≤85/	≤85/	≤85/	-
Crit Press PC					

LINE

Pipe Line Size   In 2½" SCH 80
& Schedule   Out 2½" SCH 80
Pipe Line Insulation

VALVE BODY/BONNET

R1

*Type	GLOBE
*Size	2" ANSI Class 300 #
Max Press/Temp	450 PSIG / 300 DEGF
*Mfr & Model	COPES VULCAN CV
*Body/Bonnet Matl	ASTM A216 WCB
*Liner Material/ID	
End | In	SW
Connection | Out	SW
Flg Face Finish	
End Ext/Matl	
*Flow Direction	UNDERSEAT
*Type of Bonnet	BOLTED
Lub & Iso Valve	Lube
*Packing Material	TEFLON
*Packing Type	
*Hydro Pressure	

R1

TRIM

R2

*Type	ANTI CAVITATION HUSH TRIM w/CASCADE PLUG
*Size	HUSH/CASCADE Rated Travel 1"
*Characteristic	LINEAR
*Balanced/Unbalanced	BALANCED
*Rated Cv	18.5 FL XT
*Plug/Ball/Disk Material	420 ST.ST
*Seat Material	420 ST.ST
*Cage/Guide Material	420 ST.ST
*Stem Material	316 ST.ST
ANSI/FCI Leakage Class	IV

R3 | R2 | R1

CONTROLLER

Type	Action
Measuring Element	Material
Range	Output 4-20 mADC
Mounting	
*Mfr & Model	BY OTHERS

ACTUATOR

*Type	SPRING LOADED DIAPHRAGM
*Mfr & Model	CVI 600-60RA
*Size	60 SQ.IN Eff Area

[ ] On/Off   [X] Modulating
Spring Action Open/Close   CLOSE
*Max Allowable Pressure   100 PSIG
*Min Allowable Pressure   80 PSIG
Available Air Supply Pressure :
Max 100 PSI   Min 80 PSIG
*Bench Range   /
Act Orientation
Handwheel Type
Air Failure Valve CLOSES Set at
Input Signal

R1
R1

POSITIONER

*Type	ELECTRO-PNEUMATIC
*Mfr & Model	BAILEY AP-9
*On Incr Signal Output Incr/Decr	INCR
Gauges	By-pass
*Cam Characteristic	

R1

TRANSDUCER

*Mfr & Model	N/A
Input Signal	
Output Signal	

SWITCHES

Type	DPDT Quantity 2
*Mfr & Model	NAMCO EA170II100
Contacts/Rating	
Actuation Points	OPEN & CLOSE

R1

AIRSET

*Mfr & Model	BELLOFRAM 51FR
*Set Pressure	
Filter	YES Gauge YES

R1

PILOT SOLENOID VALVE

*Mfr & Model   ASCO 212631-4U
[ ] Normally Open   [ ] Normally Closed   [X] Universal
Voltage Rating 125 VDC

R1

POSITION XMTR

*Mfr & Model   N/A

*Information supplied by manufacturer unless already specified

NOTES: 1) VALVE STRONG TIMES 7-8 SECONDS.   3) BOOSTER RELAY-MOORE 614
       2) MINIMUM GUARANTEED CONTROLLABLE CV<4.

R1 |R3

FIGURE 3.12 Control valve data sheet. Copyright © Ebasco Services, Inc., 1992.

3.18 CHANGE CONTROL

The most frustrating part of a PE's work can be maintaining control over **changes** initiated by design personnel. Engineers by their training are encouraged, and feel a professional responsibility, to initiate improvements and enhancements to designs. Unless this process is carefully managed, the PE will find that the project is out of control and that schedules and cost estimates are meaningless. A number of methods have been tried to handle this, the choice depending on the project, its organization, physical layout, and so forth. The usual method of control is to require that the discipline design leader authorize any changes. A more drastic method is to physically lock up the original drawings (tracings) so that changes cannot be made to them. When CAD is used in lieu of traditional drafting, this is more difficult, although it is possible to deny access to certain drawings by suitable access code application. Other methods that can be used are restricting access to calculations or requiring that all changes be processed through a change board, who review the need for the change before authorizing it.

The best way to deal with the problem may be to assure by **indoctrination** that all members of the design team recognize that the project has constraints of both schedule and cost and that refinement of the design beyond a certain point is counterproductive. Since the feeling of need for change or improvement normally originates with a member of the design team, if that person can initially exercise more responsible judgment, fewer changes are likely to occur.

Another way to avoid changes is to establish at the beginning of the work design margins which are sufficient to allow for some natural development of the work. This avoids the situation where a change is necessary because of lack of capacity, cramped space, excessive ambient temperatures, and the like. Design margins are effective, but to set them, one needs considerable experience because the trade-off between the quantifiable added initial cost and the mere possibility or uncertainty of design development is a difficult one. In general, unless the project is on a highly strained budget, some allowance should be made for these developments. After the project is completed the wisdom of providing the margins in the design will be apparent. The problem of enhancing the design by adding more detail, redundancy, and the like, is more difficult to control. It can only be identified and dealt with where strict attention is paid to detailed overview of design by discipline supervisors on a day-by-day basis.

In some highly structured work, for example, in the defense aerospace industry, it is common for formal Change Control Boards and **configuration management** programs to be instituted. These require that any change from the approved design be submitted for review and approval prior to being implemented. The approved design may be a conceptual one, a detailed one, a manufacturing plan, operational or performance criteria, or any similar controlling document. The structure of these boards is often quite formal and includes senior representatives of the several technical departments involved, quality assurance, project management, and the customer. Their meetings are scheduled on a regular basis, although additional meetings are held if required. The party initiating the request for change is required to present its case, including the rationale behind the change, the advantages and disadvantages of making it, the economic impact, scheduling impact, contractural effect, potential reaction by users or other groups, and any external factors such as public reaction. Where the change is minor, it is often approved immediately. Where it is more complex or far-reaching, it is not unusual for the decision by the board to take a significant amount of time or for additional study to be requested. Where the design is evolutionary, it may be necessary for the board to meet over much of the life of the project. Where the design uses proven concepts and technology and the effort is more one of detailing and optimizing the design, the board will meet fairly frequently in the initial stages of the project and then, as the project proceeds, less frequently.

During the design execution phase it may be necessary to institute a **"Design Freeze."** A common reason for a Design Freeze is to permit the design as it stands at that point in time to be "packaged" for a cost estimate to be prepared. This may occur several times during the life of the project depending on the type, number, and timing of estimates. Then the Design Freeze usually lasts a few days so that drawings and necessary data can be reproduced and gathered to support the estimator's work.

A more critical Design Freeze may be initiated to prohibit further changes and complete the design so that it can be released for fabrication, construction, or purchase. This is a significant step since the release that follows often involves a major financial commitment. Changes beyond that point not only can be very costly but may introduce unacceptable schedule delays. The Change Control Board mentioned earlier is often involved in this matter.

CHAPTER 4

ENGINEERING ORGANIZATION

Much has been written on the subject of group **organization** for complex work. While engineering organizations fall into this category, there is surprisingly little written on how to best organize an **engineering** group. An oversimplification would be to take the view that it matters very little what form the organization takes so long as the organization is staffed with good people.

Having good people in the organization is important, but it overlooks the critical concern of **efficiency** and **morale** when organizational lines and responsibilities are not clearly stated. In today's world of integrated highly technical work, individuals need to feel not only that they are making a contribution to the work but also that they are **empowered,** that is, that they have certain authority over decisions taken in their technological area. (This is discussed in more detail in Chapter 12.) The experienced working level engineers typically know their work and the best way to find an appropriate solution. The trick is to integrate their expertise into a team operation so that they can influence, and in some cases direct, the outcome of at least some of the work. It is for this reason that engineering organizations take usually one of only two or three forms. These organizational arrangements are well proven and have not only met the test of time but, more important, the test of performance. They can with only slight modification be adapted to most any type of project.

As with most all matters in engineering, there is no hard and fast rule regarding organizational structure. All of the basic types can be effectively used, the determining factors being scope definition of the project and form of contract, size of engineering organization, experience and skill level of design staff, client preference, and project type. Whichever organizational structure is used, some form of management oversight is essential to ensure proper performance.

4.1 FUNCTIONAL ORGANIZATION

The **functional** organization is the predominant form used in engineering work. Typically the technical groups stand alone and process work in their own design specialty, with a modest group of "project" or management personnel providing overall direction and guidance. Each technical group is under the direction and control of an experienced senior engineer, often holding the title, "Chief (Electrical, Civil, . . .) Engineer." If the technical personnel are not located in a common project area, work packages are sent to them from the PE. If the group is housed in a common work area, the project members work in a more integrated manner. Figure 4.1 depicts such an arrangement.

The functional organizational form has the main advantage that the design staff is technically expert and has significant related experience. For lump sum or fixed price work this type organization can be extremely **efficient** and **cost** effective. It can produce design work that is highly sophisticated and at the cutting edge of the technology.

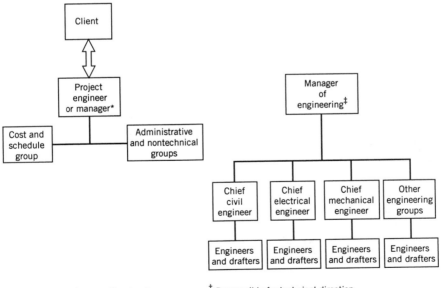

* Responsible for operational and
  administrative direction.

‡ Responsible for technical direction.

FIGURE 4.1   Functional organization.

The main disadvantage of this organizational type is that it may be difficult to obtain performance on a predictable schedule. The technical groups may work on several projects concurrently and usually perform the work in the order in which it is given them. As a result **priority** work may not be acted upon promptly. This results as the technical personnel have no project responsibilities, apart from the technical adequacy of their work. To overcome this, project personnel often must intervene with the head of the technical group to reschedule or expedite the work. This is effective if only a few projects are in progress. When there are numerous ongoing projects, there arise conflicting priorities. The functional arrangement can create difficulties for the project personnel charged with the dual responsibility of getting the work done correctly and on time. If in addition the project assignments are "packaged" and sent to a central technical group, a significant potential for omission or lack of coordination exists.

Personnel **training** and **development** is effective in the functional organization, although it will be more narrowly focused within the technical specialty of the group. But, while desirable from the standpoint of the group, this may limit the development of the more generalist-type engineers who often later become project engineers and managers. With group-oriented professional development, rotation of personnel to other groups is more difficult, since they often do not want to take on an engineer who is not a specialist in their discipline.

In the end a company using strictly the functional organization arrangement may be forced to hire from the outside to obtain the generalists needed for Project Engineer and Project Manager positions. The negative side of this often has adverse effects on the morale of the staff who feel that this violates the generally perceived concept of promotion from within the company.

On the positive side, the functional groups are likely to be more current on **technological developments** in their speciality. They are often called upon to serve on professional committees of standards organizations and trade associations. The functional groups also tend to maintain long-term relationships with their peers in other companies and with major suppliers in the field, thus promoting the interchange of technical information. This also provides for employment opportunities and referrals based upon direct knowledge as well as through technical or trade association activities.

Functional organizations do not tend to provide as many opportunities for **promotion** as do project organizations. This is caused by the hierarchical structure within the individual design groups. To compensate for this, many companies have set up salary and classification structures that provide improved pay and classifications to encourage technical personnel to stay in the design groups rather than seeking what may be higher paying jobs as Project Engineers or Project Managers.

The operational requirements of projects in this sort of organization are pursued by the Project Engineers and Project Managers charged with the responsibility of getting the work out. The difficulty for them is that they have responsibility for the schedule and cost performance of the group but have no direct control over the design engineers. As a result they must often convince, cajole, persuade, and in some cases perhaps even threaten the functional staff engineers to give their project the proper amount of attention. As the functional staff design engineers try to accommodate these requests, it is usually at the cost of other projects. Often because of their experience, the functional staff design engineers will downplay the importance of any one project request, when seen in the context of all the requests they are dealing with. To the Project Engineer or Project Manager their project is of maximum importance and should be given priority. This problem, while annoying, is normally not damaging, and when it does arise, senior management can be called upon to provide the necessary direction.

Functional organizations, in order to keep their personnel busy, will sometimes take on work for outside companies. If not handled carefully, this can create conflicts over whether the work in-house or for the outside contract gets **priority.** This is usually decided in favor of the outside contract and can further exacerbate any tensions that may exist between the project and functional staff personnel. While not a frequent occurrence, it is of sufficient concern that care should be taken when considering outside work. At a minimum the timing of the outside work should be coordinated with likely project requests for assistance.

In the functional organization the Project Engineer and Project Manager are heavily involved with administrative and management matters, and the **technical evolution** of the project may be left largely in the hands of the specialist groups. It is particularly important that the project scope be well defined and that, for large or complex projects, there be instituted a review mechanism of technical development, such as the Boards of Review or Control described in Chapter 2. In addition design freezes can be used to maintain control.

Small organizations that use the functional organization form do not suffer from many of these problems. Their small size permits close coordination and sharing of personnel, and a technical group leader often serves as Project Engineer or Project Manager. Unfortunately, small organizations often do not have sufficient personnel to provide as much flexibility as larger ones, and problems of work overloads are common.

Despite the potential problems discussed above, the functional organization is effective and widely used. With goal-oriented management it can perform at a high technical level with relative economy. In terms of understanding the organization and the specific responsibilities of the members, this type structure is simple and tends to maintain relatively clear lines of authority and responsibility.

4.2   PROJECT ORGANIZATION

Another widely used form of organization is the **project organization,** where groups of engineers are brought together as an engineering team, working full time on a single project. An example of this organization is shown as Fig. 4.2.

The main advantage is the organization's focus and responsiveness; it exists for the sole purpose of working on one project, and thus the project goals and objectives are kept constantly in mind. The personnel on the team share the common goals of the project and tend to be somewhat more responsive to secondary project requests. As in the case of the functional organization, these can be considered nuisances and may not be given sufficient attention. Usually the PE is responsible both for the **technical adequacy** of the work and for the operations of the group. As such, he or she is normally assisted not only by group leaders of the various design disciplines but also by one or more administrative personnel, and in some cases by planners, schedulers, and cost engineers.

The members of the project team will have varying amounts of experience, and thus there is sometimes the concern that the design may not be as sophisticated as that of a functional group. Some members of the design team may give too much weight to the project cost or schedule concerns and thus neglect some technical matters. To solve this problem, the design

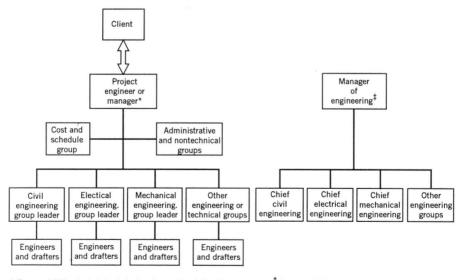

* Responsibilities include technical and operational direction.   ‡ Responsibilities include assigning personnel to project(s)—essentially administrative.

FIGURE 4.2   Project organization.

work usually undergoes periodic reviews by highly qualified technical personnel who are responsible for providing comments and advice on the technical adequacy of the design. This has the advantage of providing a form of check and balance on the design process and has proved to work quite well. It does require a cooperative relationship between the PE and the nonproject Chief Engineers, since the PE is responsible for the technical adequacy of the design and will likely be signing drawings and specifications. Normally the PE is not knowledgeable in all the technical areas and must rely to some extent upon the comments and recommendations of the chiefs. This is troublesome when the recommendations of the chief are contrary to what the client or a member of the project team feels is appropriate. The only solution in this case is to maintain close and continuing involvement with the chief and the client or team member through special meetings and, if necessary, with additional studies to arrive at a proper solution. In some cases this problem is overcome by making the chief responsible for the technical adequacy and requiring the chief to sign off on the design.

This type of organization can be subject to **abuse** if the PE is too headstrong or is concerned only about one or two engineering disciplines, or if schedule or cost matters are given too much weight. It places a premium upon the ability of the PE to understand the importance of the work of each of the technical groups and to seek a middle ground between an exquisitely correct technical solution and the demands of cost or schedule. The selection of a capable PE is therefore very important if this system is to work. The PE needs to have demonstrated technical **competence** and balanced **judgment.** The use of Boards of Review (see Chapter 2) is a good way to compensate for variations in the ability of PEs and project team members. If periodic design reviews are held, the design can be developed with most of the potential problems with this type of organization avoided.

A particular strength of the functional organization lies with projects where the scope may not be initially well defined. Because of its **responsiveness,** it is able to accommodate client requests and design developments more readily than the functional organization. It is important to remember, however, that with this type work the inherent **cost** of the design effort will likely be higher and that the very responsiveness of the organization may make scope definition more difficult.

In terms of training the project organization tends to produce more **generalist** engineers than the specialists produced by the functional organization. It provides a reservoir of experienced personnel for the future roles of Project Engineer or Project Manager. By greater exposure to the overall demands of the project, the engineers also become more skilled in making the trade-offs between cost, schedule, and design development.

From the **client's** point of view, the project organization is preferred because the client deals with a relatively constant group of people who have no other responsibilities and who are fully dedicated to the project. This permits improved understanding of the client's requirements and facilitates the development of communication between counterparts in the client and the design organizations.

One difficulty with the project organization is the constant **change** in the organization size and personnel responsibilities. Because of the importance of the early design decisions, it is usual to initially staff the project with the technical discipline leaders and to build the organization around them as the work progresses and the number of personnel on the project team grows. The project may begin with only four or five technical personnel and then grow to perhaps several dozen, or even a hundred or more personnel, as the project is fully staffed. As the project grows, personnel often assume the responsibilities formerly held by other staff members. This is less a problem during the growth phase of the project engineering group than during the later stages of the project when the staff size is declining and there is a consolidation of duties to fewer personnel. Care must be taken to ensure that there is adequate coordination to avoid omitting important requirements. Even more care must be taken to ensure that the person assuming the responsibility is not permitted to redesign the work merely because he or she had no part in its earlier development.

Staffing of project organizations requires balancing the project require-ments with the availability and skill of qualified personnel. While contract (outside) personnel can be hired for limited time periods if the in-house staff is insufficient, it can be a problem where there is a surplus of personnel. Engineering organizations can normally only carry a small percentage of personnel not actively engaged on and charging their time (and consequently cost) to project work. In addition to the operational requirements for staff size changes, the problem arises as to what to do with personnel who are surplus to the particular project organization. Few engineering organizations want the reputation of being a hire/fire company, so there can be some pressure to use personnel in areas where they are not experienced or for duties at a subprofessional level. To avoid this, the use of **contract** personnel, "jobshoppers" is prevalent. These personnel are often paid on the order of 30% more than the permanent staff but usually have no benefits such as medical insurance and vacation time. They work on the project for only as long as needed and then are terminated and return to their agency for reassignment. The skill level of the contract personnel varies somewhat, but in general because of the variety of work they have performed, they are usually able to rapidly integrate themselves into the organization and perform effectively.

While personnel from one design discipline may not be fully interchange-able with other design disciplines, in times of personnel shortages or to levelize staffing, such steps should be considered. A large part of the time of design engineers is spent writing specifications, preparing somewhat routine calculations, and other similar operations, a major portion of which can be satisfactorily handled by a competent engineer in another design speciality. This is best arranged by temporarily assigning and relocating the engineer to the other technical group, with adequate provision for coaching and review of the work.

4.3 MATRIX ORGANIZATION

The use of the **matrix** organization is becoming increasingly common. This arrangement offers the benefits of the project-type approach while retaining technical experts. With the matrix organization, everyone has two supervisors: one for the operational aspects of the work, the **what** and **when,** and another for the technical aspects of the work, the **how.** An example of a matrix organization is shown as Fig. 4.3. Note the dual reporting relationships.

Each person in the organization works for a Chief Engineer who provides guidance and direction in the execution of the technical aspects of the work while also taking operational or project direction from a Project Engineer. This could establish a condition where the individual feels that instructions are being given from two directions. While theoretically a problem, in practice it is rarely so. To exercise this responsibility, the chief is deeply involved in all the significant design decisions made on the project and provides active

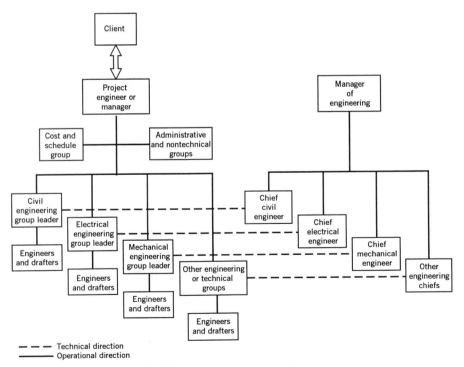

FIGURE 4.3 Matrix organization.

guidance to personnel. The involvement can be in the form of periodic meetings with the designers, review of technical approaches, calculations, work on the drawing boards, and of specifications. In some cases the chief will actually approve and sign off on these documents. Design Control Checklists, discussed in Chapter 3, ensure that important design documents are reviewed.

The design personnel also work for an operational supervisor, the PE. While he or she has a major responsibility to ensure that the design is prepared on time and within budget, in the matrix organization the PE has responsibility for **design adequacy** as well. With the PE's major role in the technical side of the design development, this is a responsibility shared with the chief but ultimately borne by the PE. Since a fundamental principle of engineering is to develop a workable design that does not endanger the health and safety of the public, the PE must temper operational concerns with an appropriate regard for a design that is safe as well as technically adequate.

Since the PE and the chiefs are part of the same company and have usually worked together for some time, there should be a satisfactory working relationship between them, so little formal structure is needed to ensure balance. When conflicts arise between the chief and the PE, senior management almost always makes the decision in favor of the chief, not to imperil the design. But in practice this is rarely a problem, although often minor decisions made in favor of an improved design are vexing to the PE. Where a PE is unable or unwilling to adjust to these type of decisions, a change in personnel is often made.

Theoretically design personnel might seem to have difficulty knowing where their primary loyalty lies since they work day to day under the PE and may only see their chief infrequently. In practice this is seldom a problem because design personnel are usually on the staff of the chief and only on loan to the PE while assigned to a project. The chief handles all personnel matters related to his or her design personnel, including salary administration, promotions, assignments to projects, sponsorship in technical societies, and so forth. As a result the longer-term career concerns of the engineer are handled by the chief and provide a powerful incentive to give careful attention to the instructions of the chief.

To ensure that a sufficient amount of authority remains with the PE, it is common to have the **performance evaluation** of the personnel on a project initiated by the PE, with additional comments by the Chief Engineer before review with the individual. This is intended to reflect day-by-day performance and to give the engineer incentive to respond to the directives of the PE. The performance evaluation is thus really a composite document reflecting both the technical and operational performance of the individual. In rare cases where a major evaluation difference arises between the chief and the PE, their common manager intercedes to resolve the problem. Typically the PEs and the chiefs interact continually, are at the same organizational level, and develop fairly close friendships, so such extreme problems are avoided.

Often PEs become chiefs, and vice versa, with the result that there is normally no significant difference in the way they view the work and its requirements.

Even on relatively large engineering projects there may not be sufficient work in some specialties to justify setting up one or more persons as full-time project **specialists.** For this reason even very large engineering organizations will maintain some groups of specialists who provide services to all the projects, much as in the case of the functional organization. For example, such specialist groups might work on piping stress calculations, water treatment, process design, heat balance, advanced fluid flow problems, dynamic control, and newer technologies such as acoustics, OSHA, environmental protection, and geotechnology.

All but the smallest firms typically end up with a **composite** type of organization—a project-oriented engineering team of full-time engineers covering disciplines requiring continual effort on a project with support from specialist groups who are located off project but who perform specific services as needed. This organizational form can provide the best of both worlds, the responsiveness of a project team approach with the technical excellence of a functional organization approach.

4.4 BUSINESS LINE ORGANIZATION

In large firms, where a number of **business lines** are pursued, it is common for the engineering organization(s) to follow the business lines. Typically this takes the form of a small number of full-time specialist engineers assigned to the business line, with additional staff drawn, as needed, from a pool of general engineers. The business line engineers are the experts in the technology of the industries served, and it is from them that the PE and the technical group leaders are chosen. The nonspecialist engineers are usually organized by disciplines and take both operational and technical direction from the business line specialist engineers.

The effect of the business line organization is to add a functional relationship that somewhat complicates the organizational structure, and this effect is felt more at the senior levels. Because the functional and project organizations are combined, the organization is less flexible. The project design, however, benefits because there are more expert personnel while retaining the responsiveness of a project-type arrangement.

4.5 LIFE OF A PROJECT ORGANIZATION

While functional organizations remain relatively constant over time, project organizations are more dynamic, starting with only one or two persons and gradually adding staff until a project reaches its maximum staffing, generally when the design effort is about 60% complete.

Initially the PE and the group leaders are assigned, followed by personnel in the engineering discipline groups depending on their immediate workloads. Thus the personnel who scope and define the work, the process, and the basic design are among the first to be assigned, followed by personnel in other groups as the work expands. Often shortly after the middle of the project, personnel of the civil group may be coming off the project while the electrical group may be adding personnel as firm design information becomes available. As a result there is a continual **change** occurring in the size of the groups and the responsibilities of the staff members. This requires that the leaders of the individual technical groups on the design team continually review staffing requirements and projections of work load for their personnel. Personnel forecasts are used or formal progress reports include effort expenditure and forecasting.

The **forecasting** work effort can be difficult because the tendancy is to underestimate the remaining effort to complete the work. This is particularly troublesome during the late stages of a project. Older, more experienced group leaders will do a better job of forecasting than less experienced personnel, who often fail to allow for the always present but undefined minor added items of work.

On **reimbursable** or **time-and-material** work the project's organization takes on a life of its own. The requests of the client and the progress of the work is tracked very closely. One troublesome area is the reassignment of **"key"** personnel to other duties when their workload declines. While this is not a major difficulty with the functional organization, it can be of some concern in project or matrix organizations, particularly if the work is of the reimbursable type. Often contracts permit the client right of approval of key personnel and changes to them. While reassignment of personnel within the groups generally is not a problem, certain key personnel become so valuable to the project that their reassignment is not feasible, since both the project and client relations would suffer. A good rule of thumb is that the truly key personnel, the groups leaders and the PE, should not be reassigned until it can clearly be shown that to continue them in their project jobs would be unnecessary. Although somewhat subjective, this mandates that these personnel will be kept on project for some time after needed.

Because of the usual loose ends, it can become costly and difficult to maintain a full project team to completely **close out** a project. On reimbursable work at some point it may be in the best interest of the client to obtain their agreement that the project is essentially complete and that personnel can be reassigned and the project team disbanded. The remaining work may be assigned to a single team member or a small project group which will close out the final work items. For firm price work there is no client involvement, and the close out can proceed under the control of the design firm. It is not unusual on design/construction work for the final work items to remain open over a period of months while vendor, construction, or other late data are obtained.

4.6   TASK FORCES AND TIGER TEAMS

Task forces or **tiger teams** are sometimes organized to expedite the solution to a project problem which would otherwise divert the design team from its basic job of executing the project design. These names are used interchangeably, the concept being that the group is set up for one purpose only, has a short life, and is composed of personnel who are not only technically competant but who are **aggressive** in pursuing the work. The group is normally comprised of people not on the project, with perhaps one or two project design people to provide some technical coordination. The charter of the group is to solve a **specific,** well-defined problem. Examples of such problems are a major subsidence problem with the foundation of a principal building on a project job site, significant development problems with a process control computer, or bankruptcy of a major equipment supplier requiring changing the sourcing. For each of these type of problems, it is critical that there be a satisfactory solution developed as soon as possible. Normally all the resources of the firm are made available, and the task force/tiger team will report to the PE and senior management as well.

In some cases, where the progress of the design is itself the problem, a block of design work may be assigned to the team. Then the design is developed relatively independently from the project design organization.

Because of the importance of the work, a report is prepared to set forth the findings and recommendations of the team. The team report will not always fully detail the solution but will develop it sufficiently so that the project design team can carry it forward. After acceptance of the report, the team is usually disbanded, its purpose having been accomplished. The life of these teams is on the order of a few weeks or perhaps months at most.

4.7   AUTHORITY

As mentioned earlier, with the functional organization the authority and responsibility for design decisions lies with the specialist groups, while with the project or matrix organization this authority lies with the project team. In both cases, however, major design decisions are subject to review by senior management. In the case of reimbursable work, or where provided for in the contract, the client (or owner) may reserve the right to make these decisions. In these cases the design group may conduct suitable studies or evaluations to define the advantages and disadvantages of each alternate and present them for a decision. Unless prohibited, the study should be accompanied by a recommendation to ensure that the client (or owner) is told in writing which alternate is preferred. This reduces the chance for misunderstanding and protects the design group from a later allegation if problems result when a less desirable alternate is selected.

4.8 JOINT VENTURES

Joint ventures are typically formed where two or more firms are needed to staff a project or where there may be significant risk which the firms wish to share. Usually one of the venturers is named the leading joint venturer and acts as the PE or Project Manager. A **contract** is developed between the venturers to define areas of responsibility, degree of liability, the resources each firm will provide to the work, and revenue sharing. Since the preparation of the contract defines the work scope, there is normally no problem establishing the principal responsibilities of each. The problems that can arise occur where the scope definition is poor or where the design development requires additional work.

The joint venture can either establish an integrated design group, where each venturer furnishes personnel in some of the same disciplines, or a design team, where each venturer provides all the personnel of a particular discipline. The latter arrangement is the more common one. With the integrated arrangement differences in design philosophy are more apparent since personnel are working side by side; with the nonintegrated arrangement there is a good chance that the design will be executed somewhat differently or to a different degree or depth of detail by the separate groups. In either case the design development must involve the management—the Project Engineer or the Project Manager.

Since the life of the joint venturer is limited to a single project and longer-term relationships may be only secondary considerations, some strains may arise between the team members. Usually these are minor and can be readily resolved. Nevertheless, the PE must be aware of them and exercise caution to ensure that the decisions made and the actions taken are evenhanded and not influenced by the organizational loyalties of the team members.

4.9 SPACE REQUIREMENTS

Space requirements always are a subject of concern to the project team. With the functional organization there is no significant problem, since the functional group members have their own usual working areas and added personnel are not normally used. Their space requirements then are relatively constant and vary but little over the life of the work.

It is quite a different matter with the project team approach. The project team is normally housed together in its own area. With the normal buildup of personnel at the beginning of the work and the decline in personnel in the later stages, space requirements vary greatly. Since large projects have a life of a year or more, it is common to see a project grow from a few people to a hundred or more within a year, with the different design groups peaking in size at different times. As a result a project that may initially need only space for 15 people may later need space for 135, and in its late stages come down to a relatively stable staff of about 20. In addition to the question of physical space, projects that use integrated computer networks for calculation work, data base development, and CAD must allow for expansion as the project grows.

The actual space required by the project will vary depending on the degree to which CAD is used. For projects with traditional drafting equipment and only a modest CAD usage, space allowances of 125 to 160 sq. ft. per person are adequate. This includes reference tables, drafting areas, space for desk-type design engineers, aisles, stick file and working file areas. CAD usage normally can be accommodated in the same space for each drafter using manual drafting methods. In addition space for offices for the PE and other senior personnel, plus conference areas, must be provided. These facilities can be evaluated and added by project and anticipated staffing requirements. Special facilities such as laboratories, shops, and testing areas may need to be considered as well.

Design groups often are housed in a large common area (sometimes called a "bull pen") to aid communication. The advantage to this arrangement is that personnel are forced to communicate and thus to coordinate their work. The disadvantage is the lack of privacy and sometimes high background noise which can be distracting for some people. But, unless the noise level is excessive, the personnel can readily adjust to it. For meetings or other activities, nearby conference rooms are useful.

Persons of a common group are usually housed together. This facilitates the work of the group supervisor and permits the use of common stick, vendor print, and data files. With the growth of the project and space limitations, it may be necessary to move one of the design groups to another area or even another floor or building. Although not desirable, the solution is to relocate the group that can best function relatively independently. This may not be only a question of the role of the group but may also reflect the ability and experience of the group leader.

4.10 ORGANIZING FOR QUALITY

While the establishment and maintenance of a quality design is itself an integral part of the engineering process, in certain circumstances and for particular types of work it is necessary to establish a formal **quality** organization to provide assurance that the quality of the design is proper and that the design organization is performing in accordance with generally accepted quality principles. Some of these principles were described in Chapter 2.

The quality organization is charged with the responsibility for **quality assurance (QA),** the term used to cover overall **programmatic** requirements. In principle the quality organization reports to a sufficiently high level that its findings and recommendations can be acted on even if disputed by the organization causing the noncompliance. Thus it is normal for the quality organization to report directly to the executive who has responsibility for the organizations that the quality group audits and reviews. To avoid establishing too large and unwieldy a group, the control of the work product is normally left in the hands of the operational organization, which establishes within their own group a **quality control** (QC) organization. This has the advantage of responsiveness and self-policing but suffers from the possible lack of diligence on the part of the organization, hence the need for the periodic audits by the QA organization. Perhaps the best arrangement is when personnel from the QA organization are assigned to work with the operational people on a continuing basis and are accepted by them, since they assist in the avoidance of quality problems. For this arrangement to work, however, it is essential that the QA personnel assigned have proper credentials and experience in the technical areas to which they are assigned. It is important that these personnel have good **interpersonal** skills to permit them to work effectively in what is in many respects a hostile situation.

PART III

PROJECT OPERATIONS

CHAPTER 5

SCHEDULING AND FORECASTING

Scheduling and **forecasting,** as the terms imply, deal with the question of **time** and only indirectly with cost. Cost and other factors come into play when the effects of schedule performance are considered. Both scheduling and forecasting involve **durations** to complete activities, the **sequencing** of activities, and assessments of time required to accomplish the work, or significant portions of the work. Cost forecasting and its associated activities are dealt with in Chapter 6.

5.1 TASKS AND DELIVERABLES

In recent years increased emphasis has been placed on the concept of **tasks** and **deliverables.** Tasks is the term used to identify activities necessary to produce the finished product or project; deliverables refer to specific pieces of work that are completed. A task might be the analysis of a bid, while a deliverable could be issuance of a drawing or specification. In general, tasks relate to internal activities or activities largely under the PE's control, while deliverables are completed units of work issued, presented, or to be presented to others.

The first step in preparing a schedule is to establish what the major tasks and deliverables are to be. In a contract deliverables are often called **line items** or **pay items.** In general, the scheduling work proceeds from the deliverables to tasks. That is, the deliverables are first established, and from them the tasks necessary to achieve them are identified. For example, if a deliverable is to issue a set of piping drawings to a contractor, a series of tasks necessary to support that deliverable would be establishment of the equipment piping and service requirements, development of the P&IDs piping specifications, line designation tables, calculations of line sizes, and material selection studies. All of these tasks would in an **interrelated** way provide input to the final activity of preparing the piping drawing. When the piping drawings are complete, the deliverable has been achieved.

While it may seem strange to proceed backward in drafting the schedule, deliverables are the critical **milestones,** so compliance with them is essential. Since tasks are the implementation aspects of a project, their durations can only later be developed to support the deliverables. Deliverables tend to include many tasks. It is far easier, and much less likely to introduce errors, if the scheduling work proceeds from the general to the detailed rather than the other way around. In addition few people can itemize all of the tasks required at the first writing of a task list. The list is thus subject to revision not only to include overlooked items but also those not previously thought necessary. Other tasks such as special studies or investigations may be identified as the work progresses. Despite the fact that engineering work may only represent 7 to 10% of the cost of a project, it will usually set the schedule for the accomplishment of downstream activities. Thus it is desirable to develop as accurate and complete a list of the tasks as possible so that the initial schedules will be accurate. Whether performing lump sum or fixed price work, R&D or reimbursable work, the same principles apply, since an accurate preliminary schedule and the estimates derived from it are essential.

Having decided upon the tasks and deliverables or the activities, the next step is to establish the correct **logic** or sequence of operations. Following that, durations are assigned to each of the activities, and a draft schedule is developed. Next come the reviews, sometimes called **sanity** checks, to ensure that the schedule is reasonable. Lastly the draft schedule is resource loaded. With this information, schedule adjustments are performed, and the schedule is given a final review, signed off, and formally issued for use. The schedule then becomes the official control document for the project. Figure 5.1 indicates the steps required to develop a schedule.

From an operational point of view it is necessary to select the **graphical** method of presentation and determine whether some form of computer-aided scheduling will be employed. Since it is not possible to keep in mind the many elements of even a simple schedule, much less their interrelationships, some form of graphical representation is essential. Similarly, with the wide availability and low cost of computer-aided schedule programs, virtually all projects use the computer to some extent in preparing and maintaining schedules.

When preparing schedules for major projects, a **hierarchy** of schedules is established. Three levels of schedules are often utilized: the **milestone, intermediate,** and **detailed** levels. Figure 5.2 shows the development and relationships among these schedules for a large engineering and construction

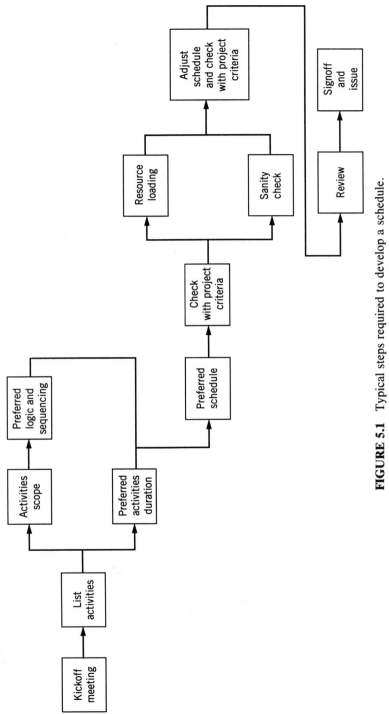

FIGURE 5.1 Typical steps required to develop a schedule.

Schedule hierarchy-EPMC projects

FIGURE 5.2 Schedule hierarchy.

Milestone Summary Schedule
- Unit
- Facility
- Category
- Activities

Project Summary Schedule (large projects only)
- Unit
- Facility
- Category
- Systems
- Bulk quantities

*Contract Summary Schedule
- Category
- Activities

Contract Requirement Schedule
- Contract scope and interfaces
- Major work activities

Project Intermediate Construction Schedule
- Facility
- Module
- Category
- Commodity
- Activity

Project 6-month Field Schedule
Interface contractor detailed activities

Project 3-week Construction Schedule
Interface contractor daily work operations

Intermediate Engineering Schedule
- System
- Category
- Type of document
- Activity

Detailed Engineering Schedule
- Drawing
- Specs
- Tasks

Interface Contractors' Major Activities

Contractor Schedules

Contractor proposal schedule
Contract scope and major work activities

Contractor master schedule
Equivalent to applicable ICS level of detail

Contractor 6-month field schedule
Detailed activities

Contractor 3-week field schedule
Daily work operations

Legend
→ Schedule development flow

Notes:
1. Front-end schedules (and in some cases start-up schedules) are prepared on EPMC projects, but are not considered part of the basic hierarchy.

2. Contract identifiers will be included on EPMC project schedules.

3. EPMC is engineering, procurements, and management of construction.

136

project. The general approach would be the same for any project, regardless of the industry. The main concern is to develop schedules that are working tools for the management of the work, and to that extent they are detailed only to the appropriate level for the work to be controlled.

5.2   SCHEDULING METHODS

There are three general types of graphic presentations of schedules: **bar** charts, **arrow** diagrams, and **precedence** diagrams (often called **critical path method**). Figure 5.3 compares the general characteristics of the three basic types of schedules.

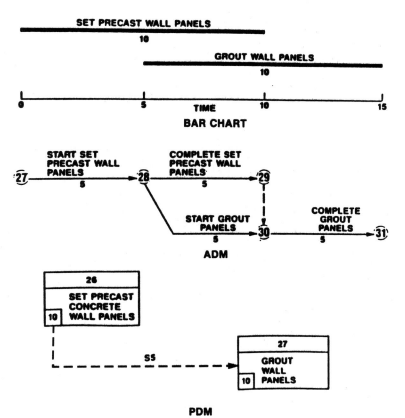

FIGURE 5.3   Comparison of schedule types.

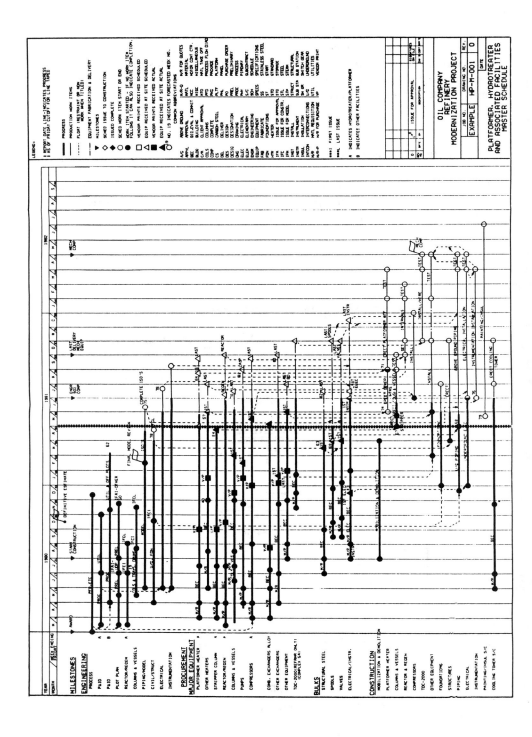

The most basic form of scheduling is the bar chart. The bars are proportional to the time required for the activity, with the beginning of the bar placed on a time scale at the beginning of the activity and the end of the bar placed at the end of the activity. This method has the advantages of ease of preparation and overall simplicity. It suffers from the major disadvantage that interrelationships, usually called **restraints,** are not apparent, although variations on this method can provide some of this information. It permits the highlighting of significant milestone dates, which is another major advantage. Speed of preparation and ease in reading are further advantages, and the bar chart can be useful for simple projects or for an overview of complex work. The later typically utilize additional more sophisticated schedules at the intermediate and detailed level(s). It is common for larger projects to use a **summary** or executive level schedule of the 20- to 50-line bar chart type that summarizes detailed schedules of perhaps 2000 or more lines at the detailed level.

Often bar charts will be embellished with triangles, circles, and other symbols to indicate specific events such as release dates, commitments, restraints, and dependencies. Because of differences in the use of these graphic symbols, it is good practice to include a legend on every schedule. The legend should be repeated on every sheet so that each can stand alone if necessary.

An example of a project bar chart schedule is shown in Fig. 5.4. Note particularly how milestone dates are highlighted by symbols. In the figure some restraints are shown as well.

The **arrow diagraming** method of scheduling utilizes arrows, representing the various activities connected head to tail to preceeding and following activities. Its primary advantage is simplicity and ease in displaying the logic or **dependency** of activities. In one form, arrow lengths are proportional to the durations required for the activities, and thus the diagram becomes a hybrid of both the bar chart and the precedence diagram, with absolute time as well as time intervals presented on it. The real value of arrow diagrams is to serve as an intermediate step between bar charts and true precedence diagrams by establishing the logic relationships between activities and displaying the sequence and interdependence of activities. For modest work of medium to low complexity it may be suitable for controlling the work without resorting to precedence scheduling. Figure 5.5 displays a typical arrow diagram.

On **precedence** diagrams, activities are shown as a network of lines whose lengths are proportional to the required time, with each activity having a node at its beginning and end. The nodes are connected to indicate **dependency** or interrelationships. This is the most widely used scheduling display method

FIGURE 5.4   Project bar chart.

ARROW DIAGRAM

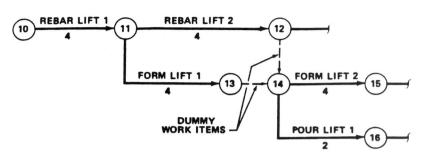

FIGURE 5.5   Arrow diagram.

because of its clarity, simplicity, and the large amount of information it conveys. Figure 5.6 shows a simplified precedence diagram.

From this basic method a variety of scheduling techniques have developed, all of them being similar and many of them utilizing computer programs to determine minimum times to complete the work. The path that this minimum time requires is called the **critical path,** and the Critical Path Method (CPM) approach was derived from it. In CPM scheduling it is possible to determine not only the schedule controlling activities but also secondary activities which with minor schedule changes, such as delays, can themselves become critical. The concepts of **float, early** and **late starts, early** and **late finishes,** and **restraints** all apply to this technique and are discussed below.

With both the arrow diagram and the precedence diagram, the **accuracy** of the logic (dependencies) is critical. The entire analysis and practicality of the scheduling work, and ultimately control of the project, depend on this. It is important to also note that there are frequently alternate ways to pursue the work and thus alternate logic chains may be constructed. As with all other operations, to avoid confusion, it is important to identify and segregate the alternates being considered.

Since the CPM method can utilize computer programs for analysis, it is possible to deal with extremely large and complex networks. Similarly it is relatively easy to perform studies varying certain critical activities to determine the impact on schedules. These alternate analyses ("what if" studies) are extremely useful, since no project proceeds exactly as initially planned and some changes and adjustments are always necessary. It is even more useful when significant variations occur, for often changes in schedule logic (the sequence and dependencies) are necessary. Lastly activities can be assigned resource requirement values **(resource loaded)** such as labor or cost to determine staffing or cash flow requirements.

Because of earlier schedule delays projects often are faced with trying to make up time. The CPM method is extremely helpful in preparing these schedule recovery programs. When coupled with resource loading the limitations on implementation become apparent.

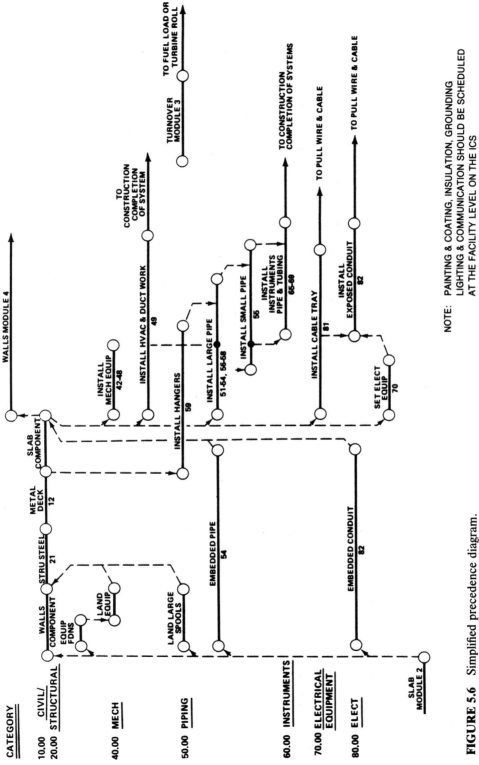

FIGURE 5.6 Simplified precedence diagram.

5.3   DURATION

Duration is the time required to perform an activity. To facilitate its definition, nodes are used to define the beginning and end of the activity. The duration may be long or short depending on the overall length of the project and the level of detail desired in the schedule. When determining durations, a balance needs to be maintained between those that provide a reasonable degree of progress reporting (and hence control) and the tendency to ask for too much detail. Excessive detail becomes burdensome to report and tends to impare the usefulness of a schedule. Where the project is relatively short, say, 90 days, the durations could be on the order of five days. For longer projects, say, two years, durations of 10 to 20 days might be more suitable.

Each node is given a unique number or letter identification to facilitate computerized processing. The numeric system uses lower numbers for earlier activities and higher numbers for later ones. Once an identifier is assigned it should not be reused, if for some reason it is discarded. It is not necessary that all the identifiers be utilized where blocks of them have been allocated. In activities of long duration control of work is improved if the durations are broken up into several short ones with nodes assigned for measurable events. For example, the preparation and issue to a fabricator of three dozen drawings could be broken into several smaller packages: the first 10, next 10, next 10, and final 6 drawings. This would permit monitoring progress more closely and identifying problems earlier, and thus there would be a greater probability of schedule compliance. Figure 5.5 shows a portion of a schedule using the arrow diagram method with node identification and internal nodes for a long duration activity, such as forming concrete lifts.

5.4   FLOAT

Float is defined as the amount of free time available not required for the work. It can be slack time available either prior to beginning an operation or after its completion. These cases are called **late start** and **early start,** respectively. If, for example, 25 weeks of schedule time are available for performing an activity requiring 16 weeks, 9 weeks of float are available. The schedule could show the activity starting immediately (early start) or at the end of the 9-week interval (late start) or any time in between. None of these cases would affect the critical path and hence the final completion of the project. Because of the large number of variables and external factors over which there is no control, it is better to plan on an early start approach to the work. This conserves the float for other possibilities. Float is often shown as a dashed (rather than a solid) line.

5.5 RESTRAINTS

Restraints are predecessor activities that must be completed prior to undertaking another activity. As such they interrelate the activities on a schedule and it is through these interrelationships that a critical path or minimum required schedule time is established. Restraints can be of many types: completion of a work activity, receipt of materials, receipt of information, issuance of permits, and so forth. The significant factor is that succeeding activities cannot proceed without their completion. Restraints need to be clearly indicated to avoid ambiguity and confusion. The precedence diagraming discussed earlier is a method that displays the linkage and dependence between predecessor activities.

Figure 5.6 shows the restraints typically found on a schedule. Note that the slab component must be completed first, for it restrains the walls, installation of mechanical equipment, HVAC and duct work, large pipe, cable tray, and the setting of electrical equipment. Note also that certain overhead work, such as hanger installation, is not dependent on and restrained by the slab completion and can be scheduled in parallel.

5.6 CRITICAL PATH

As described earlier, any precedence-type schedule will yield a sequence of activities that will complete the work in a minimum time. This **critical path** represents the shortest period required for the work. However, it is normally only one of several alternate paths having very similar time requirements. As a result, if there are delays on the other minimal **alternate** paths, it is likely that one of them will become controlling and hence the new critical path. Despite the best management on medium to large projects, say, $5 million and up, there are a large number of events that can affect the schedule many of which are outside the control of the project personnel.

Clearly it is necessary to periodically rerun the CPM network to determine shifts in the critical path and the effect on the schedule. Reruns should identify the alternate critical paths and their early and late start completion dates. With this information it is possible to determine the effects of **work arounds** and other sequence adjustments taken to compensate for schedule problems. All of the computerized CPM scheduling programs perform this critical path analysis and identification.

5.7 LEAD TIME

Lead time is the interval required from the placement of an order until the item or material is delivered to the point of use. In general, it is applied not only to equipment and materials but also to other activities such as the granting of permits and time required to resolve legal issues.

Lead times for major equipment will vary depending on the type and complexity of the equipment, the number of suppliers in the market and their current shop load. Some fabricated equipment such as large piping assemblies require relatively short lead times on the order of 10 to 20 weeks. Others such as the manufacture and delivery of a large paper machine require lengthy fabrication time of perhaps 50 to 60 weeks. Because of these wide variances accurate current lead time information is necessary when developing a schedule.

There are also numerous instances where manufacturing capacity has been rapidly filled, significantly lengthening lead times and affecting schedules. It may even be necessary in some instances to make an **early commitment** to ensure space in the supplier's manufacturing schedule and thus to protect delivery dates. An additional advantage of early commitment is that it often protects the pricing of the item from later increases.

In the case of custom-designed equipment, special tooling, and so forth, early commitment may be necessary if the supplier is to develop engineering data for the designer to integrate into the design. Some specialized activities, such as the preparation of environmental impact reports, may also require work to begin early.

For the major commodities or equipment normally purchased, lead time data are available in trade publications. More precise and specific information can be obtained by telephone surveys of suppliers. Where a large field expediting or inspection organization exists, it is possible to obtain these data from personnel who visit the fabrication facilities on a routine basis. Finally, where exact data are required, formal inquiries can be made to suppliers to determine their capability, shop loading, and backlog for particular items. To ensure accuracy, all lead time data used for major schedule decisions should be confirmed and updated at the time the schedule is prepared.

In preparing schedules, it is common to develop standard subroutines for **repetitive** operations such as procurement activities and client approval cycles. Repetitive operations are noted together with the lead times discussed above to establish overall durations for the procurement of major equipment. The subroutines are run as an integral part of the scheduling operations. An example of a subroutine is shown in Fig. 5.7.

5.8   PREPARATION OF SCHEDULE

As mentioned earlier, the preparation of the schedules for a project proceed from the general to the specific—that is, from the milestone summary schedule, through the intermediate level schedules, to the detailed schedules. Since all the information to prepare the schedules at succeeding levels of detail is not available at the start of the project, developments are **sequential** and

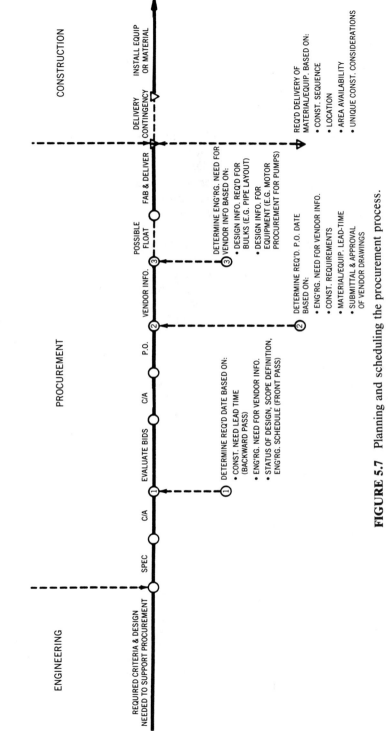

FIGURE 5.7 Planning and scheduling the procurement process.

145

depend on the definition of the work as it evolves. Sometimes on long-duration projects intermediate level schedules may not be available until four to six months after the project is initiated, with detailed schedules perhaps several months later. Because early items on the intermediate and detailed schedules may be worked on in the meantime, such schedules are often developed incrementally.

As mentioned earlier, in preparing the schedule, one works backward—starting with the deliverables and allowing time for the preceding activities to determine the starting date for the activity. For long lead items the beginning of the activity is the lead time required for the item, so early commitment may be necessary for some components. If it appears that there is insufficient time to complete a project as contemplated, senior level decisions must be made to change either the durations or sequence of activities (with its accompanying degree of risk) or the scope in order to shorten the schedule. One or a group of senior level technical managers or officers could make the major decisions so that protracted studies can be avoided. The **risk** can be significant, since the decisions may become invalid if insufficient data is available or external factors change. This can place a large part of the work at risk. The work may need to be repeated at a later date with an even longer delay in the schedule; otherwise, the design solution may end up being a poor one.

One way to shorten the schedule is to break the work up into parallel activities. Where personnel limitations are not a problem, **concurrent engineering** is often used to reduce the elapsed schedule time. This may increase costs somewhat, but it usually results in considerable time savings that offset the costs. Personnel limitations may require that the work be placed out of the house.

Scope adjustment is often possible if it can be shown that the work as contemplated cannot be completed on time. This is a sensitive question. It may involve contractural commitments and must be taken up with the highest management levels of the project.

5.9   SCHEDULE TYPES

Schedules fall into several types depending on their purpose. The principal features of the main types of schedules were discussed earlier. In practice project schedules of the bar chart type of the 20- to 40-line variety are used because they require only one page. They are particularly useful for control purposes as project status can be readily displayed and easily understood. Where an additional level of detail is needed, the 200-line configuration is often used. These charts are normally enhanced to show intermediate milestones and principal restraints, with each bar having four to eight events shown along its length. Such schedules typically show from 100 to about 1500 events.

For more complex work fully detailed precedence-type schedules are used. The 20- to 40-line bar chart schedule is retained for control and overview and supplemented with the 200-line schedules. For large, complex projects they are usually supplemented with CPM-type schedules often in great detail and running numerous pages. The detailed schedules are used for planning the actual work and can be arranged by work group to outline day-to-day operations. Because of their detail these schedules permit work progress to be monitored with great precision, and they can be used to allocate resources or to determine any remedial work. Because of their exhaustive detail they may only be run for a modest time window eg. two weeks.

Specialized schedules are often developed for R&D work, startup and operations, and test activities. Each of these categories has a different emphasis, a different level of uncertainty, and requires varying margins for specific activities. The chosen schedule presentation will therefore vary—but a good rule of thumb is to keep the schedule as simple as possible so that it is readable and thus usable.

5.10 RESOURCE LOADING

Virtually all of the current computerized scheduling programs provide the capability to **resource load** the activities and then to level the use of the resources, both personnel and financial, to reduce the effects of peaking. While the levelizing normally takes place within available float, care needs to be taken when using this feature to avoid overlooking the effect on dependent activities if insufficient resources cause earlier activities to be delayed. The ability to level personnel requirements is useful if only to stabilize the work force and thus avoid rapid swings in the number of assigned personnel. To maximize the benefit of the personnel resource leveling programs, it may be necessary to run alternates on an iterative basis until the results fit the general restraints of personnel availability.

Resource loading the schedule with **cash flow** requirements can yield significant financial dividends where the cost of money or the value obtained from alternative investment is factored in. This is particularly the case with advanced payments for equipment purchases and materials, which for very large orders may include escalation clauses tied to materials and labor indexes.

5.11 MEASURING PROGRESS

Measuring progress can be done in several ways. Regardless of the method used, it is critical that the original schedule be retained to provide a baseline against which progress (and departures) can be evaluated. Often schedules are revised because of changes in scope or an evolution in the design. In these circumstances the key original dates should be displayed with the necessary milestones changed to conform to the agreed changes.

Perhaps the easiest way to measure progress is the use of actual progress plotted against the project bar graph schedule. In this method the actual progress bar or milestone is plotted adjacent to the scheduled duration or event. This permits an easy comparison of the actual with the forecast progress. Figure 5.8 shows a portion of a project schedule with state of progress indicated.

Because of their simplicity and ease of use, **envelope** curves are often used to measure progress. It is a simple matter to compare a variable such as cumulative hours expended or drawings issued to that forecast to determine progress. Because of the wide variations in the elements of a project these curves are difficult to generate with precision and thus can only furnish broad guidance. The curves should be developed using only a maximum of three to five milestones. More than that could introduce confusion in correlating data. Considerable care should be taken when developing the curve(s) to remove, or account for, the effect of secondary items. While these curves only provide an approximate answer to the progress question, they do give an overall indication of the status of the work and can be used for general guidance. The usual problem in using the curves is a tendency to **overestimate** the degree of completion by assuming that progress for the project has proceeded more efficiently than normal. The use of these curves should be restricted to generalizations only.

One envelope curve that is frequently used in measuring overall progress is one that measures the rate or cumulative dollars expended. Often projects will appear to be making sufficient progress but will really not be making commitments or expending labor at a rate high enough to ensure completion of the project on schedule. Then the rate or cumulative curve based on **expenditures** would be a useful guide. When using this type of curve to forecast completion, one divides the expenditures remaining by typically 80 or 90% of the time remaining to determine the new rate required to complete and thus identify whether the project needs to be significantly accelerated. The use of the 80 or 90% factor provides a **contingency** to increase assurance of on-time performance. Usually during the later stages of a project the factor may be reduced to 60% or less of earlier progress to compensate for the effects of crowding, increased complexity, and so forth. This analysis, if performed early enough, will prevent the situation where the project cannot be completed on time due to limitations of space or personnel. One of the four curves shown in Chapter 8 can be used for this depending on the distribution of resources and effort versus time. The four curves considered there are back loaded, front loaded, bell curve loaded, or, in rare cases, straight-line (uniform) loaded.

One useful way of estimating both design effort and progress is the **hours expended** per drawing method. If a detailed drawing list is prepared, the values assigned to the individual drawings can be used to provide a reasonably

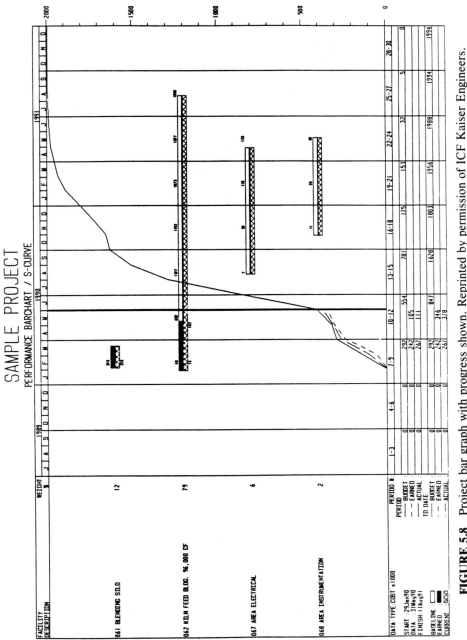

FIGURE 5.8 Project bar graph with progress shown. Reprinted by permission of ICF Kaiser Engineers.

149

accurate estimate of the engineering and drafting effort. Taking the total work hours used for design (engineering, drafting, and administration) and dividing those hours by the number of drawings to be produced provides a fairly consistent value that can be applied to the design work. Breaking that down by the various work groups provides further precision and is often used as a basis for the overall estimate of the design effort. Since each industry has a different set of design standards, it is not possible to utilize the per-drawing data from, say, the chemical process design industry to the aircraft design industry. Even within an industry design effort will vary significantly depending upon the complexity of the design and the standards applied. Figures 5.9a, b, c, and d show a set of work hours standards for design drawings in the engineering and construction industry. The hours shown for CAD are workstation operating hours only. All engineering, designer, drafter, and workstation operator effort are shown under labor. For partial progress assessment, Table 5.1 shows a typical percent complete chart for various types of construction design drawings broken down by the usual drawing milestone dates. Similarly for the engineering work required to support the preparation of requisitions and the **procurement** process, Table 5.2 lists percentage complete values that can be used to estimate progress. The example given is for a petrochemical facility.

One refinement often tried as a way to measure efficiency is to evaluate the hours expended versus actual progress. This entails the concept of "hours earned/hours expended." Here careful records must be kept of hours expended versus actual progress as defined by the deliverables, and a value is assigned to each of the deliverables to establish its worth in hours. As the deliverables are completed, the earned hours are then compared to the expended hours to determine progress. An advantage of such a system is the ability to forecast the hours required to completion. It is possible to readily extrapolate to completion the expended versus earned work hours if allowances are added for the lower productivity in the later stages of the design work. This method, however, creates the potential for a major problem, since there is a natural tendency to use these data as a kind of **efficiency** measure. Although not the stated purpose of the system, it is sometimes viewed that way, and it introduces the hazard that achieving a favorable set of earned numbers becomes an end in itself.

Another slightly different form of progress measurement is to calculate the **unit rate** that applies to the work. This is useful if the work is repetitive and of a reasonably similar type. For example, a designer detailing reinforcing steel for a fabricator prepares bar-bending sheets and schedules that govern the fabrication process; the work is repetitive, and there is little change from one project to another. The rate of progress made and a forecast of completion of the work can be readily evaluated by applying unit rates to both the schedules and the sheets. Often these designers are paid on a unit price basis, so much per sheet or per ton of rebar detailed. For those cases where the rate is based upon the tonnage of rebar detailed, it is essential to establish

TYPICAL ENGINEERING HOURS BY DRAWING TYPE

DISCIPLINE DRAWING TYPE	MANUAL			2D CAD						3D CAD					
				SIMPLE		NORMAL		COMPLEX		SIMPLE		NORMAL		COMPLEX	
	SIMPLE	NORMAL	COMPLEX	LABOR	CAD	LABOR	CAD	LABOR	CAD	LABOR	CAD	LABOR	CAD	LABOR	CAD
CIVIL															
General Arrangements	100	138	175	87	27	119	37	152	47	40	13	55	18	70	23
Rough Grading	100	125	150	87	27	108	33	130	40	40	13	50	17	60	20
Yard Piping	75	125	125	69	12	115	20	115	20	30	6	50	10	50	10
Finish Grading and Drainage	75	113	138	67	16	101	24	123	29	30	8	45	12	55	15
Fencing	75	100	125	67	16	89	21	112	27	30	8	40	11	50	13
Maps	50	63	75	45	11	56	13	67	16	43	9	54	11	65	14
CAD Model (See Note 2)										27	20	38	29	46	34
STRUCTURAL CONCRETE															
Building Foundations	100	125	150	92	16	115	20	138	24	90	14	112	17	135	21
Equipment Foundations	113	138	163	104	18	127	22	150	26	101	15	123	19	146	22
Slabs	100	125	150	92	16	115	20	138	24	90	14	112	17	135	21
STRUCTURAL STEEL															
Building Framing	113	138	163	104	18	127	22	150	26	45	9	55	11	65	13
Stairways and Platforms	100	125	150	92	16	115	20	138	24	90	14	112	17	135	21
Hoppers, Bins and Other Platework	100	125	150	92	16	115	20	138	24	90	14	112	17	135	21
CAD Model (See Note 2)										36	27	44	33	52	39

(a)

FIGURE 5.9 Typical engineering hours required by drawing type. Reprinted by permission of ICF Kaiser Engineers (*continues*).

TYPICAL ENGINEERING HOURS BY DRAWING TYPE

DISCIPLINE	MANUAL			2D CAD						3D CAD					
				SIMPLE		NORMAL		COMPLEX		SIMPLE		NORMAL		COMPLEX	
DRAWING TYPE	SIMPLE	NORMAL	COMPLEX	LABOR	CAD	LABOR	CAD	LABOR	CAD	LABOR	CAD	LABOR	CAD	LABOR	CAD
ARCHITECTURAL															
Site Development	75	100	125	69	12	92	16	115	20	67	10	90	14	112	17
Plans	100	125	150	92	16	115	20	138	24	40	8	50	10	60	12
Sections and Elevations	100	125	150	92	16	115	20	138	24	40	8	50	10	60	12
Detailed Floor Plans	100	138	175	89	21	123	29	156	37	40	11	55	15	70	19
Interior Elevations	75	100	125	67	16	89	21	112	27	30	8	40	11	50	13
Reflected Ceiling Plans	75	100	125	67	16	89	21	112	27	30	8	40	11	50	13
Door and Window Schedules	88	113	138	76	23	98	30	119	37	35	12	45	15	55	18
CAD Model (See Note 2)										29	22	37	28	46	35
MECHANICAL															
Material Balances	100	125	150	92	16	115	20	138	24	90	14	112	17	135	21
Heat Balances	88	113	138	81	14	104	18	127	22	79	12	101	15	123	19
Flow Sheets	100	138	175	87	27	119	37	152	47	83	23	114	31	145	40
General Arrangements	125	150	175	108	33	130	40	152	47	50	17	60	20	70	23
Mechanical Arrangements	100	125	150	87	27	108	33	130	40	40	13	50	17	60	20
Conveyors	100	125	150	89	21	112	27	134	32	40	11	50	13	60	16
Chutes and Hoppers	88	113	138	78	19	101	24	123	29	35	9	45	12	55	15
Dust Collection	100	125	150	89	21	112	27	134	32	40	11	50	13	60	16
CAD Model (See Note 2)										33	25	41	31	49	37

(b)

FIGURE 5.9 (continued). Typical engineering hours required by drawing type.

152

TYPICAL ENGINEERING HOURS BY DRAWING TYPE

DISCIPLINE / DRAWING TYPE	MANUAL			2D CAD						3D CAD					
	SIMPLE	NORMAL	COMPLEX	SIMPLE		NORMAL		COMPLEX		SIMPLE		NORMAL		COMPLEX	
				LABOR	CAD	LABOR	CAD	LABOR	CAD	LABOR	CAD	LABOR	CAD	LABOR	CAD
HVAC															
Flow Sheets and Air Balance	100	125	150	87	27	108	33	130	40	83	23	104	29	124	34
P&ID's	113	150	188	98	30	130	40	163	50	93	26	124	34	155	43
Equipment Schedules	100	125	150	92	16	115	20	138	24	40	8	50	10	60	12
Roof and Floor Plans	100	125	150	89	21	112	27	134	32	40	11	50	13	60	16
Sections and Details	100	125	150	89	21	112	27	134	32	40	11	50	13	60	16
Control Diagrams	88	113	138	81	14	104	18	127	22	79	12	101	15	123	19
Plumbing	88	113	138	81	14	104	18	127	22	35	7	45	9	55	11
Fire Protection	75	100	125	69	12	92	16	115	20	30	6	40	8	50	10
Isometrics	25	38	50	20	11	30	16	39	21	10	5	15	8	20	11
CAD Model (See Note 2)										26	20	33	25	41	31
PIPING															
Material Balances	100	125	150	92	16	115	20	138	24	90	14	112	17	135	21
Heat Balances	88	113	138	81	14	104	18	127	22	79	12	101	15	123	19
Process Flow Diagrams	100	138	175	87	27	119	37	152	47	83	23	114	31	145	40
P&ID's	113	150	188	98	30	130	40	163	50	93	26	124	34	155	43
General Arrangements	125	150	175	108	33	130	40	152	47	50	17	60	20	70	23
Tanks and Vessels	100	125	150	89	21	112	27	134	32	40	11	50	13	60	16
Piping Arrangements	113	138	163	98	30	119	37	141	43	45	15	55	18	65	22
Piping Details	100	125	150	89	21	112	27	134	32	86	18	108	23	129	27
Pipe Support Details	88	113	138	78	19	101	24	123	29	76	16	97	21	119	25
Line Designation Tables	75	100	125	67	16	89	21	112	27	30	8	40	11	50	13
Isometrics	25	38	50	20	11	30	16	39	21	10	5	15	8	20	11
CAD Model (See Note 2)										28	21	35	26	42	32

(c)

FIGURE 5.9 (continued). Typical engineering hours required by drawing type (*continues*).

153

TYPICAL ENGINEERING HOURS BY DRAWING TYPE

DISCIPLINE	MANUAL			2D CAD						3D CAD					
				SIMPLE		NORMAL		COMPLEX		SIMPLE		NORMAL		COMPLEX	
DRAWING TYPE	SIMPLE	NORMAL	COMPLEX	LABOR	CAD	LABOR	CAD	LABOR	CAD	LABOR	CAD	LABOR	CAD	LABOR	CAD
ELECTRICAL															
One Line and Elementary Diagrams	100	138	175	87	27	119	37	152	47	83	23	114	31	145	40
Power and Control Layouts	113	125	163	98	30	108	33	141	43	93	26	104	29	135	37
Lighting and Communication Layouts	75	113	150	65	20	98	30	130	40	62	17	93	26	124	34
Interconnection Diagrams	38	63	88	30	16	49	27	69	37	27	14	45	23	64	32
Computer Control Diagrams	38	63	88	34	8	56	13	78	19	32	7	54	11	76	16
INSTRUMENTATION															
Control Systems and Loop Diagrams	75	100	125	65	20	87	27	108	33	62	17	83	23	104	29
P&ID Input to Piping and Instrument Data Sheets	75	100	125	59	32	79	43	98	53	54	27	73	37	91	46
Block Diagram and Control Room Layout	100	125	150	84	32	105	40	126	48	79	27	99	34	119	41
Installation Details	75	100	125	63	24	84	32	105	40	60	21	79	27	99	34
Logic Diagrams	100	125	150	84	32	105	40	126	48	79	27	99	34	119	41

Notes:

1. The engineering hours by drawing type include all project engineering, design work, specification and procurement of equipment, and minor coordination with the construction contractor. It does not include plant data books, operating manual preparation or operator training.

2. The estimate of hours to produce the CAD Model is determined by multiplying the engineering hours by drawing type in the CAD Model row times the number of drawings to be extracted from the model. The drawings that will be extracted from the CAD Model are the drawings in the drawing type column shaded dark.

(d)

FIGURE 5.9 (continued). Typical engineering hours required by drawing type.

TABLE 5.1 Typical drawing percent complete table				
Drawing Type	Start	Issued for Review	Issued for Approval	First Issued for Construction (or Fabrication)	Final
Process flow diagrams	5	35	50	75	100
P&IDs	5	30	50	75	100
Utility flow diagrams	5	65	75	85	100
Plot plans/equipment arrangements	5	50	65	75	100
Piping	5	65	75	85	100
Civil	5	60	70	80	100
Structural	5	65	75	85	100
Architectural	5	65	75	85	100
Logic diagrams	5	65	75	85	100
Instrument Loop	5	65	75	85	100
Electrical	5	65	75	85	100
Vessel	5	35	50	70	100

a level of complexity for the work to reflect major tonnage differences for significantly different bar sizes.

For projects where field or fabrication activities are significant, curves of bulk materials delivery and installation curves of materials using the same principles described above are useful.

5.12 EXCEPTION REPORTS

To analyze the performance of a project and to identify specific areas for improvement, it is useful to run **exception** reports that identify those areas of the project that are not proceeding according to plan. Because of the bulk of information, it is necessary to develop suitable criteria so that only the most significant items are presented at any one time. Exception reports are a way to avoid being inundated by this mass of data and allows management to focus on the particular items that need attention. Exception reports are of many types depending on the project and the type of activities being controlled. One important type that has proven to be helpful is to run reports based on deviations from early start/late start and early finish/late finish criteria. In this way, not only items of specific schedule delay, but other activities that threaten the schedule can be identified.

Typically exception reports are run weekly for a one-year project and biweekly for longer projects. The report is **limited** to a specific number of items—often ten. Since some of the items will require reassignment of work priorities, working with a short list permits the items to virtually be kept in

TABLE 5.2 Typical material requisition percent complete table

Material Requisition Commodity	Start	Issued for Review	Material Requisition for Quotation	Bids Received	Bid Summary to Client	Material Requisition for Purchase	Delivered
C Columns and vessels	5	30	35	40	60	70	100
D Tanks	5	30	35	40	60	60	100
E Exchangers	5	35	40	45	75	85	100
F Fired heaters	5	20	25	30	55	65	100
G Pumps	5	30	35	40	65	75	100
H Vacuum equipment	5	30	35	40	65	75	100
J Instruments	5	40	50	55	70	80	100
K Compressors	5	20	25	30	55	65	100
L Piping	5	30	35	40	45	50	100
M Structural steel (Fab)[a]	5	25	30	35	40	50	100
N Insulation	5	30	35	40	45	50	100
P Electrical	5	30	35	40	65	75	100
Q Foundation materials	5	40	45	55	60	65	100
R Architectural	5	40	45	55	60	65	100
S Sitework materials	5	40	45	55	60	65	100
T Special equipment	5	30	35	40	70	80	100
V Packaged equipment	5	30	35	40	70	80	100
X Paint	5	40	50	55	75	85	100
Y Miscellaneous equipment	5	30	35	40	70	80	100

Note: This table includes workhours for vendor print review and technical coordination of material fabrication. If the vendor print effort log is utilized, workhours for vendor print review should be transferred from the MR control log and the percent complete table adjusted accordingly.
[a] Includes review of shop drawings.

FIGURE 5.9 (continued). Typical engineering hours required by drawing type.

mind. For projects that involve fabrication or construction, many of the activities are performed by others and thus are not under the direct control of the project personnel. For these a report of the delivery of critical items is an important additional tool. Where numerous activities of this type exist, routine exception reports can be extremely useful.

5.13 COMPLETION FORECASTING

Completion forecasting is perhaps the most poorly understood aspect of scheduling and forecasting. The mistakes most commonly made are to underestimate (or not allow enough time for) unforseen events that will delay the project and to assume that the work force can continue to produce at the same unit rate as during the period of maximum productivity.

While it is not possible to develop a laundry list of all the items that can delay a project, a seasoned PE will have a feeling for items that have a good chance of going wrong and jeopardizing the completion of the project. Some allowance should be made for these, while recognizing that all of them are not likely to happen but some of them certainly will. One approach is to perform a formal **risk analysis** with values of risk assigned to the key identifiable events. Because of the difficulties of assigning risk values or probabilities, these studies are often viewed as being of little value and without sufficient historical data are difficult to defend. One way to overcome this is to assign risks as **high, medium,** or **low.** This avoids the problem of actual quantification. Apart from such difficulties, however, an advantage of the studies is that merely raising the problem of risk in a formal way provides visibility. While the studies may never even quantify the risk, the mere act of performing the risk **identification** is useful in avoiding surprises and provides some warning to the prudent manager.

Unit rates of production traditionally fall off during the final phases of a project. There are numerous reasons for this, including the requirement to clean up the loose ends from earlier work. There is also the need to deal with a larger body of information and design development, which often constrains the solutions to design problems and may require more analysis and study. Further the loss of personnel with specific knowledge of the project may increase the difficulties in closing out the work. With fabrication or construction activities, crowding, personnel limitations or availability of fabrication space, or tooling often are limiting factors.

As with other forecasting, no simple formula or percentage can be assigned to cover this situation. In general, the rate of completion of the project describes an S curve. The curve significantly flattens in the latter stages. Affecting this rate, for example, are reductions in the number of personnel working on the project and reductions in efficiency. The combined effect of the two may be to reduce the productivity per hour to perhaps as little as one-quarter of the maximum rate achieved at the peak of the project.

Experience on past projects is the best guide to likely performance. If these data are available, they can be used to develop the boundaries of the problem and to assign upper and lower values to the forecast. Otherwise, best judgment must be used, and though this may seem inaccurate, it is a far better approach than not evaluating the problem.

5.14   CHANGE TRACKING

Change control administration represents the main way a firm price contract can be adjusted to compensate for increases or decreases in scope, changes in time permitted (schedule), work originally excluded from the contract, or other factors such as labor disputes and delays in supply of materials and equipment. Some system to document and identify changes should be in place. The system can be as simple as a diary or day book kept by the PE, or it can be an elaborate system with standardized forms that automatically initiates estimates and schedule forecasting efforts. Whatever system is used, it must be **rigorous** to ensure that all change requests are identified and evaluated. Key to being able to make such a system work is to ensure that all members of the project team who might receive requests for changes are aware of the system and understand their responsibility for initiating a change document.

Where a formal system is in place, it will typically utilize a standardized form that provides for each project group to evaluate the effect of the proposed change upon its work as well as to estimate the cost to the project, including additional design and capital costs. The effect on the schedule for completion of the work, both design and fabrication/construction, must be evaluated as well.

Changes should be sequentially numbered, and one person or group appointed to log and maintain the change order requests and the administrative data resulting from them. In no case should a change be implemented without the **formal approval** of the proper authority, usually Project Engineer or Project Manager. It is important that the project team members understand this. Often well-meaning personnel will institute changes on their own authority. Usually this is done in an attempt to maintain client goodwill, without realizing the consequences of such changes and the potential embarrassment that can result from a later reversal of the informal agreement.

Contractors will contend that all changes cost them money, and even if the change calls for a reduction in the scope of the work, they are still entitled to an extra charge. They base this request upon the concept that any change destroys their planning and requires them to proceed differently than originally contemplated. To a large extent this is true, and it must be borne in mind that minor reductions in scope may in fact cost more than had the work proceeded as originally scoped. If, however, the change makes significant reductions in scope, the contractor should show a reduction in price. Usually the reduction offered appears small, and the contractor is then required to furnish backup data to justify the estimate. It is good practice and saves time to require that this backup data acompany all requests for extras in order to permit proper evaluation of the change. More detail on the estimating of changes and their evaluation is found in Chapter 6.

5.15 SCHEDULING FOR MANUFACTURE

Scheduling for manufacture involves all the same steps described earlier. Because much of a typical manufacturing project involves **buyouts** incorporated into the product, and often the purchase of special tooling, the front end of the schedule needs particular care to ensure that sufficient time is allocated to permit this.

Similar to the time required for buyouts is the time required prior to production for the development and shakedown of tooling, including fixtures. Since this work is normally performed by the company itself, it is usually more predictable than the time required for buyout development. The production of the **first article** is of great value in validating the systems and production methods to be used. To the greatest extent possible they should be those contemplated for the volume production that is to follow. After the production of the first article, the schedule must also allow time for production of a sufficient quantity of early articles, including initial adjustments to the production processes, to ensure that the quality and quantity requirements of the order can be met when full production begins.

With increased emphasis on bringing products to market faster, more manufacturing scheduling is being done using the **setback** method. In this method the date when the product is available (required) to market is established, and the schedule is developed to arrange the predecessor steps to achieve this goal. Thus the schedule can be said to be completion driven. Calculations are performed for each assembly or part, depending on the amount of detail and on when the information or material is required by manufacturing to prepare its capacity for building the product. The requirement date depends on the **cumulative** times needed to acquire, build, and test, for example, the tools and parts prior to the next level of assembly. The result of this tells the design team when their work must be available. With this information, resource loading of the schedule takes place and the necessary personnel, tools, and facilities put in place to meet the schedule dates. There are some parts or buyouts as well as special tooling that may have long lead times, and these are identified when the schedule is prepared. They must be given special handling and early procurement, and design work must be authorized for them to protect the schedule. A frequently used technique to shorten the design time is to utilize concurrent engineering which, as described earlier, provides for engineering work to be done in parallel rather than sequentially.

Order entry is of course, the beginning of the manufacturing cycle, and considerable effort should be made to reduce the time required to enter the orders to the absolute minimum. The use of **real time** order entry systems and the avoidance of batch processing of orders at day's or week's end should be paramount considerations. For products having other than a very short production cycle, decentralizing the process to allow customers or sales personnel in the field to enter orders directly should be tried and implemented wherever possible. With this type arrangement order confirmation can be performed after the order is in the system in parallel with initial manufacturing steps.

Increased recognition of the importance of time and the small amount of time a product receives **added value,** as compared to the time spent as **work in progress** or awaiting decisions of one sort or another, can make a fundamental difference in the profitibility of a company. The following case study illustrates the problem and the effect on the company profits and market share of an aggressive solution:

Consider the remarkable example of Atlas Door, a ten-year-old U.S. company. It has grown at an annual rate of 15 percent per year in an industry with an annual growth of 5 percent. In recent years, after-tax earnings were in excess of 10 percent of sales, about five times the industry average. Atlas is debt free. In its tenth year of existence, it achieved the number one leadership position in the industry.

The company's product: industrial doors. These doors involve considerable variety, with many possible choices of width, height dimensions and material. This variety limits the effectiveness of responding to customers through inventory. Most doors must be manufactured to order.

Historically, the industry has needed on average 12 to 15 weeks to respond to an order for a customized or out-of-stock door. Atlas' strategic advantage is time. It can reliably respond in three to four weeks because it has structured its order entry, engineering factories, and logistics to move information and product quickly and reliably.

First, Atlas built just-in-time factories. These are fairly simple in concept: extra tooling and machinery to substantially reduce changeover times; and a fabrication process organized by product and scheduled so that most all of the parts needed to fulfill an order for a door can be started and completed at about the same time. However, the performance of the factory, while critical to the company's overall responsiveness, consumes only about two-and-a-half weeks of the complete cycle.

Second, Atlas compressed time at the front end of the system, where the order is received and entered into the process. Traditionally, when a customer, distributor, or salesperson called a door manufacturer with a request for price and delivery, he or she might have to wait a week or more for a response. If the door was not in stock, not in the schedule, or not engineered, then the request had to be kicked around the supplier's organization before the answers were known. Atlas automated its entire order entry, engineering, pricing, and scheduling processes. Today 80 percent of all incoming orders can be priced and scheduled while the caller is still on the telephone. Special orders can be engineered quickly because the amount of re-engineering has been substantially reduced by preserving the design and production data of all previous special orders.

Third, Atlas controls logistics tightly so that a complete order can always be shipped to construction sites. An order requires many components. Getting them together at the factory and making sure they are with the correct order can be a time-consuming task. Getting the correct parts to the job site if they missed the initial shipment is even more time consuming. Atlas developed a system to track the parts in production and the purchased parts for each order to ensure that all parts arrive at the shipping dock in time and at the customer site at the same time.

Early in the company's life, Atlas' salespeople were often rebuffed when they approached new distributors. The large, established, and attractive distributors already carried the door line of a larger competitor and saw no reason short of major price concessions to switch suppliers. As a startup company, Atlas was too small to compete on price alone. Instead, the company positioned itself as the door supplier of last resort. It was the company people came to if the established supplier could not deliver or missed a key date.

Of course, with industry lead times averaging 12 to 14 weeks, the company was likely to receive some calls. When the company did get a call, it was able to command a higher price because of its faster delivery. Not only was its price realization high, but its streamlined and effective processes were lower cost. The company had the best of both worlds. In the short span of ten years, Atlas replaced the established door suppliers for 80 percent of the distributors in the country. Now the company could be selective when asked to become the house supplier and could acquire the stronger distributors.

Atlas's competitors are not responding at all effectively. The conventional view at one major competitor is that the company is a "garage shop operator" that cannot sustain its growth. In other words, the competitors expect the company's performance to degrade to the industry average as it grows larger. But this response—or nonresponse—only reflects a fundamental lack of understanding time as a source of competitive advantage in business. The competitors' delay in responding often allows the timebased competitor to build a lead that is insurmountable or at least very expensive to close.*

5.16 REPORTS

As with all other phases of the work, without **quantified** data it is impossible to control the design process. Several forms of reports that provide overviews

---

* Reprinted with the permission of The Free Press, a Division of Macmillan, Inc. from *Competing against Time* by George Stalk, Jr., and Thomas M. Hout. Copyright (c) 1990 by The Free Press.

of the work and that also list the most critical items to be dealt with are usually required.

Weekly or biweekly updates of schedules are useful, so is a ten-item critical item action list run biweekly or more frequently. In addition special runs of early and late starts as well as **rolling schedules** for, say, the next two weeks or a month can be useful in identifying potential problem areas. When the critical path changes, it will usually be necessary to rerun the entire remaining schedule and make a distribution to all parties. When this is done, it is important, as mentioned earlier, to retain the **original** schedule for reference.

As with all of this work it is important that the schedules and their reports be kept as simple as possible and that the scheduling personnel analyze and reduce the data for management use.

CHAPTER 6

ESTIMATING AND COST CONTROL

The development and utilization of accurate cost information is a primary tool for the management of a project and can provide both rapid and accurate data for decision making. The forecasts prepared from these data provide leading indicators of the direction of the project development and permit early corrective action of deviations.

6.1 PROJECT COST CONTROL

The **cost control** process includes inputs from a variety of sources, both in-house and outside sources, and estimates their effect on the project cost (or budget). The process also often includes methods that generate **cash flow** data and forecast expenditures on both near- and long-term bases. By capturing both actual cost data and estimated expenditures and commitments as they occur, the system will permit the rapid development of these data with a minimum of effort. To provide these data a cost or **estimating function** is established, having as its primary goal the gathering of cost and **commitment** data and the preparation of **budgets,** estimates, and **forecasts.** The personnel performing this work are usually cost engineers with extensive experience. Many of these engineers are able to estimate final costs with a surprising degree of accuracy, from very preliminary or sketchy data.

The person or group performing the cost or estimating function must be given sufficient authority and support to be able to draw on all the resources and experience of the organization in preparing their work. Preferably they are housed together in a convenient area and have access to data-processing equipment, historical records, communications equipment, administrative support, and other services to facilitate their work. The size of the group will vary, with a core of people assigned for the duration of the project but augmented by additional temporary personnel for short, labor-intensive activities such as the preparation of estimates, bids, or special reports.

A suitable project cost control plan uses the definition of the deliverables described in Chapter 5, defines the responsibilities of the particular groups of project personnel, identifies the interfaces with other groups, and establishes the methods by which the various cost control activities will be performed.

The **plan** establishes who is responsible for the preparation and input of basic data (engineering, construction, accounting, etc.), the **frequency** of input (daily, weekly, etc.), responsibility for review and use of data, and the relationships among parties. A block diagram, together with a table, is a convenient way to display this information, and it should provide a clear picture of the roles and responsibilities of the project team members. The block diagram is developed in concert with the functional departments to ensure that the plan fits their normal mode of work and that special requests are minimized. A sample block diagram is shown in Fig. 6.1.

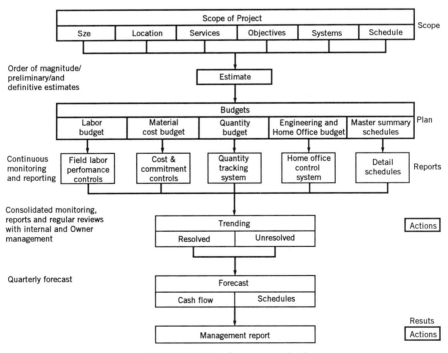

FIGURE 6.1 Cost control plan.

It is useful to keep in mind the significant difference between **reporting** and **controlling.** The first—reporting—is essentially passive and historical in nature. Reporting consists of **accumulation** and processing of data on the costs and commitments as defined by the contract and implemented through project decisions, design, procurement commitments, labor reports, subcontracts, and other project documents. The second—control—represents the use of these data in a pro-active way to control and direct the destiny of the project and thus to effect control over the expenditures and the ultimate overall cost of the work.

With the wide availability of computerized programs for the accumulation and integration of financial data into cost control programs, it is important to ensure that at the outset of a project the cost-reporting systems are, if not integrated, at least compatible. This permits data to be entered once in the system, and after that the data are available to all authorized users. The system should utilize to the maximum extent the documents and practices of the firm and thus avoid the need to establish special systems or methods. Doing so reduces the need for special training of personnel, permits their ready integration into the project cost group, and greatly reduces the number of input errors that occur when special, one of a kind, systems are employed. Many projects will in addition require some form of special reporting—often because of the mode of financing. Thus frequently some added special systems or subsystems may be required.

The overall nature of the cost control function, cuts across all the project activities, and the cost personnel will need to be knowledgeable of and have access to all the project controlling documents. Table 6.1 lists the major documents usually identified in the project cost control plan or procedure.

TABLE 6.1 Project Cost Control Plan Documents	
Contract	Cash flow and payments curves
Scope document	Definitive project schedules
Procedures manual	Engineering performance
Milestone summary schedule	Procurement status
Material assignment schedule	Critical items reports
Change control system	1 week schedule (or equivalent)
Project review reports	Engineering release curves
Weekly/monthly project progress reports	
Weekly cost reports	Construction installation curves
Subcontract cost reports	Manpower curves
Commitment register	Start-up schedule
Craft performance data	Material receiving and control
Nonmanual personnel schedule	Accounts payable ledger
Quantity tracking program	

6.2   COST CODE AND ACCOUNTING SYSTEMS

A system of **cost codes** is an essential tool in controlling costs and developing data for use in future estimates. It also assists in cost identification and classification both for **accounting** and **tax reporting** purposes. The project cost control system should be established at the very beginning of the work, and its categories and classifications used for all activities generating costs. Most firms have existing cost code systems that form the basis for the project system.

A principle usually followed is that all costs that can be identified as being incurred for the benefit of the project are to be charged to the project. This is applied to costs shared between projects as well as those incurred only for the benefit of the single project. As a result such things as rental of an automobile for the exclusive use of a project should be charged totally to that project, while casual automobile use, such as pool cars, for the benefit of the project should be prorated to the project depending on actual use. Unless such a policy is followed, it is never known just what a project and, more important, its various elements, did in fact cost, and the historical data developed are invalid. As a result there is no basis for accurately estimating the costs of future work. Another added advantage is that by identifying costs to the specific project, less cost is found in various overhead or pool-type accounts that must be later distributed to the projects. This reduces the overhead rate and permits more accurate control of activities of only limited value to the projects. In some cases the project has the choice of purchasing services from the parent firm or going to an outside vendor for them. In these cases firms have required that overhead-type activities be bid to the project. This is done to try to limit the charging to the project of costs that may not be truly necessary or that appear to be excessively high.

The basic **accounting** system used by the firm is the starting point for assignment of cost codes because it permits ready capture of costs by the normal accounting system. With their linkage to the basic cost accounting system, these codes are standardized for each company, although by their nature they permit customizing for a particular project. Cost codes are of two types: those that cover the activities in a design or project management office and those that cover construction and/or fabrication activities. The categories of typical cost code for office activities are given in Table 6.2.

For construction or fabrication work the cost code used is more complex. It consists of a series of numbers and letters to identify each portion of the work, usually on a facility basis together with additional information on the type of work. Because of the wide variety of work a firm may undertake, the codes can be very long, often consisting of eight or more alpha or numeric fields with their identity established by the type of work performed. Thus for a multiple building complex, two digits might identify the building, three others the type of work, with additional ones used to identify materials or services used. Thus a cost code might be 14.627.E26. In this case translating

TABLE 6.2   Cost Code for Office Activities
100 Salaries and wages 200 Payroll additives 300 Other employee related expenses 400 Travel, subsistence, and entertainment 500 Materials, supplies, tools, and office expenses 600 Equipment and facilities 700 Communications 800 Outside services and subcontracts 900 Computer and miscellaneous services

into, Plant 1, Level 4, Local Control Boards 627, Electrical E, Raceways 2, Low Voltage 6.

For design office activities, labor and similar charges are automatically captured by the accounting system. Other costs are coded onto the origination document, such as a travel request or voucher, a telephone bill, or a computer services invoice. The accounting department picks these up when billed and enters them in the accounting system, charging them to the appropriate project.

For material purchases, use of the system begins with the responsible engineer assigning the cost code to the **material requisition.** Cost codes are similarly assigned to labor activities, subcontracts, service contracts, and all other sources of charges to the project. Once assigned, that cost code will not normally be changed, and the costs that flow from that part of the work will be captured automatically and reflected in the various cost and budget reports. Frequently certain categories of cost are exempt from state or federal tax or may be given preferential tax treatment. The cost code must reflect these categories, and proper instructions must be provided to all personnel who assign cost codes or classify costs. Certain regulated industries require that cost and other data be furnished using a specific set of cost codes such as electric utility industry Federal Power Commission Code of Accounts. The regulatory cost code is often adopted by the regulated company and used for their normal operations as well. Where the cost code for reporting is different from that used for the normal accumulation of cost data, a simple computer program can be used to prepare the conversion.

Costs for field **labor** are tracked based on timekeeping records. These records establish the employee's badge number, hours worked, and type of work performed. Often the foreman will be consulted by the timekeeper to advise the correct work category to be assigned since crews normally work on a variety of tasks that have different cost categories. Following the identification of the employee's badge number and time charged, the timekeeper/field accounting department will extend the costs by the hours expended for the appropriate labor rate for the individual, including fringe benefits and overheads, and thus develop the labor charge for the activity code.

Assignment of accurate cost codes is important not only to provide good budget control, cost forecasting, and future estimating but to ensure that the most favorable tax treatment is obtained. It is good policy to conduct an occasional review to ensure that cost codes are being properly assigned.

One basic problem with the use of cost data for budget control purposes is their lack of **timeliness.** Accounting data are generally available only on a monthly basis, and in the case of outside services the actual commitment when billing, entry, and reporting times are added together may have been incurred as much as two months earlier. As a result, though valuable, some care needs to be taken that when such cost data are used, this **delayed reporting** must be recognized.

6.3 DEVELOPMENT OF ESTIMATES, BUDGETS, AND FORECASTS

There is a significant difference between estimates, budgets, and forecasts, each being used to provide a different type of cost information. **Estimates,** as their name implies, can be used to predict the cost of the entire project or some portion of it. Estimates are converted into **budgets,** which are the controlling cost limitations established for the work or portions of it. Periodically, in order to obtain an updated view of the financial performance of the project, **forecasts** are prepared. Forecasts are an estimate updated and adjusted for actual circumstances and trends such as labor performance/productivity, commitments for materials and subcontracts, scope changes, design evolution, testing, and start-up. Forecasts are an extension of an estimate but do not replace it, rather being used to **predict** costs without going through the effort of preparing a complete estimate.

Estimates are prepared for a variety of purposes, the end purpose of the estimate establishing not only the degree of detail and accuracy of the estimate but also the effort involved and time allowed (or required) for its preparation. Broadly speaking, estimates are one of the following types: rough or order-of-magnitude estimates, conceptual estimates, preliminary estimates, definitive estimates and bid estimates. In addition to the purpose for which the estimate is used, the **risk** involved affects the degree of effort put into it and the accuracy and timing of the work. Thus, if the estimate is for lump sum or fixed price work, the estimate will be prepared with a higher degree of precision than for work where the scope is not well defined and which would be executed using some form of reimbursable contract.

Timing of estimates is often dictated by outside factors such as bid preparation dates, budget establishment dates, option expiration dates, and—of increasing importance in recent years—requirements of regulatory bodies for use in hearings or permit applications.

The number and timing of estimates are also affected by contract type as work progress. In the simple case where firm prices can be bid a single initial estimate (for bid purposes), and the final cost report reflecting the ultimate price, including adjustments may be all that are required. It is fairly rare that such simple circumstances exist, however. Thus for most projects more detailed and sophisticated methods are used.

Typically a **series** of estimates are prepared over the life of the project. These estimates increase in accuracy from the initial one, which may literally be on the back of an envelope, to final definitive estimates that may be precise as those for fixed price work.

Because of the cost and time required to prepare estimates, forecasting is used to update estimates and reduce their frequency. This is particularly useful on large projects spanning a number of years. In general, it has been found that about 80% of all project costs are determined in the first 20% of the life cycle of the project. Clearly the importance of reasonably accurate early estimates can not be overstated.

The difficulty of defining the scope of the work may affect the type of contract used and, together with the usual interest by the owner in maintaining a short schedule, often lead to very different estimate approaches. Where the scope is well defined and enough time is allowed, an accurate estimate can be prepared, so the risk to the bidder is minimal. In this case fixed price or lump sum contracts can be used effectively. Where it is not possible to accurately scope the work because of the developmental nature of the project, uncertain licensing and permitting requirements, weather, availability of labor or materials, and other similar factors, some form of **reimbursable** contract is used. (Chapter 11 discusses these contract types in more detail.)

Estimates for fixed (or firm) price work are prepared wherever possible with extensive backup material, since as their name implies, the price(s) are **fixed** and, barring changes in scope, are not subject to adjustment. The bidder assumes the **risk** for the accuracy of the estimate, with cost underruns representing a profit and overruns representing a loss.

While estimates for work—whether of the fixed or reimbursable type—are prepared to the highest level of detail available, bidders on fixed type work will normally expend added time and effort, at their own expense, to add definition to the scope and to firm up pricing because they assume a significant financial risk. As a result fixed price bid estimates are usually of the most definitive type.

Where fixed price bids are prepared for lump sum proposals, as is often the case, it is rare for the definitive detail in them to be made available to the client. A major exception exists where a significant **error** has been made in the bid and there is a request for reconsideration of the bid amount. But, when such a request is made, the bid may be automatically rejected. Where reconsideration is permitted, it is not unusual for the appropriate portion of the bid, including the working papers, to be presented in support of the request. The same procedure may be followed where a claim for additional work is filed, as a way to support the request. For some elements and often the entire project scope of reimbursable work, it is more common for the estimating data to be made available to the client when requested.

For work where the scope is not well defined, different approaches are taken. If the owner insists upon a fixed price bid, the bidder will usually define the scope in the bid response in such a way as to limit the exposure and to create a suitable **definition** (and limitation) of the work. This building of a fence around the scope of the work is essential to limit the risk that the bidder assumes. If the bidder recognizes the problem and wishes the entire scope of the work undertaken, even though it is not possible to definitively establish it at the outset of the work, it is usual to perform the work under a reimbursable contract that can be converted later in the project to fixed price work, such as a lump sum or **target maximum.**

Where it is not possible to agree on either a fixed price or reimbursable contract, work is often performed on a **time and materials** basis, which permits the work to proceed while assuring the contractor reimbursement for costs of labor, materials, subcontract work, and so forth, together with suitable allowances for overheads, distributable costs that can not be identified with specific activities, bonds, and, most important, for profit. One problem with this approach is the difficulty in reaching agreement on the **allowances** for overhead, distributable costs, and profit.

The number and timing of the estimates will vary with the type of work, its complexity, and the size of the project. There may also be outside factors that require the preparation of estimates such as lending institution requirements and permitting agencies.

In general, only one or perhaps a few Order-of-Magnitude (OOM) estimates are prepared. These are done to support economic feasibility studies and basic management decisions. They are no longer required once the basic decision to undertake the project is made. Supplementary OOM estimates may be made over the life of the project where major changes in scope are contemplated or where significant supplemental work may be considered. Again, given their very generalized accuracy, on the order of plus or minus 20% (or even more with developmental work), they are useful only for the broadest type of decision making and cannot really be used for project control activities.

Conceptual estimates are often prepared where designs are not yet available and only the most preliminary information exists on capacity, performance requirements, and similar data. While similar to OOM estimates, their accuracy will vary widely depending on the degree of definition of the work and the skill and experience of the estimator. Some can have an accuracy of plus or minus 10%, while others, where project definition is less precise and historical data lacking, can vary by plus or minus 50%.

Bid estimates are used for fixed price work. They are usually prepared only once during the bidding phase of the project, prior to contract award. If the project later requires scope additions, estimates are prepared for the added work and, if the scope can be suitably defined, can have the accuracy of a bid estimate. Usually bid estimates will have accuracies as high as plus or minus 3 to 5%, since they are the product of an intensive effort that includes obtaining firm pricing on as many of the project components and activities as possible.

Preliminary estimates are often prepared for work where either the scope cannot be fully established and detailed or, in some rare cases, where there is insufficient time to prepare a suitably detailed estimate. Because of the variation in the definition of the project, preliminary estimates can vary in accuracy from those of the order-of-magnitude estimates to bid estimates. Usually they fall at about the midpoint, with an accuracy of about plus or minus 10%.

With reimbursable work a project will often be started using a preliminary estimate that is later converted to a **definitive** estimate. To support the preparation of the definitive estimate, the project's scope is fully defined, sufficient engineering is completed to detail the design, and, most important, sufficient experience is acquired to establish the **productivity** of field labor. The estimate will typically have an accuracy of plus or minus 5%, about that of a bid estimate, and it forms the basis for negotiating the final contractural cost of a project—if it is converted from a reimbursable to a fixed price basis.

Additional special estimates will be prepared for engineering costs, tooling, special additions to the scope of projects, and many other requirements. They may fall into any of the estimate types listed above depending on the accuracy required.

Where lending institutions are involved in providing loans, the degree of accuracy required in the estimates will be higher than required for in-house decisions. Generally lenders require estimates having accuracies in the range of 3 to 5%, as with fixed price bidding. For developmental projects of a competent, established firm, with a track record of satisfactory performance where the work does not represent more than say 30% of the net worth of the company, the lending institutions may be willing to accept estimates with accuracies of plus or minus 30%. This is very **subjective** and, besides those factors noted above, depends on the general economic climate at the time of the loan request, the cost of money, the duration of the project, likely regulatory involvement, the degree of participation by the borrower, guarantees by outside parties, and other similar factors.

Where drawings and specifications are incomplete, as is often the case, it may be necessary to perform some conceptual estimating. This occurs where an experienced estimator is given only very sketchy data and, using historical data and past experience, makes assumptions regarding quantities, complexity, material types, labor productivity, and so forth. Thus the estimate is developed based largely upon overall insight and **historical data.** Usually conceptual estimating is employed only for parts of a project, and not the overall job itself. For example, for a chemical processing plant an on-site electrical substation may not yet have been engineered and the estimator will, based on experience, make certain allowances in the estimate for the required equipment, foundations, structures, control systems, and so forth, sufficient to price that portion of the work. Significant amounts of preliminary engineering probably will not be involved apart from perhaps a brief discussion with the design personnel.

To support the different types of estimates, varying amounts of data are necessary. Table 6.3 summarizes the degree of definition required for each of the estimate types. Note that slightly different names are used for some of the estimates. For work in the engineering/construction industry, Table 6.4 presents the various types of estimates and the scope of the design as well as the construction development that supports each estimate.

In the manufacturing sector, **capital cost** estimates are often prepared to determine feasibility of projects or to establish funding requirements for agreed changes. These estimates have the same characteristics as those discussed earlier, with the added need to evaluate the effect of the **learning curve** and the economic consequences of the action. Factors such as market risk, return on investment, alternatives, life cycle cost, and warranty service become significant here.

The learning curve—where successive units are produced more efficiently than earlier ones—has an important effect on any estimates involving repetitive operations. While the learning curve is often difficult to evaluate, there are generally accepted methods used to assist in this evaluation. Two general approaches are used, based on either the **unit** or **cumulative average** theory. In general, the unit theory tends to predict somewhat lower costs but is widely used because of its simplicity. This approach is based on the concept that predicts a constant percentage reduction in the cost of successive units using the formula

$$Y = (AQ)^x,$$

where

Y = cost of the Qth unit,
A = cost of the first unit,
Q = number of the Qth unit,
x = learning curve exponent.

TABLE 6.3 Types of Estimates	
Type	Typical Accuracy
Conceptual	± 10–50%
Order of magnitude estimate	± 20
Feasibility estimate	± 20
Bid estimates (often partial fixed price)	3–5
Preliminary estimates	± 10
Definitive estimates	± 5
Special estimates (engineering, tooling, etc.)	Varies

TABLE 6.4 Comparison of Capital Cost Estimate Types

	Type 1 MAGNITUDE: Study	Type 2 CONCEPTUAL: Feasibility/ Budget	Type 3 PRELIMINARY: Budget/ Control	Type 4 DEFINITIVE: Control	Type 5 ENGINEERS: Control	Type 6 BID: Competitive Bid
Typical Use of Estimate						
Engineering						
Plant product and cap.	First cut	Approximate	Fixed	Fixed	Fixed	Fixed
Geographical location	Usually known	Approximate	Exact	Exact	Exact	Exact
Topographical maps	Not Available	Rough	Rough/detailed	Detailed	Detailed	Detailed
Soil/geology reports	Not Available	Outline/draft	Draft	Complete	Complete	Complete
Process flow sheets	Diagram	Diagram	Detailed	Complete	Complete	Complete
Equipment selection	None	Draft equipment list	Complete listing/sized	Sizing refined	Complete	Complete
General arrangement drawings	None	Minimum	Complete	Complete	Complete	Complete
Civil structural drawings	None/sketches	Typical	Sized	Semicomplete	Complete	Complete
Architectural drawings	None/sketches	Exterior elevation	Sized	Semicomplete	Complete	Complete
Piping/HVAC drawings	None	P&ID	Sized	Semicomplete	Complete	Complete
Electrical drawings	None	One-line drawings	Sized	Semicomplete	Complete	Complete
Criteria/specifications	None	Outline criteria	Outline specifications	Draft specifications	Complete	Complete
% engineering complete	0–2	2–5	8–15	35–45	75–100	100

TABLE 6.4 (*Continued*)

Typical Use of Estimate	Type 1 MAGNITUDE: Study	Type 2 CONCEPTUAL: Feasibility/ Budget	Type 3 PRELIMINARY: Budget/ Control	Type 4 DEFINITIVE: Control	Type 5 ENGINEERS: Control	Type 6 BID: Competitive Bid
Implementation Planning						
Construction work plan	None	Not required	Desirable	Desirable	Desirable	Mandatory
Construction contract configuration	None	Not required	Desirable	Draft	Final	Final
Construction schedule	None	Not required	Outline	Draft	Detailed	Detailed
Capital Cost Estimates Basis						
Estimates prepared by	Estimating department	Estimating department	Estimating department	Estimating department	Estimating department	Estimating department
Estimate form	Unit cost	Unit cost or spread	Spread	Spread	Spread	Spread
Site visit by project estimator	Unnecessary	Sometimes desirable	May be required	May be required	May be required	Mandatory
Equipment	Historical	Major: letter quote	2 letter quotes	Vendor proposals	Vendor proposals	Firm vendor proposal
Civil structural work	Historical	Takeoff major items	Partial takeoffs	Partial takeoffs	Complete takeoff	Complete takeoff
Architectural work	Historical	Takeoff major items	Partial takeoffs	Partial takeoffs	Complete takeoff	Complete takeoff

Piping/HVAC work	Historical	Takeoff major items	Partial takeoffs	Partial takeoffs	Complete takeoff	Complete takeoff
Electrical work	Historical	Takeoff major items	Partial takeoffs	Partial takeoffs	Complete takeoff	Complete takeoff
Labor rates	Not evaluated	Current rate schedule	Current rate schedule	Current rate schedule	Labor contract	Labor contract
Labor productivity	Not evaluated	Assumed productivity	Evaluated	Evaluated	Evaluated	Evaluated
Construction equipment usage	Not evaluated	% of Labor	% of labor/crew method	% of labor/crew method	% of labor/crew method	Crew method
Material pricing	Not evaluated	Historical	Telephone quotes	Some letter quotes	Some letter quotes	Firm written quotes
Subcontract pricing	Not evaluated	Historical	Telephone quote	Written quotes	Written quotes	Firm written quotes
Contractors' OH&P	Included in unit cost	% of direct cost	% of direct cost	% of direct cost	% of direct cost	Detailed estimate
Engineering and management	% of constructed cost	% of constructed cost	Broad estimate	Broad estimate	Detailed estimate	Detailed estimate
Escalation	Often excluded	%/Year to midpoint	%/year to midpoint	Detailed evaluation	Detailed evaluation	Detailed evaluation
Contingency	Single %	Broad evaluation	Broad evaluation	Detailed evaluation	Detailed evaluation	Detailed evaluation
Relative Cost of Estimates	1	4	10	14	19	23

Reprinted by permission of ICF Kaiser Engineers.

The result is a curve having the form shown in Fig. 6.2.

The percentage reduction is usually applied to a doubling of the units, thus the learning curve might be calculated and plotted for 1, 2, 4, 8 . . . units at which points the value of the cost can be determined. An example is shown in Fig. 6.3.

Market risk may be the most important factor in these estimates, since the ability to win market share (and to retain or expand it) is crucial to the production contemplated and hence the number of units over which fixed costs are spread. The problem is somewhat more complex with nondurable goods because their acceptance by the public can be more uncertain than durable goods. As a result these estimates inherently carry more risk than those in other market situations. In market risk it is wise to prepare both a high and low estimate and thus to bracket the range of possibilities.

Return on the investment can be treated more accurately once the market risk is established. As expected, the higher-risk options should carry a higher return, and the markup on nondurables may be several hundred percent, as compared to the markup on durable goods, which may be only on the order of 100%. Return on the investment can be calculated in a number of ways depending on the uncertainties and the degree of accuracy required. For the high-risk alternatives discussed above, the calculation may be a simple percentage per month, or year, return on the investment. The number of times per year the inventory is **turned over** may also be a factor, although it is usually less significant than in the durable goods sector.

For more durable goods the return on the investment, presuming proper market research, can be calculated with more assurance. Typically the calculation should evaluate the life cycle costs, including the present worth of the investment, the present and future capital requirements, scrap or salvage value of the investment, comparison of alternate investments, cash flow, change in market share and potential pricing changes. In addition, since **durable goods** often have a longer cycle from conception to marketing, changes in labor and material costs need to be considered. With a longer useful life **warranty** provisions and exposure need to be carefully evaluated. If the product also has a potential impact on the public or the environment, some provision, perhaps insurance, may need to be provided.

Internal operations, such as the number of times per year the inventory is turned can provide a clue to the efficiency of the operation and can be used as a rule-of-thumb measure as well. Since inventory represents monies tied up and not earning either interest or earnings available in alternative investments, the smaller the cost of inventory, or the more frequently it turns over, the more efficient are the internal operations. Typically in manufacturing operations annual inventory turns of five to ten are goals. Some caution needs to be taken, however, since the efficiency of internal operations is only one of many factors affecting the cost, and hence return, on the investment.

The consideration of **alternatives** is an important factor with any capital

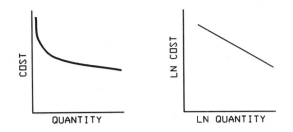

FIGURE 6.2 Learning curve (general form). From *Cost Estimating* by Rodney D. Stewart. Reprinted by permission of John Wiley & Sons, Inc. Copyright © 1991 by John Wiley & Sons, Inc.

cost analysis. Typically the minimum alternates considered are potential earnings from investment in other internal operations of the firm, investment in low- to moderate-risk securities, and retirement of internal debt. When doing a comparative cost study, it is fairly simple—by using the appropriate assumptions—to **skew** the results to reach a preferred conclusion. As a result it is important to treat the alternates uniformly to ensure that the study's findings are both objective and accurate for the subsequent decision making.

In addition to the factors mentioned above, the **tax law** that applies to the investment and its potential changes may need to be considered. Large uncertainties often arise in this area, since these decisions are beyond the control of the enterprise. Examples include potential changes in the capital gains tax law, changes for tax credits for participation in various programs, credits for export activities, and so forth.

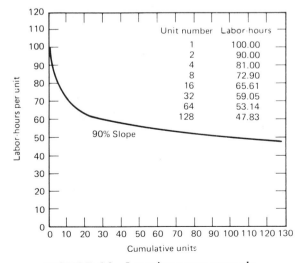

FIGURE 6.3 Learning curve example.

6.4 PREPARATION OF ESTIMATES

When preparing the estimate, it is important to account for the variables that affect the effort, and hence the cost, required to perform each of the individual elements of the work. It is useful to develop a detailed listing of these **factors** and to assign values of them as multipliers of the standard estimating values for each of the elements of the estimate. These lists of factors can be highly detailed; in other cases they may be used at the summary level. One such estimating factor form is shown in Fig. 6.4.

Most companies have a standardized way in which costs are defined and categorized and estimates prepared. One method widely used categorizes costs based on **location** and whether they are **direct** or **indirect** in their support of the project work. For each of these categories, subdivisions of the cost are used to break them down by facility or type. Then the costs are estimated or accumulated by type of activity, for example, labor, materials, equipment rental, and subcontract. Because of their unique nature, escalation, contingency, and fee or profit are usually considered as a separate category.

For example, in practice engineering costs in connection with the design work are performed in the office and, since they are directly project related, are considered a direct cost. Field engineering in support of construction activities is considered a direct field cost, while engineering support activities in each location, not directly identifiable to the project are considered indirect costs, again with the location establishing whether they are considered office or field indirects.

Indirect costs can be recovered by some form of overhead percentage, which is usually developed from historical data. In most cases the indirect costs are further categorized into those expended in support of project activities, but not identifiable to a particular project, and those of an overall nature, which support the general operations of the firm and which are considered **overhead** expenses in the more usual sense of the word. Indirect costs are often called **distributable** costs from the manner in which they may be allocated to the various projects underway.

Both direct and indirect costs are broken down into several categories, which for a construction project are often titled as noted in Table 6.5. The categories in the table are still too general to be useful except for the smallest, most simple project, and they can be broken down again into more specific topics. They are normally broken down even further and classified according to the work item, for example, Air Compressor #2 Foundation, Liquid Nitrogen Tankage. These items of work are then given an accounting numerical code to permit cost accumulation by facility.

Home office costs need to be included with the estimate. They will reflect the type and degree of involvement of these groups. Generally engineering, procurement and project management functions are performed at least initially in the home office, while support is provided to these groups by the legal, personnel and similar departments. Later in a project it is common

JOB NO _____

DATE _____

BY _____   JOB TITLE _____

JOB SECTION O □ ▷

ESTIMATING DEPARTMENT
JOB FACTOR FORM

← MINUS ———— % ———— PLUS →

LABOR

Factor	Low descriptor	MINUS		%	PLUS			High descriptor
SKILL, WORKER	EXPERIENCED	10	5	0	5	10	15	UNTRAINED
UNION	CO-OPERATIVE	10	5	0	5	10	15	DISPUTES
AVAILABILITY	CHOICE	10	5	0	5	10	15	SHORTAGE

SITE

Factor	Low descriptor	MINUS		%	PLUS				High descriptor
LOCATION	LOCAL	10	5	0	5	10	15		REMOTE
TEMPERATURE, HUMIDITY	IDEAL	5	0	5	10	15	20	25	EXTREME
WIND, DUST, MUD, SNOW	NEGLIGIBLE	5	0	5	10	15	20	25	SEVERE
PORTAL TO PORTAL — MINUTES PER DAY		15 / 0	30 / 5	45 / 10	60 / 15				
MULTI LEVEL — NUMBER OF LEVELS		1 / 5	2–3 / 0	4–9 / 5	10–15 / 10	16–20 / 15			
SCAFFOLD — HEIGHT	NONE / 10	5 FT / 5	10 FT / 0	15 FT / 10	20 FT / 15	25 FT / 25			
TRENCH — DEPTH		2 FT / 5	4 FT / 0	6 FT / 5	8 FT / 10	10 FT / 15	12 FT / 20	14 FT / 25	
WORKING SPACE		YARD / 5	BUILDING / 0	EQ. RM. / 10	TUNNEL SHAFT 15 / 20				
COMPLEXITY	FAMILIAR / 10	20	15	5 / 0	5	UNUSUAL 15 / 20	25		
HOISTING — FEET		5	10 / 0	30 / 5	50 / 10	80 / 15	100 / 20		
OTHER									

WORKING CONDITIONS

Factor	Low descriptor	%	PLUS				High descriptor
TRAFFIC & UNLOADING	SMOOTH	0	5	10	15		OBSTRUCTIONS
MATERIAL STORAGE	SUFFICIENT AREA	0	5	10	15		INADEQUATE AREA
MATERIAL & ENGINEERING SUPPLY	AS REQUIRED & UP-TO-DATE	0	5	10	15	20	SHORTAGES & REVISIONS
INTERFERENCE—OTHER CREWS	NONE	0	5	10	15	20	SEVERAL CRAFTS
OCCUPANTS	NONE	0	5	10	15		WORKING PLANT
QA/QC	NO SPECIAL REQUIREMENTS	0	5	10	15	20	TIGHT SPECIFICATIONS
SECURITY	NONE	0	5	10	15		STRICT
NOISE	NONE	0	5	10	15		EXCESS
LIGHTING	DAYLIGHT	0	5	10	15		POOR, ARTIFICIAL
VENTILATION	OUTSIDE	0	5	10	15		POOR, ARTIFICIAL
FIRE HAZARDS	NONE	0	5	10	15		DANGEROUS
SAFETY REGULATIONS	NORMAL	0	5	10	15		STRICT
OTHER							

SCHEDULE

Factor	Low descriptor						High descriptor
OVERTIME — HOURS PER WEEK		48 / 5	50 / 10	60 / 15	65 / 20	70 / 25	
SHIFTWORK — NUMBER OF SHIFTS			2 / 10	3 / 20			
EXCESS WORK FORCE			25% / 10		50% / 25		
DURATION OF PROJECT	SHORT	0	5	10	15	EXTENDED	

PROJECT TYPE

TOTAL %

TOTAL % ÷ 100 + 1 = LABOR ADJUSTMENT FACTOR

FIGURE 6.4 Estimating job factor form. Reprinted by permission of ICF Kaiser Engineers.

TABLE 6.5   Cost Categories

Office Direct Costs
Engineering: Primarily labor, including payroll additives (employer's portion of Social Security, sick leave, vacations, employee benefits, etc.)

Procurement: Labor and material commitments, and purchases

Project Management: Primarily labor

Other directly identifiable costs: Legal, personnel, data processing, reproduction, supplies, etc.

Office Indirect Costs
Non-project-related activities: Public relations, training, general legal, security, general business licenses, insurance, etc.

Field Direct Costs
Process facilities

Non-process/yard facilities

Ancillary facilities

Start-up and initial operations

Field Indirect Costs
Construction facilities

Construction support

Construction supervision

Engineering

Insurance, taxes, duties

Other Costs
Escalation

Contingency

Fee or profit

---

practice to transfer the final engineering and project management work to the field location. Any remaining procurement work is also picked up by the field procurement organization at the job site.

All of these costs must be included in the estimate. To develop them, an estimate is needed for each group. Where the work cannot be defined by output documents, experience must be used to develop an estimate of the time and effort required. A **staffing bar chart** is easily prepared and will provide an in depth assessment of the required staff.

Home office labor activities must be included. Typically there are two types of home office costs: payroll additives and overhead or burden costs.

Payroll additives, include the employers portion of social security payments, sick leave and vacation allowances, medical and life insurance, and so forth. These can range from 25 to 35% of direct payroll costs, and if retirement or other benefits are included, they can be even higher. Overhead costs include rent, insurance, miscellaneous business fees, utilities, security, and similar non-project-related costs. These costs can range from 35% to as much as 100% of the cost of the direct payroll plus payroll additives, depending on the type of space provided, its utilization (square foot per employee), and the type of organization and complexity of the firm.

The estimate of **engineering** costs is normally based on a detailed account of hours required to produce the various documents, including drawings, specifications, administration of procurement, resolution of field questions, preparation of studies, and preparations of reports including environmental impact statements. In other cases where the work may be fairly conventional and well defined, an estimate can be prepared from the number of drawings to be produced, with an overall figure for the hours required to prepare each drawing (which includes all of the engineering activities listed above). Of the two methods the detailed approach is preferred because it forces a better definition of the work to be developed as well as a more precise set of values for the components of the work. It also permits the work to be tracked in more detail, enabling early warning of deviations in performance and thus early corrective action. The detailed approach, however, can suffer from excessive optimism and should always be checked by a quick overview, namely the hours-per-drawing method.

Field direct costs usually include labor, materials, equipment, and subcontract charges. The quantities of labor required are developed by applying **labor productivity** factors to a takeoff of the materials on the drawings and in the specifications and the requirements of the specifications as to methodology and quality of work. Labor productivity is best obtained from historical data for the same work. Most firms have this type of data in their records. Where these data are not available, the normal estimating guides should be consulted, with particular care given to the description of the work and such factors as crowding, working above the ground, and temperature or weather effects. Typically productivity factors are given as decimal values, which are applied to the first estimate of hours to obtain adjusted hours. Thus for a productivity factor of 1.08, a work item requiring 150 hours under standard or normal conditions would require 162 hours (150 multiplied by 1.08). These productivity factors are adjusted as experience is gained and can be projected, using either the cumulative or last-period productivity. Of the two methods the last-period approach, if a reasonable amount of experience is included, yields a more accurate value.

A precise and detailed **takeoff** of materials and quantities establishes the labor content of estimates. Labor is usually the largest controllable cost item and thus underlies the overall estimating effort. The importance of accurate and complete takeoffs cannot be overestimated.

To estimate field direct costs, the scope of the work is broken down into its parts and subparts, with accounting codes assigned to each. The work items are listed vertically down the left side of the estimate preparation sheet. Across the top of the form, a series of cost categories are noted, for example, labor, materials, subcontract, and equipment rental. For each work item the next step is to assign the proper costs to the item and subitem of work and then to sum up all the types of costs that can be incurred for the item.

Suppose that the item of work is a diesel generator foundation. It would be subdivided into its several construction activities, such as excavation, formwork, rebar and embeds, and concrete placement. Within each of these categories, the work is quantified by the units of measure for the item such as cubic yards for excavation, square feet for formwork, pounds or hundred-weight for rebar and miscellaneous iron, and cubic yards for concrete. In this example, the overall block of work is found in work package 1, civil/structural work, subdivided into a multiple digit cost accumulation breakdown. Figure 6.5 displays this type of information.

When all the subitems of work are accounted for, they are added up, and the direct costs associated with each item are broken down by labor, materials, subcontract work, and finally equipment rental. In this way, from the cumulative costs of subitems, the total cost of the item is obtained, and it can then be entered against the cost code identifying the particular item. For the entire estimate the procedure followed is the same and the work sheets when taken together give the specifics of the estimate.

The cost of **materials** (as distinct from equipment) can be estimated by using historical data or by obtaining verbal or written quotations from vendors. The labor necessary to install the material is calculated using installation rates per hour based on either historical data from previous projects or data from standard estimating references.

To these figures productivity factors are applied to reflect actual working conditions in the field. Crowding, significant overtime, the use of extended workweek for more than a short time, adverse weather, lack of experience with the materials and processes, use of less-skilled labor on portions of international work and similar factors—all tend to reduce the effectiveness of the labor employed. The effect is to lower the installation rates, which not only increases the numbers of hours required but also extends the schedule. Even if crowding, significant overtime work, or the other factors mentioned above are not present, the reduction in productivity may require **acceleration** to make up for the schedule delays that would otherwise occur. As a check on the work, it is important to review the effect on the schedule and adjust the estimate to reflect the most likely labor requirements. In adjusting the estimate, costs that relate to units of work such as direct labor are increased proportional to the extent that the productivity is decreased—materials and consumable items such as small tools are not normally increased. The one exception to this occurs where labor rates change

ESTIMATE DETAIL BY WORK PACKAGE/FACILITY

--WORK BREAKDOWN--- FACIL.STANDR.BID	DESCRIPTION	QUANTITY	MANHOURS	LABOR	EQUIP USAGE	MATERIAL	SUB-CONTRACT	EQUIP-MENT	SALES TAX	TOTAL DOLLARS
WORK PACKAGE 1	CIVIL/STRUCTURAL									
1	BASECASE - 3 DIESEL GENERATORS									
11	MODIFY EXISTING CHILLER ROOM									
11 .0211000.1	JACKING AND PROPPING SYSTEM	1 LS	216	6461	1234	1650	0	0	0	9345
11 .0213433.1	DEMOLISH CONCRETE CURBS AND CORBEL	2 CY	12	319	37	0	0	0	0	356
11 .0214322.1	DEMOLISH CONC. SLAB ON GRADE W/ REBAR (AIR EQUIPMENT, EXCLUDING DISPOSAL)	3 CY	12	319	40	0	0	0	0	359
11 .0214342.1	CHIP OUT CONCRETE BEAMS, FOR KEY AND TO ROUGHEN SURFACE	3 CY	69	1835	221	0	0	0	0	2056
11 .0214521.1	SAW CUTTING - CONCRETE SLAB W/ MESH (SF=LENGTH X DEPTH)	140 SF	12	300	52	0	0	0	0	352
11 .0214551.1	DRILL, FOR ANCHOS, DOWELS	70 LF	19	475	85	0	0	0	0	560
11 .0214556.1	CORE DRILL, 6" DIA	150 LF	110	2752	491	0	0	0	0	3243
11 .0214744.1	DEMOLISH STUD DRYWALL PART'N	1000 SF	32	800		0	0	0	0	800
11 .0214900.1	DISPOSAL OF DEMOLISHED ITEMS AND EXCAVATED MATERIAL	27 CY	3	81	58	184	0	0	0	323
11 .0241300.1	EXCAVATION STRUCTURAL - HAND	11 CY	32	801	0	0	0	0	0	801
11 .0299110.1	EPOXY COAT EXIST BEAM SURFACE	600 SF	14	381	5	360	0	0	0	746
11 .0302100.1	REINFORCING STEEL	6000 LB	86	2511	418	1452	0	0	0	4381
11 .0303100.1	EMBEDDED METALS, ANCHORS,NUTS	200 LB	14	383	5	330	0	0	0	718
11 .0314100.1	CONCRETE-FOOTINGS & PEDESTALS 3000 PSI	12 CY	21	559	105	594	0	0	0	1258
11 .0345100.1	CONCRETE FINISH - SLABS	150 SF	3	88	2	25	0	0	0	115
11 .0361200.1	FORMS - BEAMS	900 SF	324	8866	104	1020	0	0	0	9990
11 .0364200.1	CONCRETE - BEAMS	20 CY	55	1463	296	990	0	0	0	2749
11 .0511100.1	STRUCTURAL STEEL	9 TON	233	7035	1226	14603	0	0	0	22864
TOTAL	MODIFY EXISTING CHILLER ROOM		1,267	35,429	4,379	21,208	0	0	0	61,016
TOTAL	BASECASE - 3 DIESEL GENERATORS		1,267	35,429	4,379	21,208	0	0	0	61,016
TOTAL WORK PACKAGE 1			1,267	35,429	4,379	21,208	0	0	0	61,016

FIGURE 6.5 Estimating detail by work package/facility. Reprinted by permission of ICF Kaiser Engineers.

due to a labor contract expiration. In this case it is necessary to reflect the new, generally higher, labor rates in the estimate from the date of effect of the new agreement. Where work is in **high-risk** areas, such as roofing or tunneling, Workman's Compensation insurance premiums can become substantial and have a major effect on field costs. Items that are **time related** such as field engineering, field supervision, equipment rental, and insurance are increased proportional to the extension in the schedule. Home office costs, which are time related, such as general supervision, project administration, and legal services would be increased in the same way. Those functions that provide direct support to the field and relate to the rate or number of units installed would not be increased.

To develop the cost for supervision and field engineering, a field staffing plan is prepared, usually in bar chart form. The staffing plan lists each position, its start and end dates, and the labor rate to be applied to the position. The labor rate often includes payroll additive costs, incentives, and allowances for increases if the project life is lengthy. From this detailed listing it is possible to develop not only the overall staffing requirements but the specific costs likely to be incurred for supervision, field engineering, first aid, security, and so forth. Figure 6.6 shows a simple field staffing plan together with extensions and costs.

The cost of installed **equipment** is fairly straightforward to estimate. As mentioned earlier, quotations are obtained from vendors or recent historical data can be used, and a base equipment cost is established. To this must be added transportation to the point of use as well as installation and start-up supervision. **Spare parts** for a specific period of time (often one year) are normally also included. In some cases operating or **consumable** supplies are included as well. Where equipment is large, heavy, or delicate, it is often necessary to arrange special transportation and handling, and this can be expensive and time-consuming. The estimate would reflect these costs as part of the equipment cost. The furnishing and installation of each piece of major equipment is usually a line item on the estimate and may include several categories of both direct and indirect costs.

For indirect or distributable costs, the situation is different. These costs usually affect many operations, and not merely a single block of work. Generally they are expressed as a percentage of direct labor, even though many of the items are not in themselves labor activities. A typical listing of distributables is shown in Table 6.6.

The usual method of developing both the type of items and the quantity included as indirect costs is to utilize **historical** data. Since field conditions are never the same from one project to the next, wherever possible, the historical data should be supplemented by more precise estimates of its components and adjusted accordingly. The many estimating guides in print today offer some guidance as to the proper values to be used where historical data may be lacking.

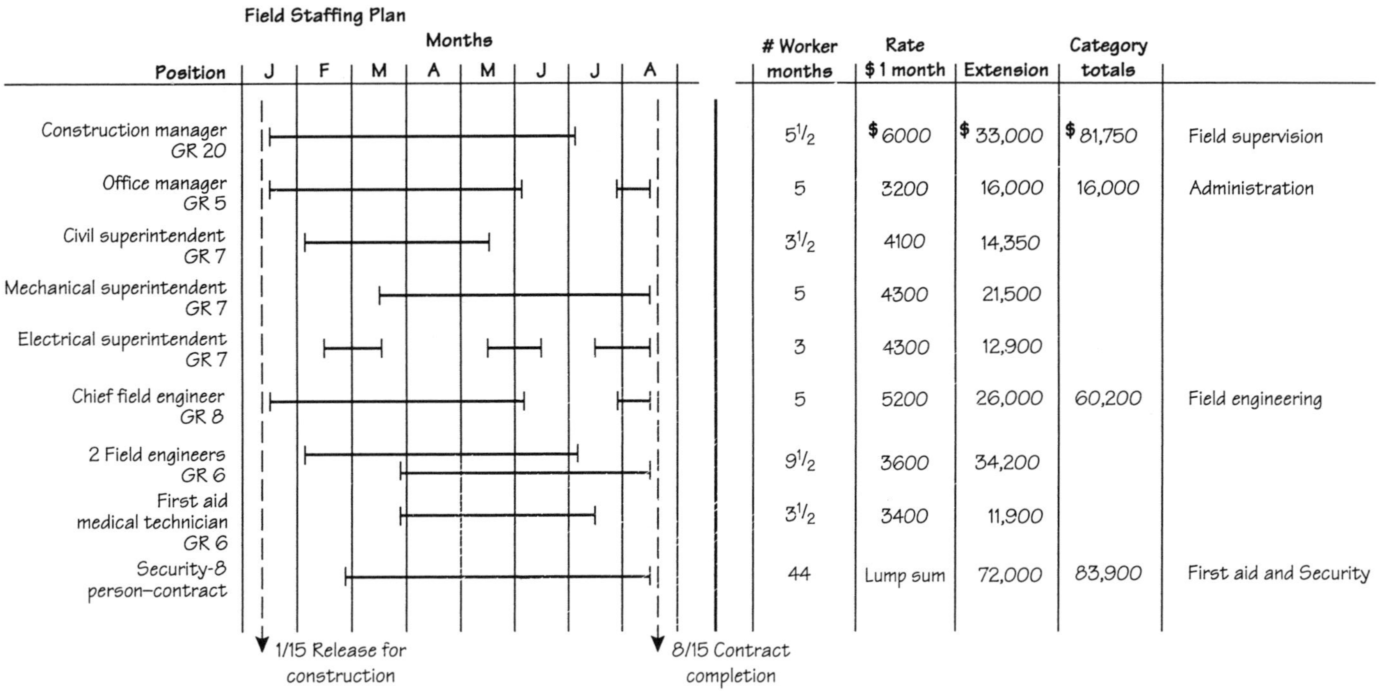

FIGURE 6.6 Field staffing plan.

TABLE 6.6 List of Typical Distributable Costs

Construction Facilities
  Work areas/bays
  Roads, walks, parking areas
  Temporary buildings
  Construction plants
  Temporary utilities
    Power, light, communication
    Water, fire, air, steam, sanitary
    Temporary fuel
  Transportation facilities
  Weather protection
  Scaffolding, cribbage, dunnage (general purpose)
  Minor temporary construction

Construction Support
  Cleanup (final)
  Material handling and warehousing (general)
  Craft training and testing
  On-site services
    Security
    Testing
  Other miscellaneous construction facilities
  Unallocated service labor
  Surveying
  Transportation
  Pre-op testing
  Manual labor allowance

Construction Camps
  Camp construction
  Camp operation and maintenance

Construction Equipment, Tools, Supplies
  Construction equipment (purchase/salvage/rental)
  Tools
  Maintenance of equipment and tools
  Fuels and lubes
  Consumables
  Medical and safety supplies
  Purchased utilities

Material Shipping
  Duties
  Freight (ocean/air)
  Shipping agents
  Lightering/stevedoring
  Loading/offloading vessels
  Receiving/clearing expenses
  Demurrage/container rentals

Project Expenses
  Insurances
  Licenses and taxes

Often overlooked, but always necessary, is some field engineering. Together with insurance and taxes upon field operations, field engineering costs can be significant. They can be estimated as previously described.

For those states with **Workman's Compensation Programs** the amount of insurance premium paid to the carrier can represent as much as 30 to 40% of direct payroll costs and can become a significant cost factor. Depending on the historical safety and claims record of a company, the premiums quoted by the same carrier for the same work may vary between one company and another by as much as a factor of four or five. An effective safety program that reduces injuries will more than pay for itself with the reduction in the required insurance premiums. Specific quotations should be obtained from prospective Workman's Compensation insurance carriers because their rates vary widely.

Taxes will depend upon the locality in which the work is performed, the type of work, and similar factors. To estimate these values, inquiries need to be made to determine the taxation policies of the locality, plus any special permits or fees that may be assessed because of the operations.

In addition to the foregoing, some provision must be made for costs that are **not recovered.** While this applies principally to reimbursable work, certain activities such as entertainment, and selling and bidding costs are categorized as nonrecoverable costs for fixed price work as well. For work in the private sector bidding and selling costs can typically be as much as 5 to 15% of the cost of the project, while for work in the **public sector** these values are often only half as much. For either case the basic cost of estimating and bid preparation is similar, with the principal difference being in the effort of selling. With public sector work, the effort to be included on the bid list is usually fairly modest, while in the private sector it may require extensive cultivation of the client and developing a feeling of confidence by the client prior to receiving a bid request. It is important to ensure that this cost category is considered because these costs can, in some cases, be significant.

The contract may establish certain categories of cost as nonrecoverable. Thus, while care must be taken to ensure that such costs are not billed, they must be paid for from project revenues, and their identification and tracking is important.

After all of the foregoing costs are estimated, it is necessary to establish an estimate of **escalation, contingency,** and finally the **profit** planned from the project. Escalation may not be a consideration if the project is a short one with a life of only a few months. Since, however, most projects are longer, it is likely that some escalation will occur. Two basic methods are available to evaluate this depending on the degree of detail in the estimate. For estimates that use a generalized approach, or where details may not be available, **overall factors** can be applied to general categories, such as craft labor, engineering labor, management, materials, and project equipment, and the totals for these categories adjusted accordingly. Overall factors are available from a variety of sources including the Bureau of Labor Statistics, contractor associations such as the AGCA, and trade publications such as *Engineering News Record*. Where detailed information is available, the labor rates are adjusted for **each craft** type and classification, using the expiration dates of the labor agreements and the anticipated succeeding labor rates. Where quotations have been obtained, project equipment costs would normally include escalation, but for nonquoted items, or general materials, an escalation allowance will still be necessary. Adjustments for other costs would be made as cited above for the generalized case unless special factors permit better definition.

Virtually every estimate includes some degree of contingency. The amount of contingency to be included is a responsibility of senior management and must be reserved for them. When preparing the estimate, all values must be estimated as closely as possible **without** including margins or allowances for the unexpected. If this is not done, the senior management, when reviewing the estimate, will have no way to know how much margin exists as it is buried in each line item and the estimate will invariably be high.

To assist in making such a decision, the estimating group often prepares **risk analysis** calculations. A risk analysis defines the percentage of work in different categories, and the degree of confidence in their definition and accuracy. Included are recommendations on the amount of contingency that would be appropriate. With these data, senior management can make an overall judgment about the completeness and accuracy of the estimate, and the likelihood of difficulty with the project, and then assign a value to **contingency.** An example of a contingency analysis is shown in Table 6.7.

Profit or **fee** is the final category of cost that must appear in the estimate. As with contingency it is a subjective judgment, although it is possible to establish certain guidelines based on risk, utilization of assets, dollars per work hour, or percentage of the value of the work in place. This is discussed in more detail in Section 6.8.

A major difficulty with fixed price bids is their **time** requirement. Usually the time allowed to prepare the bid is very short, and as a result there is a flurry of work at the last minute to incorporate subcontract and equipment quotations and establish values for contingency and profit. Frequently the estimating team will work very long hours, and as the bid due date approaches, even through the night to prepare the final figures. Sometimes the final figures are not ready until minutes before the bid due date and time, and often we see a bid delivered at the very last moment—just prior to the bid opening. In the case of public bid openings, timeliness is particularly important because public agencies will not accept late bids, and if the bid is but one minute late, it will not be accepted, so the entire bid preparation effort is wasted and opportunity to perform the work lost.

Bids should be prepared only to the level of detail necessary to support the requirements. To the extent practical, they should be of a relatively uniform level of accuracy, although it is widely recognized that variance of accuracy will occur among the different bid items.

TABLE 6.7 Contingency Analysis Example

Contingency (recommended)	Base $M	Contingency % 1	2	3	$M Minimum under 1	$M Most Probable 2	$M Maximum over 3
Low risk							
Major equipment on POs	1200	1	2	3	12	24	36
Awarded subcontracts	749	1	2	3	7	15	22
Quoted major equipment	190	2	3	4	4	6	8
Medium risk							
Estimated major equipment	194	2	5	8	4	10	16
Quoted subcontracts	788	3	5	7	24	39	55
Field distributible subcontracts	162	5	6	8	8	10	13
Home office costs	1302	5	5	7	65	65	91
High risk							
Estimated bulks	2436	4	8	11	97	195	268
Direct manual labor costs	2798	4	8	11	112	224	307
Distributible manual labor costs	478	5	8	13	24	38	62
Estimated subcontracts	1248	8	9	12	100	112	150
Field nonmanual costs	574	8	15	18	46	86	103
Field distributible materials	805	6	9	12	50	72	97
Schedule	—	—	—	—	98	141	159
Total	12,924				651 (5.1%)	1,037 (8.1%)	1,387 (10.8%)

Note: The estimator should recognize that the project scope definition for offplots may not be of the same quality as that for onplots.

6.5 UNIT PRICING

Unit pricing follows the same approach as described earlier, with the exception that the costs to be recovered must be spread over a number of work units. Care needs to be taken to ensure that **one-time** costs such as mobilization, equipment, and tooling purchases are recovered either in the early unit costs or through an additional lump sum price. It is normal for the unit pricing to be structured with the early units priced higher than later ones to permit recovery of these front end costs and also to provide some funding for the later work. Thus early units would carry higher overhead and similar charges and later units few or none of these charges. Often bid request documents will establish the approximate quantities of items for unit price bidding, and the bidder will have to provide pricing in that fashion. More information on this is found in Chapter 10.

6.6 SOURCES FOR ESTIMATES

As mentioned earlier, **historical** data or quotations for the materials or work are the best sources of data for estimates. Historical data are important because they not only reflect the work to be performed but include the way the company would organize and accomplish the work. For most work there are some differences between the historical data and the requirements of the work contemplated. For this reason it is common to apply factors to the data to reflect the anticipated differences. A good deal of judgment must be exercised to ensure that the factoring neither overstates nor understates the differences and thus skews the estimate.

For **subcontract** work or the purchase of materials and equipment, it is possible to obtain quotations that are fixed price and thus can be used in the estimate with no allowance for contingency and in many cases escalation. The only caution needed is to ensure that the timing and quantities of work are accurate.

Estimating the amount of labor required is more difficult if no historical data are available. For these cases it is necessary to refer to published data. Typically such estimating guides utilize a specific format. The format will vary from project to project, and considerable care should be taken when using them. Some references widely used for the construction industry are *The Richardson Rapid System* (a multiple volume series), *Means Construction Cost Data, Walker's Building Estimator's Reference Book,* and *The Dodge Construction Cost Data* series published by McGraw-Hill Information Systems, Princeton NJ.

Feedback data are extremely important for estimates or forecasts prepared after some experience is available on a specific project. As mentioned previously, the effect of location, climate, and weather—not to mention the availability and productivity of the local labor force—will all effect productivity and hence cost. Of all the sources of information, actual cost feedback data are the best and should be used wherever possible.

6.7 PROFIT AND FEE

Earlier it was pointed out that profit and fee determinations were often established from factors other than the basic interest in maximizing the profit on the particular project. If the work is of the fixed price variety, this may be a particular concern. For reimbursable work the problem is a bit simpler, since the competition for work is somewhat reduced and thus the margins that can be obtained are higher. Further, since reimbursable contracts normally establish the profit or fee as a separate item, with most all other items reimbursed, there is less pressure on the profit to cover unforeseen items or items underestimated in the bid. This is usually recognized by the client, and, though there may be extensive negotiations, the concept of profit is normally acknowledged.

In some cases the calculation of fee or profit may be a composite with more than one method used. On a large project, where, for example, both construction work and start-up services are provided, it might be appropriate to base the profit on the construction work as a **percent** of the value of the work in place, while the profit on the start-up services could be based on a flat value of dollars to be earned per hour of start-up services expended, with the profit **included** in this unit rate. The profit or fee for **management**-type work is derived from an overall assessment of the value to the client. If calculated on a per-hour basis, the amount earned will seem extremely high, since relatively few personnel are directly involved in the management of the work. These fees can run as high as several hundred percent of the total labor cost. If, however, the management fee or profit is viewed as the amount that could be earned with the same number of management personnel running an equivalent direct hire project, it can often be shown that the management fee is actually modest. Profit determination is more subjective than contingency calculation because projects are sometimes undertaken for reasons of positioning, expanding market share, learning new technology, and so forth, so a purely mathematical approach to profit may not be in the best long-term interests of the firm. All of these factors must be considered when setting the value of profit, while bearing in mind that a project taken at a low or zero profit should as a minimum offset the cost of or the potential return on an investment of the money employed in the work. In general, firms do not undertake work that promises 6% or less fee or profit, since that could be earned from alternate, relatively risk-free investments. Fees in the range of 8 to 15% represent the normal range desired. Where work is high risk, in new technologies or under adverse geographic or weather conditions, the size of the fee is customarily increased.

Chapter 10 presents the fee curves developed by the American Society of Civil Engineers for design services. These curves can be used as a guide to suitable values for this type of work.

6.8   OTHER OPERATIONAL REQUIREMENTS

Cash flow schedules are often needed to permit suitable financial planning. They are prepared from the estimate with the added dimension of time, and they require significant supporting data—namely the size and duration of work crews and their productivity. Placement of subcontracts and orders for purchased materials and equipment must be estimated (including the breakdown of payments for engineering, material procurement, and fabrication), and the entire effort is then converted into a time-phased cost schedule. Usually, where there is uncertainty when payments are required, the cash flow estimate will assume an **earlier** date to ensure that funds are available when needed.

Fortunately cash flow requirements are not bid documents in themselves, and thus they are not required to have the same degree of accuracy as a bid. If they are set too low, however, there may be an absolute funding limit established and because of this limitation, **restraints** on the work that can be accomplished.

In the manufacturing sector, inventory turnover is often used as a measure of the efficiency of the operation and will have a direct effect on the cash flow requirements of the organization. Where the firm is operating at a profit, this is usually not a significant problem because the cash flow requirements are more than offset by the revenue from sale of the product.

6.9   REPORTING

Periodic reports of actual performance against budgets, forecasts, or estimates are required. Because of the cost and time needed to perform an estimate, they are prepared only infrequently. On a project with a long life, say, five years, a full estimate may be prepared two or three times, with forecasts used to provide updated information between estimates. Since they only look at the **changes** from the last estimate and with the extensive computer capability available today, forecasts can be prepared much more easily. As a result it is now common to update the forecast as frequently as every month.

An important part of a forecast is a reconciliation of the changes since the last forecast or estimate. This permits management to operate on the "management-by-exception" principle and to concentrate on areas where the planned performance has not been achieved. An example of such a reconciliation is shown as Fig. 6.7.

ESTIMATE RECONCILIATION

PROJECT   _8723_

Jason Products

	Orig. Estimate	Fcst. #1		
Dated	_6/9/93_	_11/15/93_		
Eng. % Complete	_55_	_87_		
Const. % Complete	_14_	_42_	Diff.	Cause
Total Cost ('OOO's)	_$10,570_	_$10,406_	⟨164⟩	
Eqpt.	4,026	4,142	+116	_INCR. IN BOILER COST +110,MIS. MATL'S +6_
Labor	2,148	2,098	⟨50⟩	_IMPROVED LABOR PERF._
S/C's	1,024	1,139	+115	_INCR. QUAN. ROCK EXCAV._
Eng.	397	438	+41	_DETAIL FIELD RUN PIPE_
Other _Permits_	510	376	⟨134⟩	_3 VARIANCES REC'D_
Legal	95	105	+10	_INCR. TO OBTAIN VARIANCES_
Admin.	240	255	+15	_ADD'L TEMP. OFFICE ADMIN._
Fld. O'Hd	310	315	+5	_ADD'L HRS.–SECURITY_
Contingency	1,210	928	⟨282⟩	_REV. PER 10/25 RISK EVAL._
Fee/Profit	610	610	—	

Remarks _SCHED. PERF IMPROVING_

Prepared By   _RCN_   Date _3/9/94_
Reviewed ___AT___ Date _3/11/94_

FIGURE 6.7   Cost reconciliation.

Detailed reporting is often prepared for areas that have a major impact on project cost performance. Typical of these are labor performance, quantity tracking, subcontract monitoring, and change orders and claims.

Labor performance is a major concern on most projects and takes the form of productivity reporting. The actual performance hours are compared to the hours predicted in the budget (derived from the estimate), and the comparison is used to determine actual performance. Using these data, it is possible to determine the performance level. Some care needs to be taken to ensure that the true cause of the labor performance is known, for often variations in the quantities of work to be performed result from engineering and other changes. Because of the learning curve it is common for the hours required for initial installation units to be much higher than those performed later. This difference can be as large as 45%, particularly where tooling or other developments can significantly improve productivity. Figure 6.8 illustrates this type of calculation.

In addition to the concern with the efficiency of labor performance, there is usually need for information on the rate at which the work is being completed. Some of the same data are used to produce installation performance data. For ease of understanding this is normally shown graphically, with tabular data for backup. Such a graph is shown in Fig. 6.9.

Because of the impact of changes on the size or capacity of work items as well as on the quantity of work items, most projects will institute some form of **quantity tracking** to identify and monitor these changes and their effects on the project's cost and schedule. The system for monitoring these changes can range from a simple memorandum to continuous updating of quantities and extensive computer programs for recording and predicting the effects of the changes. With the importance of maintaining accurate and current information and the ready availability of integrated systems for project management, including cost and schedule data, it is becoming more common to maintain a continual updating system.

The quantity-tracking systems will track items as they evolve in the design process in engineering, material procurements as purchase orders and contracts are placed, and finally the actual quantities of work units performed by the field forces. Where the scope is well defined and no significant changes occur, only the field portion of the tracking system needs to be main-

FIELD LABOR PERFORMANCE

ITEM NO.	ERECT STEEL FRAME (UNIT OF MEASURE)	BASE ST		APRIL	MAY	JUNE	JULY	AUG	SEPT
1	BUDGET QTYS	90		9	26	45	65	82	90
2	BUDGET MHRS	3780		378	1096	1890	2722	3440	3780
3	FORECAST QTYS	95		9.5	28	47.5	68	86	95
4	FORECAST MHRS	3990		399	1157	1995	2873	3631	3990
5	INSTALLED QTYS			6	20	50	70	90	95
6	BUDGET UNIT MHRS	42.0							
7	EARNED MHRS (5 X 6)			252	840	2004	2940	3780	3990
8	SPENT MHRS			289	914	2067	2897	3776	4005
9	PERCENT COMPL (STD EXP C		0.0%	10%	29%	50%	72%	91%	100%
10	PERCENT COMPL (ACT) 7 / 4		0.0%	6%	21%	50%	74%	95%	100%
	PRODUCTIVITY (MHR PF) 8 / 7			1.15	1.09	1.03	0.99	1.00	1.00

FIGURE 6.8   Labor performance calculation.

EXPENDITURE CURVES

PERCENT COMPLETE OVER DURATION
SCHEDULE VS. ACTUAL

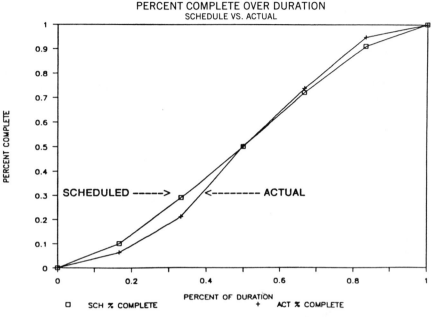

FIGURE 6.9 Expenditure curve.

tained—although some firms still employ the entire system to ensure management control over the entire project.

Feedback from the quantity-tracking work is essential to determining the effect on the project schedule, particularly where the quantities have increased. It is important to bear in mind that even without scope changes there is almost always a growth in the quantities for a project.

As a part of developing the base estimate for the project, a definition of the project quantities should be established to the highest level of detail and precision available at the time. From that point forward all quantity changes are measured against that baseline, and the baseline is not changed until the next definitive estimate, when the quantities are again taken off and a new baseline is established.

On a typical construction project engineering costs, though significant, are not a large percentage of the cost of the project. Thus changes in them have only a small effect on the overall cost of the project. The greater effect may be on the procurement and construction activities, which can be affected if engineering fails to complete its work on schedule and delays issuance of requisitions, specifications, and drawings.

While **subcontracts** are often bid and awarded on a fixed price basis, the competance of the subcontractor, size of the organization, management, labor, or organizational problems, or changes in quantities, may significantly affect completion dates for the subcontractor's work. Not only delays in the specific work of the subcontractor but also problems of access to or crowding may result in certain areas, with an impact on other activities and even the project overall. Since subcontract work is not under the direct control of the parent organization, it is vital that detailed close **monitoring** of the performance of the subcontractor be maintained and that corrective action be taken immediately when performance problems are identified. Part of the monitoring of the subcontractors may require independent verification of crew sizes, equipment availability, and installed quantities.

Change orders are an important source of revenue, and some contractors bid work on the basis of no profit planning instead to "make it up on the change orders." The idea of making excessive profit on change orders is fairly common, since the owner may have little choice but to give the added work to the contractor and has little negotiating leverage. Maintaining a reporting system for change orders provides **visibility** of them and permits some alternatives to be considered, such as the use of time and material's arrangements in lieu of merely adding the work to the existing contract. In some instances it may be possible to break the work out of the basic contract and award it to another contractor, possibly including competitive bidding. Regardless of the overall approach taken, early identification of change orders and prompt negotiation to reach agreement or pursue an alternate approach are important to maintaining schedule.

Claims are similar to change orders and can affect revenue in a similar fashion. Usually claims are instituted either by the contractor against the owner or from subcontractors or suppliers for changes or extra work. Subcontractor or supplier claims are often charged against the contractor rather than the owner. A similar category are **backcharges** against suppliers and subcontractors for the correction in the field of incorrect or incomplete material shipments or unsatisfactory workmanship. In contrast to the other type of items, backcharges represent potential revenue to the project.

All these categories—change orders, claims, and backcharges—are potential sources of heated **negotiations,** and it is essential that, when prepared and presented, they be sufficiently documented to clearly demonstrate the need, the authorization, and the cost. Backcharges in particular require more backup than a normal claim, since the charge is against a supplier not present at the job site and negotiations are rarely conducted on a face-to-face basis. Photographic evidence and drawings of the conditions both before and after are often used in documenting the basis for backcharges and should be considered in every case. In addition basic data furnished should be included with the backcharge. This includes the time records of the personnel, including the supervision and administrative personnel involved in the work and the labor rates that apply to each person, or category, such as payroll additives,

---

FIGURE 6.10   Project financial status report.

ACME CORPORATION
PROJECT FINANCIAL STATUS REPORT
($US and Hours in 1,000's)

I. PFSR NUMBER & DATE		
PFSR No: __ as of:	DD-MMM-YY	
Payroll Cutoff :	DD-MMM-YY	
Date Prepared :	07-Dec-92	

JOB # _____

DESCRIPTION _____

II. JOB DATA

Job No:
Project: Proj Mgr: ____
Client: R/O Mgr: ____
Location: B/L Mgr: ____
Scope:
Contract Type:
Business Line:
Regional Office:
"As Sold" GM/Jhr:

III. APPROVALS & CONCURRENCE

Prepared by: Date   Intial   Date   Intial
Proj Cntls Supv: ____   ____   ____   ____
Proj Cntls Mgr:

JOB HOURS	TOTAL ORIG BUDGET (A)	TOTAL CURRENT BUDGET (B)	TOTAL CURRENT FCST (C)	ACTUAL I-T-D MM/YY (D)	CHANGE THIS PERIOD (E)	ACTUAL PRIOR YEARS (F)	ACTUAL Y-T-D MM/YY (G)	CUR YR OPER PLAN (H)	CUR YR TOTAL FCST (I)	CHANGE THIS PERIOD (J)	NEXT YEAR FCST (K)	REMAIN YEARS FCST (L)
1 Permanent Office Jhrs												
2 Project Office Jhrs												
3 Construction Office Jhrs												
4 Total Non-manual Jhrs	0	0	0	0	0	0	0	0	0	0	0	0
5 Normal NR Jhrs												
6 Contractual NR Jhrs												
7 Manual (Direct Hire)												
8 Subcontract (Manual & NM)												
REVENUE (EARNED)												
9 Agency/Client Furnished												
10 Principal												
11 Interest on Receivables												
12 Fees												
13 Incentive/Participation												
14 Contingency												
15 Other												
16 TOTAL CONTRACT REVENUE	0	0	0	0	0	0	0	0	0	0	0	0
COST												
17 Agency/Client Furnished												
18 Normal Non-Re Costs												
19 Contractual Non-Re Costs												
20 Charge for Working Capital												
21 All Other Project Costs												
22 Contingency												
23 Other												
24 TOTAL PROJECT COSTS	0	0	0	0	0	0	0	0	0	0	0	0
25 TOTAL PROJECT GROSS MARGIN	0	0	0	0	0	0	0	0	0	0	0	0
26 Other Performing ___ GM	0	0	0	0	0	0	0	0	0	0	0	0
27 Total Sponsoring ___ GM	0	0	0	0	0	0	0	0	0	0	0	0
28 PROJECT GM/NM JHR	0.00	0.00	0.00	0.00	0.00	0.00	0.00	0.00	0.00	0.00	0.00	0.00

PERSONNEL FCST AT QUARTER END & QUARTERLY GROSS MARGIN WORKOFF	CURRENT YEAR				NEXT YEAR
	1Q	2Q	3Q	4Q	
29 Permanent Office					
30 Project Office					
31 Construction Office	0	0	0	0	0
32 Total Non-manual					
33 Total Manual (Direct Hire)					
34 Total Subcont (Man & NM)					
35 GM Workoff by Quarter					

> markers appear at rows 4, 16, 24, 25, 28, 33, 35

IV. PERCENT COMPLETE (Basis: _____)
                                    Sched   Actual
1. Engineering   ____   ____
2. Engr & Home Office   ____   ____
3. Construction   ____   ____
4. Total Project   ____   ____

V. PROJECT PERFORMANCE (1.00=Cur Bgt)
1. Engineering (ITD)
2. Construction (ITD)
3. Total Cost (CF/CB)
4. Schedule-Wks Ahead (+), Behind (-)
5. Quality Assessment (+,0,-)
6. Customer Satisfaction (+,0,-)
7. Construction Safety (+,0,-)

VI. TRENDS (Not included in Cur Fcst)
                    Jhrs   Rev   GM
1.
2.
3.
4.
5.

VII. GROSS MARGIN EVALUATION
                    Cur
              Min   Fcst   Max
Current Year
Total Project

VIII. KEY MILESTONES
                          Sched Fct/Act
1.
2.
3.
4.
5.
6. Duration (in months)
* Liquidated Damages Apply

IX. RECEIVABLES & WORKING CAPITAL
                          As of MMM-YY
1. Unbilled   0
2. Billed 0-30 days   0
3. Billed 30-90 days   0
4. Billed 90-365 days   0
5. Billed over 365 days   0
6. Total Receivables due   0
7. Advanced Payments   0
8. Project Working Capital   0
9. Payment Terms:

X. COMMENTS/SIGNIFICANT CHANGES

applicable indirect costs, equipment utilization records and rates, and POs or material issuance records for materials used. Since backcharges are normally only a small part of the value of the project work, a fee is not usually charged.

Registers are usually maintained for each category of costs, and they form the basis for controlling them. Since the items are all negotiable, it is normal that some portion of them will become uncollectable and be ultimately written off.

Reporting of the overall **financial** status of a project is critical to the ability to effectively manage and control the project. Most projects on a monthly basis prepare some form of financial report that covers all categories of cost and includes forecast data as well. In addition, because of the importance of trend data and the effect of change orders and claims, these reports usually have at least a brief narrative covering these items as well as purely numerical data. Each firm will have a slightly different form of report tailored to its type of work and reporting requirements. An example of one form of such a report is shown in Fig. 6.10.

6.10 BILLINGS

Billings, while essential to the financial management of the project, are prepared by the accounting organization and will not normally require the involvement of the estimating group. The exception to this is where a contract provides for **advance payments.** Then it is necessary to estimate the costs to be incurred prior to the next billing. The estimating group will provide information on costs to be incurred, including not only labor but equipment and subcontract costs, commitments and progress payments for equipment, payments for materials, insurance, and other indirect costs as well as an allowance for a pro-rata share of profit (or fee) to be earned.

CHAPTER 7

MATERIAL ACQUISITION

As the complexity of materials and components increases, a significantly larger portion of the cost of the project is devoted to their economic procurement. With the exposure to unsatisfactory materials, inadequate procurement practices can have a major impact on project schedules and cost.

7.1 MATERIALS MANAGEMENT

The procurement of materials and parts is generally called **materials management.** This system includes not only actual purchasing but also the planning, shipping, storing, and issuance of materials. The functions of expediting and inspection of these materials are normally considered part of this process as well. **Procurement** personnel work side by side with the design engineers to develop requirements, set standards, assign inspection levels and shipping standards, and in general become an extension of the engineering organization. In many cases the procurement functions, because of the technical complexity of the materials, or the activities to procure them, are managed by technical personnel.

Although still often called the **purchasing** department, the more common term used today is the procurement or materials management department to denote the broader responsibilities of the group. The department is usually staffed with personnel who have extensive experience in the procurement function and internally may further be specialized into such groups as purchasing, planning, contracts, expediting, inspection, traffic, materials handling, warehousing, and inventory control. In smaller organizations these functions are performed by fewer persons possibly include engineers, and in some instances by only one person.

Materials management involves an entire continuum of the process, from identification through issuance for use, with the final stage being that of asset recovery (the sale of surplus material and equipment). In addition it includes the functions of materials planning, inventory control for buyouts, and vendor materials management. The process begins in the engineering office where materials are first identified and concludes at a point of use, often thousands of miles away where a worker receives and utilizes the material.

The process as shown in Fig. 7.1 is the normal one with bids taken, analyzed, a Purchase Order awarded, fabrication performed, inspections (including release for shipment, transportation to the job site, warehousing), and ultimately issuance to the using organization. In many cases these steps are more extensive; in others some may be omitted, the decision depending on the value or complexity of the material or its importance. In some cases schedule, cash flow requirements, or other external factors will affect the steps taken.

7.2 SCOPE ESTABLISHMENT

Table 7.1 lists many of the typical questions needed to establish the **scope** of procurement. Because projects often include procurement activities that may call for client involvement, the checklist contains several instances where this occurs. In addition, since these activities involve basic operational decisions, they should be put in place as early as possible.

7.3 DEFINING THE MATERIALS

The first decision in materials management is to determine the scope of the work. This involves a **description** of the materials and equipment to be purchased and differentiating between what is to be supplied by others, such as subcontractors, and what, for companies with fabricating or manufacturing capability, is to be produced in house. On many projects most of the cost of the project is in purchases from outside sources, thus the make or buy decisions can be critical to the project's success.

The **make or buy** decision, while seemingly applicable only to large firms, does have some application to smaller companies and their projects as well. For example, a project of only modest size may have the option of producing certain concrete precast elements on the company's premises or purchasing them from a vendor. A large project, however, would consider this as almost a matter of course, particularly if the parent company or its affiliates has significant manufacturing or fabrication capability.

On a recent project to produce a large number of serially built ships, the

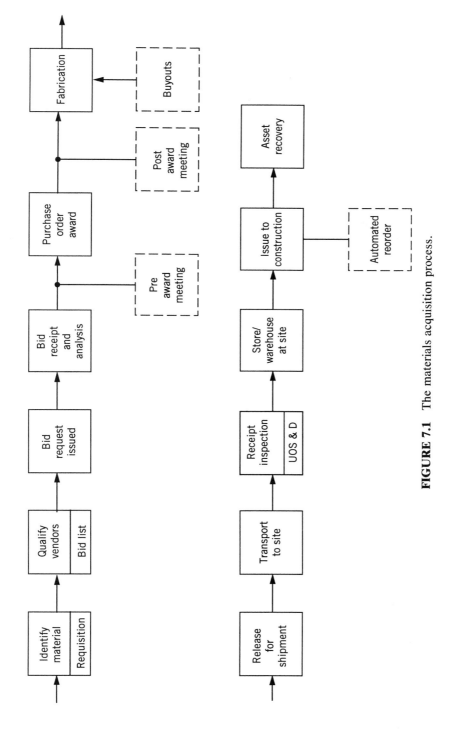

FIGURE 7.1 The materials acquisition process.

TABLE 7.1   Procurement Scope Determination

1. If procurement is a part of the scope, does it include
   Bid list development
   Obtaining bids
   Bid evaluation
   Issuing purchase orders
   Purchase order administration
   Expediting
   Inspection
   Traffic
   Receiving, warehousing, inventory, and issuance
   Asset recovery
2. Is procurement to develop a list of proposed suppliers and subcontractors?
3. Is client approval of the bid list required?
4. Typically how many bidders are required?
5. Will the list be exclusive, or can procurement add or delete with or without client approval?
6. Are there sole source procurements?
7. Are foreign sources permissible?
8. Will procurement solicit bids using the standard bid request form, general terms and conditions, and supplemental instructions?
9. Does client require bids to be solicited from sources nearest client's marketing area or jobsite area?
10. Is client approval required prior to award? If yes, what are the specific orders or dollar limits that apply? Does this apply to both purchase orders and subcontracts?
11. What internal approval levels are required?
12. What type of transportation is contemplated?
13. Are special routing and/or preferred carriers to be used?
14. Is any proprietary data being sent in the request for quotation (RFQ)? If so, a nondisclosure clause is needed.
15. Will the design organization act as agent for the client? Will the design organization purchase order and subcontract forms be used?
16. What are the terms of payment and retention, and the progress payment standards?
17. How are prices to be quoted, FOB destination, FOB suppliers plant, FOB plant WFA to destination, FAS port, etc.?
18. Are any special tax clauses required in the procurement documents, such as tax exemptions?
19. Is a cost breakdown required—particularly for government or sole source procurements?
20. Are there special arrangements that should be applied to the procurement such as JIT (just-in-time) inventory or MRP (material replenishment programs)?
21. Are there any special insurance requirements for product liability, etc.?

parent company performed a detailed study to determine which elements of the vessel—engines, reduction gearing, shafting, major electrical machinery, pumps, and so forth—could be produced in company facilities. The study included the following factors:

1. *Estimated cost to produce versus the comparable price to purchase on the open market, the price to produce including the required return on investment, loss of income from work not taken due to shop space tied up.* Some care needs to be taken to ensure that the return on investment used includes all the investment employed as well as personnel or shop space and equipment used, awaiting use, or unavailable for other purposes because of commitment to the make decision.
2. *Comparable profit to be earned versus using that shop space for outside orders.* This is a subset of the item above, using different financial criteria.
3. *Required and available shop space, and the impact upon production of other orders in house.* Where required shop space was unavailable, the cost to obtain other shop space was considered as well as the differentials for cost and time to ship from other shop space to the point of use.
4. *Resultant delivery schedules impact on project schedules.* Where deliveries were later than required their cost impact was calculated in terms of added labor, including overtime, allowances for crowding, reduced efficiency, and so on. In some cases special delivery methods, such as charter aircraft, compensated for some of the delay, and the cost was charged against this alternate.
5. *Warranty exposure for self-supplied materials and equipment.* This was established by a direct calculation based on experience in the performance history of the same or comparable equipment. For unfamiliar equipment performance is more difficult to evaluate and protection was provided by assuming conservative (i.e., higher values) of warranty service required.
6. *Exposure to liability by virtue of equipment or components manufactured rather than purchased.* The evaluation was similar to the warranty evaluation above.

The results of the study were that only about 40%, on a dollar basis, of the materials that could be made were kept in house, the balance were purchased on the open market.

The preceding study is fairly typical where the results indicate that "make" decisions are not as profitable as initially anticipated. The one exception to this is the case of developmental work, where it is essential to link the design development with the fabrication work or where **proprietary** design or processes are involved.

A side benefit to a suitably executed make or buy analysis is that it becomes a precursor to **Activity-Based Costing.** Here both the direct and overhead costs are broken down into their contributing factors and charged against the specific operation being analyzed. This calculation obtains an actual cost value for the specific operation to be used in its review and offers a way to manage the total product cost.

In the engineering/construction industry similar decisions must be made. The make or buy decisions are less complex, but there is a need to carefully document responsibility for procurement activities. A **Material Assignment Schedule** that lists which items are to be produced in-house, purchased, subcontracted, the organization responsible for the requisition, and so on, is often used for this type of control. A sample of a Material Assignment Schedule is shown in Chapter 3.

Once the make or buy decision is made, the design process can produce the required takeoffs and **requisitions** for procurement personnel to perform their work. The takeoffs that design provides are often done on a drawing-by-drawing basis as the design develops. Usually a bill of material is shown on the face of the drawing, and the materials listed on it are then requisitioned and purchased. This has the advantage of convenience and simplicity with so clear a definition of the requirements, but it can mix the different types of items that show on the same drawing. For example, a drawing could have ball bearings, electrical switches, instruments, anchor bolts, grout, and insulation on it, leading to a requisition that reflects this wide mix of materials. Such a wide mix of material could be supplied by an industrial supply house, but typically at a higher cost than if like items were grouped, requisitioned, and purchased from specialty suppliers or manufacturers. Alternatively, the list of materials could be broken up and multiple purchase orders placed with a number of different suppliers. This tends to be costly, however, since the number of units to be purchased on any one order is small with no pricing advantages and likewise since the administrative effort required for the multiple orders is higher. For small projects the convenience of this approach outweighs the cost disadvantage, while for larger ones, particularly with long schedules, grouping and **specialty purchasing** is advantageous. For designs that only require a limited range of materials, the bill-of-material approach is generally favored and has proved to be effective.

Regardless of approach it is necessary to perform a detailed **takeoff** of the materials required. In some organizations the takeoff is not performed by the designers but rather by an estimating group, which not only calculates and counts the materials required but also estimates their cost for estimate updates. In smaller organizations the takeoff may be performed directly by the designers. When the designers perform the takeoffs, they feel more responsible for the total work and can often make adjustments to the design to reduce material costs or requirements. The disadvantage is that they cannot be expert in all matters and often do not have specific detailed knowledge of material costs and availability as would a takeoff person who is somewhat more specialized. With several of the newer CAD programs, bills of material are developed by the program and listed directly on the drawing, thus eliminating takeoffs. This automation has the advantage of reduced effort to produce the takeoff. In addition accuracy is improved because omissions, and more commonly duplications, are eliminated.

Whether or not a formal bill of material is used, the listing of material on a requisition is essential to its procurement. The usual form of requisition lists like kinds of materials using an item system with identification of the quantity of the item to be obtained plus a suitable description and any special information. The special information can be a manufacturer's or purchasing department's part number, a brief specification (if a formal specification is not used), a brief description, a cost code, delivery requirements, a tag or identification number, or other appropriate information. Often material from a series of drawings or groups of materials from the same drawing are gathered together to make up the requisition. In most cases the material requisition is one part of the package of information sent to the bidders, with standard conditions and other commercial information forming the balance of the package. Figure 7.2 shows part of a typical requisition in such a package.

In general, the materials to be procured are defined by **codes, standards,** or **part numbers.** Regardless of the method it is crucial to provide the most precise definition of the material using those standards or descriptions normal to the industry. The American Society for Materials Testing (ASTM) publishes standards covering materials' chemical and physical properties and testing methods and is the usual referenced industrial standard for the United States. For federal governmental work in the United States, each governmental entity imposes its own standards and specifications on the work. The specifications referenced most frequently are the Military (MIL) Specifications.

The listing of the requirement for drawings and data or other information from the vendor should be shown as a separate line item on the requisition.

Some companies today utilize a takeoff system that permits the takeoff person to establish the quantity of an item needed and to assign a preestablished **bar code** to the item, using a bar code dictionary. This information is downloaded into a computerized data base that automatically prepares the requisition and in some systems also bid requests and similar documents and ultimately the purchase order. This saves a considerable effort in the preparation of these documents. Some manufacturing firms have tied this system into their suppliers through data links and thus have developed a paperless system that permits virtually instantaneous ordering of material.

7.4 BID LIST DEVELOPMENT

One of the most important steps in the procurement process is the development of a suitable **bid list.** The first question to be considered is whether there will be bidding for materials or equipment. In many cases it is not appropriate to request bids, examples being where the dollar value is low

SECTION I   <u>SCOPE OF WORK AND PRICING</u>

<u>GOODS: BED ASH HANDLING SYSTEM</u>

<u>A. WORK INCLUDED</u>

Equipment to be furnished shall include the following major items together with all materials and accessories necessary to provide a fully operable system within the intent of this Purchase Order and Attachments.

ITEM NO.	QUANTITY	DESCRIPTION	UNIT PRICE	EXTENSION
A	lot	Drawings and Data in accordance with attached Drawing and Data Requirements, Form G-321-E	lot	Included
B	lot	Expediting and Scheduling Reports in accordance with Section VI, Expediting Requirements	lot	Included
C	lot	Special tools for installation, operation, and maintenance	lot	Included
D	lot	Freight Charges, Not-to-Exceed: Ocean $_____ Inland $_____	lot	Not Included
1.0	lot	Design and furnish complete Bed Ash Handling System. Tag No. JN-H-500 A and B Tag No. JN-H-501 A and B Tag No. JN-H-502 A and B	lot	_____
2.0	per ln. ft	Add or deduct price for Bed Ash Handling System per ft. of drag chain conveyor.	(Add) (Deduct)	_____
3.0	per ln. ft	Add or deduct price for Bed Ash Handling System per ft. height of bucket elevator.	(Add) (Deduct)	_____
4.0	per ln. ft	Add or deduct price for Bed Ash Handling System per ft. of conveying chute.	(Add) (Deduct)	_____

FIGURE 7.2 A typical requisition.

and the administrative costs to solicit, place, and administer the process would be excessive, or where it is necessary to obtain identical or matching equipment. This last concern is often the case where existing equipment or systems are involved as in the case of spare parts procurement. In these instances, called **sole sourcing,** a negotiation with the selected supplier establishes the price, and the order is awarded accordingly.

The usual best source of names for a bid list is previous bid lists and bidders with satisfactory past performance. These bidders are familiar with the company procedures, know the procurement personnel, and have done business with the firm in the past. Lacking that, the yellow pages of the telephone book are a good place to start. Often a few telephone calls will provide sufficient information to identify several bidders. If working in the public sector (regulated utilities, governmental bodies, etc.), there are likely to be specific legal requirements regarding the number of bidders, the makeup of the bid list, the kind of businesses given preference (whether small, and/or owned by women or minorities), and so on. There may also be requirements regarding the percentage of business that must be awarded to certain categories of businesses, as well as "Buy American" clauses, which require that the business be placed with American companies unless the cost differential is significant. Project **financing** may further require that some portion of the procurements be placed with companies in certain countries or localities. Certain clients will have previous companies whom they wish on the bidders list because of **reciprocity,** location, or other factors. For work in high-tech fields, particularly aerospace and nuclear, the Consolidated Aerospace Supplier Evaluation (CASE) register maintains listings of suppliers with prequalified quality systems who are able to provide high-quality products with comprehensive, formal documentation, while for electronics work, the *Electro Buyers Guide* is particularly helpful.

Another source of bidders names is the *Thomas Register of American Manufacturers* which lists manufacturers in the United States by product and location. It is possible to identify a bidder close to the point of use, and this can represent a significant savings in the cost of freight if the material is heavy or must be shipped a considerable distance. More recently a similar set of industrial buying guides has been developed for U.S. regions. The *Thomas' Regional Industrial Buying Guides* lists products and services by city, with A to Z alphabetical listing and cross-references. Other useful sources are chambers of commerces, builders' exchanges or trade associations, and technical publication advertisements and their periodic (usually annual) procurement indexes. For work in foreign countries their **consulates** employ commercial attaches whose job is to stimulate trade, and they are an excellent source of this information.

For most work some **prequalification** of bidders is desirable, while for high-tech work it may be essential. The prequalification would normally require demonstrating, preferably by recent past performance including quality, price, and delivery, the ability to supply the equipment or materials. In some cases of material or equipment supply, but more commonly with subcontractors, there is a concern that the bidders may not have the financial strength to carry the work forward, so some form of a financial prequalification may also be necessary. For subcontractors a **performance bond** is frequently required to provide this assurance.

For all major orders it is good practice to obtain a **financial evaluation** of the supplier. D&B (Dun and Bradstreet) and TRW (Thomas, Ramo & Woolridge) offer credit reporting that can provide information about the strength of the supplier. A sample of a D&B Business Information Report is shown in Fig. 7.3.

The financial reporting companies can also furnish a variety of supplemental reports that provide other useful data. As an example, D&B can also provide a Payment Analysis Report which provides an overview of a firm's

			SUMMARY
DUNS: 00 007 7743	DATE PRINTED:		
GORMAN MFG CO INC	OCT 30 199-	RATING	3A3
(Subsidiary of Gorman Holding Companies Inc.)			
492 KOLLER ST	COMMERCIAL PRINTING	STARTED	1965
AND BRANCH(ES) OR DIVISION(S)	SIC NO.	PAYMENTS	SEE BELOW
SAN FRANCISCO CA 94110	2752	SALES F	$13,007,229
TEL: 415-555-0000		WORTH F	$2,125,499
		EMPLOYS	500 (150 HERE)
		HISTORY	CLEAR
		FINANCING	SECURED
		FINANCIAL	
		CONDITION	FAIR

CHIEF EXECUTIVE: LESLIE SMITH, PRES

SPECIAL EVENTS 10/20/9- On Oct. 13, 199-, the subject experienced a fire due to an earthquake. Damages amounted to $35,000, which was fully covered by its insurance company. The business was closed for two days while employees settled personal matters due to the earthquake.

PAYMENTS REPORTED (Amounts may be rounded to nearest figure in prescribed ranges)

	PAYING RECORD	HIGH CREDIT	NOW OWES	PAST DUE	SELLING TERMS	LAST SALE WITHIN
10/9-	Ppt-Slow 90	1000	500	-0-	N30	1 Mo
09/9-	Ppt	250	100			4-5 Mos
	Ppt-Slow 30	2500	2500	1000		1 Mo
	Slow 30	500	500			2-3 Mos
	Slow 30-60	70000	70000	65000		1 Mo
08/9-	Disc	2500	1000			1 Mo
	Disc-Ppt	25000	25000	-0-	2 10 Prox	1 Mo
	Ppt-Slow 15	1000	500	250		1 Mo
	Ppt-Slow 30	15000	10000	5000		1 Mo
	Ppt-Slow 30	1000	-0-	-0-	N30	4-5 Mos
07/9-	Ppt	250000	250000	-0-		1 Mo
	Ppt	7500	250	-0-	N15	1 Mo
	Ppt	500	-0-	-0-	N30	6-12 Mos
	Ppt	100	50	-0-	Regular terms	1 Mo
	Ppt-Slow 30	100000	100000	40000		1 Mo
	Ppt-Slow 30	70000	70000	50000	2 15 Prox	1 Mo
	Slow 30	7500	-0-	-0-		1 Mo
	Slow 30		-0-	-0-	N30	6-12 Mos
06/9-	Disc-Slow 30	30000	30000	7500		1 Mo
05/9-	Ppt	250	-0-	-0-		6-12 Mos
	Ppt-Slow 60	200000	200000	90000		1 Mo
04/9-	(022)	100	100		N30	

* Payment experiences reflect how bills are met in relation to the terms granted. In some instances payment beyond terms can be the result of disputes over merchandise, skipped invoices etc.
* Each experience shown represents a separate account reported by a supplier. Updated trade experiences replace those previously reported.

CHANGES 03/17/9- Subject moved from 400 KOLLER ST. to 492 KOLLER ST. on March 11, 199-.

UPDATE 08/17/9- On August 17, 199- KEVIN J. HUNT Sec-Treas stated for the six months ended June 30, 199- profits were up compared to same period last year.

FIGURE 7.3 Sample Dun & Bradstreet Report. Copyright 1992 by Dun & Bradstreet, Inc. All rights reserved. Reprinted with permission.

Dun & Bradstreet. Inc.

	This report has been prepared for:
BE SURE NAME, BUSINESS AND ADDRESS MATCH YOUR FILE.	ANSWERING INQUIRY
	Page 2

THIS REPORT MAY NOT BE REPRODUCED IN WHOLE OR IN PART IN ANY MANNER WHATEVER.

FINANCE 09/11/9-		Fiscal Dec 31,1989	Fiscal Dec 31, 1990	Fiscal Dec 31, 1991
	Curr Assets	4,643,821	4,825,611	5,425,125
	Curr Liabs	3,595,821	3,625,000	4,125,718
	Current Ratio	1.3	1.3	1.3
	Working Capital	1,048,000	1,200,611	1,299,407
	Other Assets	1,468,291	1,485,440	2,191,690
	Worth	1,879,451	1,912,112	2,125,499
	Sales	9,321,118	10,325,582	13,007,229
	Net Income	32,661	213,387	26,014

Fiscal statement dated Dec. 31, 1991:

Cash	$ 925,000	Accts Pay		$ 2,125,114
Accts Rec	1,725,814	Notes Pay		450,000
Inventory	1,643,311	Bank Loan		1,100,000
Other Curr Assets	1,131,000	Other Curr Liabs		450,604
	— — — —			— — — —
Curr Assets	5,425,125	Curr Liabs		4,125,718
Fixt & Equip	1,667,918	L.T. Liab-Other		1,365,598
Other Assets	523,772	CAPITAL STOCK		50,000
		RETAINED EARNINGS		2,075,499
	— — — —			— — — —
Total Assets	7,616,815	Total		7,616,815

From JAN 1, 1991 to DEC 31, 1991 sales $13,007,229; cost of goods sold $9,229,554. Gross profit $3,777,675; operating expenses $3,751,661. Operating income $26,014; net income before taxes $26,014. Net Income $26,014.

Submitted SEPT 11, 199- by Leslie Smith, President. Prepared from statement(s) by Accountant: Ashurst & Ashurst, PC. Prepared from books without audit.

--0--

Accounts receivable shown net less $12,586 allowance. Other current assets consist of prepaid expenses $64,471 and $1,066,529 of a loan from an affiliated concern. Other assets consist of deposits. Bank loans are due to bank at the prime interest rate, are secured by accounts receivable and inventory, and mature in 3 years. Notes payable are due on printing equipment in monthly installments of $37,500. Other current liabilities are accrued expenses and taxes. Long term debt consists of the long term portion of the equipment note.

On SEPT 11, 199- Leslie Smith, president, submitted the above figures.

Leslie Smith submitted the following interim figures dated JUNE 30, 199-.

Cash	$ 1,011,812	Accts Pay		$ 1,932,118
Accts Rec	1,932,118	Owe Bank		1,100,000
Mdse	1,421,112	Notes Pay		350,000

Sales for 6 months were $7,325,001. Profits for 6 months were $103,782.

Projected annual sales are $14,000,000.

The financial condition at DEC 31, 1991 was fair. Total debt was heavy in relation to the net worth. Total debt exceeded net worth by 2.6 to 1 vs an industry median of 2.3 to 1.

Current assets were centered in a slow moving inventory, while profitability has been below average the past three years. The median return on sales for this industry is 1.9%. Return on sales for the past three years has been 0.2%, 2.1% and 0.4% respectively.

Total equity represented 27.9% of the total capitalization which compared unfavorably to the industry average of 37.7%. On SEPT 11, 199-, Leslie Smith, president stated profits were below average due to heavy price competition in the industry, higher operating expenses, and decreased advertising budgets following the nationwide move towards cost containment. Net worth increased due to additional paid in capital.

FIGURE 7.3 (continued). Sample Dun & Bradstreet Report (*continues*).

The following data is for information purposes only and is not the official record. Certified copies can only be obtained from the official source.

PUBLIC FILINGS

If it is indicated that there are defendants other than the report subject, the lawsuit may be an action to clear title to property and does not necessarily imply a claim for money against the subject.

* * * SUIT(S) * * *

DOCKET NO.:	21211	STATUS: Pending
SUIT AMOUNT:	$1,000	DATE STATUS ATTAINED: 03/25/199-
PLAINTIFF:	MAZZUCA & ASSOC.	DATE FILED: 03/25/199-
DEFENDANT:	GORMAN MANUFACTURING CO. INC.	RECEIVED BY D&B: 03/31/199-
CAUSE:	Goods sold and delivered	
WHERE FILED:	SAN FRANCISCO, CA	

* * * UCC FILING(S) * * *

COLLATERAL: Accounts receivable - Inventory including proceeds and products

FILING NO.:	86188586	DATE FILED: 07/24/199-
TYPE:	Original	RECEIVED BY D&B: 10/04/199-
SEC. PARTY:	A.C. Paper, Palo Alto, CA	FILED WITH: SECRETARY OF STATE/
DEBTOR:	Gorman Manufacturing Co., Inc.	UCC DIVISION, CA

There are additional UCC's in D&B's file on this company available by contacting 1-800-DNB-DIAL.

The public record items contained in this report may have been paid, terminated, vacated or released prior to the date this report was printed.

BANKING 09/9-

Account(s) averages high 6 figures. Account open over 10 years. Loans granted to low 7 figures on a secured basis. Now owing low 7 figures. Collateral consists of accounts receivable and inventory. Matures in 1 to 5 years. Borrowing account is satisfactory. Overall relations are satisfactory.

HISTORY 09/11/9-

LESLIE SMITH, PRES   KEVIN J. HUNT, SEC-TREAS
DIRECTOR(S): THE OFFICER(S)

BUSINESS TYPE: Corporation - Profit   DATE INCORPORATED: 05/21/1965
AUTH SHARES - COMMON: 200   STATE OF INCORP: California
PAR VALUE - COMMON: No Par Value

Business started May 21, 1965 by Leslie Smith and Kevin J. Hunt. 100% of capital stock is owned by parent company.

SMITH born 1926. Married. Graduated from the University of California, Los Angeles, CA in June 1947 with a BS degree in Business Management. 1947-65 general manager for Raymor Printing Co., San Francisco, CA. 1965 formed subject with Kevin J. Hunt.

HUNT born 1925. Married. Graduated from Northwestern University, Evanston, IL in June 1946. 1946-1965 general manager for Raymor Printing Co., San Francisco, CA. 1965 formed subject with Leslie Smith.

RELATED COMPANIES: Through the financial interest of Gorman Holding Company Inc., the Gorman Manufacturing Co. Inc., is related to two other sister companies:
1. Smith Lettershop Inc., San Diego, CA; commercial printing, started 1972.
2. Gorman Suppliers Inc., Los Angeles, CA; commercial printing, started 1980.
Intercompany relations consists of loans.

OPERATION 09/11/9-

Subsidiary of Gorman Holding Company Inc., Los Angeles, CA, started 1965 which operates as a holding company for its underlying subsidiaries. Parent company owns 100% of capital stock. Parent company has 2 other subsidiaries. Intercompany relations: consist of loans and advances.

A consolidated financial statement on the parent company dated Dec. 31, 1991 showed a net worth of $4,125,112, with an overall fair financial condition.

Commercial printing specializing in advertising posters, catalogs, circulars and coupons.
Net 30 days. Has 175 accounts. Sells to commercial concerns. Territory: United States. Nonseasonal.
EMPLOYEES: 500 including officers. 150 employed here.
FACILITIES: Rents 55,000 sq. ft. on first floor of one story cinder block building in good condition. Premises neat.
LOCATION: Central business section on well traveled street.
BRANCHES: Subject maintains a branch at 1073 Boyden Road, Los Angeles, CA.

FIGURE 7.3 (continued). Sample Dun & Bradstreet Report.

payment habits as reported to D&B, a two-year trend, and a comparison to industry payments; a Government Activity Report which provides governmental transaction as well as private sector transaction data for firms working in both fields; and Dun's Financial Profile Reports which depicts financial statement data and numerous key financial ratios for up to three years. In addition analytical computer software is available as well as special reports.

An important reference book, also available from D&B is the *Dun & Bradstreet Reference Book of American Business* which contains summary business information on more than three million businesses throughout the United States, Puerto Rico, and the Virgin Islands. Updated six times a year, the listings include the D&B Rating or Employee Range designation, line of business, an indication of the availability of the Payment Analysis Report and/or the Dun's Financial Profile Report, telephone number and year of establishment.

Of increasing importance is the requirement that suppliers have in place appropriate quality assurance programs to control their product and processes and to conform to standards found in the industry. Where significant quantities of material are to be furnished, ever more purchasers are requiring that fairly extensive programs including **statistical process control** (SPC) be used. As a result firms that do not comply often find themselves prevented from being placed on bid lists. It is often necessary to be **qualified** by a quality audit prior to being placed on a bid list, and the existence and implementation of a suitable quality program is an essential part of today's business operations. As a result it is common to require that bidders provide copies of their programs to qualify as a bidder, or with their proposals to comply with bid request requirements. Chapter 2 discusses some of the usual quality programs used in both manufacturing and service-type activities.

The **number** of bidders, while not usually a problem, must also be given some specific consideration. It is costly to the bidder to prepare a bid, and bids should not be solicited merely to keep someone in line or to act as a check on someone else's pricing. Thus it is important to establish a principle that all solicited bids will be given proper and fair consideration. The minimum number of bids usually desired is three, with up to five or in some cases seven or so considered the preferred upper limit. Beyond five the effort to evaluate all but the most simple bid becomes costly and rarely changes the apparent low bid. For work in the public sector, certain regulations may require advertising and the acceptance of all bids received from a large number of bidders, which can greatly expand the analysis work. The number of bidders may also vary with the dollar value of the work. Often the number of bidders is limited by the complexity of the product or the schedule requirements for delivery. Frequently because of high administrative costs **low-value** orders do not justify competitive bids.

In the case of **sole suppliers** or bidders, a cost breakdown of the bid can provide enough data to determine if the pricing is reasonable and can save the time and effort required for a fully detailed evaluation. Often where pre-engineered/standardized materials or equipment are being purchased, bids can be considered competitive on the basis that bidders have published price lists to which their proposal conforms and the price list accepted as the bid document.

7.5   THE BID PACKAGE

An initial decision in establishing the bid package is whether the work is to be implemented with a **purchase order** or whether some form of **contract** or **subcontract** will be used. The material that follows deals with the purchase order type procurement in which competitive bidding is utilized. (Contracts and subcontracts are dealt with in Chapter 11.)

A request for quotation, or a request for proposal (RFQ or RFP) is the document that initiates the process and establishes where and when the bids are to be delivered, the number of copies required, whether the bids are to be specially handled and will include special labels or other similar items or provisions. The bid package included with it consists of a series of interrelated documents which includes both the **commercial** and **technical** requirements of the procurement. A typical bid package includes a set of instructions to bidders and a pro forma purchase order with attachments. The instructions to bidders can be extensive. A typical set for a reasonably complex procurement is shown in Fig. 7.4.

The purchase order will normally include the following sections:

- Scope of work and pricing, including general conditions
- Price policy
- Seller's promised milestone and delivery schedule
- Correspondence
- Quality surveillance requirements
- Expediting
- Shipping instructions
- Invoicing instructions
- Technical notes
- Attachments

A major factor to consider when developing the bid package is the question of whether or not the plans and specifications are to be complete or purposely less complete, and thus place a burden of completion or development on the successful bidder. This approach is often taken where the work or performance of the component is broadly defined, the detailed design is purposely left somewhat incomplete. It reduces the purchasers cost of the initial design work and transfers the responsibility for final completion of the design and

INSTRUCTIONS TO BIDDERS

THE INSTRUCTIONS CONTAINED HEREIN ARE IMPORTANT

PLEASE READ THEM THOROUGHLY

ALL REPRESENTATIONS, CERTIFICATIONS, AND FORMS

PROVIDED WITH THIS BID REQUEST

MUST BE COMPLETED, SIGNED AND

INCLUDED WITH BIDDER'S PROPOSAL

NOTIFY WITHIN FIVE DAYS OF RECEIPT OF THE BID

REQUEST OF YOUR INTENT TO QUOTE

CRITICAL NOTES:

1. Late bids received after the Bid Due Date may not be considered but may be promptly returned unopened to the Bidder.

2. Bidder shall submit four (4) copies of the proposal and all subsequent correspondence. (Original plus three (3) copies).

3. BIDDERS are encouraged to ensure proposals are technically and commercially their best offer and that all drawings/data necessary to permit a complete technical evaluation of the offer is included with the proposal. Because of the late release to proceed with this project and the ensuing impact on jobsite schedule, the proposal which most closely meets specification, offers competitive pricing and takes fewer, if any, commercial exceptions will be selected for further evaluation and negotiation.

FIGURE 7.4 Instructions to bidders (Page 1 of 6 pages).

INSTRUCTIONS TO BIDDERS

These instructions are a part of   Bid Request requirements and must be followed in the preparation of your proposal. Throughout this document   may be referred to as the Buyer.

1.   **Inquiry Documents**

      Bidder's attention is directed to all the documents which form a part of these instructions. All forms and documents must be completed and submitted with the Bidder's proposal.

2.   **The Proposal**

    2.1   The Bidder's proposal and the Bidder's equipment and/or material are to be in strict conformance with the Buyer's requirements as set forth in the Bid Request letter, pro forma Purchase Order, the Material Requisition (if applicable), Technical Form of Proposal (if applicable), the Specifications and all attachments thereto. If there are any variations from the Buyer's requirements, the Bidder shall itemize all variances by attachment to the proposal. All variances must be correlated to the appropriate section of the Bid Request.

    2.2   Quote firm Unit Prices and Total Net Extended Prices based on the Shipping Terms shown on the attached pro forma Purchase Order.

    2.3   Shipping charges are to be quoted as maximum amounts. Any actual shipping charges in excess of the quoted maximum will be to the Bidder's (SELLER'S) account.

    2.4   All prices are to be quoted in U.S. dollars.

    2.5   Bidder shall use a copy of Section I of the pro forma Purchase Order to record Unit Price and Total Net Extended Price for each line item listed thereon. When applicable, customs duty, sales or use tax and shipping costs shall be stated as separate items.

          All figures must be typed. If you are not providing a price for a listed item, type "Not Quoted" in the extension column. Add your company name and proposal number at the top of each page of Section 1 and include the required number of copies with your proposal.

    2.6   Insert the following information in the appropriate blanks on the pro forma Purchase Order:

         a.   Origin of shipments   - Cover Sheet

         b.   Maximum, NOT TO EXCEED,   - Section I
              Shipping Charges

FIGURE 7.4 (continued). Instructions to bidders (Page 2 of 6 pages).

	c.	Unit prices and extensions	- Section I
	d.	Recommended spare parts and prices	- Section I
	e.	Sellers promised milestone and delivery schedule	- Section III
	f.	Per diem rates for on-site Seller's representative	- Special Conditions Article SC-4

2.7   The following information is to be provided elsewhere in your proposal.

     a.   Period of time for which bid is valid. (See paragraph 6).

     b.   Shipping gross weight.

     c.   Shipping gross cube.

2.8   The attached Bidder Information Sheet (Att. 1) is to be completed and included with Bidder's proposal.

2.9   The attached Commercial Terms Checklist (Att. 2) is to be completed and included with Bidder's proposal.

2.10   The attached Technical Form of Proposal (Att. 3) is to be completed and included with Bidder's proposal.

3.   Alternates

Bidders are encouraged to submit alternate proposals when they consider the alternate to be an improvement or more economical. However, base proposals must meet Specifications. (See Specifications for considerations to be used in selecting equipment). Consideration shall be taken to provide maximum interchangeability of parts for units quoted. When such consideration causes a significant increase in lost, or loss of operating efficiency, alternates may be submitted.

Bidders are also encouraged to submit commercial offerings such as extended warranties, performance guarantees, financing, etc. which may enhance their bid.

4.   Brand Names

Brand names and figure numbers, when specified, are illustrative of an approved type and the substitution of an equivalent type by another manufacturer shall be proposed by Bidder for approval. Bidder shall give complete description of equipment, including materials of construction.

FIGURE 7.4 (continued).   Instructions to bidders (Page 3 of 6 pages).

5. **Delivery**

   It is essential that delivery of the equipment and/or material specified herein be met. Promised ship date quoted by the Bidder will be a key consideration in making award. Bidders should therefore quote best shipment based on a realistic production schedule. In determining Bidder's promised ship date allow three (3) weeks for Buyer's approval and return of drawings required by attachments to the Bid Request.

6. **Proposal Validity Period**

   Bidder's proposal shall be firm for a period of 120 days from the Bid Due Date. Bidder shall clearly state that his proposal is firm for 120 days and will not be withdrawn within that time.

7. **Basis of Award**

   This is a request for Bid; however, Buyer reserves the right to accept or reject any proposal with or without prior discussion with the Bidder. Buyer may either:

   o   Make award on the basis of proposals received, without discussion of proposals. Consequently, proposals should be submitted initially on the most favorable terms from a commercial, price and technical standpoint that the Bidder can submit.

   o   Select one or more proposals for further negotiation.

   o   Reject all proposals or modifications received after the Bid Due Date unless it is determined by the Buyer that failure to arrive on time was solely due to delay in the mail for which the Bidder was not responsible.

   o   Buyer reserves the right to procure all or part of the quantities specified herein from one or more Bidder.

   o   Not make an award based on this request.

8. **Payment Terms**

   General Conditions, Article 3, "PRICE AND PAYMENT", and Special Conditions, Article SC-2, "Payment," set forth the required payment terms in this procurement action. Progress payments, if proposed, shall be considered as an exception to these conditions. If progress payments are proposed, the Bidder's proposed payment terms must:

   o   Define the percentage of payment and the provisions under which these payments are to be made.

   o   Tie each proposed progress payment to a milestone(s) or actual work performed and must be verifiable by the Buyer.

FIGURE 7.4 (continued).   Instructions to bidders (Page 4 of 6 pages).

o   Define milestone in detail, i.e. "major components" must be identified.

o   Show each milestone date as a duration period either from date of order or prior to the shipment date.

o   Incorporate a schedule identifying the milestone date to either date of award or date of shipment in those cases where a progress payment is contingent upon the completion of a milestone event, i.e. delivery of your materials to the Seller.

Additionally, if progress payments are proposed, the Bidder must enclose audited financial statements for the Buyer's review.

9.   **Escalation**

Firm price quotations are preferred and will be preferentially evaluated.

10.   **Progress Reports**

Regular periodic production schedules and progress reports will be required from Seller for materials covered by a Purchase Order resulting from Bidder's proposal. Unpriced copies of Seller's Purchase Orders to sub-suppliers for major components must be supplied to Buyer.

11.   **Drawings**

As required in the Bid Request, the Bidder shall furnish typical outline, arrangement and sectional drawings complete with Bidder's description of equipment, and a priced recommended spare parts list. Bidder shall complete and return   Data Sheets with proposal. Bidder shall submit with the proposal either a statement of concurrence with the time allotted for preparation and submittal of drawings and other documents to the Buyer, or the number of calendar weeks required by the Bidder to prepare and submit these documents after receipt of a Purchase Order.

12.   **Documentation**

As applicable, complete documentation in accordance with Quality Assurance requirements stated in the Specification must precede and/or arrive with the material at the jobsite. Any shipment arriving at the jobsite without proper documentation will not be received and will be subject to return or hold on invoices pending corrective action. Any costs incurred in connection therewith shall be to the Bidder's account.

13.   **Tests**

Proposal shall state the extra cost, if any, for witnessed performance tests. If the Bidder does not have adequate facilities for performance testing (i.e., power supply, motors for pumps, steam, etc.) he shall so state in his proposal.

FIGURE 7.4 (continued).   Instructions to bidders (Page 5 of 6 pages).

14. **Palletization**

    As applicable, shipments shall be palletized to the maximum extent practicable to preclude damage during shipment. Bidder shall state in his proposal the extent of palletization. Additional costs included for palletizing and export packing shall be stated separately.

15. **Tools**

    Bidder shall submit a list of the type, quantity and unit price of special tools required for operation and maintenance of the equipment that Bidder proposes to furnish, as a part of the proposal.

16. **Proprietary Information**

    This Bid Request and all drawings, designs, specifications and other data appended to or related to are the property of   and are provided only for the purpose of enabling each potential Bidder to prepare and submit a proposal. The information contained or referred to in the Bid Request, or appended to it, is not to be disclosed or released for any other use or purpose and must be returned to   if requested.

17. **Mechanics' Lien Waiver Agreement**

    In the event of award, Bidder will be required to execute the Mechanics' Lien Waiver Agreement included   as an attachment to the pro forma Purchase Order. Bidder shall acknowledge in the proposal it's acceptance of the terms of the Mechanics' Lien Waiver Agreement.

FIGURE 7.4 (continued). Instructions to bidders (Page 6 of 6 pages).

detailing to the contractor or subcontractors. Since the successful bidder will have to complete the plans and specifications, this cost is in the bid price. As a result the savings in the initial design work will be somewhat offset by the cost the supplier includes for the design work. As this approach parallels that of the performance specification (below), there is also a risk that the work, at least in its details, will turn out differently than expected. Where these factors can be accepted, this form of bid package has proved effective.

The bid package, and later the PO, includes the documentation described earlier. For special cases it may also include the **basis** for bid evaluation, whether pre- or postaward meetings are contemplated, and so forth.

The technical portion will include the drawings, specifications for the particular item of work, and referenced **standard specifications** such as painting, special requirements for testing, qualification of processes, and providing of spare parts. It will include the requirements for the submittal of drawings and data as well as other documentation of a technical nature such as test results, code certificates, and quality control procedures. A copy of a standard form used to request these data is shown in Chapter 2.

Perhaps the most important principle when preparing a specification is the requirement for **clarity** and **specificity.** The use of ambiguous phrases or wording can, and almost certainly will, create problems during the life of the project and perhaps beyond. The choice of phrases and the use of language must be such that there is no question as to what was intended. Further the description of the item must be such that there are **acceptance criteria** established and the question of whether the work meets the specification requirements is not a matter of personal opinion but rather can be demonstrated by measurement.

As an example, on a recent project large steel plates were to be field assembled into a circular tank about 90 ft in diameter, the entire tank then being moved by a crane into an excavation. A concern was the out of roundness that could occur in the fabrication and handling. As a result the specification in its first draft contained a phrase that "excessive out of roundness shall be corrected by the contractor at his cost . . . ." Because of the potential for disagreement with the general language used, out of roundness was defined and the specification revised and issued with the phrase "out of roundness of more than two inches in diameter in two feet of circumference of the tank shall . . . ."

As can be seen by this example, the improved language defined and quantified the requirement in such a way that there would be no disagreement upon what was intended and what was acceptable. It also established a standard for the contractor who could then ask for a modification before award if it was too restrictive. In this case a modification was requested, and the requirement was adjusted to 3 inches out of roundness.

For certain types of work it is not possible to properly define the acceptance criteria in the specification. Two examples are weld finish and contouring where decontamination is a requirement and unique paint or fabric colors. In these instances, it works out well if the bidder supplies **samples** of the end products with the bid, and upon approval they become the acceptance standard. Alternatively, samples may be provided by the purchaser to the bidder.

Precedence of documents is another item that can create difficulties if not clearly specified. The normal precedence of documents is from the specific to the general. In the event of conflict, the specific documentations govern and take precedence over any general ones. While the principle is clear, it is nevertheless useful to include, usually in the general requirements a statement on precedence. For example, a listing could include a statement that ". . . in the event of conflict, the precedence of documents shall be as follows with the higher listed documents having precedence over the lower ones":

- Equipment or component unique specifications
- Basic design or data sheets
- Process requirements
- Project, equipment, or component drawings
- Standard specifications
- Standard drawings

- Reference drawings
- Industrial standards

When writing specifications it is necessary to determine whether they are to be **performance** or **design** specifications. In a performance specification, the performance requirements are established together with any other external or environmental requirements, and the **supplier** assumes complete responsibility for offering the best package of features to meet the specification. For instance, a performance specification for an engine to drive a generator would spell out the required torque or horsepower, the duty cycle, the type of fuel to be used, the overall dimensional or weight limitation (if one exists), and the type of coupling or drive arrangement. It would leave to the bidder the questions of the number of cylinders, valving, fuel system, starting system, electrical system, and so on. This gives the bidder the maximum flexibility to offer the best and least expensive piece of equipment and provides the owner with the opportunity to see different pieces of equipment offered, some of which may have significant advantages. The disadvantage with this type of specification is that the equipment offered will vary somewhat.

The other form of specification is the design type, which not only describes the performance required but also establishes many design features. This approach ensures that the equipment, as offered, will be exactly what the owner wants, but it can expose the specifier to claims or future problems. The design-type specification significantly increases the **liability** of the specifier, since, if the equipment does not perform correctly, it is quite possible that the fault lay in specifying incorrect features or design elements. Another disadvantage of this approach is that it **limits** the offering of alternates and new or different items and does not tend to be at the state of the art for the particular service intended. It should only be used where the equipment features are well established, and thus a reduced exposure to inadequate performance.

In general, it is good practice to permit the offering of **alternates.** This allows the bidders to offer something that, while not precisely meeting the specification, may in fact be a better item for the intended service. It also permits improved pricing, for bidders can often offer their standard products which may be close to the specification requirements, cheaper, and more readily available—an advantage when the schedule is tight. To ensure that the bidders are free to offer alternates, language such as "proposals for alternates are encouraged" and "bidders should consider offering alternates if they can substantially meet the requirements of the specification(s)" can be used.

Where it is necessary to change any of the provisions of the specifications or of any other data in the bid package, a suitable **addendum** should be prepared and issued to all of the bidders. It is not sufficient to merely advise bidders by telephone, suitable written confirmation must be sent as well to ensure that there is a clear definition of what is required.

7.6 BID ANALYSIS

When **evaluating** bids, it is often necessary to perform not only a commercial or priced bid analysis but a technical one as well. Regardless of which approach is used, the basis for the award needs to be firmly established. It may not be in the best interest of the project to award the order on the basis of low price alone, for other significant factors may be present. For this reason a policy widely used is to award the order to the lowest **evaluated** bid. This requires that the bid evaluation process consider other pertinent factors in the bid(s) that affect the total price.

The preparation of the **priced** bid analysis is straightforward, although it may be necessary to consider more factors than merely the base cost of the item as offered. Factors such as weight (particularly heavy items to be shipped some distance), requested payment terms, past performance, delivery schedule, cost of spare parts, erection services, or extended warranties can all swing the order from one supplier to another. These factors can be readily evaluated by assigning values to their differentials and adding them to the bid price. This yields an evaluated price which, while different from the bid price, more accurately represents the true cost to the project. Sometimes, for complex items with long delivery times, protection against cost increases to the bidder are found in the bid. These often take the form of tying the price offered to a standard index, such as the Bureau of Labor Statistics index of cost for metalworking industry labor, with the material cost tied to a Department of Commerce standard for, say, copper wire, if that is the principal material component. The price analysis must in this case evaluate the likely escalation against another bid that may be higher but offers a firm price. This of course assumes that the bid request does not exclude nonfirm pricing. While priced bid analyses are often prepared by design personnel, they can in many cases be prepared equally well by nontechnical personnel who are experienced in the procurement function. An example of a priced bid analysis is shown in Fig. 7.5.

Technical bid analyses are more complex than priced ones and are normally performed by the technical design personnel. An important feature of them is an orderly listing of the principal requirements of the technical specification and an indication of the compliance with or deviation from the requirement. All deviations are not equally undesirable, and when preparing the analysis, this must be kept in mind. One useful way to approach this evaluation is to separate the essential items in the specification from those that are merely desirable. Often differences in performance of the offered items must be evaluated not only on the basis of the feature itself but also on the cost of operations. Typical of this is a bid where the low bidder offers a unit that has a lower efficiency than those of the other bidders. When energy costs are included or when capitalization or other factors are applied, the additional energy required exceeds the quoted price savings, and the second low bidder whose unit has a higher efficiency is in fact less expensive.

CONFIDENTIAL

BID TABULATION

JOB NO:	PROJECT: COGENERATION — M/R NO: M
MATERIAL: HEAT RECOVERY STEAM GENERATOR	

DESCRIPTION	1	2	3	4	5	REMARKS
BIDDER						1. BASE BID = $4,351,075 FURTHER EVALUATION NOT PERFORMED
PROPOSAL DATE	7/19/90	7/20/90	7/19/90	7/27/90	7/25/90	2. FURTHER EVALUATION NOT PERFORMED
REFERENCE NO.	3900.03	3900.03	1893.REV5	004-16.REV3	910534JDH	
BID PRICES:	$3,587,000	$3,496,000	$3,276,800	$3,480,000	$3,638,700	
RECOMMENDED OPTIONS:			$258,530	REMARK 2	REMARK 2	SEE PAGE 3 FOR DETAIL
NET PRICE:	$3,587,000	$3,496,000	$3,535,330	$3,480,000	$3,638,700	
TAXES:	$269,025	$262,200	$265,150	$261,000	$272,903	
FREIGHT CHARGES:	$160,000	$160,000	$192,000	$134,000	$165,000	NOT TO EXCEED
VENDOR REP.:	$12,600	$12,600	$10,500	$10,920	NOT PROVIDED	BASED ON 21 DAYS STRAIGHT TIME
ESCALATION:			$104,292			
PERFORMANCE BOND:						
TOTAL COMMITMENT:	$4,028,625	$3,930,800	$4,107,272	$3,885,920	$4,076,603	
ADJUSTMENTS FOR COMPARISON:	$373,007	$371,187	$0	$419,704	$435,478	SEE PAGE 2 FOR DETAIL
TOTAL COMPARABLE EVALUATION:	$4,401,632	$4,301,987	$4,107,272	$4,305,624	$4,512,081	
DIFFERENTIAL:	$294,360	$194,715	$0	$198,352	$404,809	
PROMISED DELIVERY:	5/1/91	5/1/91	4/15/91	5/18/91	5/18/91	REQUIRED DELIVERY/COMPLETION DATE: 4/1/91
PAYMENT TERMS:	NET 30	NET 30	NET 60	NET 30	NET 30	
PAYMENT TYPE; LS, UP, T&M:	LS	LS	LS	LS	LS	RECOMMENDED BIDDER:
ESCALATION TERMS:	NONE	NONE	NONE	NONE	NONE	CURRENT BUDGET: $4,111,050 (INCLUDES SCOPE CHANGES)
PROGRESS PAYMENTS:	YES	YES	YES	YES	YES	MAXIMUM COMMITMENT AMOUNT & FUND AUTHORIZATION
ADVANCE PAYMENTS:	NO	NO	NO	NO	NO	LIMITED TO: $4,107,272
F.O.B. POINT:	JOBSITE	JOBSITE	JOBSITE	JOBSITE	SHIPPING PT.	
SHIPPING WEIGHT:	1,953,000	1,963,000	1,600,000	1,770,000	NOT AVAIL.	
EXPIRATION DATE:	8/20/90	8/20/90	8/10/90	8/25/90	8/24/90	CREDIT CHECK.: NOT PERFORMED, 50% BOND TO BE PROVIDED

	ORIGINATOR	RECOMMENDATION		APPROVALS	
		PROCUREMENT	PROCUREMENT	DIVISION	-POSTED COST/SCHDLE
SIGNATURE					
DATE	8/10/90	8/10/90	8/10/90	8/10/90	8/10/90

FIGURE 7.5 Priced bid analysis.

Another somewhat qualitative approach is to assign a weight from 1 (smallest) to 20 (largest) to each deviation in terms of its importance, with the bidder having the lowest score offering the most acceptable bid. An example of a technical bid evaluation appears in Fig. 7.6.

The management of the material acquisition process should include a system of internal review and delegation of **approval** authority for purchase or commitment. As a part of the review process, the individual bid analyses and the bid tabulation forms (summaries of the individual bid analyses) should be checked by an independent party who approves them for correctness both from a technical and pricing standpoint. These steps normally occur within the design group, although for extensive or complex pricing arrangements, including escalation, the cost-estimating group or the finance department may review labor, burden, or general and administrative rates, particularly for sole source offerings. Depending upon the size of the organization, the analysis can then be approved for release for purchase by the signatures of designated personnel. One system for larger projects has the levels of approval for bid tabulations presented in Table 7.2.

For complex items, where final prices are subject to negotiation, it is wise to establish **negotiation** objectives early. These can be based on previous prices paid or on an independent estimate of the pricing. With these values it is then possible to have a strategy rather than to merely respond to the flow and style of the other parties.

Conflict of interest is always a concern but it can be particularly damaging if personnel with important procurement responsibilities have conflicting involvements. Full disclosure by personnel is essential to avoid a situation where relatives may be involved in potential vendor companies or where stock ownership or an equity position may be held with a vendor. In such cases the personnel must advise their management of this potential conflict and abstain from involvement either in the analysis or placement of the order. Since order administration is less sensitive, conflict of interest is usually not an issue, unless significant change orders are contemplated in which case personnel with this problem should disqualify themselves.

7.7 TYPES OF ORDERS

Orders for materials and equipment take one of several forms or represent variations of these: purchase orders, letters of intent, letter orders, contracts, or subcontracts.

Purchase orders (POs) are almost universally used. They are convenient, and small firms can purchase preprinted standardized forms available in office supply stores. The purchase order is an authorization to provide materials or services in accordance with a prior offer, with payment to be made upon execution of the conditions of the order and the presentation of invoice(s) in accordance with the terms of the order. The prior offer can be a formal

	MTM	PROSPER	FLJ	SPEC
PERFORMANCE				
RATED FLOW GPM	✓	✓	✓	6900 GPM
EFFICIENCY @ 6900 GPM	76.5%	83.0%	73.0%	—
DESIGN FLOW GPM	✓	.	✓	8400 GPM
DESIGN TEMP. °F	✓	✓	✓	112°F
SPEC. GRAVITY @ 112°F	0.991	✓	0.991	—
DIFF. HEAD @ 8400 GPM	✓	✓	✓	1080'TDH
PRESS. @ MIN. FLOW PSIA	710	670 ②	610	675 PSIA
MIN. CONT. FLOW GPM	800	700	800	—
FLOW@ RUN OUT GPM	10,000	9400	11,500	9400GPM
NPSH REQ. @ 9400 GPM	15'	21.4'	15'	—
NPSH REQ @ DESIGN (8400GPM)	14.5'	17.6'	14.5'	—
BHP @ 6900GPM HP	2700	2600②	2650	—
BHP @ 9400GPM HP	2900	2860	3100	—
SPEED RPM	1180	1180	1180	900/1200
1ST CRITICAL SPEED RPM	NS	>9500	NS	—
N° OF STAGES	7	9	6	—
WEIGHT ROT. PART LBS	NS	3700	3100	—
REVERSE FLOW RPM@100%HEAD	<125%	1250 RPM	<125%	—
DESIGN				
SUCTION : SIZE/FLANGE RATING	24"/150#	30"/150#	24"/150#	24"PREF.
DISCHARGE: SIZE/FLANGE RATING	18"/300#	18"/300#	18"/300#	16"PREF.
DIAM (O.D.) OF SHELL	42"	44"	42"	—
LENGTH OF SHELL	18'-6" ①	22'-5½"	20'-8"	—
WEIGHT OF PUMP (EACH) LBS	43,000	46,500	40,000	—
WEIGHT OF MOTOR LBS	26,000	23,000	26,000	—
SUCTION : DESIGN PRESS PSI	1.5xMAX.SUCT.	50	1.5xMAX.SUCT.	—
DISCHARGE : DESIGN PRESS. PSI	1.5xSHUT-OFF	650	1.5xSHUT-OFF	—
SUCTION : TEST PRESS. PSI	1.5xMAX.SUCT.	75	1.5xMAX.SUCT.	—
DISCHARGE:TEST PRESS. PSI	1.5xSHUT-OFF	975	1.5xSHUT-OFF	1.5xMAX.SHUT-OFF

① AS PER LETTER, MARCH 9,1970; INCREASE IN AVAILABLE SUCTION
HEAD BY 8 FT. ENABLES MTM TO SHORTEN LENGHT OF BARREL
FROM 21'-6" TO 18'-6".
② ADJUSTED TO PROSPER LETTER MARCH 6, 1970
NOTES : ENCIRCLED ITEMS ARE UNDESIRABLE
A CHECK ✓ INDICATES COMPLIANCE WITH SPECS

SUMMARY OF BIDS	JOB No XXXX	
CONDENSATE PUMPS	P.O.NO. M-23	REV.
XYZ POWER COMPANY	SHT. 2 OF 4	0

(Left margin vertical text) XXX DATE · APPROVALS · MATL · SUPV · CHK · OR · ENG · ISSUED FOR CLIENT APPROVAL · REV. DESCRIPTION

FIGURE 7.6 Technical bid evaluation. From I. J. Karassik and William C. Krutzch, *Pump Handbook* 2d ed., copyright 1985. Reproduced with permission of McGraw-Hill, Inc.

TABLE 7.2	Bid Tabulations Approvals and Authorization for Purchase		
	Project Engineer	Chief Engineer	Vice President
Under $25,000	Signs	N/R	N/R
$25,000 to $200,000	Signs	Approves	N/R
Over $200,000	Signs	Approves	Approves

bid or offering, an unsolicited bid, a telephone conversation, or even a handshake. Caution should be used with these less formal forms of commitment, however, for they almost invariably result in misunderstandings and disagreements. Thus there should be some system for confirming the agreement in writing using one of the documents mentioned above.

Regardless of the basis for the agreement, it is important that the PO spell out in sufficient detail the item(s) to be furnished. For informal agreements it is even more important to spell out the agreement to avoid later difficulties. The PO must be **accepted** by the vendor to become a valid contract with the purchaser. As a result an acceptance phrase or section is typically included on the face of the document. Figure 7.7 includes the facing sheet and selected portions of a typical purchase order.

The actual preparation and issuance of a PO often takes a significant amount of time, varying from a few days to a week or more for complex or expensive orders. For orders that are totally conforming with no exceptions taken and where automated purchasing systems are in force, this can be reduced to same day issuance of orders. For most large or complex orders some changes to the specification are almost always necessary to **conform** it to the items selected for purchase. Normally these are minor in nature and relate to secondary items such as slight differences in performance, accessories, or even paint color, type of packing for shipment, and so forth.

When items of relatively low value are being purchased (e.g., less than $5000), formal bidding procedures and purchase orders are often not used. In this instance, in one system a purchase order number is assigned to and written on the face of the requisition, which then becomes the document of purchase. To maintain control, the PO number is listed in the commitment or Purchase Order Register, and thus overall control of the PO and the commitment is maintained. When the material is furnished, this revised requisition is used as the controlling document, and receipt of the materials and authorization for payment are made against it. This procedure has the advantage of simplicity, and it significantly reduces the cost and effort to administer small orders. In addition, to further reduce the work in administering small orders, some firms waive competitive bidding, and they may award orders based merely on a telephone quotation.

PURCHASE ORDER

POWER PROJECT

Purchase Order No.:_____

Revision:_____0_____

Date:_____June 4, 1991_____

Required Jobsite Delivery

Start:_____February 1, 1992_____

Complete:____March 2, 1992_____

Earliest Acceptable Delivery:__February 1, 1992_____

TO:

(SELLER)

This Purchase Order confirms Buyer's Notice of Award leter dated June 4, 1991 and Buyer's Notice of Change letter dated July 3, 1991.

Telephone: **DO NOT DUPLICATE**

FAX:

In accepting this Order, Seller agrees to furnish the goods specified in full accordance with all conditions stated in this Purchase Order package, contents of which are listed on page 2, or any revisions thereof.

GOODS: **BED ASH HANDLING SYSTEM**

Total Value: _$_____

Ship & Consign To:

Shipping Terms: FOB Jobsite Prepay Freight and Add to Invoice

Origin of Shipment(s): _____

By:

_____

Title: _Project Procurement Supervisor

Telephone: _____

Fax: _____

ACKNOWLEDGEMENT

Seller hereby accepts all terms and conditions contained herein.

Accepted By: _____

Title: _____

Firm: _____

Date: _____

Please sign and return this page within 10 days. If Seller fails to acknowledge this Purchase Order, Buyer reserves the right to withhold final payment until acknowledgement is received.

FIGURE 7.7   Some aspects of a typical purchase order (Page 1 of 7 pages).

TABLE OF CONTENTS

SECTION	DESCRIPTION
I	Scope of Work and Pricing
II	Price Policy
III	Seller's Promised Milestone and Delivery Schedule
IV	Correspondence
V	Quality Surveillance Requirements
VI	Expediting
VII	Shipping Instructions
VIII	Invoicing Instructions
IX	Technical Notes
X	Attachments

FIGURE 7.7 (continued). Typical purchase order (Page 2 of 7).

SECTION I **SCOPE OF WORK AND PRICING**

GOODS: BED ASH HANDLING SYSTEM

A. WORK INCLUDED

Equipment to be furnished shall include the following major items together with all materials and accessories necessary to provide a fully operable system within the intent of this Purchase Order and Attachments.

ITEM NO.	QUANTITY	DESCRIPTION	UNIT PRICE	EXTENSION
A	lot	Drawings and Data in accordance with attached Drawing and Data Requirements, Form G-321-E	lot	Included
B	lot	Expediting and Scheduling Reports in accordance with Section VI, Expediting Requirements	lot	Included
C	lot	Special tools for installation, operation, and maintenance	lot	Included
D	lot	Freight Charges, Not-to-Exceed: Inland $_____	lot	Not Included
1.0	lot	Design and furnish complete Bed Ash Handling System. Tag No. JN-H-500 A and B Tag No. JN-H-502 A and B Total Value (Excluding Freight)	lot	_____ ▬▬▬▬

B. UNIT PRICES

1.0	per ln. ft	Add or deduct price for Bed Ash Handling System per ft. of drag chain conveyor.	(Add) (Deduct)	_____ _____
2.0	per ln. ft	Add or deduct price for Bed Ash Handling System per ft. height of bucket elevator.	(Add) (Deduct)	_____ _____
3.0	per ln. ft	Add or deduct price for Bed Ash Handling System per ft. of conveying chute.	(Add) (Deduct)	_____ _____

FIGURE 7.7 (continued). Typical purchase order (Page 3 of 7).

ITEM NO.	QUANTITY	DESCRIPTION	UNIT PRICE	EXTENSION
		C. OPTIONAL PRICING		
1.0	lot	Dust collector system complete with collector, ductwork and hoods, stand alone control panel.	lot	_____

Prices are valid to:_____

FIGURE 7.7 (continued). Typical purchase order (Page 4 of 7).

SECTION II   **PRICE POLICY**

All prices shown in the Purchase Order are firm and not subject to price adjustment for the duration of the Purchase Order.

SECTION III   **SELLER'S PROMISED MILESTONE AND DELIVERY SCHEDULE**

Seller promises to meet the milestone and delivery dates outlined below:

Milestone	Submittal Date
Submittal of calculations, outline, general arrangment with bill of materials, final foundation requirements:	8-5-91
Submittal of complete Spare Parts List and cost:	11-18-92
Submittal of all engineering documentation:	8-26-91
Start of fabrication:	9-30-91*
Completion of all fabrication:	2-24-92
First shipment to jobsite:	1-27-92
Arrival of first shipment at jobsite:	2-3-92
Last shipment to jobsite:	2-24-92
Arrival last shipment at jobsite:	3-2-92

* Based on 4 weeks to process all front end drawings and 4 weeks thereafter to start fabrication.

SECTION IV   **CORRESPONDENCE**

Send _1_ copy of correspondence relating to price, schedule, or other commercial provisions of this purchase order to the attention of the Buyer. Send correspondence relating to the technical aspects of the purchase order and design drawings to the attention of **Project Engineer**. The address for all correspondence is shown on Page 1. Purchase Order Number, Item and Equipment Number must be referenced on all such correspondence.

SECTION V.   **QUALITY SURVEILLANCE REQUIREMENTS**

Manufacturer's standard.

FIGURE 7.7 (continued).   Typical purchase order (Page 5 of 7).

SECTION VI   <u>EXPEDITING</u>

In addition to the Expediting requirements described in Section 6 of the General Conditions, the following statements are applicable:

Seller shall furnish to   no later than <u>two (2) weeks</u> after commitment, a complete schedule forecasting or acknowledging receipt and entry of the order, engineering schedules (drawings, data, manuals, spares list, etc.), material acquisition, manufacture/fabrication detailing dates to start machining, assembly, test, cleaning/painting, etc. (include all witness and hold points for quality verification).

In addition, the Seller shall furnish a progress report to   every 2 weeks in sufficient detail to allow a realistic evaluation of scheduled events towards contract/order completion.

Immediately after placing your order for MAJOR COMPONENTS of this order, please furnish two (2) unpriced copies of your purchase order showing sub-supplier's name, shipping point, order number and promised shipping dates to

The requirements of the above paragraphs shall be addressed as follows:

Attention: Project Procurement Manager
            Job

SECTION VII   <u>SHIPPING INSTRUCTIONS</u>

A copy of these instructions must be included in each of your purchase orders to sub-suppliers making shipments direct to Buyer's jobsite.

PREPAY FREIGHT CHARGES FOR ALL SHIPMENTS.

All items for shipment shall be prepared and packaged so as to prevent damage during transit and handling. Seller shall indicate on shipping containers any special storage, hoisting, and handling requirements.

Purchase order number, item number and equipment and/or tag number(s), must appear on all containers, shipping papers and packing lists.

One copy of PACKING LIST must accompany each shipment.

FIGURE 7.7 (continued). Typical purchase order (Page 6 of 7).

ON DAY OF SHIPMENT mail one (1) copy of Packing List and Bill of Lading to each of the following:

(1)   (2)   Construction Company

Attn: Project Procurement Manager
Job   Attn: Field Procurement Manager

SECTION VIII   **INVOICING INSTRUCTIONS**

Invoices must show freight, taxes and duties as separate items, if applicable.

Three (3) copies of your invoice and the original signed bill of lading <u>MUST</u> be mailed to:

Attn: Accounts Payable

All discount periods will be calculated from the time <u>complete</u>, <u>accurate</u> invoices are received by

Payment of invoices will be by   Buyer will require Seller to execute the Mechanics' Lien Waiver Agreement, Attachment 8, prior to any payment being made under this Purchase Order.

SECTION IX   **ATTACHMENTS**

<u>No.</u>	<u>Description</u>	<u>Rev.</u>
1	General Conditions	0
2	Special Conditions	0
3	Technical Specification for the Bed Ash Handling System,	1
4	General Project Requirements,	1
5	Attachment Specification for Induction Motors,	1
6	Technical Specification for Electrical Requirements for Packaged Mechanical Equipment,	2
7	Technical Specification for Controls and Instrumentation for Packaged System,	1
8	Mechanics Lien Waiver Agreement	0

FIGURE 7.7 (continued). Typical purchase order (Page 7 of 7).

Where schedules are tight and it is necessary to begin work immediately, a **letter of intent** is sometimes used. This is a written notification, and it can be sent by fax or telex if necessary. Typically the letter of intent would read as follows: "It is our intent to issue to you a Purchase Order to furnish the . . . per our bid request . . . . Please begin engineering and/or the purchase of materials and commence fabrication immediately and confirm your acceptance of this order." After issuance of the letter of intent it is important to follow up and issue the confirming formal purchase order at the earliest date.

Where only engineering information is needed, the RFQ will normally require a breakout of engineering costs, and orders are often issued with the phrasing, "**Released for engineering only,** no material is to be procured or fabrication to take place without specific prior authorization."

Where there is a sole source that has offered a **proprietary** design, and lack of supply from that vendor could shut down the operations of the purchaser, it is prudent to set up an escrow account with the vendor and a reputable bank. In this arrangement the bank, or other agreed agency, holds copies of the drawings and manufacturing data so that in the case of bankruptcy, fire, and other such incidents, the information can be given to the purchaser to arrange alternate sources of supply until the vendor can once again resume production. Naturally the purchaser and the alternate manufacturer, which produces the item during the unavailability of the vendor, would have to abide by a suitable **secrecy** agreement to avoid future infringement on the design and methods of the vendor.

To facilitate the ordering of **bulk**-type items such as small valves, electrical fittings, wire and cable, and bulk cement, it is often advantageous to place an **open** order or **blanket** purchase order with a supplier. This order establishes the pricing for the items based on estimated quantities that permit the vendor to offer pricing levels for the general quantity anticipated but leaves the exact quantity and delivery dates open. It authorizes someone, typically at a jobsite or plant to request **releases** (shipments) against the order as needed. This arrangement has the advantage of avoiding the need to accept a large shipment and thus warehouse all the materials at one time, and it permits the jobsite to establish the final quantities and delivery schedules of the items needed.

In many manufacturing facilities, where there is confidence in the quality of the product produced by the supplier, the recent approach is to issue blanket purchase orders and require that the supplier deliver to and maintain an appropriate minimum quantity of the items in bins on the shop floor. This avoids the need for receiving, and the parts are invoiced and paid for as used.

Subcontracts or contracts used to purchase materials or equipment or to undertake on-site work are normally not handled as purchase orders. More information on them is found in Chapter 11.

7.8 TERMS AND CONDITIONS

Standard terms and conditions, often called **general conditions,** are found on the PO form and are frequently referred to as "boilerplate." These are usually the fine print on the back of the PO that establishes the standard commercial and legal conditions that apply to all orders. While these vary among businesses, most have at least the following provisions:

1. A statement that the entire agreement is found in the documents and that changes will be reduced to writing.
2. Changes must be proposed and agreed, and only specific parties may authorize changes.
3. Price and payment terms are defined.
4. A statement regarding title and risk of loss.
5. Expediting may be applied to this order.
6. Quality responsibility rests with the supplier; shop inspection may be applied to the order.
7. A statement regarding warranties and guarantees.
8. A hold harmless clause regarding patents and processes.
9. A statement requiring compliance with all applicable laws and regulations.
10. A statement regarding assignment.
11. A provision permitting suspension of the order.
12. A provision permitting termination for convenience or default.
13. A statement regarding the applicable law under which the agreement is to be construed.
14. A list of the various executive orders and federal codes that apply to the order.

7.9 ON-SITE INSPECTION

On-site **inspection** is often used for fabricated items particularly if custom designed or of high cost. To provide this capability, design personnel are often sent to the supplier facility. Alternatively, an in-house or separately hired outside inspection group can be used. Whatever the case, the inspection needs to be conducted in accordance with a written inspection **plan** that incorporates the technical requirements and the acceptance criteria set forth in the specifications. The advantage in using the design personnel for these inspections is their familiarity with the specification requirements and the design requirements for the component. They are able to make informed judgments on the spot and can resolve many minor difficulties rapidly. The difficulty is that these inspections take the design personnel away from the design office and often slows down the design work. There is a further danger that the designer will waive certain requirements and make no record of the waiver, and other design personnel will then not benefit from the thinking that led to the waiver. It also provides an opportunity for the designer to conceal errors in the specification or design and again not permit management review of these changes. From a commercial point of view where a **waiver** or a lessening of the specification requirements is permitted, a price reduction should be obtained, but this is often not the case unless there is a careful review of the changes permitted.

The planning of the inspections necessary to verify that the work meets the specification should be done in the design group. Where there is an in-house inspection department, their personnel should work closely with the designers and assist them in integrating testing and inspection requirements into the technical specification. From this work an inspection plan emerges that spells out the inspections required, their frequency, and, most important, whether these inspections constitute **witness** or **hold** points. A witness point is one that the inspector should observe particularly early in the fabrication cycle while a hold point is one beyond which manufacture cannot proceed without the required inspection. It is important to make the supplier aware that inspections by others do not relieve it of responsibility for the correctness and accuracy of the work.

For manufacturing facilities, inspection of materials received is often performed on a routine basis. Samples—or in some cases all—of materials received are inspected by the receiving facility to ensure that the materials received conform to the specification requirements. With increased stress being placed on supplier quality, more of this responsibility is being shifted to the supplier to permit the manufacturer to place received materials directly into production without this additional step.

Even with this reduced inspection mode, extensive inspection is always given to the initial item produced or received. This is called "**first article inspection**" and is a critical step in ensuring product quality. An example of an inspection plan and a completed inspection report are shown in Figs. 7.8 and 7.9.

With the increasing complexity of procurements, the importance of the documentation furnished with the order at, or just before, time of shipment is becoming ever more important. The documentation includes not only the normal shipping documents but also the record of tests and inspections performed during the fabrication process. **Certified material test reports** (CMTRs) and **certificates of conformance** (CofCs) are also included. The careful review of these documents prior to shipment is essential to prevent nonconforming material from being shipped. For custom designed equipment there is almost always some deviation in the documentation that must be resolved prior to shipment. As a result it is a normal requirement that for

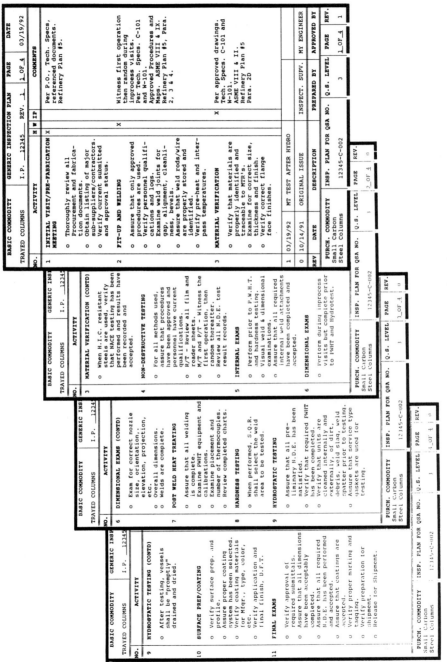

FIGURE 7.8 Inspection plan.

```
Item    0720705                92/11/08        16:06
================================================================================
                        QUALITY SURVEILLANCE REPORT
QSR NUMBER :    15                  FORM PSQ-221
================================================================================

I. GENERAL
ASSIGNMENT NUMBER    : 12345-C-002              SUB:
P.O. NUMBER          : 12345-C-002-HAC          REV: 4
TECHNICAL SPEC NUMBER : P.O. Section 2          REV: 4
MATERIAL ON ORDER    : Small C.S. Columns
LEVEL OF SURVEILLANCE : 3
SHOP ORDER NUMBER    : 2214, 2347 & 2397
PRIME SUPPLIER : Any Tank Co.           LOCATION: Small Town, USA
SUPPLIER       : Same                   LOCATION: Same

================================================================================
II. PRODUCTION STATUS (Show only applicable)

X   IN PROCESS ENGINEERING
X   IN PROCESS PROCUREMENT
X   IN PROCESS MANUFACTURING
================================================================================
III. QUALITY SURVEILLANCE ACTIVITIES PERFORMED:
     (Include type (Progressive/Final), date, Supplier contacts)
     Final       03NOV thru 07NOV92                - Manager Quality Control
================================================================================
IV. MATERIAL RELEASED FOR SHIPMENT WITH THIS REPORT:

     P.O. ITEM#          DESCRIPTION       TAG#           QUANTITY
     ==========          ===========       ====           ========

        1                DESULFURIZER      C-2420            1

================================================================================
V.  Supplier NonConformances:  None
================================================================================
VI. SUMMARY OF SURVEILLANCE ACTIVITIES PERFORMED:

     The writer visited the supplier's facility at Small Town, USA to perform
the following surveillance:

Hold Point - Final Inspection

Final inspection was performed on S.O. 2214, Vessel Tag No. C-2420.  The
following activities were performed:

 *  The writer inspected the installation of trays, and installation was found
    to be satisfactory to the following        Approved      Dwgs.:

    Tray 23        -       Dwg. 35834-E-02   Rev. A
    Tray 1-17 Odd -             35834-E-05   Rev. A
    Tray 2-16 even-             35834-E-05-1 Rev. A
    Tray 18                     35834-E-05-3 Rev. A
    Tray 19 & 21                35834-E-05-2 Rev. A
```

FIGURE 7.9 Inspection report (Page 1 of 2 pages).

0720705 21311-C-002 MS Tank QSR #15   PAGE 2 of 2 PAGES

    Tray 20   35834-E-04   Rev. 0
    Tray 22   35834-E-03   Rev. A
    Downcomer & Seal Pan of   35834-E-05-4 Rev. A
    trays 1-19

*   Final Paint check. Paint was checked for total DFT and was found to average
    11 mils, which meets the Cleaning and Painting Requirements Sheet for this
    vessel, which has been approved Code 1. Note that the skirt inside and
    outside and only the top portion of this vessel requires painting.

*   Lettering and nameplate were checked and verified to be correct.

*   Documentation Review was completed and signed off for these vessels. A copy
    of the documentation package will be mailed to the site to the attention of

*   Prep to ship was performed in accordance with the requirements of the
    Preparation for Shipment Sheet approved for this vessel and was found to be
    satisfactory.

*   The extra set of gaskets were boxed and identified for the vessel, and is
    being shipped with the vessel.

*   The writer verified that the vessel internal was dry and clean and allowed
    the supplier to close the vessel for shipment.

*   The writer witnessed the loading of the vessel on two railcars, with no
    damage noted. However, Nozzle N3 which extends thru the skirt was found to
    extend past the shipping clearance envelope of 13'6", with the vessel
    oriented such that the tailing lug was in the upright vertical position.
    The vessel was rotated so that nozzle N3 was within the shipping clearance
    envelope. This placed the tailing lug approx. 8" away from the vertical
    upright position. The writer, along with Mr.   from Traffic
    and the supplier believe that this slight re-orientation will not alter the
    unloading sequence at the site. Also, 4 sets of ladder clips for Ladder
    #1 were found to be outside the shipping clearance envelope. Three of the
    sets are located on the skirt and one set on the vessel shell. The writer
    had the supplier to heat the clips just above the weld and bend the clips
    using as long a radius as possible to bring the clips within the shipping
    clearance envelope. The site will have to heat and straighten these clips
    after receipt.

This vessel was released for shipment on 07NOV92. An inspection release
Notification form was completed and signed by the Q.C. Manager and the writer
with notification provided to the project and the site. Mr.   from
Traffic will verify the final tie down on the railcars.

Surveillance to continue.
==
PREPARED BY: S.Q. Rep.   DATE: 07NOV92

REVIEWED BY: DATE:

DISTRIBUTION:
          SQA File

FIGURE 7.9 (continued).   Inspection report (Page 2 of 2 pages).

complex or expensive equipment, shipment by the vendor is not permitted without a release for shipment by the inspector or other authorized representative.

7.10 EXPEDITING

A similar but different function is that of **expediting.** The role of the expeditor is to ensure that the entire procurement and manufacturing cycle is the shortest possible. Expeditors will function both with regard to in-house and external activities. When expediting internal activities, the expeditors will work with the procurement and design personnel to ensure that the specifications and bids are received as scheduled and that they are reviewed, analyzed, and the order(s) placed promptly. They will prepare written reports on the status of the work on an order-by-order basis and in the case of critical items even on an item basis. An example of an internal expediting report is shown in Fig. 7.10.

The largest part of an expeditor's work is expediting the work of vendors. While some status information is available by telephone calls or facsimile messages, **visits** to the vendor facilities are essential to confirm the availability of shop space, receipt of materials, fabrication progress, testing, preparation for shipment, and so forth. Shortages of materials or labor in the shop are often critical, and this is reflected in the expeditors reports to permit changes in priorities if necessary. Often order entry is a problem, and a delay at this point affects delivery of the entire shipment, so the work of the expeditor starts from the very issuance of the PO. Because the expediting function is not related to inspection milestones, visits can be scheduled in a more routine fashion and reports issued in a more uniform way. An example of an expediting report is shown in Fig. 7.11. Note that each item on the order is tracked.

With the increasing importance of the development of drawings and data by the vendors, expeditors are becoming more involved in obtaining this information for the design office. As a result reports on the documentation called for by the order are increasingly found in expediting reports.

Expeditors can provide useful information regarding shop loadings, and these data when analyzed can indicate overall trends in various industrial facilities such as pressure vessel manufacturing, large valve manufacturers, and heavy transformers. From these data **lead time** reports can be developed. These reports are useful in planning and scheduling not only the procurement activities but the overall project schedule as well.

At the management level a summary expediting report on the status of the main materials for the project is usually prepared. A suitable cutoff level is established in terms of dollars or some other measure, and a periodic report is issued based on the data developed by the expeditors. For large projects running over several years, the report would be issued monthly; for projects with shorter schedules, weekly or biweekly issuance may be

```
--------------------------------------------------------------------------------------------------------
                                    SPECIFICATION SCHEDULE                                    PAGE   4
                                                                                             ISSUE  18
E.O. NO. COGT  1608                                                                   STATUS DATE 11/01/91

Organization - ELECT ENGRG                                                             RUN DATE   11/22/91
--------------------------------------------------------------------------------------------------------
   SPEC.      |--Latest----|                      |PCT| START | ISS SPEC | BID EVAL | AWARD    |
  Number      |REV | Date  |       Title          |CMP| SPEC  | INQ MEMO | & SELECT | P.O.     | COMMENTS / OPEN ITEMS
--------------|----|-------|----------------------|---|-------|----------|----------|----------|----------------------
EE-402        | 0  |07/23/90|345KV GIS SUBSTATION |100|04/02/90|04/06/90 |04/30/90 |05/07/90 |
TASK    :EE2300|   |        |                     |   |04/02/90A|04/06/90A|04/30/90A|05/07/90A|
--------------|----|-------|----------------------|---|-------|----------|----------|----------|
EE-403        | 1  |07/23/90|MAIN POWER           |100|03/22/90|03/28/90 |04/24/90 |05/07/90 |
TASK    :EE2100|   |        |TRANSFORMERS         |   |03/22/90A|03/28/90A|04/24/90A|05/07/90A|
--------------|----|-------|----------------------|---|-------|----------|----------|----------|
EE-405        | 1  |10/20/90|15 & 5 KV, 600V      |100|06/26/90|08/14/90 |10/12/90 |10/24/90 |
TASK    :EE3500|   |        |N. S. BUS DUCT       |   |06/26/90A|08/14/90A|10/12/90A|10/24/90A|
--------------|----|-------|----------------------|---|-------|----------|----------|----------|
EE-406        | 2  |10/01/90|AUX TRANSFORMER      |100|04/09/90|07/13/90 |10/01/90 |10/11/90 |
TASK    :EE2500|   |        |                     |   |04/09/90A|07/13/90A|10/01/90A|10/11/90A|
--------------|----|-------|----------------------|---|-------|----------|----------|----------|
EE-407        | 1  |10/19/90|4.16KV METAL CLAD    | 99|06/04/90|07/20/90 |10/22/90 |10/26/90 |
TASK    :EE3100|   |        |SWITCHGEAR           |   |06/04/90A|07/20/90A|10/22/90A|10/26/90A|
--------------|----|-------|----------------------|---|-------|----------|----------|----------|
EE-409        | 2  |12/11/90|480V SWITCHGEAR      |100|05/21/90|07/20/90 |10/30/90 |11/01/90 |
TASK    :EE3300|   |        |& TRANSFORMERS       |   |05/21/90A|07/20/90A|10/30/90A|11/01/90A|
--------------|----|-------|----------------------|---|-------|----------|----------|----------|
EE-410        | 2  |03/06/91|480V MOTOR CONTROL   |100|05/21/90|07/20/90 |12/07/90 |12/28/90 |
TASK    :EE2900|   |        |CENTERS              |   |05/21/90A|07/20/90A|12/19/90A|12/26/90A|
--------------|----|-------|----------------------|---|-------|----------|----------|----------|
EE-412        | 1  |01/31/91|BATTERY & CHARGER    | 98|08/06/90|10/18/90 |12/21/90 |01/02/91 |
TASK    :EE4100|   |        |                     |   |08/06/90A|10/18/90A|02/19/91A|03/12/91A|
--------------|----|-------|----------------------|---|-------|----------|----------|----------|
EE-413        | 1  |04/02/91|UPS                  |100|08/06/90|10/18/90 |12/21/90 |01/02/91 |
TASK    :EE3700|   |        |                     |   |08/06/90A|10/18/90A|04/03/91A|04/04/91A|
--------------|----|-------|----------------------|---|-------|----------|----------|----------|
EE-415        | 1  |12/28/90|ELECTRICAL EQUIP &   |100|07/23/90|12/28/90 | N/A     | N/A     |
TASK    :EE6100|   |        |INSTALLATION         |   |07/23/90A|12/28/90A|         |         |
--------------|----|-------|----------------------|---|-------|----------|----------|----------|
EE-416        | 0  |08/26/91|ELECTRICAL FREEZE    |100|02/12/91|03/05/91 | N/A     | N/A     |
TASK    :EE4900|   |        |PROTECTION           |   |08/20/91A|08/26/91A|         |         |
--------------|----|-------|----------------------|---|-------|----------|----------|----------|
EE-417        | 2  |05/06/91|DIESEL GENERATOR     | 93|06/11/90|08/14/90 |12/20/90 |01/08/91 |
TASK    :EE2700|   |        |                     |   |06/11/90A|08/14/90A|02/25/91A|03/14/91A|
--------------|----|-------|----------------------|---|-------|----------|----------|----------|
EE-419        | 1  |02/25/91|MEDIUM VOLTAGE       | 99|09/06/90|10/29/90 |12/14/90 |01/09/91 |
TASK    :EE4300|   |        |POWER CABLE          |   |09/06/90A|10/29/90A|03/12/91A|03/15/91A|
--------------|----|-------|----------------------|---|-------|----------|----------|----------|
EE-420        | 2  |05/13/91|600V POWER, CONTROL  |100|09/06/90|11/15/90 |12/21/90 |01/16/91 |
TASK    :EE4700|   |        |& INSTRUMENT CABLE   |   |09/06/90A|11/15/90A|05/16/91A|06/03/91A|
--------------|----|-------|----------------------|---|-------|----------|----------|----------|
EE-420        | 2  |05/13/91|600V POWER, CONTROL  |100| N/A   | N/A      | N/A     |06/06/91 |SPLIT AWARD (BULK P.O.)
TASK    :EE4750|   |        |& INSTRUMENT CABLE   |   |       |          |         |06/06/91A|
--------------|----|-------|----------------------|---|-------|----------|----------|----------|
EE-421        | 1  |05/15/91|DC SWITCHBOARD       | 96|08/06/90|12/14/90 |01/30/91 |02/06/91 |INQUIRY ISSUED AFTER
TASK    :EE3900|   |        |                     |   |08/06/90A|03/01/91A|05/21/91A|05/29/91A|BATTERY BID EVALUATION
--------------------------------------------------------------------------------------------------------
1st Date Line = Target ("N/A" indicates milestone not applicable)
2nd Date Line = Current Schedule / Actual (if an "A" suffix appears)
START SPEC =  10%, ISS SPEC INQ MEMO =  20%, BID EVAL & SELECT =  20%,
```

FIGURE 7.10 Expediting report (internal). Copyright © Ebasco Services Inc., 1992.

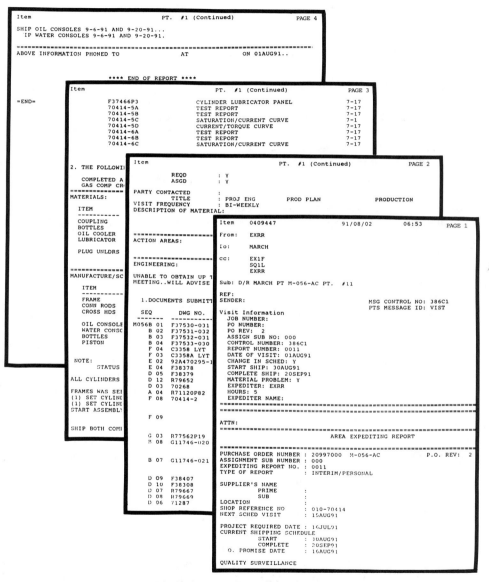

FIGURE 7.11 Expediting report.

necessary. With the computerization of these records, the actual preparation of the report takes less effort than formerly but depends on the ability of the expeditors to update the information prior to report issuance. An example of a portion of such a report is shown in Fig. 7.12.

7.11 TRAFFIC AND FREIGHT

For most projects it is not enough to merely order the materials, there must be some thought given to the methods of **transportation** to be used and a transportation plan developed for all major pieces of equipment. For domestic projects, rail, motor freight and in some cases barges or air freight are used to move the product to the point of use. Each has advantages and disadvantages shown in Table 7.3.

Where the items to be shipped are large or unusually heavy, it is wise to begin working with the freight carrier early in the design phase. It is often possible to design the materials and equipment to fit within normal size or weight limits and thus avoid significant extra freight costs and delays. On a recent pipeline project, for example, the designers rather than using only one size of pipe utilized three sizes of 36, 38, and 40 inches, which permitted the pipes to be nested within each other when shipped and saved about 60% of the transportation cost.

With the increased use of **containers,** loss by theft has been reduced and handling made significantly more convenient and rapid. The ports of the world, however, tend to fall into only one of two categories, those that are highly mechanized to handle containers or those where the cargo must be moved by the ship's tackle, which is slower and generally does not permit the use of containers. Unfortunately, only about 10% of the worlds ports are suitable for container handling, and thus when planning for a significant amount of freight movement, the capabilities of the port and alternate ports need to be evaluated. Where containers are used, if not owned by the shipper, they can be provided by the carrier. Since demurrage is charged for the containers after a few days of being unloaded from the ship, if customs clearances are slow or inland freight distances long, it is often more economic to own the containers even if they are only purchased for a single project and sold at its conclusion.

Most companies have standing **blanket** agreements with major carriers to take advantage of price discounts negotiated at the time of placement of the blanket order. These are often supplemented by project agreements which may require special provisions or modes of transport unique to the project and the location of the work. Table 7.4 lists some typical **freight** rates for the same commodity for the principal methods of shipment. The data may vary significantly depending on the rates negotiated with a carrier; however, the relative differences will remain fairly constant, thus the basis for selection of one mode over another will normally not change. In particular, note that the units of weight vary widely between the modes.

For international work any of these modes might be used, often in connection with each other. In this circumstance it is common to set up a single location within the host country where all goods are received. Since most shipments have significant weight, they are usually made by water. Thus the foreign location is often a port that has the advantage of good communications facilities and in-place weight-handling equipment. Centralization of all receiving activities allows the development of good working relations with the customs officials of the receiving country and avoids spreading the few expatriates into several locations. The single receiving location also permits the centralized management and control of the in-country traffic and freight function. It also becomes the point of transshipment to the point of use, for shipments can be broken down and regrouped to project requirements and the transportation mode used. Where shipments are made by air, they should be routed to the central receiving location if convenient.

Transshipment modes will depend upon distance, terrain, infrastructure, and so on. It is not unusual in less-developed nations to have to build roads, ports, or airfields at jobsites, particularly in remote areas. Water freight using barge or coastal steamer is common, and on projects in some remote areas passenger ships have even been leased and moored at the jobsite to house and feed the work force. Special **heavy lift ships** are available to move large and heavy pieces of equipment, but they are usually booked months or even years in advance. Hence long-range front-end planning may be necessary for their use.

With the potential for damage at each handling, the suppliers transportation plan should be reviewed and **transshipment** of material minimized. On a recent project a complex hydraulic assembly that included ASME Code piping was shipped across five states by truck-trailer. The material was transferred from one trailer to another several times. In doing so, the hydraulic equipment was slightly damaged, but more important, several code piping welds were cracked. As a result the vendor had to send a certified welder several thousand miles to repair the piping under its quality program and issue a weld repair certificate. All of the foregoing could have been avoided had the transportation plan of the vendor been reviewed and multiple transfers of the load avoided.

Each transportation mode has limitations. Typically **air** transport is the most restrictive as regards both weight and length. A project in northern Canada had a major building with columns 32 ft high but required that all materials be transported by air. When the columns arrived at the airfield to be loaded, it was found they were several feet too long to fit within the aircraft which had a 24-ft interior length limitation. As a result they were cut to length on the spot. Later when welded back together at the jobsite there was significant distortion and fit-up difficulty, requiring a large amount of extra field labor and a delay in erecting the steel for the structure. Had the project's engineers been aware of this limitation, they would have had the vendor manufacture the columns in lengths shorter than 24 ft and build splices into them to permit easier and quicker erection in the field.

JOB NUMBER:
RUN DATE: 09APR92
RUN TIME: 15:56
LAST RPT :

PAGE: 7
SORTED BY:
PO NUMBER

STATUS OF MAJOR MATERIALS & EQUIPMENT AS OF 09APR92
EXPEDITING OFFICE:
PO NUMBER :

DOCUMENT DESCRIPTION	SUPPLIER INFORMATION	EXPEDITING INFORMATION
PO NUMBER : 20710-210121 DESCRIPTION: LEVEL CONTROL VALVES	SUPPLIER :	PLACEMNT OFC :
REV/DATE : 0 03DEC91	SUPPLR ID :	EXPDTNG OFCS :
MR NUMBER : R14-033R	CITY/STATE :	EXPEDITOR :
	MFG ORIGIN :	BUYER :
SHIP TO :	CONTACT :	ENGINEER :
BILL TO :	TELEPHONE:	AREA EXPEDTR :
COMMODITY :	TELEX :	SUP QUAL REP :
	FAX :	

MILESTONES	SCHEDULE	FRCST/ACT	EVENT DESCRIPTION	WKS VAR	% EARNED
	03DEC91	03DEC91	PO AWARD TO VENDOR	0	
	31DEC91	20DEC91A	DESIGN DATA DUE	2	
	27JAN92	03FEB92A	DESIGN DATA APPROVED	-1	
	25FEB92	26FEB92A	FAB/ASSEMBLY START	0	
	17APR92	17APR92F	FINAL TEST / INSPECTION	0	
	24APR92	24APR92F	EXIT WORKS	0	
	05MAY92	05MAY92F	FIRST LIQ DAMAGE DATE	0	
	15APR92	15APR92F	REQUIRED JOBSITE	0	

DOCUMENTS CONTROL NUMBER	DESCRIPTION	PROJECT SCHEDULE	VAR	NO	QTY	REV	SUBMITTALS REQUIRD	PROMISED	VAR	RETURNED	CODE	CODE 1 COUNT	CODE 2 COUNT	CODE 3 COUNT	CODE 4 COUNT	CODE 5 COUNT
	GENERAL ARRANGEMENT DRWGS/DATA	20DEC91					31DEC91	20DEC91A	2	03FEB92A						
	WELD PROCEDURES	27FEB92						27FEB92A								
	ASSEMBLY DWGS	27FEB92						27FEB92A								
	OP/MAINT. MANUALS	20MAR92						20MAR92A								
	FINAL ASSEMBLY DWGS	30MAR92	-2					14APR92F								

244

REMARKS | | | DATE CHANGED

ALL MAJOR BUYOUT COMPONENTS ON-HAND. PACKING DUE 4/13. | 09APR92

REMARKS | | DATE CHANGED

REQUEST FOR SPEC DEVIATIONS TO ROCKWELL HARDNESS AND — 08APR92
YIELD STRENGTH REQUIREMENTS HAVE BEEN APPROVED. REQUEST — 08APR92
FOR SPEC DEVIATION TO BALANCED TRIM FOR ITEM NOS. 1 AND 2, IN LIEU OF — 08APR92
UNBALANCED HAS BEEN APPROVED. SPEC REVISION IN PROGRESS. — 08APR92 / 25MAR92

ADVISES THAT CONSIDERABLE IMPROVEMENT OF CURRENT — 09APR92
SCHEDULE IS UNLIKELY DUE TO P.O. REQUIREMENTS AND CURRENT SHOP LOAD, — 09APR92
BUT IS REVIEWING WAYS TO ADVANCE CURRENT SHIP DATES - WILL ADVISE — 09APR92
FEASIBILITY OF SCHEDULE IMPROVEMENT BY 4/10. — 09APR92

ACTION ITEMS	DATE REQUIRED	DAYS LEFT	RESPONSIBLE GROUP	ACTION REQUIRED	FOLLOW-UP INDIVIDUAL	PREVIOUS SCHEDULE	ORIGINAL SCHEDULE	TIMES RESCH
	17APR92	8		FINAL TEST / INSPECTION	SDP			

VALVES

ITEM NO.	MATERIAL ID.	PACKAGE NO.	QTY	UN	ITEM DESCRIPTION	START FAB	COMPLETE FAB	READY FOR INSPCTN	ORIGINAL PROMISE	EXIT WORKS	REQUIRD DELIVRY	FORECAST DELIVERY	WKS VAR	MRR
1	HD-LCV-11A2		1	EA	LEVEL CONTROL VALVE, 6", 900#	25FEB92A	17APR92F	17APR92F	24APR92	24APR92F	15APR92	01MAY92F	−2	
1	HD-LCV-11B2		1	EA	LEVEL CONTROL VALVE, 6", 900#	25FEB92A	17APR92F	17AP492F	24APR92	24APR29F	15APR92	01MAY92F	−2	
1	HD-LCV-7A2		1	EA	LEVEL CONTROL VALVE, 6", 900#	25FEB92A	17APR92F	17APR92F	24APR92	24APR92F	15APR92	01MAY92F	−2	
1	HD-LCV-7B2		1	EA	LEVEL CONTROL VALVE, 6", 900#	25FEB92A	17APR92F	17APR92F	24APR92	24APR92F	15APR92	01MAY92F	−2	
2	HD-LCV-10A2		1	EA	LEVEL CONTROL VALVE, 300#	25FEB92A	17APR92F	17APR92F	24APR92	24APR92F	15APR92	01MAY92F	−2	
2	HD-LCV-10B2		1	EA	LEVEL CONTROL VALVE, 300#	25FEB92A	17APR92F	17APR92F	24APR92	24APR92F	15APR92	01MAY92F	−2	
2	HD-LCV-8A2		1	EA	LEVEL CONTROL VALVE, 300#	25FEB92A	17APR92F	17APR92F	24APR92	24APR92F	15APR92	01MAY92F	−2	
2A	HD-LCV-8B2		1	EA	SAME AS 2 WITH 304L S.S. OUTLET NPLE	25FEB92A	17APR92F	17APR92F	24APR92	24APR92F	15APR92	01MAY92F	−2	
3	HD-LCV-9A2		1	EA	LEVEL CONTROL VALVE, 300#	25FEB92A	17APR92F	17APR92F	24APR92	24APR92F	15APR92	01MAY92F	−2	
3	HD-LCV-9B2		1	EA	LEVEL CONTROL VALVE, 300#	25FEB92A	17APR92F	17APR92F	24APR92	24APR92F	15APR92	01MAY92F	−2	

FIGURE 7.12 Status report of major equipment and materials.

245

TABLE 7.3 Transportation Modes	
Advantage	Disadvantage
Rail	
Low cost	Sensitive and perishable cargo at some
Dependable	risk
Particularly suitable for full carloads	Difficult to predict exact delivery
Can handle extremely heavy loads	Appropriate loading/unloading site
Unaffected by weather	necessary
	Some size limitations
Motor Freight	
Modest cost	Heavy loads may be troublesome
No special facilities required	Size limitations exist
Can deliver directly to point of use	
Flexible	
Useful for small shipments	
Reasonably rapid	
Barge/Ocean Freight	
Lowest cost	Limited to water access areas
Virtually no size or weight limitations	Ocean freight may have tonnage
Containers provide increased security	minimums
	Slow
Air Freight	
Very fast	Very high cost
	Limited to high value or critical
	cargo
	Both weight and size limited
	May be affected by weather
	Landing field/airport required

Where it is in the national interest, the U.S. government has permitted the charter of military aircraft. Since they offer larger size and heavier weight capacity than commercial aircraft, their use has been advantageous. Following the Korean war, for instance, a core for a large power transformer for a U.S. government-sponsored project was flown from eastern United States to Korea in an air force plane, the only one which could accommodate it, thus saving about six weeks of delay by ocean freight shipment.

Extremely heavy or wide loads also require suitable planning. For extremely **heavy** loads, planking of roads, reinforcement of bridges and culverts, and perhaps building segments of roads is necessary. All work must be coordinated with and reviewed and approved by the jurisdictional authority. This often requires several months. A similar problem arises when very large

TABLE 7.4	**Typical Freight Rates**			
Mode	Shipment Size	Routing	Rate	Insurance
Air freight	500 lb	San Francisco–Chicago	$1.20/lb	Add 1%
Motor freight	500 lb (min)	Same	$37.50/cwt	Included
Rail freight	75,000 lb (min)	Same	$5.50/cwt	Included
Water freight	40-ft container	San Francisco–Taipei	$270/long or measurement ton	Included

Note: For large, heavy, or hazardous materials, special rates apply.

prefabricated process plan modules—on the order of 300 tons, and perhaps 75 ft wide by 50 ft high by 60 ft long—were to be moved inland from a port to a jobsite. Not only do roads, bridges, and so forth, need to be strengthened but also telephone or power poles and lines raised or relocated. In addition sometimes buildings need to be moved or alternate roads built around towns to permit movement. One very useful way to work out the difficulties that may arise is to prepare a **dummy** load, including a three-dimensional envelope (template) of the load and move it to the site using the same routing and transport equipment. This permits not only the checking of the physical problems encountered but allows for working out the timing of the move, the effects on local traffic, and similar problems. This should take place several months before the actual move is scheduled to permit revision of the plan, added strengthening of culverts, and so forth.

Another type of transportation problem arises when cargo sensitive to **shock** must be moved. It is useful to fit out the material with accelerometers (usually in three planes) to indicate and record the g forces that the load receives. In some cases, for particularly delicate and costly items, shipments may be accompanied by couriers to ensure that there is no mishandling.

Hazardous cargo presents its own special problems in both loading and handling. Chemical, nuclear, and explosive cargos require special permits, special handling and storage provisions, and special routing and escorts. Their movement and routing need to be carefully coordinated with the public safety agencies at both the state and local level. For air or water freight shipment overseas, it may be necessary to load the carrier in a special area and to separate materials onto different vessels or aircraft; explosives and their detonators are usually handled this way. Because of the hazard these materials are usually the last to be loaded on the carrier and the first to be unloaded.

Rates for freight movement by rail, truck, air, and ship are virtually all negotiable. Due to international competition ocean freight rates are subject to negotiated reductions that can be significant. In addition the freight tariff for the commodity will vary depending on its classification, and considerable savings may be realized by careful classification of the material. While published tariffs for such movement still exist in part, these rates can usually be improved upon, often significantly if the quantity of material to be moved is of importance. On a large, approximately $300 million, road building project in the Mediterranean area recently, negotiations of more favorable ocean freight rates saved the project approximately $600,000. If a company does not have expertise in this type work, the use of a shipping agent is essential to obtaining the best rates and handling the large amount of administrative work resulting from a project, even of modest size.

For projects performed for the federal government, Government Bills of Lading are available. These authorize the payment of the transportation costs directly by the government at a significant savings. As a result it is not necessary for the supplier or the purchaser, if the purchasing is done on behalf of the government, to pay for the transportation, although these parties still have to arrange for the transportation.

7.12 FIELD PROCUREMENT

For all but the smallest projects a jobsite will find it necessary to **procure** some materials and supplies locally, to purchase small tools and certain bulk materials, perhaps by issuing releases against open orders as described earlier. For these reasons it is usual to establish a field procurement organization that will not only handle the procurement functions listed above but will also receive materials, warehouse or store them, and issue them to the using organization. In addition the **field procurement** organization will perform final acceptance of materials and approve vendor invoices for payment. The actual moving and storing of the material is performed by personnel from the field organization who work under the direction of the procurement personnel. Thus the procurement personnel may only comprise a handful of people who will direct the work, issue purchase orders, and maintain records and so forth. For small projects that do not justify such an organization, it is still necessary to specifically designate someone to be responsible for such activities.

7.13 RECEIVING

Material receiving is the last link in the chain where problems in material fabrication can be detected and readily corrected. Materials received at jobsites should be given a suitable receiving inspection, including the issuing of a receiving report. The material inspection normally includes two aspects, the commercial one for quantity, shipping documents, and so forth, and the technical one to be sure the material is that ordered and that it conforms to the requirements of the PO. To perform this inspection, the procurement personnel include a field engineer. A satisfactory receiving report forms the basis for the payment of the final invoice or retention unless there are extended or special warranty provisions.

Where the material is conforming and no problems exist, the material can be warehoused or stored until ready to be issued for use. Where the material is nonconforming or damaged, it will be quarantined and claims against the freight carrier, backcharges to the vendor, or negotiations with the vendor to correct the problem instituted. A typical receiving report is shown in Fig. 7.13.

To keep track of the location of materials and to maintain an accurate inventory, some form of **computerized** data base system is often used. In such a system the materials are coded by type and often PO number, and a location coding system is utilized. As withdrawals are made from the inventory, it is possible to maintain an up-to-date status of the material by type and location, and in the case of bulk materials with preestablished reorder levels, automatically prepare POs for signature and issuance to the vendor. The data base permits sorting by type of material, PO number, vendor, and other similar characteristics so that it can be accessed for additional requirements.

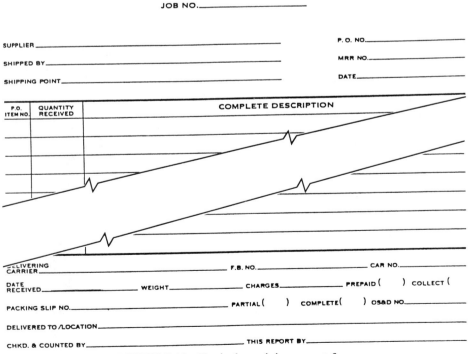

FIGURE 7.13 Typical receiving report form.

Where materials are excessive, damaged, or missing an **OS&D** (over, short, and damaged) report should be issued. If materials are obviously damaged, as with a crate with the sides caved in or severe water damage to packing, it is essential to open the packing to ensure that damage exists. Where there is no damage, it is still useful to open the packing and at least count the items within to check the count is as stated on the packing list. This also provides an opportunity to confirm the description of the material against the packing list and eventually the order.

Where materials are sensitive or delicate, the proper packing of these materials by the vendor is critical, and particular pains should be taken to ensure that the bid request includes the requirement for proper packing. In some cases it may not be wise to open the packing on receipt because this could lead to damage or perhaps void a warranty. For these cases the materials should be left in the packing and when later issued for use, their condition noted. A typical OS&D report form is shown in Fig. 7.14.

Certain materials and equipment such as instruments are too **delicate** to be subject to dew-point changes or swings in temperatures in uncontrolled storage and are normally warehoused in separate areas that have temperature and sometimes humidity control. As a result it is wise to provide some portion of the warehousing facilities with suitable environmental controls for these type of materials. Some equipment that is sensitive to dew-point changes may be equipped with internal **heaters** to permit storage in nonenvironmentaly controlled spaces. Then provisions must be made for electrical power distribution to the storage location. A similar problem arises with equipment shipped with gaseous **nitrogen blankets** to prevent the admission of ambient air. While temperature control is not a problem, the maintenance of the nitrogen blanket is essential, so special provisions must be made for the monitoring of nitrogen pressure.

The use of **bar coding** to identify and permit automated data entry of materials location and other data into the storage data base is becoming more widespread. With this system, as each shipment is received, a bar code tag is applied to it. The code then is read by an optical scanner to feed the significant data into the data base, with a location code being noted as well. With suitable bar coding the vendor, type of material, size, special handling or storage requirements, reorder points, and other important information can be entered, and the data base is sorted for exception reports and other management uses.

7.14 ISSUANCE

Field organizations require a formal system of paperwork to control material **issuance.** Material can be diverted or carelessly handled, and thus possibly misapplied, if there is not a rigorous system to control its issuance and use. This is particularly important if there are formal quality programs requiring

REPORT OF UNSATISFACTORY-OVER-SHORT-AND DAMAGED MATERIAL

SUPPLIER			JOB NO.		OS & D NO.	
ADDRESS			P.O. NO.		MRR. NO.	
CITY & STATE			RECEIVED AT			
CARRIERS/ROUTING					POINT OF ORIGIN	
DELIVERING CARRIERS F/B NO.			CAR/VESSEL/TRUCK/TRAILER/NO.			
ORIGINAL B/L NO.		SHIPPING ORDER NO.			SUPPLIERS ORDER NO.	
WEIGHT					JOBSITE ARRIVAL DATE	
GROSS	TARE		NET			

QUANTITY			COMPLETE DESCRIPTION
ORDERED	SHIPPED	RECEIVED	

NATURE AND EXTENT OF OVERAGE-LOSS-DAMAGE OR UNSATISFACTORY CONDITION OF MATERIAL AND REMARKS:

TYPE OF CONTAINER	CONTAINER FILLED TO CAPACITY?		TYPE OF PACKING USED	
EXCEPTION NOTED ON FREIGHT BILL	CARRIER NOTIFIED	DATE		
WAS DAMAGE CONCEALED?			BEFORE ☐ AFTER ☐ UNLOADING ☐	
INSPECTED FOR CARRIER BY:	DATE		CARRIER INSPECTION REPORT NO.	

ATTACHMENTS:
CARRIERS INSPECTION REPORT ☐   PHOTOGRAPHS ☐   CARRIERS F/B ☐   ORIGINAL B/L ☐

THIS REPORT IS FOR MATERIAL: OVER ☐   SHORT ☐   DAMAGED ☐   UNSATISFACTORY ☐
INSPECTING FIELD SUPT./ENGR.   DISPOSITION RECOMMENDED

I CERTIFY THE ABOVE REPORT TO BE TRUE AND IN ACCORDANCE WITH THE CONDITION OF THE GOODS UPON RECEIPT.

FIGURE 7.14 Typical OS&D report form.

that the status and condition of the material be continually controlled and that the final act of transfer to the using organization document that the proper material was issued.

In the manufacturing sector material issuance, movement, and control effort is considered "non-value-added" work. That is, in itself it adds no improvement in value of the product, and the customer has no interest or concern how it is handled. As a result most organizations are trying to organize material flow and staging to require as little effort as possible. This requires the maximum use of open stock at the point of use, with the virtual elimination of material transactions—namely the use of requisitions and similar paper to authorize the movement or release of material. In its place the local inventory is controlled by production demand using a bill-of-material or similar form as the operation is completed and the work moved ahead ("kanban"). This places the control of inventory in the hands of the production personnel and drastically alters the role of the material organization. The net result of these systems is to reduce the quantities of material required, lower the cost of material handling, decrease space requirements, and reduce the frequency of shortages.

7.15 REORDER POINTS

Earlier it was noted that automatic **reorder** points can be readily included in a suitable computerized storage/inventory data base system. For manufacturing operations such order points may be essential in maintaining operations. This is normally a part of the **Material Requirements Planning (MRP)** system for a typical manufacturing facility, in which a computerized system is used for tracking withdrawals and automatically issuing reorder documents to the supplier(s). It becomes of increased importance if the firm is using **JIT (Just-In-Time)** inventory techniques for reducing the quantities of materials and in-process operations.

7.16 CONTROL DOCUMENTS

To control the issuance of requisitions, a Requisition Register is maintained. This provides a running listing of the requisitions issued and their status. As mentioned earlier, when requisitions are canceled, their number should not be reused, and the register must carry a suitable entry.

Similar to the Requisition Register, a Purchase Order Register is used to maintain control over the issuance of POs, which may include the values committed.

Frequently a separate Commitment Register with more extensive data, such as timing of major payments and applicable cost codes, is used to keep track of the funds committed and to provide a measure of procurement progress on a dollar basis. The register is useful not only to the procurement personnel but also to the estimating personnel who will use its information for progress reports and for forecasting cash flows and preparing similar financial estimates.

With the widespread use of computerized systems to maintain the different procurement registers, it is common to have the data found in the several control documents in a single data base from which the various reports can be obtained.

7.17 PARTNERING IN MANUFACTURING

Of increasing interest in the manufacturing industries is the concept of **partnering** between the purchaser and supplier. In this arrangement the purchaser is able to establish a long-term relationship with a supplier who can provide the desired material at a fair price, on the desired schedule, and, most important, of the required or higher quality. Sometimes also called a **strategic alliance,** the advantages of partnering also include the development of long-term sources of product and the opportunity to actively improve the quality of the product supplied and eventually to reduce or eliminate receiving inspection of the product. Some firms call this practice ''supplier-to-line.''

In the supplier-to-line concept the supplier becomes an extension of the purchaser, and resources are often shared between the two. Thus the purchaser may provide personnel on a loan basis to the supplier to solve production or quality problems, or the supplier may work with the purchaser's personnel to revise product design to facilitate production. Even cost data may be shared to determine where operations can be performed for optimum cost. Over time, while the two firms retain their identity, on an operational basis the distinction between them blurs, and much of the paperwork for production releases, changes to product design details, and so forth, are handled less formally.

With long-term contracts the supplier is able to justify the investment in improved personnel training, plant, and equipment to produce the product, and the purchaser can avoid the need to periodically repeat the bidding, evaluation, and award cycle.

The difficulty with partnering is that the price paid for the product may not always be the lowest and that some monitoring of cost needs to be undertaken periodically to ensure that the pricing is reasonable. Normally this is a minor concern and easily resolved. The benefits of partnering are so substantial that it is widely growing in acceptance.

Most firms only begin partnering discussions with suppliers after some experience with them and then proceed in a stepwise fashion until their reliability is demonstrated. Because of the far-reaching nature of the partnering agreements, they are usually complex and lengthy and worked out in great detail, since it is necessary to protect the interests of both parties. For example, partnering agreements typically include the requirement for usage of common cost codes and scheduling systems to facilitate electronic interchange of information.

In the engineering and construction industries and other areas in the private sector, partnering is increasingly being used, while for work in the public sector it is much less frequently used because of legal prohibitions and the requirement for competitive bidding.

7.18 GOVERNMENTAL PROCUREMENT

Governmental procurement takes a variety of forms depending on whether the entity is at the city, county, state, or federal level. With the sophistication and structure of the federal regulations and practices, they have become the model for most other public sector work. Because of this the requirements of federal procurement, in particular by the Department of Defense, are widely applied. The Defense Department procurement activities are governed by the federal procurement regulations. These are voluminous and highly structured. The regulations provide for procurement by a variety of contract types for virtually any form of goods and services. When receiving an RFP from a governmental agency, it is prudent to carefully review their contractural requirements prior to bidding.

CHAPTER 8

PROJECT MANAGEMENT

Because of scope and contractural differences the role of the Project Manager will vary from industry to industry, company to company, and even from project to project within a company. Despite these variances there are a family of activities for which the Project Manager is normally either directly responsible or has a leading role in implementing. This chapter will describe the PM's common responsibilities and attempt to answer the questions: "Who is the PM?" "What does he or she do?" "How is it done?"

8.1 PROJECT MANAGER'S RESPONSIBILITIES

Table 8.1 lists the PM's responsibilities in the groupings most frequently used. The PM is responsible for establishing the project **objectives,** the project **plan,** and the project **budget** and **schedule.** The PM provides the necessary direction to the project team to ensure satisfactory performance by the functional departments through all the phases of the project including closeout and final report preparation. By the approach taken and dedication to the objectives of the project, the PM leads the project team by example.

Being in overall charge favors a person with a background in the technology of the project—chemical plant engineering, banking, metal fabrication, and so forth. The requirement, however, of providing overall direction and balance to the work tends to also favor someone who is more of a **generalist** than a specialist. A large part of the duties of the PM is to set priorities and to choose between conflicting requirements for the same resources—people, time, or dollars. As a result the PM should maintain a balanced impartial

TABLE 8.1   Project Manager's Responsibilities
Major Responsibilities
Overall performance of the project team, including design, procurement, construction/manufacturing, test and operations
Contract administration
Client relations
Cost control and financial management
Schedule performance
Quality assurance and control
General project administration
Change control
Safety
Secondary Responsibilities
Environmental compliance
Licensing and permits
Public relations
Security
Other Responsibilities
Labor relations
Housing
Camp operations
Automotive equipment

position vis-à-vis the team members to establish mutually agreed goals and to ensure execution of the project plan.

The selection of the PM is always a **subjective** one. While it is desirable for the PM to have performed the work before, it is not essential, and many first-time PMs have performed very well on difficult assignments. Ideally the new PM is started on smaller less complex projects that have low risk and less costly potential errors. From these, he or she progresses to larger, more complex work. While the selection of a PM normally favors someone with a technical background, the more important question is the **overall strength** and **balance** of the project team. A new or weak PM can be effective if the team is strong and experienced and if sufficient coaching and oversight is provided. For work in isolated areas or with a high public interest, more experienced and skilled PMs are required.

Since the PM is required to make decisions in technical areas, the normal pool from which the PMs emerge is the technical groups on the project. Once in the role of the PM, however, the PM will function as more of a generalist, with a balance required between the technical and operational requirements of the project. Because of this, and as mentioned above, it is possible for successful PMs to come from backgrounds other than the technical project groups. Problem-solving ability, interpersonal skills, decisiveness, and a commitment to the project goals are all essential **attributes** in the PM. Other characteristics that are useful are energy, stamina, resourcefulness, and flexibility. One characteristic that can often create difficulties in the performance of a PM is too large an ego. As part of the management process, it is essential that the client and individual project personnel be given a great deal of credit for the work. If the broad goals of the project are kept in mind and the long view maintained, the PM will be judged by the overall performance of the project, and there is no need for self-promoting actions or statements.

From the foregoing it can be seen that the attributes of successful project managers includes a wide spectrum of personality traits, technical and management skills, and attitudes toward the work. As a result, while there is no simple formula to select personnel who will be successful managers, candidates having a reasonable mix of the characteristics described above stand a good chance of being effective and successful.

The PM is usually defined as the highest ranking person associated with the work who has **overall responsibility** for execution of the work. Very often he or she is the most senior person working on the project on a full-time basis. By definition, the PM is responsible for the overall direction of all the activities of the project including engineering, procurement, construction (or manufacturing), financial administration, start-up, test, and initial operation. Unless the project involves sales or product development, PMs are not normally included in these responsibilities.

On management versus "direct hire" projects, the organizational factors will be somewhat different, including the responsibility of the several entities involved, but the role of the PM will be essentially the same.

Normally each department is responsible for the quality and technical execution of its work within the schedule and budget established for the function. The PM assumes overall responsibility for the **integrated performance** of the project team, with primary responsibility for the performance of the departments remaining with their functional managers (or management). The PM is also the focal point contact in dealings with the client and thus represents the organization to the client and the client to his own organization. As noted above, the PM's primary emphasis on a project relates to matters of project direction with a strong emphasis on schedule and cost performance.

The functions of engineering, construction, manufacturing, for example, are normally stand-alone functions that can be considered line activities of the company. The functions of cost and schedule, however, are really staff functions, much like an accounting function in a manufacturing company. Since cost and schedule functions do not in themselves produce a final product, they are primarily monitoring and control activities. In a typical project organization they become the extension of the PM and provide the most important **management tools** that the PM uses to run the project to monitor, detect trends, and predict directions and results. They also provide a way in which the PM can quantify the performance of the "line" groups. Through budget allocations and the establishment of milestone dates, the PM can control the performance of these groups to achieve the project goals.

It is usual for organizations to have engineering, procurement, and manufacturing or construction departments of long standing with extensive capability. The role and function of the PM is a newer one and does not normally intrude upon the internal operations of these established groups. Rather, the PM's responsibility is focused on the performance of the project and interactions related to the execution of the project by the aforementioned groups.

The cost group usually reports directly to the PM. The PM then is responsible for both the budget and the control of the expenditures of the project. In effect the PM becomes the senior manager for the cost group. These responsibilities involve the PM more deeply in the day-to-day work of preparing costs and schedules than in engineering, procurement, and manufacturing or construction work.

Through the cost and schedule groups, the PM can measure, in a **quantified** way, the performance of the various project groups and their effect on the project overall. The overall performance is measured by comparison of the completion of items against the schedule milestones previously established and by measuring expenditures against cash flow, revenue, and forecast estimates. Other measurements of cost and schedule may be in the form of days of schedule time, dollars committed or expended, forecasts of these values, and so on. Quantification is the means by which the PM can establish both near and long-term measurable goals for the project groups.

An important aspect of establishing project goals is the participation of the groups themselves. For example, if the PM is trying to set a date for issuance of drawings to a vendor, the engineering department should be asked for its best estimate of the date and the number of drawings to be issued and its commitment to the date or a negotiated date obtained. At this point the schedule for the item becomes truly engineering's, and not some abstraction. By participating, the group will feel more responsible for the item's fulfillment. The PM can make periodic reviews of the work in progress during the preparation period to ensure that the schedule will be met and that the drawings, when issued, will contain sufficient information to permit the vendor to proceed. This last point is important. Often groups are tempted to put out incomplete work so that they can get credit for meeting a milestone date (e.g., 38 drawings issued on June 14th).

One way to prevent the problem is to schedule the work in smaller increments. Thus, for the preceding example, if the PM has the schedule broken down into, say, four groups of six drawings and two of seven (or some other logical work package grouping), each group of drawings can be more easily tracked. This realistic approach also seems to reduce sensitivity to criticism since the problem is treated more **objectively.** But, more important, it has the advantage of permitting earlier identification of problem areas and thus improving the chance to correct them by reallocating personnel or making schedule adjustments.

The principal factor affecting the PM on **management**-type assignments, as distinct from direct hire work, is that the operating departments may only be indirectly under his or her control and may not be from the same organization and thus be unfamiliar. While the relationships of the groups in this circumstance are somewhat different, the role and responsibilities of the PM are essentially unchanged. Due to the less direct authority available to the PM, more vigilance and care must be exerted, for the capabilities and practices of the groups may be significantly different than expected. Often there are private long-standing relationships about which the PM may know nothing, complicating the political environment in which he or she must operate. This may be further exacerbated by the fact that after the project is completed, the PM will be gone but the different groups will likely continue to do business together, and there is often a feeling of temporizing on their part. There will then sometimes be a feeling of reluctance on the part of some to follow the direction of the PM. The usual effect of this is to frustrate the PM and increase the difficulty of the job. Recognizing this as the work proceeds can be helpful in maintaining a proper perspective, including the role of each group. By careful handling it is sometimes possible to utilize the long-term interests of these groups to assist in the conduct of the work.

In recent years with increased competitive pressures and with vastly more complex external factors impinging on the project, the role of the PM has been **broadened** and become more **complex.** As a result the PM often spends more time on external factors such as licensing and financial management than on internal ones. With a seasoned project team this is not a big problem, but where the project team is less experienced, Assistant Project Managers may be required to supervise one or more of the operational functions to compensate for the reduced involvement of the PM.

Contract administration is perhaps the most critical area of the PM's responsibility. If possible, the PM should be involved in the original contract negotiations, including the development of the principal contract provisions. The PM should also be informed of the discussions and negotiations that led to each provision. Only with complete background knowledge such as this will the PM be able to effectively administer the contract.

It must be constantly borne in mind that the contract is a binding legal agreement between the parties. Regardless of other factors, the overall performance of the project team and the PM will be judged by contract **compliance.** Even if the project is superbly controlled and is on time and within budget, if the contract is not properly administered, major problems can and likely will arise. The PM rarely is a lawyer but some knowledge of general legal principles is helpful. A survey course in business or contract law is very useful to PMs. The PM also needs access to suitable legal support and advice. Larger firms may have legal staffs, while smaller ones typically use outside legal firms for this purpose.

In recent years there has been a trend to write contracts in less formal language, making them easier to read and understand. Despite this, certain provisions, such as liability and default are still written in a legal style and thus subject to legal interpretation. The PM should avoid falling into the trap of attempting to perform his or her own legal interpretations. Contract interpretation is a critical matter that can lead to complex legal disputes.

Contract administration will often take up a large part of the PM's time. The PM should recognize that the efficient and effective administration of the contract is a primary duty. Where significant problems arise it is prudent to check with legal counsel on interpretations, precedents, and so on, to ensure that there is a clear understanding of the background of the requirement and its implication. The administration of the contract can affect such major concerns as scope and schedule and such minor ones as the timing of the submission of automobile insurance certificates. The PM will be involved in a full range of concerns that depend on an intimate knowledge of the contract.

The contract exists to expedite the work and to establish a clear understanding between parties. **Strained interpretations** or extensive arguments over minor points can destroy the basic purpose of the agreement and poison the working atmosphere both for immediate and future activities. In working with the client, the larger purpose of the project should always be kept in mind. There is nothing more infuriating to a client than, for example, to be presented with a claim in which many items are clearly not proper and which are included merely for negotiating advantage. Similarly, to inundate a client with trivia in order to mask other activities is neither ethical nor wise.

Where the contract provides for actions on the part of the client, it is fitting and proper that these issues be raised if these responsibilities are not being discharged. A PM gains nothing by avoiding what may be difficult issues. Sooner or later all issues must be resolved. The best approach is to present the concerns affirmatively with the appropriate degree of tact while maintaining a professional and objective position.

A fair, consistent, and equitable interpretation of contract provisions goes a long way to maintaining goodwill between parties. The benefit of this becomes obvious when there is a request to change or negotiate new contract provisions because of changed circumstances. Ideally, if the contract administration can result in a "win-win" situation, both parties can feel they have been fairly treated.

Client relations are considered by some to be almost the most important area of activity for the PM. Regardless of the outcome of the physical work, if the relations with the client have become severely strained or broken down, much of the work of the project team will have been wasted. In such an environment, no matter what has been done correctly or positively, the overall impression left with the client (and the client's organization) will be unsatisfactory, and the likelihood of future work will be impaired. In addition there may be severe and long lasting damage to the reputation of the firm and potential clients may learn of the dissatisfaction and avoid working with the organization.

Thus, in developing and maintaining favorable relations with the client, the PM must make an effort to understand the client's needs and instill in the client an understanding of the needs and confidence in his or her organization. Paramount to this is developing an appreciation that the two organizations are allies and that what is being done is for the benefit of both. As mentioned earlier, a successful project benefits both parties, whereas if only one party benefits, "wins," the project cannot be considered successful, for in the long run both parties lose. The closer the relationship between the parties, the smoother the work will proceed, and the easier it will be to resolve areas of differences.

One of the more sensitive areas is that of **cost control** and financial management. In theory cost control should proceed automatically if the functional departments or project groups, including the cost estimating and cost control personnel, are performing correctly. In reality, however, there are always outside factors that alter the cost control environment. Typical of these are unanticipated cost overruns due to lower worker productivity. This may be due to scope changes, a poor initial estimate of productivity, changed working conditions, or unforseen conditions such as the presence of underground water when performing an excavation. In addition cost or rate-of-expenditure limitations sometimes exist, and they affect the rate of work or commitments that can be made. The cost control requirement is thus often more complex than merely reviewing progress and costs to date and forecasting further requirements. The factors cited above create a more dynamic situation and require, not only the development of accurate historical data, but forecasts and projections that take into account such external circumstances.

When performing work in the United States, letters of credit are rarely used, and a dedicated fund may not be available for the work. When this is the case, it may be necessary to adjust the work rate to suit the cash flow available to support the work. This is often the case with major new projects where the parent company (client) itself is financing the new expansion. Where full financing comes from other sources, this is a problem only if the scope expands or if productivity problems or unforseen conditions arise.

For reimbursable work, if the client is kept fully informed of the progress and problems, it reduces the surprise of added funding, should that become necessary. Financial solutions can then be developed with full knowledge and understanding of the client, and thus the timing for resolution of the problem and its concurrent uncertainties can be significantly reduced. Often financial problems can be resolved by extending the schedule, with the expectation that the anticipated additional costs can be covered by later funding or future cash flows.

For fixed price work cost control is just as essential as for other types of work. Maintaining cash flow requirements within forecast values holds down overall project costs, since the interest charges to use lines of credit can be high. Thus a rigorous program of cost control will pay large financial dividends. It is always wise then to maintain an overview of the current status of the line of credit, funds committed, and funds potentially needed to avoid the problem of cash needs in excess of the line of credit. If no advance action is taken, and an increase in the line of credit is necessary on short notice, there is no negotiating advantage left to the borrower, and the accompanying charges and interest rates are often much higher than they need be.

Payments are always a source of concern. The sole purpose of performing the work is to generate payments and to perform the work at a profit. As such, it is essential that the payments for the work be made promptly and in accordance with any contractural requirements. Requests for payments or billing must be **accurate** and **timely.** Since billings are rarely prepared directly by the PM's staff, it is wise to have them checked before presentation to the client. After checking and correction of errors, the billings should be presented to the client by the PM who can answer any initial questions and facilitate the preliminary approval of them by the client. Once billings have been rendered, there must be an orderly program for follow up and resolution of additional client questions until payment has been received. On a large project the responsibility for this follow-up should be given to an individual who reports directly to the PM. This may take as much as half of the time of this person.

Where billings are complex, some **disputed** items may be present. When this happens it is useful to ask that payment be made, setting aside the disputed items for the next billing in which they may be deleted if they are not valid. Such a practice will both maintain a steady cash flow from the project and permit the client to feel not overwhelmed by billings to which it may not have fully agreed.

For fixed price work the billing problem is usually much simpler, although where payment is based upon quantities of work completed there may be differences in evaluation of the **quantities.** In this circumstance it is useful, even if the contract does not provide for it, to establish either a client observer to participate in the review of the quantity measurement(s) or, in extreme cases, to set up a joint group to perform or certify the measurements.

As previously mentioned, **schedule** performance is a primary responsibility of the PM. Poor schedule performance will affect the completed cost of the project because of the cost of added labor and supervision, escalation in the cost of the materials, and possible loss of an equipment production slot with schedule delay. In some cases the client may have to fulfill delivery contracts for a facility product. Recently a petroleum refiner built a facility to produce liquified natural gas (LNG). The refinery completion was delayed about six months because of a materials problem in the construction of the facility. As a result of the delay the owner had to purchase LNG on the open market to honor contracts based on the original completion date of the facility. It is interesting to note that while the cost to make the repairs to the plant was on the order of half a million dollars, the cost to the owner to purchase LNG to fulfill previous supply contracts approached fifty million dollars, indicating the enormous financial impact of this delay. While all delays do not have such a dramatic financial effect, they all affect the financial and operational well-being of the company. Even if no obvious financial problem results, the lost opportunity cost of the funds tied up or the unavailability of the staff to pursue other profitable work can be significant.

Last, and perhaps most important, poor schedule performance tends to damage the **reputation** of the firm, thus significantly weakening the potential for future work. Most clients will object to cost overruns on reimbursable-type work or if there are major claims on fixed price work. They will particularly be damaged if the project is completed significantly later than planned. The old axiom that "time is money" is never truer than in project work. While cost increases may in time be forgotten, projects that experienced major delays tend to remain in everyone's memories and can never seem to be put to rest. Of course the underlying reason is that while money can be recovered by the facility or company, time once lost can never be regained.

While each functional department is responsible for the quality of its work, the PM must be conscious of the integration of the work and responsible for the **quality** of the project. The term "quality" as used here refers to the quality of the project in all its aspects and not a narrow view of quality in terms of the number of defects of a particular type, and so on. This quality concern begins with the relationship with the owner and runs through the entire project and all its activities, both internal and external. It can also be considered a concern for **excellence.**

While measurement of the level of quality achieved is difficult because it can be subjective, many areas and activities can be measured and thus evaluated and corrected if difficulties exist. Normally the day-by-day measurement, evaluation, and correction is left to each functional group, since these groups are in control of their own work and are responsible for it. A quality assurance or quality control group reports directly to the PM to provide independent information on the work. This group conducts **audits** and **reviews** of the work of the functional departments, determines deviations from standards, and assists the departments in taking corrective action to revise their systems or procedures to prevent recurrence. The audits take two forms: system or procedure audits and technical audits. While both are necessary and useful, if the organization has reasonably well-developed procedures, the technical audits tend to be more useful in pinpointing problem areas. If the personnel in the quality group are primarily quality systems oriented personnel, as is often the case, they have sufficient expertise to perform the system audits by themselves, though for the technical audit work they need to be augmented by technical personnel. Usually they are personnel not assigned to the project but sufficiently knowledgeable to understand the technical activities underway. In establishing the quality group and in reviewing and evaluating their work, the PM must constantly keep in mind that the purpose for their activity is to help the functional groups and not to act as police. It is very easy to fall into the trap of counting the number of this or that without recognizing the impact of the item. The basic purpose is to assist the project groups in their performance to meet the schedule and cost goals established. Thus the type and seriousness of deviations found and corrective action taken must be viewed from that perspective.

Another major area of responsibility often assigned to the PM is overall administration of the project. This relates to the **business** side of running things. Someone needs to have the broad overview to control what is being done and determine whether there are improvements or changes needed. These are the tasks that facilitate the work and without which the work might not get done. Many of the PM's tasks are housekeeping functions that, though mundane, are vital to being able to operate a business. They can range from simple tasks such as office furniture procurement to complex ones such as establishing secure communications for highly classified projects. For an international project, as described in Chapter 9, the PM's tasks become even more important. The preparation of a **procedures manual** is a good example of this type of activity. Every project, whatever its size, requires some documentation that defines who does what and how it is to be done. A procedures manual tailored to the specific project, including the client, contract, jobsite, local conditions, and governmental requirements, is usually prepared for all projects of significant size and provides a framework for the overall operations of the project team. An example of the contents of a project procedures manual for an engineering/procurement/construction project is given in Table 8.2.

Change control is an area of major interest and responsibility for the PM. As mentioned earlier, change control ensures that the project proceeds in the planned manner. Without it, focus and direction is lost, and excessive overruns in both time and cost can result. Change control should be maintained internally by departments as a part of their ongoing work, with periodic reports to apprise the PM of the progress of their work. With large or complex projects a project change control board is often established. This board reviews and approves any change prior to implementation. The PM of course attends, and may even chair, the meetings and demonstrates the importance of change control and impresses upon the operating departments that they bear the responsibility for the first level of change control.

With increased emphasis upon **safety,** the PM often is called upon to assume responsibility for this area as well. Safety issues arise not only from obvious humanitarian and schedule considerations, but also from the cost of Workman's Compensation Insurance premiums. Since these premiums are set based on the claims history of the company, there is a major cost savings if these premium rates can be kept low. The difference between two contractors doing the same work can be as much as $5 or more per hour of field labor expended and thus can represent a very large savings to the company. In a very real sense a dollar spent on safety equipment, toolbox meetings, the salary of a safety engineer, and similar measures is returned many times over by the savings in insurance.

TABLE 8.2   Project Procedure Manual Table of Contents

1. General
   1.1 Key considerations
   1.2 Scope
   1.3 (Organization) responsibilities and organization
   1.4 Contract
   1.5 Communications
   1.6 Overall project reporting
   1.7 Management controls
   1.8 Government and community relations
   1.9 Staffing
   1.10 Legal

2. Project Control
   2.1 Organization
   2.2 Centers of operation
   2.3 System implementation
   2.4 Frequency of reports and program outputs
   2.5 Cost control programs
   2.6 Schedule control programs

3. Engineering
   3.1 Process design and coordination
   3.2 Engineering scope
   3.3 Engineering organization
   3.4 Functional engineering group responsibilities
   3.5 Quality assurance
   3.6 Jobsite office
   3.7 Preassembly/subassembly approach
   3.8 Construction/preassembly support
   3.9 Compliance with governmental regulations
   3.10 Engineering contracts to local firms
   3.11 Interfaces with local infrastructure

4. Procurement
   4.1 Scope of work
   4.2 Organization
   4.3 Execution plan
   4.4 Logistic plan

5. Construction
   5.1 Purpose and objective
   5.2 Home office construction management
   5.3 Field organization
   5.4 Preassembly/subassembly
   5.5 Industrial relations
   5.6 Subcontract plan
   5.7 Quality assurance
   5.8 Field controls
   5.9 Field procurement
   5.10 Housing
   5.11 Rigging

TABLE 8.2 *(Continued)*
5.12 Safety
5.13 Security
5.14 Mechanical completion
5.15 Precommissioning and startup
6. Quality Assurance Program
6.1 Program management
6.2 Program description

8.2 SECONDARY RESPONSIBILITIES

There are several secondary responsibilities usually assigned to the PM that have a lower level of activity but can nevertheless affect the project, if not properly discharged.

Environmental compliance and information is an area of increasing importance that can substantially affect the project. We all know of examples where misunderstandings or ignorance have led to major public issues and violations of regulatory statutes. The PM will usually assume direct responsibility for and control of these activities and will assemble a staff to deal with them. As with the other PM responsibilities, one person should be designated for this responsibility and given sufficient resources to not only deal with problems when they arise but to monitor developments in these areas and take preventive action to avoid major difficulties. Of particular importance for large projects with significant public impact or interest is the establishment and implementation of a suitable program of public information, including, where appropriate, project tours, public speakers, guest appearances on radio and television programs, articles, interviews, and so on.

The PM is usually responsible for obtaining and maintaining the necessary **permits** and **licenses** for the project. As the number and complexity of these increase, more attention is required, and the PM will typically assign one or more staff members to work on permitting. As with environmental matters, relationships with the general public and the government agencies that issue permits are critical. This applies not only to identifying and resolving the need for a permit, but to the issuance of permits on a timely basis. In many cases significant effort by the project team is required to prepare environmental impact reports, special studies and drawings for permits, and similar data.

Paralleling the emphasis of these areas is the general topic of **public relations.** Since it is significantly broader in its scope and area of public interest, it is likely to have a greater impact on the project, and thus the subject of public relations should not be left to chance. Again affirmative programs to establish and maintain favorable public relations should be undertaken with particular emphasis on the media, both print and electronic. Almost as important are less formal activities such as the establishment of a bureau to provide speakers for civic and other local groups, development of favorable relationships with governmental authorities, including mayors, planning commissions, and labor or other groups in the area. Providing lecturers to schools and colleges in the area and participation in civic activities can assist in forming a favorable public image of the project and the personnel involved.

All of these activities should be part of a plan prepared for the project with clear objectives as to the goals desired. The type and number of contacts and articles or speeches given can be used as ways to measure the effort and, at least partially, the effectiveness of the program.

Security, because its requirements cut across all the project functions, is usually a responsibility assumed directly by the PM. In general, it can be divided into the physical and personnel areas. The requirement for **physical** security can be provided through contract with any number of competent security agencies. They can provide both armed and unarmed personnel depending on the type and degree of the security required. Coordination with local law enforcement authorities is essential, and they can assist in establishing suitable coordination and mutual assistance provisions. The local authorities can also assist and provide counsel in evaluating the security companies prior to soliciting proposals from them. For work in remote or small communities, more complete, self-standing arrangements are necessary, while for work in urban areas, the existing police forces can play a larger role. In any event, care should be given to selecting the security firm and checking out their reputation to ensure that they are competent to provide the level of security required. After security operations commence, it is useful to conduct surprise inspections at odd hours and to attempt to smuggle weapons or unauthorized materials past the guards to ensure that they are performing satisfactorily. To avoid misunderstandings, a neutral party such as the local authorities should be advised of these tests before they are attempted.

Personnel security is a more complex question, for it involves not only the actions of the personnel but their living and travel habits as well. Generally it is poor practice to publish directories of senior personnel that list their home addresses; the listing of their office telephone numbers is sufficient. If there is a concern about kidnapping or terrorist activity, it is wise for these people to vary the time and the route of their travel to their office so that their movements are not predictable. In some cases, if drivers are used for these people, the drivers can be trained in evasive methods.

When traveling, particularly by air, it is good practice to arrange itineraries so that senior personnel do not all travel on the same aircraft and thus possibly perish in a common accident. While this is not always practical, it should be utilized where appropriate. Similarly it is useful to maintain a low profile when traveling to avoid calling attention to oneself and thus becoming a target for possible action. Luggage tags, for example, should give merely the owner's name and a telephone number (it can be the office number) rather than the title and in some cases the home address. This has the obvious advantages of not identifying the traveler's position and also protects the location of the home in the traveler's absence. Finally, it is good practice to restrict the distribution of travel itineraries to only those persons having a **need to know.** Others who need to contact the traveler can do so through his or her office.

8.3   OTHER RESPONSIBILITIES

Many other responsibilities may be assumed by the PM depending on circumstances. Those that follow in this section are indicative of the more common ones.

Labor relations is a responsibility usually reserved for the head of the construction or operational group rather than the PM; however where the labor relations practices or agreements affect more than one group, the PM will often take the lead. This is particularly the case where several contractors are working on a project and an agreement by one could set a precedent and become binding upon the others. In other cases there may be agreements with the national headquarters of the unions involved which take precedence over local practices and thus need to be dealt with on a different basis.

Housing is an area that can contribute to the productivity and stability of the project work force. For work in the United States it is rarely a big problem, but on international projects (Chapter 9) it often requires major planning. Usually the work force is left to find its own housing, although in remote areas it may be necessary to establish a **camp** operation. Sometimes housing may be leased or purchased by the project and made available to the personnel, the rent being deducted from their wages. At the conclusion of the project the housing is sold, and the company takes a gain or loss depending on the changes in the value of the property. The preferred approach is not to take an ownership position but rather to merely locate property and advise personnel of its availability. One helpful arrangement is to work with one or two real estate agents in the area and let them locate suitable housing for the personnel, with the company having no economic interest in the housing question. A complaint frequently voiced is the inflation of local real estate values when a large project is undertaken nearby. This problem occurs more in smaller communities where, upon completion of the project, values return to normal. There is really no way to avoid this problem, for it is simply the law of supply and demand—coupled with the fact that often the wages paid to the project work force are higher than the typical local wages and thus more money is available for housing. Finally, there is an element of personal preference for the size and type of housing, its features, its decoration, and numerous other factors—all of which can generate an innumerable series of minor problems and irritations for the PM. If it is at all possible to minimize the role of the PM in housing questions, it will permit more attention to the problems of the execution of the work and will remove a significant source of distraction.

Camp operations may become a responsibility of the PM in certain circumstances. Usually this is the case where operations of more than one department are involved and the PM is the senior project representative in the area. Camps are a more difficult operation than merely providing and maintaining housing for personnel. They involve food supply, schooling for dependents, recreation, emergency services such as fire and health, garbage services, sanitation, water supply, security, transportation, and other similar services. As such they represent the full range of municipal sevices found in a city. Since camps are located in remote areas, it is difficult if not impossible to provide all the services ordinarily found in urban areas. Perhaps the most important factor in running a satisfactory camp operation is the quality, quantity and variety of **food** provided. Some firms develop reputations for running outstanding camps based largely on the food service provided. They rarely have trouble recruiting personnel for camp-type locations. Since camp operations are more common in international locations, more information on them is found in Chapter 9.

Often the PM will be responsible for the **automotive** equipment assigned to the project. The best course of action is to delegate this to one person for administration. Some useful thoughts include minimizing the number of vehicles assigned to individuals unless there are enough for virtually every person or family to be assigned one, using buses to transport personnel where possible, and minimizing the need for frequent trips to stores and recreation by providing a form of shuttle or scheduled vehicle runs. Responsibility for the automotive operations involves maintenance and repair of the equipment, with the usual problem of spare parts availability, maintenance of the safety of vehicles, and rigorous periodic safety checks. In some cases it may be more convenient to lease the vehicles rather than purchase them. This avoids the resale value question when the vehicles are no longer needed. It may be possible to lease vehicles on a complete basis—to include not only the vehicle cost but fuel for operations and all normal maintenance items such as brakes, hoses, coolant, and lubrication. As an adjunct to this it may be possible to have the lessor carry all necessary automotive insurance coverage as well, the type and levels of coverage being set by the PM.

8.4 PROJECT PHASES

The role of the PM and his or her involvement and primary emphasis will change with the various phases of the project. Initially, in the **marketing** phase, the primary emphasis will be on determining the prospects for the project and its feasibility. Often clients are not aware that a particular project may be to their advantage, and a PM may be assigned to assist the sales personnel, develop the project concept, and establish a strategy for presenting it to the client and gaining approval. This activity combines the function of sales or business development and sufficient technical work to ensure that the project is both technically and financially practical. It also includes the establishment of working relationships with the client to demonstrate a need for the project and the advantages that the client will realize not only from pursuing the project but also by contracting with the PM's organization.

This work may involve the development of sources of **financing** for the project and may include arrangements with banks, financial consortia, public agencies or private investors. The role of the PM is more than merely putting the deal together, it involves the areas above and includes a willingness to follow the work through to completion when the project is completed and working; thus the PM has a deeper and more long-term relationship with both the client and the project than merely being an orchestrator. Of course the PM should begin early to cultivate working relationships with the client's personnel, being careful to keep them cordial and reciprocal. The development of **reciprocal relationships** is likely to be the most difficult part of the task, however. Usually the PM is not known to the client, so it may take some time and effort to demonstrate to the client that its best interest is at heart and that the proposed project is not merely a device to enrich the PM's company. This does not mean that the PM should transfer his or her loyalty to the client but rather that the PM has an **ethical** responsibility to represent the project fairly and accurately to the client so that, when the project is approved, it will be viable. This point will eventually become obvious to even the most poorly informed client, and thus the true basis for the project will be known to the client with severely adverse results if deceit or trickery has been used to justify it.

Often the client will accept the idea of the project but will still wish to have the work competitively bid or negotiated. While it is disappointing if the firm has put a lot of development effort into the feasibility evaluations to then have to compete for the work, at least the firm will get a chance to bid. One hazard often considered is the fact that the firm knows perhaps too much about the project, having developed the initial concepts and thus may build excessive contingency into its proposal. While this can be a disadvantage, the detailed knowledge of the client provides added insight and usually balances out this concern. If normal bidding practices are followed as regards contingency development (see Chapter 10), this should not be a problem.

Assuming the bidding has been successful and a contract is to be negotiated, the PM can play a major role in contract **negotiations.** Primary among these is the requirement that the PM will have to execute the project in accordance with the terms of the contract and thus will be more concerned than legal personnel about the operational requirements of the contract. Such things as progress reporting, measurement methods for progress payments, warranties, approval of personnel, reports, and in some cases, housing, transportation, foreign currency exchange rates, and a myriad of other items that arise during the negotiations are of great concern to the PM and must be considered from an operational point of view. Contract negotiations provide an opportunity for the PM to start to develop good working relationships with senior client personnel and with his or her counterparts. Wherever possible the PM should include in the contract negotiations engineering and other senior personnel who will have to execute portions of the contract. Naturally before bringing them into the meetings, they should be briefed on the issues to be considered and the most effective way to deal with the client—remembering that if the discussions can result in a "win-win" situation all parties will benefit in the long run. Chapter 11 deals with the content of contracts and contract considerations more extensively.

The first operational task of the PM will be to develop a suitable **scope document** for the project and thus to begin the planning effort. For reimbursable work the scope document should be developed in conjunction with the client to the greatest extent possible. Following that, the initial scope document—typically over the first half of the project—will undergo more definition. To ensure client agreement, periodic reviews should be held, as the scope evolves.

For firm price work the scope is normally fixed, and the details of the scope implementation are turned over to the PM and the project team. The major reason for establishing a detailed scope is to provide adequate cost control. In effect the scope builds a fence around the work and establishes the level of quality of the equipment and components in the project. This prevents the gratuitous expansion of the work by well-meaning, but uninformed, project personnel. Lack of proper scope control can easily cost an additional 25% or more and make the difference between the project being highly successful or suffering a major loss.

Because of the long lead time required for applying for and obtaining the various **permits** required for most projects, owners will often start this work early in the project, in many cases even before the negotiation phase. In cases where the applications are particularly complex and require environmental studies, preparatory work on them may be started several years in advance and treated as a separate project. For less complex situations involving reimbursable work, the owner may have started this earlier and then turned the responsibility over to the design or construction firm. On firm price work the owner often takes no active part in obtaining permits.

The responsibility of the PM here is no different than that for other phases of the project, except that the issuance of permits is usually a pacing event affecting the entire schedule and thus the completion of the project. As such the PM has a major interest in timely issuance of permits and will usually take deep personal interest in how this progresses.

The marketing phase of the project usually concludes with **agreement** on the contract and its **execution.** At the moment of signing the contract the parties are typically euphoric, and a heightened sense of goodwill and cooperation exists. As the work proceeds, differences will arise, and the relationships will begin to suffer if a conscious effort to maintain them is not established. Thus the PM has the responsibility to maintain the initial goodwill both within and external to the project team. As the leader of the project it is essential that he or she set the example of cooperation and understanding for the project team.

Much has been written about the PM's responsibilities during the execution phase of the project, and they will not be repeated here except to stress the **leadership** role of the PM. No one else on the project has the breadth of responsibility or the understanding of the overall needs of the project. The PM is the only one who can provide a balanced view of the operations of the project, that is, in contrast with the more partisan view often taken by the various project groups. The PM will have to make many decisions that are certain to be unpopular with one or more of the groups but that is what the job is all about. The PM earns his or her pay by being the arbiter of the conflicting needs and views of the various groups. Nothing can more quickly destroy morale or dedication of project team personnel than an indecisive PM—or worse, one who tries to please everyone and thus does not provide the leadership and direction required.

Management of large complex projects usually follows the principle of management by **exception.** That is, the major attention of the PM and the senior managers on the project is given to those items or operations which are not proceeding according to plan. Exception reports are normally prepared on this basis by the cost/estimating and scheduling personnel on the PM's staff. To ensure a broad overview and avoid omitting a major item, the PM should also independently review the progress of each of the operating groups on the project including their major internal activities. It will also permit the PM to utilize his or her experience as well as knowledge of external factors, such as client attitudes and public acceptance, to be factored into project decisions. Thus with review of the progress of all the activities for the project and the identification of deviations and the development of approaches to recover from their effect, the likelihood of successful execution of the work is greatly improved.

Where there are schedule problems, **collapsing** the schedule, or **schedule recovery,** are always concerns of the PM, particularly where the project includes bonuses for early, or penalties for late, completion. In theory the CPM (critical path method) schedule for the project represents the minimum time sequence for its completion. To be sure, there often are other work sequences with a shorter elapsed time that can be substituted for the original one but that may be inefficient in terms of personnel requirements or cost. Although less desirable, alternatives must be considered when a significant schedule reduction is needed. Typically they involve the assignment of blocks of work to other organizations, the use of quick conservative design estimates to avoid the time needed for the development of more economic detailed design, payment of premiums to fabricators to shorten material delivery schedules or the development of alternate suppliers, the assignment of field work to additional subcontractors, and the offering of productivity incentives. If the project has a board of review or a similar control group, these people should assist in such decisions, for they have probably encountered similar situations and can provide helpful guidance.

In some respects the last 10% of the project is often the most difficult and may seem to take as much effort as the preceeding 90%. The reasons are fairly clear: It is no longer possible to defer design decisions, for the numerous details that had previously been deferred must now be resolved; delivery of the final materials and equipment must be expedited to meet field requirements; and in the field there is crowding and reduced crew efficiency, and as more work is in place access becomes more difficult. Then on some projects there may have been a slowdown in the effort of the field labor force as they see their work ending. Because of the shorter remaining time to completion, the PM may reduce the interval between repetitive reports, and may do away with some reports and replace them with "flash" type reports in particular areas of concern.

As the project approaches completion, the interest of the owner will increase, particularly with respect to **start-up** and **initial operation** of the project. Clients will usually use this period to train and familiarize their personnel with the plant and its operations. One of the important responsibilities of the PM is to ensure that planning for this phase is started early and, where the client personnel participate, that the plan as finally developed is agreed to by the client. The planning should start as early as possible, but no later than 18 to 24 months before the beginning of start-up for complex high-tech facilities and 12 months for typical lower-tech facilities. (These classifications can be defined by the number of P&IDs required, with the complex high-tech facilities commonly having from 75 to 100 or more, while the lower-tech facilities having perhaps no more than 30 or 40 P&IDs.)

The selection of the client **operational personnel** is not generally a concern of the PM, except in international work. In developing countries the client will want assistance in both selecting operating personnel and setting up a training program for them. There are often special reasons, internal to the client organization, for selecting personnel for this type of training. As a result it is wise to limit the PM's role in the selection process to merely providing test score data and similar information to the client so that the decision regarding who is selected remains with the client.

The program might involve classroom work at the design office, at local universities or technical schools, and specialized training at the major equipment vendor's plants. If the personnel are not fluent in English, special language courses (beginning with both diagnostic and performance examinations) may be required. If the training time in the United States is brief, there may not be time to train personnel in English, and thus English fluency must be made a selection requirement prior to identifying personnel for the program.

The PM is usually in charge of the overall program, and as with other responsibilities, it is important that another individual be designated to be in charge of the day-to-day activities of the program. The establishment and administration of the program will require significant time of project personnel and will often involve one or more persons on a full-time basis. If the trainees are in the United States for more than a month or so, or if the group is more than a handful of people, one trainee of the group should be placed in charge and most administrative matters taken up with him or her. There will be a multitude of other problems, ranging from housing, to salary payments in dollars, and to recreation, for the senior member to handle and reduce the burden on the U.S. project staff.

In most cases the client will want to integrate plant operational personnel into the **start-up** group to provide additional orientation. This is extremely useful and helps reduce the size of the contractor's start-up crews. Because of contractural liabilities, the client personnel should either not be permitted to actually operate the equipment, or the contract, and the contractor's all-risk insurance policy should reflect that and the assumption of responsibility by the owner for actions of client personnel.

In certain cases, for example, the communications industry, it may be necessary to cut in the new system over a very short time, perhaps a day or two. In these cases it is common to build the new system in parallel with the existing one, even to duplicating equipment to facilitate making the cutover. Some of the factors to be considered are these:

- What is the maximum time permitted for the cutover?
- How much parallel effort is required to ensure cutover without a significant exposure to major problems?
- What are the consequences of major cutover problems? Is there a fall-back position?
- What is the cost of the equipment to be duplicated? Can it be reused or salvaged, or must it be scrapped?
- Are there any space or other restraints that limit the ability to build the parallel system?

Service representatives are usually provided by the vendor for complex equipment, and they are resident at the project during important construction operations. In addition they often return for the final calibration, start-up checkout, and actual start-up to ensure satisfactory performance. Where jobsite performance test requirements are found in the purchase order, it is common for the representatives to witness the testing as well. They can be integrated into the start-up crews using the vendor's equipment, and they can be of great assistance to the field personnel, both construction and start-up.

As a part of this work a procedure for **tagging** of equipment and systems that are completed by the construction department is required. Usually a computerized data base is established, with each system and major piece of equipment (valves, sensors, controls, microprocessors, etc.) within the system identified.

The preparation of an overall outline is usually the first step in setting up the start-up operations. This outline establishes the organization of the start-up procedures as well as normal administrative controls on documentation, including revisions, control of test equipment, safety considerations, and so forth. The next step is establishing the standard procedures with which to perform the testing, and the final step consists in preparing a start-up procedure for each specific system.

The **test procedures** are prepared for components that may be tested in a generic way, regardless of the system in which they are found. Following that the start-up procedures for the systems (and sometimes subsystems) are prepared. The start-up procedure for each system is a stand-alone document that prescribes the proper sequence for start-up and includes, for each mode of operation, the performance parameters to be achieved (pressure, flow, temperature, etc.) for each instrument or measurement location.

The starting point is the identification of all systems and the establishment of their **boundaries.** Each system is given a unique identifier, and its P&ID is marked to indicate the system boundaries. Within a system all components are given a unique number that carries the system identifier, a subsystem identifier, a component-type identifier, and finally a component-specific number. For example, a control valve might bear the number D2-VAC-CV-6; that is, in the #2 drain system, for its vacuum subsystem, the component is control valve 6. When properly arranged, such a system can be used with various computerized sorts not only to prepare the punch lists of open items but for status reports and developing priorities for completion work.

Normally the procedure includes steps to check out the wiring and piping of the system to ensure that the basic elements of the system are correctly installed. With the large number of items to be demonstrated during start-up, the procedures become extensive and rather voluminous, although many of them may be routine and fairly simple.

A typical index of start-up procedures for a complex project is given in Table 8.3.

Using much of the engineering data—including the P&IDs, control and logic diagrams, wiring diagrams, ladder diagrams, piping design drawings, flow diagrams and manufacturers start-up recommendations for the system, subsystems, and components—start-up procedures are developed. The sequence is to first check out and place in operation the components, followed by the subsystems, finally building up to the overall plant systems themselves. In process plants, because of their importance, the basic **utility** systems

TABLE 8.3   Index of Start-up Procedures

1. Organization, both jobsite and home office
   Relationships with client and other personnel, vendor representatives
2. Estimating and cost control
3. Records and reports
4. Materials and equipment
   Test equipment control, wetting agents, chemical cleaning procedures, strainers
5. Personnel safety
   Tagging and clearance, radiation exposure, noise control
6. System cleaning
   Mechanical cleaning, chemical cleaning, system blowdown
7. Electrical component tests
   Testing of individual components, meters and relays, motors, switchgear, circuit breakers, motor operated valves, major electrical equipment, etc.
8. Mechanical component tests
   Testing of individual components, valves, rotating equipment, relief valves, etc.
9. Instrument component tests
   Same as above, including control valve and control loop testing, calibration
10. Electrical operations
   Megger testing, energization, battery, control systems, low voltage, medium voltage, high voltage, cathodic protection, instrument power, major equipment testing
11. Mechanical operations
   Hydrostatic testing, leak detection, rotating equipment, vibration, noise, water hammer, major equipment testing, system testing
12. HV&AC testing
13. Special testing
14. Preoperational and acceptance testing
15. Component and system turnover

Note: Within the electrical and mechanical operations sections, each system or major piece of equipment will have a specific procedure. Their number and complexity will vary with the type of facility and its operational requirements.

are generally started first. They include the primary and secondary voltage electrical systems, control, power, lighting and communications, compressed air, vacuum, process water, drainage, and so forth. When operational they permit the plant process systems that draw on them to be started.

Status tags are usually used for tagging components and systems to indicate **status** for the completion and start-up activities described above. Often three tags are used for each component: one to indicate completion by the construction department, a second to indicate turnover to the start-up group, and a third to indicate that checkout is satisfactory and the component is ready to operate. In some cases multiple-step tags are used to permit recording the status as it evolves.

With a large complex plant it may not be possible to turn the plant over to the client in a single action. Instead the client would take over and assume responsibility for systems as they are started up and demonstrated to be in conformance with the design and the contract. Then the contractor will set up a procedure to record the **turnover** to the client, with the client signing for the receipt of each system or component. Since from that point the client assumes responsibility, it is common practice for the client to establish controls for the operator/attendants who are to be in charge of the system after turnover. To avoid any question of responsibility, the contractor may subsequently be controlled or even locked out of entering certain areas of the plant. This can be awkward if some common functions occur in these spaces, and special arrangements may be needed to permit the contractor to continue with necessary work.

Since there are always cases where more work needs to be done on a system, either after turnover by construction to start-up or after turnover to the owner, there needs to be a procedure established to turn a system back to either the construction or the start-up groups for **remedial** action. This must be included in the written start-up procedures. From the contractual standpoint it is important that turnover of systems to the owner, and any subsequent turnbacks and remedial work, be **documented** in writing so that there is a clear record of responsibility for the system and the work or operations performed on or with it.

8.5 SCHEDULES

Much has been written earlier on schedules, but from the PM's point of view the schedules represent a primary tool in controlling the project and predicting its ultimate completion. The usual 20- to 40-line bar charts, with a few restraints, are excellent for an overview of the entire project.

For actual management control of the project, two different types of schedules are needed. First is a series of highly detailed **rolling** schedules of the CPM (precedence) type showing a window of perhaps two weeks to three months. The disadvantage of rolling schedules is that they do not look too far ahead and decisions noted on them are necessarily short term and tactical in nature. In addition, because of their detail, deviations can only be summarized to permit management attention. Second are CPM schedules of the 200-line variety which look ahead at least a year and thus include longer-range decisions. The rolling schedules should be run on a continual basis, typically weekly, for problem areas and periodically, perhaps monthly, for the three-month window. The full project CPM should be run every two or three months for projects having a two- or three-year duration. For shorter or longer projects the frequency can be adjusted accordingly. If the schedules are being continually updated, as they should, the rerunning of them is normally not burdensome.

In addition special schedules or schedule studies are often performed for unique conditions or events, such as loss of a ship carrying major equipment to the site, a protracted strike in a major manufacturer's plant, a change in law or regulation requiring additional permits or reviews. These studies would vary in degree of detail depending on the event considered and the potential impact on the project. In the extreme case it may be necessary to reschedule the entire project.

As a part of the schedule analysis, the schedule must be **task** oriented and **resource loaded.** Tasks, deliverables, and resource loading were discussed in Chapter 5, so they will not be repeated here except to caution that considerable care needs to be taken to ensure that when the schedules are developed at all levels, the line items or activity nodes must be based on the concept of tasks and deliverables and the precedence or dependence of activities and restraints, must be accurately reflected. Resource loading will normally follow but need not be rerun for each schedule revision. If the revision is a minor one, as are most of them, the effect on resources will be minor and not significant. As in the case of the overall project CPM, it is useful to **periodically** rerun the resource loading program to check that the cumulative effect of actual progress and changes have not significantly affected the resources required.

Lead times should be confirmed when the schedules are rerun to ensure that there are no significant changes that can impact the schedule. Again, as in the case of the basic rerunning of the schedule, if these updates are being maintained on a continual basis, the information provided will be accurate, and no major effort should be necessary to develop these data.

8.6 ESTIMATES

Estimates together with the schedules previously discussed are the two principal management tools of the PM. The form of estimate in use will depend on the type of work (whether firm price or reimbursable), the state of evolution of the project, the particular area of concern (whether the overall project or some portion of it), and the method of financing of the project. Estimates always proceed from the general to the specific, with more detail and precision added as the design and execution of the project proceeds. Estimates fall into several types depending on their purposes, as described in Chapter 6.

For firm price work the initial or baseline estimate used will have the best definition and precision that can be developed. To the maximum extent, materials and equipment pricing will be based on current quotations for the project itself. Labor data (productivity, crew makeup, wage rates, etc.) will be based on extensive investigation at the project site followed by considerable analysis. As commitments are made for materials and equipment, they would be entered into the cost-tracking system, and the accuracy of their values constantly improved. Similarly labor cost and productivity would be entered as the work proceeds. To forecast the final cost for the labor component(s) of the project is somewhat more difficult than for the materials portion, for labor productivity will vary depending on the type of work, crowding, skills of the local labor force, where the labor force is on the learning curve, introduction of labor-saving equipment and procedures, and so forth. For this reason it is important that the cost-forecasting program continually monitor the productivity of the labor employed and update the estimated labor portion of the forecast. Further, as the project proceeds, not only does the definition of the project improve, but less of the work remains to be performed. Thus the allowance for **contingency** should be reviewed and reduced accordingly.

For projects that run two or more years, a monthly update of the cost forecast is usual, while for projects that have a shorter life, the forecast updates can be made every other week. Similarly for schedule forecasting, if the latest data on materials, equipment, and labor are continually added to the cost-forecasting system, there should be no significant extra effort required to produce the periodic updates.

All of the foregoing apply as well to reimbursable work. However, by nature initial estimates are much less precise and are more subject to change, so more contingency is usually provided. It is usual to see larger changes in the forecasts for reimbursable work for which a truly definitive estimate or forecast is not available until a significant portion of the work has been completed.

A concern of the management of the project and in the case of reimbursable work, the client, is the projected **cash flow** for the work. While not complicated to perform, it requires that the forecast be coordinated with the schedule and include payments for materials, equipment, or services tied to their scheduled dates. Where the project is of the firm price type and relatively small as compared to the financial size of the firm, an approximation of the cash flow may be sufficient, but where the project is of a significant size, there will almost certainly need to be periodic cash flow estimates. On very large projects these are often routinely prepared as a part of the monthly cost forecast, while on moderate-sized projects they may be required less often.

8.7   ORGANIZATION

The organization of the project can take a variety of forms. As the leader of the project team, the PM is responsible for the organization of the work. The usual approach uses **work package breakdowns** in which, with the exception of certain common activities, each major portion of the work is organized essentially on a stand-alone basis. This work package approach normally applies only to the field or jobsite activities. In some cases the engineering of the work is split into different work packages and given to different parts of the organization, or even to different organizations on a subcontract basis. This arrangement facilitates not only the management of the work by making each work package organization responsible for its own work but also provides a convenient way to deal with any problems. For example, if the group designing the storage tanks for a major petrochemical facility is having difficulty, it can be given more personnel to achieve the schedule or if a change is necessary, the work could be reassigned as a package to another group.

The common activities usually excluded from the work package breakdown include engineering, materials acquisition, scheduling, cost estimating and forecasting, personnel recruitment and administration, labor relations, security, fire and safety, and public relations. An important concern in the work package arrangement is to clearly establish, early in the design effort, the boundaries of and interfaces between the work packages. While the boundaries are fairly easy to establish, the interfaces often are only finally established after some degree of design work.

For example, if the sludge unit of a pumping plant receives its influent from a settling chamber, the size of the supply pipeline is an important design factor. Since the engineering of the settling chamber is not complete when the engineering of the sludge unit must begin, it is necessary to assume a line size to permit engineering to proceed. As a result a quick calculation and assumption of a conservative value for the line size will permit the work to proceed, with the final line size determined at a later date. The principle to follow is to make the design assumption sufficiently **conservative** that it can accommodate later design evolution and that the design will not have to be redone when the final design interface values are established. While the design is being developed, it is wise for the different groups to exchange information so that the dependent groups will be aware of the direction of their development and will avoid last minute surprises and rework.

Subcontracting portions of the work often have a useful effect on the organization. Since subcontracting is usually done on a firm price basis and later changes are both costly and somewhat embarrassing, it forces more precise, early definition of the work including both scope and interface requirements. When working on firm price work, subcontractors also tend to be efficient and prompt in their performance. This can accelerate the work of the contracting organization. There needs to be some overview of the work, however, to ensure that quality has not suffered. If subcontractors are performing design work to carefully drawn criteria, ensuring that the design is performed under the direction of a registered engineer and that the drawings are sealed and signed off by a registered engineer provides this assurance. In any case, a brief overview and check by a qualified engineer in the contracting organization is usually sufficient to confirm compliance.

The **liason** with local organizations, such as licensing and permitting offices, chambers of commerce, environmental groups, interested citizen groups, and local governmental agencies is usually the responsibility of the PM. The development and maintenance of good working relations with these organizations is essential, and the PM must provide the necessary leadership for this. Where the project is a large one, the PM will often establish a small staff to deal with these matters, with the PM attending significant meetings with these external groups to provide the visibility and ensure personal involvement.

In a similar fashion the PM, as the most senior full-time person on the project, will be the focal point to develop and maintain satisfactory **client relations.** This does not imply that the PM is solely responsible for client relations, but rather that he or she is likely to have the most ongoing involvement with the client. It is both wise and useful to develop a strong, cordial relationship with the client's senior personnel. In effect the PM should view this role as an opportunity to develop new friendships in the industry and to learn more about the client. Positive relationships go a long way toward simplifying any work problems with the client and can mollify potentially difficult situations.

8.8 MOBILIZATION

Mobilization for the project is always a stressful time for the PM. The time required for mobilization will vary from a few days to several months in the case of remote or large-size projects. While the contractor(s) are normally left fully in charge of the mobilization, the PM must review their plans to ensure that they are adequate and compatible with future project requirements. On a recent job, for example, one subcontractor had the responsibility for large vessel and plate fabrications that were to be used in the early stages of the work. The subcontractor, through an oversight, was allocated the space immediately in front of the main building for doing prefabrication, and thus for the first six months of the job the subcontractor severely restricted access by others to this area. This required that everyone else use a side access, and this caused significant inconveniences and delays. Had there been more discussion with the subcontractor, this problem would not have arisen, as a less critical area would have been assigned for the work. A review of the contractors plans for mobilization can often also reveal whether there is a true understanding of the scope of the work and whether sufficient planning has been done to ensure its proper execution.

8.9 REPORTS

The usual reports used to control a project have in part been discussed earlier. Table 8.4 summarizes those that most projects have found useful. The frequency of the reports noted in the table is typical for projects having

TABLE 8.4 Project Reports	
Report	Typical Interval
Engineering progress	Monthly
Engineering workerhours earned	Monthly
Cost versus forecast	Monthly
Cash flow requirements next period	Monthly
Major commitments versus schedule	Monthly
Commitment register	Monthly
Status of major materials and equipment	Monthly
Schedule progress	Monthly
Schedule forecast	Monthly
Critical items action list	Biweekly
Construction progress	Monthly
Project financial status	Monthly
Project progress	Monthly

a two-year or longer duration, for shorter projects the reporting interval would be reduced accordingly. Similarly the content and level of detail of the reports will vary with the type and complexity of the project. The reports may include extensive additional data, for example, the cost versus forecast report might have several sections showing cost this period, cost to date, both versus planned, as well as similar data for the forecast. The construction report might include productivity data and compare the forecast progress with actual, establish new forecasts, and so forth. Project reports also include reconciliation of the data where appropriate as well as comments on significant events and milestones achieved or forecast.

To permit data to be readily understood, much of the summary information should be reduced to **curves** or **charts.** With the computerization of most data today, it is relatively easy to present the data in this form.

8.10 INTEGRATED PROJECT MANAGEMENT CONTROL SYSTEMS

The development of project management reporting and control systems has been a natural evolution of the increased capability in both speed of processing and capacity of computers. Among the many advantages computers afford are **single entry** of data, availability of status data on essentially a **real time** basis, and the ability to perform rapid analyses of alternate action scenarios. The cost to manipulate the data to perform such work has fallen dramatically as computer capability, speed, and memory have vastly improved in recent years. This has permitted the expansion of the computer applications from merely reporting or scheduling at a summary or secondary level of detail to reporting, data analysis, and scheduling at typically two or more orders of magnitude of increased detail. Numerous programs are available commercially to perform project management control activities, and **integrated programs** that are applicable to several aspects of the management of projects are becoming increasingly common.

ICF Kaiser Engineers, a large engineering, procurement, construction, and construction management firm has developed an integrated project management control system that incorporates the principles discussed earlier and represents a fully operational system proven by application and widely used in their work. Their description of it, which follows on pages 283–305, while specific to their system, includes the features required of any fully integrated system for the management of large or complex projects.*

> Computers and computer software have played an increasingly vital role in project management in recent years. Since the days of hand-drawn schedules and log books for expenditures and progress, great strides have been made with computers which allow project data to flow more quickly and be much more accessible to project managers. Computers have also allowed the integration of cost and schedule data more easily, thereby allowing true management control on a real time basis as opposed to merely management reporting.

> Computers increase the integrity of project data by decreasing the chance of human error. Mainframe computers allowed the data to be stored in data bases and for the first time, project personnel had quick access to a variety of report formats, although often at a considerable cost. Minicomputers arrived on the scene with a degree of portability, allowing computers to be located at or near the jobsites, cutting down on communication costs and turnaround times for reports. Finally, the introduction of the microcomputer gave project controls engineers and project managers the tools that allow them to perform their jobs efficiently and economically.

> The IBM-PC and compatibles have allowed software developers to produce the most flexible and low maintenance programs of any computer platform. Included in this group of software programs are data base systems that come with their own fourth-generation language. Examples are the Xbase products such as Dbase and FoxPro. Products like these allow companies to integrate project data by developing their own data base management systems tailored to the way they do business. Project scheduling is usually accomplished by commercial scheduling software such as Primavera, and the integration of cost and schedule data is accomplished by developing a data interface between the data base program and the scheduling program. Some software packages such as ARTEMIS provide a schedule processor within a data base language and allow the integration of cost and schedule data within the same software.

Kaiser Engineers Management System (KEMS)

> An example of a fully integrated, multi-project, multi-user project management system is the ICF Kaiser Engineers Management System, which is referred to by its acronym KEMS. KEMS was initially developed to run in the mainframe

---

* Reprinted by permission of ICF Kaiser Engineers.

environment in the early 1970s and was based on Kaiser Engineer's (KE) own methodologies of controlling projects and the practices and procedures that had been developed over many years. The mainframe system contained all of the functionality required for the management of large complex projects; however, due to limitations in computer technology it lacked the desired data integration. In the early 1980s the need for a fully integrated management system was apparent, and a project was initiated to rehost the system on a microcomputer platform. An evaluation team was first put together and developed information-gathering techniques, interviewed key staff, and thoroughly reviewed project and corporate procedures to ensure that all of the technical requirements for the desired system were defined. As a result of this process, a technical specification was developed for the system, which included a full set of prototype reports that the system was expected to produce.

When the specification was complete, the next step was to make a "build or buy" decision. A study was initiated to find a packaged software that would meet all or most of the requirements. First, all available project management software products were identified and placed on a list of candidates. The candidate products were reviewed against a list of "must have" criteria, and a short list of candidates was developed by eliminating noncomplying products. The next step was to evaluate the remaining products against a list of important, but not critical, criteria. A score was assigned for each product against each criteria item and an overall rating developed to determine the most likely candidates.

As a result of the evaluation process it was determined that none of the commercially available software afforded the flexibility that was required to control the many types of projects in the various industries that they service. It was also determined that most packages were either designed as a scheduling package or as a cost control package and that none had the full features of both that would allow the automatic integration of cost and schedule data for producing time-phased budgets, cash flows, the "what if" analyses, as well as the analyses necessary for earned-value performance measurement. Being forced to track costs by activity or scheduling by cost account was a compromise that KE was not willing to make. All of the commercially available software forced this type of compromise. The conclusion was that in order to support their philosophy of project management and implement the type of cost schedule integration that was required, they needed to develop their own computerized project management system. The decision was then made to establish a team to develop a project management and control system that met all of the criteria outlined in their specification, yet had the flexibility to change as the type of project or reporting requirement changed. The first step in this process was to go through an evaluation of development software. The evaluation process was identical to the selection of candidate-packaged software. The most important criteria for the development software was the ability to create critical path method (CPM) schedules and integrate the data with all other project information so that they could achieve true data integration in a single software package.

KE selected ARTEMIS-PC for their development tool when programming was begun in 1984. Although ARTEMIS-PC was a relatively slow data base software, it was the only relational data base with its own fourth-generation language and built-in scheduling and resource leveling programs. The fourth-generation language reduced development and maintenance time by a considerable amount when compared to programs such as S-BASIC. Additional features allowed a better user interface on screen as well as improved management graphics.

In 1986 ARTEMIS 2000 was introduced to replace ARTEMIS-PC, and the system was rewritten to work with the new data base language. In 1991 AR-TEMIS 7000/386 was introduced to take full advantage of the Intel 80386 processor for increased performance, and the system was ported to the new product. ARTEMIS 7000/386 is identical to ARTEMIS 7000, which operates in the VMS/VAX and UNIX environments, and the system can be installed on these platforms and operated without modification of program code.

Design Considerations—The Modular Approach

KEMS was designed with a modular approach for enhanced operation. Within each project data base there are separate functional modules for project organization, estimating and budgeting, scheduling, procurement, tracking installed equipment, engineering progress, controlling services costs, and controlling project cost and progress. The modular approach provides flexibility in the organization of the data, and this in turn makes it easier to meet management and client reporting requirements. Depending on the scope of work for a project, not all modules may be required. The data base and the program can be kept to a minimum size by using only the modules that apply to the scope of work for the particular project. Modules help facilitate division of work and responsibilities by the project management staff. For example, an engineer can have full control of his or her module and data. Further the modules that tend to categorize data- and project-specific customizations to the software are generally easier to make, for changes often affect only one module.

Projects logically are developed with schedule activities that do not usually have a one-to-one relationship with a cost account. The system was designed to allow cost/schedule integration with one-to-many and many-to-one relationships between schedule activities and cost accounts. This allows the estimator or cost engineer to develop project budgets without being forced to conform to the schedule logic or construction sequencing. This is significant since in reality many cost items are not related to schedule activities and many schedule activities do not have costs associated with them.

To take advantage of its integrated features, all coding of schedule activities, estimates, cost accounts, drawings, and installed equipment needs to be based on a project Work Breakdown Structure (WBS). A WBS first defines the project end item or system, and then successively subdivides it into increasingly detailed and manageable subsidiary work products. The WBS allows lower-level detailed estimates to be rolled up into higher-level cost accounts, installed equipment to be related to individual drawings and cost accounts, and activity dates to be transferred to all other modules.

To define responsibility for an activity, an Organization Breakdown Structure (OBS) is also established. The OBS defines who is responsible for a specific activity and can be entered with up to two levels of detail, the first level being the company, entity, or department and the second the specific individual. Thus the OBS, for example, allows engineering workerhours and progress to be rolled into services cost accounts.

To realize the maximum benefit, a WBS should be established before any data are entered into the system. A study should be made to establish the information needs of project and client personnel throughout the life of the project. The WBS and OBS should be designed to meet these needs. The WBS should be incorporated not only into KEMS, but all coding schemes throughout the project (e.g., document control). Normally all or part of the WBS is used in key fields. Key fields are the major or controlling codes in each module that are used for reporting data from the module as well as integrating data with other modules. Key fields within each module are the schedule activity number, cost account number, drawing number, and equipment number. Great care should be taken when designing key field coding structures. It is important that the WBS level used is appropriate for the project. The OBS should not be made part of the key field, since the personnel on any given job may change and this could cause a key field change. Changing key field codes should be avoided, since it effectively removes an audit trail. The WBS and OBS are the heart of efficient distribution of project data to craft supervisors, contractors, management, the client, and other interested parties. Effective reporting depends on the completeness of the WBS, OBS, and other user-defined sort fields. With the use of a WBS in combination with an OBS, a supervisor is able to determine if the scheduled work will be constrained and who is responsible for causing the delay. A client is able to get an accurate estimate of what a particular facility will cost, even though there are several contractors working on it, and managers can compare the unit costs, progress, and performance of one crew versus another.

KEMS System Modules

The Organization Module

The system is made up of eight modules, as shown in Fig. 8.1. The Organization Module is used to set up the environment for the project. The selection of the modules to be active for the project is made, as well as choices for project and company descriptions, WBS and OBS titles, reporting period start dates, user names with passwords and permissions, screen colors, and printers. Utilities such as archiving and restoring data, loading program changes, and editing files are included in this module. There is also an option to identify one data base for summary level reporting in multiproject environments, or with a construction management project each individual contract can be handled as a project with a summary partition or project setup for reporting the overall project. This approach allows for efficient processing of data, since it is being managed in smaller groups.

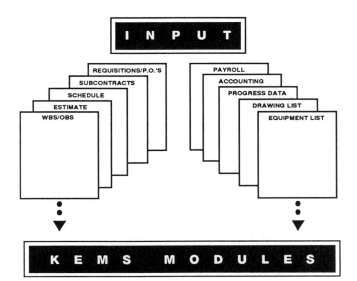

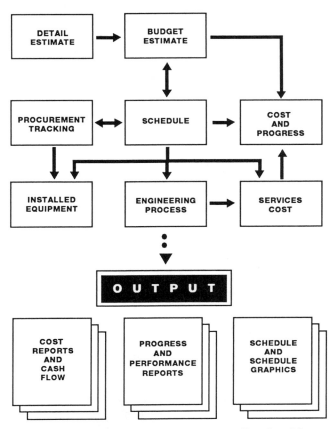

FIGURE 8.1 Kaiser Engineers management system. Reprinted by permission of ICF Kaiser Engineers.

The Budget Estimate Module

The Budget Estimate Module is used to establish the project baseline budget to be used for control. The primary dataset consists of quantity, hours, and cost data for each estimated item. Cost is broken down by labor, materials, installed equipment, equipment usage, indirects, and other user-defined categories. Each estimate item is identified by an account code, which is a unique combination of a WBS code and a work element code.

The Budget Estimate Module is designed to accept detailed estimates from other estimating software, or an estimate may be entered directly. These detailed estimates, once they are entered into KEMS, can then be combined or rolled up into cost accounts at a level of detail that is appropriate for control of project costs. After the cost accounts are established, the schedule is integrated with the budget by assigning one or more schedule activity codes to each cost account. Resources such as hours from the estimate can then be electronically transferred to the schedule, which allows the schedule to be calculated with both actual and forecast hours from the estimate. After schedule calculation is completed, the start and finish dates from the schedule can then be transferred back to the Budget Estimate Module, and the budget then time phased. The resulting time-phased budget is transferred to the Cost and Progress Module, where it provides the basic target for monitoring the performance of schedule and cost during project execution. Examples of both a budget cost estimate and a time-phased budget are shown in Figs. 8.2 and 8.3.

The Schedule Module

The Schedule Module represents a fully functional CPM schedule processor that uses precedence diagram network analysis to produce the schedule for the project. The Schedule Module is integrated with all other modules in the system, which allows for instant analysis of schedule impact when there are changes in project data from any source. If the estimated or budgeted hours change, the CPM can be recalculated based on the new resource loading. If schedule changes occur in the production of design drawings, the impact of these changes on the construction schedule can be immediately analyzed. Similarly schedule changes in the development of specifications or other procurement activities can immediately be analyzed for impact on the project schedule. A "network logic" (schedule report) is shown in Fig. 8.4.

The Procurement Tracking Module

The Procurement Tracking Module is an extension of the Schedule Module. It is designed to take some of the burden off of the CPM calculation by taking the detail of repetitive sets of activities and replacing them with one activity per set in the schedule network. For example, the procurement of any piece of equipment would usually follow the same sequence of events (prepare specifications, prepare bid documents, issue for quotation, receive quotations, analyze bids, award purchase order, etc.). For a piece of equipment, instead of

COST ESTIMATE

JOB #

KEMS 6.2 SAMPLE DATA

CONSTRUCTION PHASE

ESTIMATE #	DESCRIPTION	QUANTITY	UNIT	MANHOURS	LABOR	MATERIAL	EQUIPMENT	CONTRACTS	EQ. USAGE	INDIRECTS	TOTAL
AREA: 3000.0000	**PROJECT MANAGEMENT**										
	CONSTRUCTION MANAGEMENT	100	PC	750	62775	0	0	0	0	0	62775
AREA:	PROJECT MANAGEMENT			750	62775	0	0	0	0	0	62775
AREA:	**GAS COMPRESSION AND METERING**										
3100.1000	COMPRESSION AREA - CLEARING & GRADING	120	ACRE	150	5000	0	0	0	75000	0	80000
3100.2000	COMPRESSION AREA - FOUNDATIONS & STRUCTURAL STEEL	1800	TONS	1200	80000	750000	0	100000	230000	0	1160000
3100.4000	COMPRESSION AREA - FINISHING	100	PC	800	30000	80000	20000	0	10000	0	140000
3100.5000	COMPRESSION AREA - H.V.A.C.	1800	CY	500	20000	77000	140000	25000	0	0	262000
3100.6000	COMPRESSION AREA - PIPING	5000	LF	1200	50000	140000	0	0	20000	0	210000
3100.7000	COMPRESSION AREA - ELECTRICAL	10000	LF	2500	100000	200000	0	0	10000	0	310000
3100.8000	COMPRESSION AREA - INSTRUMENTATION	40	EA	400	16000	20000	120000	0	0	0	156000
AREA:	GAS COMPRESSION AND METERING			6750	301000	1267000	280000	125000	345000	0	2318000
AREA:	**GAS WITHDRAWAL AREA**										
3200.1000	GAS WITHDRAWAL - CLEARING & GRADING	96	ACRE	120	4000	0	0	0	60000	0	64000
3200.2000	GAS WITHDRAWAL - FOUNDATIONS & STRUCTURAL STEEL	1440	TONS	960	64000	600000	0	80000	184000	0	928000
3200.4000	GAS WITHDRAWAL - FINISHING	100	PC	640	24000	64000	16000	0	8000	0	112000
3200.5500	GAS WITHDRAWAL - H.V.A.C.	1440	CY	400	16000	61600	112000	20000	0	0	209600
3200.6000	GAS WITHDRAWAL - PIPING	4000	LF	960	40000	112000	0	0	16000	0	168000
3200.7000	GAS WITHDRAWAL - ELECTRICAL	8000	LF	2000	80000	160000	0	0	8000	0	248000
3200.8000	GAS WITHDRAWAL - INSTRUMENTATION	32	EA	320	12800	16000	96000	0	0	0	124800
AREA:	GAS WITHDRAWAL AREA			5400	240800	1013600	224000	100000	276000	0	1854400
AREA:	**LIQUID SEPARATION AREA**										
3300.1000	SUPPORT FACILITIES - CLEARING & GRADING	180	ACRE	180	6000	0	0	0	82500	0	88500
3300.1500	SUPPORT FACILITIES - CONNECT UTILITIES	100	PC	800	3000	20000	0	0	5000	0	28000
3300.4000	SUPPORT FACILITIES - PRE-ENGINEERED BUILDING	100	PC	1440	96000	12000	0	90000	93000	0	291000
3300.4000	SUPPORT FACILITIES - FINISHINGS	100	PC	960	36000	96000	14000	0	11000	0	157000
3300.5500	SUPPORT FACILITIES - H.V.A.C.	2700	CY	600	24000	92400	98000	22500	0	0	236900
3300.6000	SUPPORT FACILITIES - PIPING	7500	LF	1440	60000	168000	0	0	22000	0	250000
3300.7000	SUPPORT FACILITIES - ELECTRICAL	15000	LF	3000	120000	240000	0	0	11000	0	371000
3300.8000	SUPPORT FACILITIES - INSTRUMENTATION	60	EA	480	19200	24000	84000	0	0	0	127200
AREA:	LIQUID SEPARATION AREA			8900	364200	652000	196000	112500	224500	0	1549600
REPORT TOTALS		21800		968775	2933000	700000	337500	845500	0	5784775

FIGURE 8.2 Budget cost estimate. Reprinted by permission of ICF Kaiser Engineers.

TIME-PHASED BUDGET

AREA SUMMARY

KEMS 6.2 SAMPLE DATA
COSTS: x1000

PERIOD ENDING 31Dec92

AREA DESCRIPTION	PREVIOUS PERIODS	01Jul91 31Aug91 #1-#2	01Sep91 31Oct91 #3-#4	01Nov91 31Dec91 #5-#6	01Jan92 29Feb92 #7-#8	01Mar92 30Apr92 #9-#10	01May92 30Jun92 #11-#12	01Jul92 31Aug92 #13-#14	01Sep92 31Oct92 #15-#16	01Nov92 31Dec92 #17-#18	01Jan93 28Feb93 #19-#20	01Mar93 30Apr93 #21-#22	01May93 30Jun93 #23-#24	REMAINING PERIODS	TOTAL AMOUNTS #30
PROJECT MANAGEMENT															
PERIOD	0	139	259	276	271	261	231	163	101	83	76	67	49	41	2,017
TO DATE	0	139	398	674	945	1,206	1,436	1,600	1,701	1,784	1,860	1,927	1,976	2,017	2,017
GAS COMPRESSION AND METERING															
PERIOD	0	101	182	184	153	141	254	107	139	119	115	92	445	1,854	3,887
TO DATE	0	101	282	467	620	761	1,016	1,123	1,262	1,381	1,496	1,588	2,033	3,887	3,887
GAS WITHDRAWAL AREA															
PERIOD	0	20	123	160	141	139	316	1,128	431	175	4	4	3	2	2,646
TO DATE	0	20	143	303	443	583	899	2,027	2,458	2,634	2,638	2,641	2,644	2,646	2,646
LIQUID SEPARATION AREA															
PERIOD	0	298	345	225	183	146	146	388	1,017	885	856	662	38	1,453	6,642
TO DATE	0	298	643	868	1,051	1,197	1,344	1,732	2,749	3,634	4,489	5,152	5,190	6,642	6,642
STATION SUPPORT FACILITIES															
PERIOD	0	35	69	87	64	53	130	92	262	1,206	623	22	0	0	2,643
TO DATE	0	35	104	192	256	309	438	531	793	1,999	2,621	2,643	2,643	2,643	2,643
REPORT TOTALS															
PERIOD	0	593	978	932	812	741	1,077	1,879	1,951	2,468	1,673	847	535	3,350	17,836
TO DATE	0	593	1,571	2,503	3,315	4,056	5,133	7,013	8,963	11,431	13,104	13,951	14,486	17,836	17,836

FIGURE 8.3 Time-phased budget. Reprinted by permission of ICF Kaiser Engineers.

N E T W O R K   L O G I C

JOB #
CONSTRUCTION PHASE

KEMS 6.2 SAMPLE DATA

03Mar92-13:13
PAGE 1
DATA DATE 31Dec92

CT REL LAG	ACTIVITY NUMBER	DESCRIPTION	DURATION TOTAL	REM.	EARLY START	EARLY FINISH	LATE START	LATE FINISH	TOTAL FLOAT
AREA:	**PROJECT MANAGEMENT**								
P FF 0	0000.0003	CONSTRUCTION FINISH	22	0		24Nov93		24Nov93	0
P SS 0	3100.1000	COMPRESSION AREA - CLEARING & GRADING	15	0	A16Jul92	A14Aug92			
P SS 0	3200.1000	GAS WITHDRAWAL - CLEARING & GRADING	15	0	A20Aug92	A09Sep92			
P SS 0	3300.1000	SUPPORT FACILITIES - CLEARING & GRADING	31	0	A26Oct92	A07Dec92			
P SS 0	3400.1000	LIQUIDS AREA - CLEARING AND GRADING	15	0	A30Jun92	A20Jul92			
********	3000.0000	CONSTRUCTION MANAGEMENT *	374	241 *	A30Jun92	24Nov93	30Jun92	24Nov93 *	0
AREA:	**GAS COMPRESSION AND METERING**								
P FS 60	1100.2000	COMPRESSION AREA - STRUCTURAL DESIGN	168	0	A02Sep91	A22Apr92			
********	3100.1000	COMPRESSION AREA - CLEARING & GRADING *	22	0 *	A16Jul92	A14Aug92			
S SS 0	3000.0000	CONSTRUCTION MANAGEMENT	374	241	A30Jun92	24Nov93	30Jun92	24Nov93	0
S FS 0	3100.4000	COMPRESSION AREA - FOUNDATIONS & STRUCTURAL STEE	64	64	19May93	16Aug93	19May93	16Aug93	0
P FS 0	1100.2000	COMPRESSION AREA - STRUCTURAL DESIGN	168	0	A02Sep91	A22Apr92			0
P FS 0	2100.7001	PROCURE ENGINE COMPRESSOR UNITS	162	98	A05Oct92	18May93	05Oct92	18May93	0
P FS 0	2100.7002	PROCURE AIR COOLED HEAT EXCHANGER	28	28	A05Oct92	09Feb93	05Oct92	18May93	70
P FS 0	2100.7003	PROCURE PROPANE REFRIGERATION COMPRESSOR SKID	156	92	A05Oct92	10May93	05Oct92	18May93	6
P FS 0	2100.7004	PROCURE FUEL GAS SKID, HEAT EXCHANGER, ETC.	152	88	A05Oct92	04May93	05Oct92	18May93	10
P FS 0	3100.1000	COMPRESSION AREA - CLEARING & GRADING	22	0	A16Jul92	A14Aug92			
********	3100.4000	COMPRESSION AREA - FOUNDATIONS & STRUCTURAL STEE*	64	64 *	19May93	16Aug93	19May93	16Aug93	0
S FS 0	3100.5500	COMPRESSION AREA - H.V.A.C.	25	25	17Aug93	20Sep93	21Oct93	24Nov93	47
S FS 0	3100.5000	COMPRESSION AREA - PIPING	57	57	17Aug93	03Nov93	17Aug93	03Nov93	0
S FS 0	3100.7000	COMPRESSION AREA - ELECTRICAL	38	38	17Aug93	07Oct93	04Oct93	24Nov93	34
P FF 15	3100.6000	COMPRESSION AREA - PIPING	57	57	17Aug93	03Nov93	17Aug93	03Nov93	0
P SS 0	3100.6000	COMPRESSION AREA - PIPING	57	57	17Aug93	03Nov93	17Aug93	03Nov93	0
********	3100.4000	COMPRESSION AREA - FINISHING *	25	25 *	17Aug93	24Nov93	21Oct93	24Nov93	0
S FS 0	0000.0003	CONSTRUCTION FINISH							
P FS 0	3100.2000	COMPRESSION AREA - FOUNDATIONS & STRUCTURAL STEE	64	64	19May93	16Aug93	19May93	16Aug93	0
********	3100.5500	COMPRESSION AREA - H.V.A.C. *	25	25 *	19May93	20Sep93	21Oct93	24Nov93	47
S FF 0	3100.7000	COMPRESSION AREA - ELECTRICAL	38	38	17Aug93	07Oct93	04Oct93	24Nov93	34
P FS 0	3100.2000	COMPRESSION AREA - FOUNDATIONS & STRUCTURAL STEE	64	64	19May93	16Aug93	19May93	16Aug93	0
********	3100.6000	COMPRESSION AREA - PIPING *	57	57	17Aug93	03Nov93	17Aug93	03Nov93	0
S FF 15	3100.4000	COMPRESSION AREA - FINISHING	25	25	17Aug93	24Nov93	21Oct93	24Nov93	0
S SS 0	3100.4000	COMPRESSION AREA - FINISHING	25	25	17Aug93	24Nov93	21Oct93	24Nov93	0

FIGURE 8.4 Network logic (schedule report). Reprinted by permission of ICF Kaiser Engineers.

having each event shown as an individual activity, the Schedule Module would only show one activity covering all the events. The Procurement Module would, however, continue to show each event in the sequence, include durations and schedule dates for up to 13 events. This type of arrangement can be provided for any operation that consists of a series of repetitive sequential events such as vendor prequalification cycles and preparation of construction bid packages. Another data set holds the descriptions for the different sets of events. After filling in the durations for each of the events for an activity, the total duration is transferred to the same activity in the Schedule Module. The start and finish dates are transferred back to the Procurement Module after the network is recalculated. Subsequently the individual schedule dates for the events are also recalculated. Progress of the procurement activities can be reported for any period of time at any level of summary, and action reports can be generated that identify critical items, the responsible individual, and the action required. A procurement status and schedule interface report is shown in Fig. 8.5.

The Installed Equipment Module

The purpose of the Installed Equipment Module is to keep track of all the engineered equipment that will be permanently installed in the project. Three data sets belong to this module, one each for equipment, motors, and instruments. Each item has a unique identification number assigned to it as well as a description. Each item may be linked to an activity in the Schedule Module or Procurement Tracking Module for the transfer of delivery and installation schedule dates. Each item may also be linked to a cost account for cost information accumulation and linked to design drawings for reference purposes. Up to eight properties may be assigned to describe the equipment (weight, size, voltage, capacity, etc.). By incorporating WBS and OBS coding along with status fields, the responsible parties will be in control with timely reporting. An example of a report from this module is shown in Fig. 8.6.

The Engineering Progress Module

The Engineering Progress Module is used to monitor the design phase of the project. The primary data set keeps information on individual drawings, nondrawing documents (e.g., specifications), and tasks (e.g., meetings). Each entry includes a unique drawing (or nondrawing) number for ready identification, a description, and codes to designate whether the item is a drawing, whether it is part of the original scope of work, and whether it has been deleted from the work order. The data for any drawing that has been abandoned is retained in the data set, which allows a complete history of the design effort to be maintained. Budget, actual, and estimated hours and computer hours (CAD/CAE) may be assigned to each entry. Each item is linked to a schedule activity for the transfer of start and finish dates from either the Schedule Module or the Procurement Tracking Module. Predicted finish dates are entered for each drawing and subsequently transferred into the Schedule Module as target (or mandatory) finish dates. Drawing hours can also be transferred to the Schedule Module as required resources for the schedule activities. Progress for each drawing may

PROCUREMENT STATUS AND SCHEDULE INTERFACE

03Mar92-13:14
PAGE 1

JOB #
KEMS 6.2 SAMPLE DATA
DATA DATE 31Dec92

PROCUREMENT PHASE

PROCUREMENT

ACTIVITY #	DESCRIPTION / SCHEDULE START, VAR.	PREPARE SPECS	PREPARE BID DOC	ISSUE QUOTES	RECEIVE QUOTES	ISSUE R.F.P.	AWARD P.O.	RECEIVE VEN DWG	RELEASE FOR FAB	DELIVER TO SITE	SCHED. REQ., V
AREA:	**GAS COMPRESSION AND METERING**										
2100.5001	PROCURE PRESSURE VESSELS A23Apr92 0 DAYS	A23Apr92 10 DAYS				A07May92 18 DAYS	A19May92 26 DAYS	A29May92 48 DAYS	A30Jun92 104 DAY	A16Sep92 122 DAY	A09Oct92 0 DAYS
2100.7001	PROCURE ENGINE COMPRESSOR UNITS A05Oct92 0 DAYS	A05Oct92 22 DAYS				A04Nov92 27 DAYS	A11Nov92 32 DAYS	A18Nov92 50 DAYS	U14Dec92 132 DAY	07Apr93 162 DAY	18May93 0 DAYS
2100.7002	PROCURE AIR COOLED HEAT EXCHANGER A05Oct92 0 DAYS	A05Oct92 12 DAYS				A21Oct92 17 DAYS	A28Oct92 22 DAYS	A04Nov92 37 DAYS	U25Nov92 77 DAYS	20Jan93 92 DAYS	09Feb93 0 DAYS
2100.7003	PROCURE PROPANE REFRIGERATION COMPRESSOR SKID A05Oct92 0 DAYS	A05Oct92 15 DAYS	A26Oct92 20 DAYS	A02Nov92 25 DAYS	A09Nov92 37 DAYS	A25Nov92 45 DAYS	A07Dec92 50 DAYS	U14Dec92 72 DAYS	13Jan93 126 DAY	30Mar93 156 DAY	10May93 0 DAYS
2100.7004	PROCURE FUEL GAS SKID, HEAT EXCHANGER, ETC. A05Oct92 0 DAYS	A05Oct92 15 DAYS	A26Oct92 20 DAYS	A02Nov92 25 DAYS	A09Nov92 37 DAYS	A25Nov92 42 DAYS	A02Dec92 57 DAYS	U23Dec92 72 DAYS	13Jan93 132 DAY	07Apr93 152 DAY	04May93 0 DAYS
AREA:	**LIQUID SEPARATION AREA**										
2300.4001	PROCURE PRE-ENGINEERED BUILDING A02Jul92 0 DAYS	A02Jul92 20 DAYS	A30Jul92 30 DAYS	A13Aug92 35 DAYS	A20Aug92 45 DAYS	A03Sep92 53 DAYS	A15Sep92 59 DAYS	A23Sep92 77 DAYS	U19Oct92 143 DAY	19Jan93 173 DAY	01Mar93 0 DAYS
2300.7001	PROCURE ENGINE GENERATORS A22Sep92 0 DAYS	A22Sep92 15 DAYS	A13Oct92 29 DAYS	A02Nov92 39 DAYS	A16Nov92 47 DAYS	A26Nov92 55 DAYS	A08Dec92 70 DAYS	U29Dec92 90 DAYS	26Jan93 166 DAY	12May93 186 DAY	08Jun93 0 DAYS

FIGURE 8.5 Procurement status and schedule interface report. Reprinted by permission of ICF Kaiser Engineers.

JOB #   KEMS 6.2 SAMPLE DATA

DELIVERY SCHEDULE   DATA DATE 31Dec92

ACTIVITY NUMBER	DESCRIPTION	EQUIPMENT NUMBER	DESCRIPTION	DELIVERY DATES	CURRENT BASELINE
	A R E A : G A S C O M P R E S S I O N A N D M E T E R I N G				
2100.7001	PROCURE ENGINE COMPRESSOR UNITS	C-3100A	ENGINE COMPRESSOR UNIT 6000 BHP	A05Oct92	12Oct92
		C-3100B	ENGINE COMPRESSOR UNIT 6000 BHP	A05Oct92	12Oct92
		C-3100C	ENGINE COMPRESSOR UNIT 6000 BHP	A05Oct92	12Oct92
		F-3605	PRE-FILTERS FOR AIR COMPRESSORS	A12Oct92	12Oct92
		F-3606	PRE-FILTERS FOR AIR COMPRESSORS	A12Oct92	12Oct92
		F-3608	MOISTURE SEPARATOR	A12Oct92	12Oct92
		F-3610	AFTER FILTER	A12Oct92	12Oct92
		F-3611	MOISTURE SEPARATOR	A12Oct92	12Oct92
		F-3613	AFTER FILTERS	A28Sep92	12Oct92
		F-3716A	ENGINE/COMPRESSOR INLET AIR FILTER	A28Sep92	12Oct92
		F-3716B	ENGINE/COMPRESSOR INLET AIR FILTER	A28Sep92	12Oct92
		F-3716C	ENGINE/COMPRESSOR INLET AIR FILTER	A28Sep92	12Oct92
2100.7002	PROCURE AIR COOLED HEAT EXCHANGER	B-3104	WITHDRAWAL GAS COOLER	A16Nov92	12Oct92
		B-3115A	GAS COOLER INTERSTAGE	A09Oct92	12Oct92
		B-3115B	GAS COOLER INTERSTAGE	A09Oct92	12Oct92
		B-3115C	GAS COOLER INTERSTAGE	A09Oct92	12Oct92
		B-3121A	GAS AFTERCOOLER	A25Oct92	12Oct92
		B-3121B	GAS AFTERCOOLER	A25Oct92	12Oct92
		B-3121C	ENGINE JACKET COOLER	02Jan93	12Oct92
		B-3130A	ENGINE JACKET COOLER	02Jan93	12Oct92
		B-3130B	ENGINE JACKET COOLER	02Jan93	12Oct92
		B-3130C	WITHDRAWAL GAS COOLER	A23Nov92	12Oct92
		B-4104	GLYCOL HEATER UNIT (ELECTRICAL)	11Jan93	12Oct92
		H-3123	GLYCOL HEATER UNIT (ELECTRICAL)	11Jan93	12Oct92
		H-3124	GLYCOL HEATER UNIT (ELECTRICAL)	11Jan93	12Oct92
		H-3125	GLYCOL HEATER UNIT (ELECTRICAL)	11Jan93	12Oct92
		H-4123	GLYCOL HEATER UNIT (ELECTRICAL)	11Jan93	12Oct92
		H-4124	GLYCOL HEATER UNIT (ELECTRICAL)	11Jan93	12Oct92
		H-4125	GLYCOL HEATER UNIT (ELECTRICAL)	11Jan93	12Oct92
2100.7003	PROCURE PROPANE REFRIGERATION COMPRES	B-3208	SURGE VESSEL VAPOR COOLER	A12Oct92	12Oct92
		B-3305	REFRIGERANT SUB COOLER	A12Oct92	12Oct92
		B-3308	REFRIGERANT CONDENSER	A30Sep92	12Oct92
		B-4308	REFRIGERANT CONDENSER	A30Sep92	12Oct92
		C-3207	VAPOR RECYCLE COMPRESSOR	A30Sep92	12Oct92
		C-3601	AIR COMPRESSOR	A22Oct92	12Oct92
		C-3602	AIR COMPRESSOR	A22Oct92	12Oct92
		D-3306	PROPANE REFRIGERATION COMPRESSOR SKIDS	A30Oct92	12Oct92
		D-3420	LIQUID METER SKID	A23Oct92	12Oct92
		D-4306	PROPANE REFRIGERATION COMPRESSOR SKIDS	A30Oct92	12Oct92
		V-3307	REFRIGERANT SURGE DRUM	A12Oct92	12Oct92
		V-3701	COOLANT STORAGE	A12Oct92	12Oct92

FIGURE 8.6 Installed equipment module report. Reprinted by permission of ICF Kaiser Engineers.

294

be manually entered as a percent or can be a predetermined value depending on the phase of the drawing. Each drawing may also be assigned to a services cost account for the transfer of budget, actual, and estimated hours and a weighted percent complete value. An example of an engineering progress report and a drawing list are shown in Figs. 8.7 and 8.8.

The Services Cost Module

The Services Cost Module is used to monitor and control the hours, cost, and progress of the KE service-related tasks on the project. Invoices and invoice backup reports to the client are produced with this module. Examples of tasks covered include project services, engineering and design, procurement, quality assurance, quality control, and construction/project management. Each entry in the primary data set includes a cost account number, description, account type (labor, expense, fee), labor rates, indirect (fringe benefit, overhead, cost of money, general and administrative) and fee multipliers, and budgets, actuals, estimates, and earned values for hours, labor and expense, and cost. Labor and expense represent the bare or unloaded dollars, while cost represents the bare dollars plus indirects and fee. A baseline budget is developed by entering budget hours and allowing the labor and expense to be calculated using the labor rate and the cost to be calculated using the multipliers. Actuals can be entered in a similar fashion. Hours for an employee whose rate is stored in the employee data set may be calculated for labor and expense as well as cost amounts. Alternatively, labor and expense and cost dollars may be put on the actual transaction along with hours. Cost accounts may be cross-referenced to an unlimited number of schedule activities for the transfer of start and finish dates. Both early and late dates are moved. The budget data and schedule dates in conjunction with the reporting periods defined in the Organization Module make it possible to produce a time-phased budget.

One of four pre-defined curves can be used to shape the distribution of the budget. Front loaded, back loaded, normal bell, or straight-line curves as shown in Figs. 8.9*a, b, c,* and *d,* are available for the time phasing of cost accounts. User-defined curves may also be entered into the system for cost accounts with special circumstances. The estimate to complete can be distributed in order to develop cash flow estimates. Each account is assigned an earned-value method to establish how the physical progress (percent complete) is calculated. Earned-value methods include weighted progress from schedule activities, weighted progress from drawings, level of effort, actuals over estimate at completion, and manual entry. Each cost account in the Services Cost Module may be rolled into a cost account in the Cost and Progress Module, since project services are a part of the overall capital cost of a project.

The Cost and Progress Module

The Cost and Progress Module is used to monitor and control the capital cost and physical progress of a project. The Cost and Progress Module has the ability to develop baseline budgets, track budget changes, record and commit actual costs, evaluate estimates at completion, monitor progress, and analyze

PROGRESS AND PERFORMANCE

KEMS 6.2 SAMPLE DATA
COST

JOB #
DESIGN PHASE

DET. ACCOUNT NUMBER	DESCRIPTION	PERIOD % COMP.	EARNED	ACTUAL	INDEX	TO DATE % COMP.	EARNED	ACTUAL	INDEX	BUDGET	AT COMPLETION ESTIMATE	INDEX
AREA:	**PROJECT MANAGEMENT**											
1000.00	ENGINEERING MANAGEMENT	4.08	7783	9341	0.83	21.55	41129	46636	0.88	190836	213435	0.89
AREA:	PROJECT MANAGEMENT	4.08	7783	9341	0.83	21.55	41129	46636	0.88	190836	213435	0.89
AREA:	**GAS COMPRESSION AND METERING**											
1100.20	STRUCTURAL	5.00	0	28484	0.00	60.00	75958	91259	0.83	126596	133386	0.95
1100.40	ARCHITECTURAL	5.00	2946	10818	0.27	85.00	50086	52193	0.96	58925	67441	0.87
1100.55	HVAC	25.00	10933	1956	5.59	85.00	37173	23642	1.57	43733	46841	0.93
1100.60	PIPING	0.00	0	53113	0.00	60.00	314879	324757	0.97	524799	546953	0.96
1100.70	ELECTRICAL	30.00	21752	4949	4.40	60.00	43503	22925	1.90	72505	61860	1.17
1100.80	INSTRUMENTATION	0.00	0	4891	0.00	60.00	33836	32855	1.03	56393	57601	0.98
AREA:	GAS COMPRESSION AND METERING	4.04	35631	104212	0.34	62.91	555435	547632	1.01	882951	914082	0.97
AREA:	**GAS WITHDRAWAL AREA**											
1200.20	STRUCTURAL	30.00	26585	4028	6.60	60.00	53170	25999	2.05	88617	86949	1.02
1200.40	ARCHITECTURAL	25.00	6905	1266	5.45	85.00	23478	14962	1.57	27621	29347	0.94
1200.55	HVAC	0.00	0	4776	0.00	60.00	29002	28745	1.01	48337	48682	0.99
1200.60	PIPING	0.00	0	47704	0.00	60.00	285877	283966	1.01	476462	477843	1.00
1200.70	ELECTRICAL	30.00	7251	4891	1.48	60.00	14501	10883	1.33	24168	23708	1.02
1200.80	INSTRUMENTATION	0.00	0	5121	0.00	60.00	35493	34416	1.03	59155	58119	1.02
1200.90	GENERAL	0.00	0	4885	0.00	10.00	4316	7711	0.56	43158	47474	0.91
AREA:	GAS WITHDRAWAL AREA	5.31	40741	72672	0.56	58.09	445837	406683	1.10	767519	772122	0.99
AREA:	**LIQUID SEPARATION AREA**											
1300.10	CIVIL	0.00	0	0	0.00	100.00	152376	155253	0.98	152376	155253	0.98
1300.15	GEOTECHNICAL	0.00	0	0	0.00	100.00	18414	17839	1.03	18414	17839	1.03
1300.20	STRUCTURAL	5.00	5524	2705	2.04	35.00	38669	30097	1.28	110484	104212	1.06
1300.40	ARCHITECTURAL	0.00	0	9667	0.00	60.00	58004	57605	1.01	96674	95868	1.01
1300.50	MECHANICAL	-0.03	-50	9437	-0.01	100.00	92070	100702	0.91	92070	100702	0.91
1300.55	HVAC	4.82	11381	12660	0.90	50.14	84949	84244	1.01	169409	168028	1.01
1300.60	PIPING	26.22	62147	12487	4.98	84.17	198792	151169	1.32	236160	242547	0.97
1300.70	ELECTRICAL	0.55	1922	633	3.04	60.00	142248	78686	1.81	237080	224651	1.06
1300.80	INSTRUMENTATION	0.00	0	0	0.00	39.64	139437	136170	1.02	351707	348025	1.01
1300.95	SPECIFICATIONS	0.00	0	0	0.00	100.00	43158	43158	1.00	43158	43158	1.00
AREA:	LIQUID SEPARATION AREA	5.37	80925	47589	1.70	64.22	968116	854922	1.13	1507531	1500281	1.00
REPORT TOTALS		4.93	165080	233813	0.71	60.03	2010518	1855873	1.08	3348837	3399920	0.98

FIGURE 8.7 Engineering progress report. Reprinted by permission of ICF Kaiser Engineers.

DESIGN FORECAST

JOB #

KEMS 6.2 SAMPLE DATA

DESIGN PHASE

AREA: GAS WITHDRAWAL AREA

DRAWING NUMBER	DESCRIPTION	TYPE	REV.	PHASE	% COMPLETE	ESTIMATE M.H.	ESTIMATE TO COMPLETE C.H.	CURRENT START	SCHEDULE FINISH	PREDICTED FINISH	LAST ISSUE DATE
1200.2001	GAS WITHDRAWAL AREA - YARD FDN PLAN SHEET 1	DOE	1	C	60.00	159.00	40.00	15Oct91	12Jun92	12Jun92	
1200.2002	GAS WITHDRAWAL AREA - YARD FDN PLAN SHEET 2	DOE	1	C	60.00	144.00	48.00	15Oct91	12Jun92	12Jun92	
1200.2003	GAS WITHDRAWAL AREA - YARD FDN PLAN SHEET 3	DOE	1	C	60.00	133.00	40.00	15Oct91	12Jun92	12Jun92	
1200.2004	YARD FOUNDATIONS - SECTIONS & DETAILS SHEET 1	DOE	1	C	60.00	151.00	48.00	15Oct91	12Jun92	12Jun92	
1200.2005	YARD FOUNDATIONS - SECTIONS & DETAILS SHEET 2	DOE	1	C	60.00	133.00	36.00	15Oct91	12Jun92	12Jun92	
1200.2006	YARD FOUNDATIONS - SECTIONS & DETAILS SHEET 3	DOE	1	C	60.00	133.00	36.00	15Oct91	12Jun92	12Jun92	
1200.2007	YARD FOUNDATIONS - SECTIONS & DETAILS SHEET 4	DOE	1	C	60.00	133.00	36.00	15Oct91	12Jun92	12Jun92	
1200.4001	SKID BUILDING - FLOOR PLAN, ROOF PLAN & EXT. ELEV.	DOE	1	D	85.00	136.00	75.00	07Nov91	27Jan92	27Jan92	27Dec91
1200.4002	SKID BUILDING - CROSS SECTION & WALL SECTION	DOE	1	D	85.00	114.00	75.00	07Nov91	27Jan92	27Jan92	27Dec91
1200.5501	REFRIGERATION SKID BLDG #1 - HVAC PLAN & DETAILS	DOE	1	C	60.00	77.00	35.00	20Sep91	15May92	15May92	
1200.5521	CHILLER - P & I DIAGRAM (INPUT)	DOE	1	C	60.00	129.00	57.00	20Sep91	15May92	15May92	
1200.5522	CHILLER - PLAN AND SECTIONS	DOE	1	C	60.00	145.00	72.00	20Sep91	15May92	15May92	
1200.6001	MAIN GAS FLOW PROCESS FLOW DIAGRAM SHEET 1	DOE	1	C	60.00	206.00	39.00	27Aug91	03Aug92	03Aug92	
1200.6002	MAIN GAS FLOW PROCESS FLOW DIAGRAM SHEET 2	DOE	1	C	60.00	225.00	52.00	27Aug91	03Aug92	03Aug92	
1200.6003	GLYCOL RECYCLE STORAGE PIPING PLAN TRAIN 1	DOE	1	C	60.00	129.00	35.00	27Aug91	03Aug92	03Aug92	
1200.6004	PRIMARY SEPARATOR & AIR COOLER PIPING PLAN TRAIN 1	DOE	1	C	60.00	129.00	35.00	27Aug91	03Aug92	03Aug92	
1200.6005	GAS TO GAS H.E. & FINL SEPARATOR PIPNG PLN TRAIN	DOE	1	C	60.00	97.00	26.00	27Aug91	03Aug92	03Aug92	
1200.6006	GAS TO GAS H.E. PIPING SECTIONS TRAIN 1	DOE	1	C	60.00	129.00	35.00	27Aug91	03Aug92	03Aug92	
1200.6007	LOW TEMP. SEPARATOR & PROPANE PIPING PLAN TRAIN 1	DOE	1	C	60.00	129.00	35.00	27Aug91	03Aug92	03Aug92	
1200.6009	GLYCOL REGENERATION SKID PIPING PLAN TRAIN 1	DOE	1	C	60.00	129.00	26.00	27Aug91	03Aug92	03Aug92	
1200.6010	REF. COOLER & PROPANE SKID PIPING SECTS. TRAIN 1	DOE	1	C	60.00	97.00	26.00	27Aug91	03Aug92	03Aug92	
1200.6012	LOW TEMPERATURE SEPARATOR PIPING SECTIONS TRAIN 1	DOE	1	C	60.00	97.00	18.00	27Aug91	03Aug92	03Aug92	
1200.6013	PIPING DETAILS SHEET 1	DOE	1	C	60.00	64.00	18.00	27Aug91	03Aug92	03Aug92	
1200.6014	PIPING DETAILS SHEET 2	DOE	1	C	60.00	64.00	18.00	27Aug91	03Aug92	03Aug92	
1200.6015	PIPING DETAILS SHEET 3	DOE	1	C	60.00	64.00	18.00	27Aug91	03Aug92	03Aug92	
1200.6021	MAIN GAS FLOW P & ID SHEET 1	DOE	1	C	60.00	198.00	39.00	27Aug91	03Aug92	03Aug92	
1200.6022	MAIN GAS FLOW P & ID SHEET 2	DOE	1	C	60.00	194.00	43.00	27Aug91	03Aug92	03Aug92	
1200.6023	MAIN GAS FLOW P & ID SHEET 3	DOE	1	C	60.00	211.00	46.00	27Aug91	03Aug92	03Aug92	
1200.6024	MAIN GAS FLOW P & ID SHEET 4	DOE	1	C	60.00	227.00	50.00	27Aug91	03Aug92	03Aug92	
1200.6025	MAIN GAS FLOW P & ID SHEET 5	DOE	1	C	60.00	243.00	54.00	27Aug91	03Aug92	03Aug92	
1200.6029	REFRIGERATION COOLER PIPING SECTIONS TRAIN 2	DOE	1	C	60.00	32.00	9.00	27Aug91	03Aug92	03Aug92	
1200.6030	GLYCOL REGEN. & PROPANE REF. P & ID SHEET 1	DOE	1	C	60.00	162.00	36.00	27Aug91	03Aug92	03Aug92	
1200.6031	GLYCOL REGEN. & PROPANE REF. P & ID SHEET 2	DOE	1	C	60.00	178.00	39.00	27Aug91	03Aug92	03Aug92	
1200.6032	GLYCOL REGEN. & PROPANE REF. P & ID SHEET 3	DOE	1	C	60.00	81.00	18.00	27Aug91	03Aug92	03Aug92	
1200.6033	EMERGENCY SHUTDOWN SYSTEM P & ID	DOE	1	C	60.00	211.00	46.00	27Aug91	03Aug92	03Aug92	
1200.6060	CHILLER & REFRIGERATION COOLER PIPING PLAN TRAIN 1	DOE	1	C	60.00	129.00	35.00	27Aug91	03Aug92	03Aug92	
1200.7000	GAS WITHDRAWAL AREA SCHEMATIC DIAGRAM SHEET 1	DOE	1	C	60.00	72.00	19.00	10Jul91	26Nov92	26Nov92	
1200.7001	GAS WITHDRAWAL AREA SCHEMATIC DIAGRAM SHEET 2	DOE	1	C	60.00	71.00	36.00	10Jul91	26Nov92	26Nov92	
1200.7002	GAS WITHDRAWAL AREA SCHEMATIC DIAGRAM SHEET 3	DOE	1	C	60.00	72.00	27.00	10Jul91	26Nov92	26Nov92	

| AREA: GAS WITHDRAWAL AREA | 39 | | | | 61.04 | 5227.00 | 1493.00 | | | | |
| REPORT TOTALS | 39 | | | | 61.04 | 5227.00 | 1493.00 | | | | |

FIGURE 8.8 Drawing list. Reprinted by permission of ICF Kaiser Engineers.

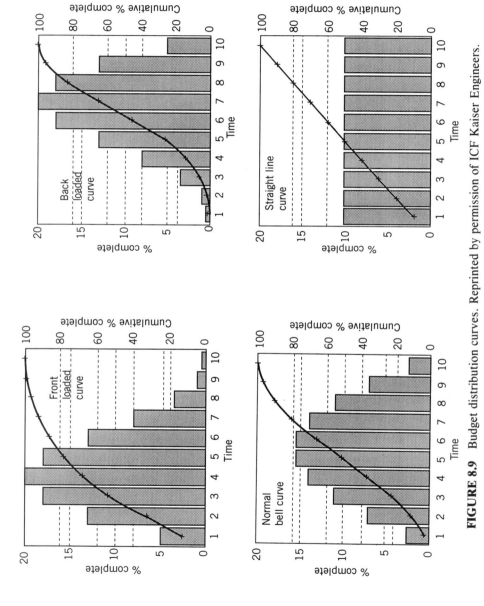

FIGURE 8.9 Budget distribution curves. Reprinted by permission of ICF Kaiser Engineers.

performance. As in the Services Cost Module, budget distributions are calculated using one of the four curves or a user-defined curve. The time-phased budget data with the earned and actual distributions allow budget envelope and performance S-curves and histograms to be plotted. An example of a bar chart/S-curve and two cost and progress reports are shown in Figs. 8.10, 8.11, and 8.12.

Estimate at completion data may also be distributed for cash flow reporting. Rules that apply payment delays of 30 or 60 days may be used to increase the accuracy of the cash flow forecasts. Earned-value methods are assigned to each cost account and include weighted progress for schedule activities, weighted progress from service cost accounts, weighted progress subaccounts, apportionment of other cost accounts, level of effort, actuals over estimate at completion, and manual entry. Based on the earned value, progress payment forms for contractors may be produced automatically.

Exception reporting is accomplished through a secondary data set that allows creation of records for cost variances that exceed defined thresholds. These records may be created at any WBS level and allow explanations to be associated with the performance data.

The Cost and Progress Module is also used for performance reporting. As a minimum, a good performance report must be able to answer the question, "Where are we now?" in a clear manner. Integration of project cost and schedule information is required to eliminate much of the subjectivity with which cost performance is determined. For example, if a project manager has to work with a report that compares actual costs to budgets and a separate report of planned versus actual schedule status, then quantifying the cost attributable to the schedule deviation is either impossible or highly subjective. If, however, budgets are related directly to scheduled increments of work, much of the subjectivity can be eliminated, and an objective report of work accomplishment can be produced to provide a basis for determining meaningful cost performance. Overall schedule performance can also be viewed in dollar terms by simply comparing budgets for completed work to budgets for scheduled work.

Because of the importance of control of the design effort, an additional report is often used. An example of this cost performance report is shown in Fig. 8.13.

Reporting

Each of the modules has a variety of reports in which the data can be summarized, selected, sorted, and grouped according to the WBS, OBS, or sort field level required. There are reports for baseline data, progress data, performance data, time-phased data, comparisons, action items, intradata set relationships, and cost variance explanations. Other special reports can be produced as needed. Reports are available for checking input and marking updates. A complete set of schedule network graphics as well as histogram, S-curve, and bar chart/S-curve graphics are available in the system. In addition KEMS is in full

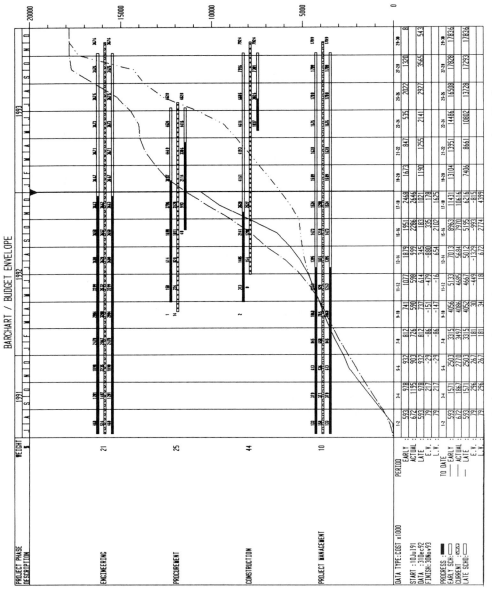

FIGURE 8.10 Bar chart/S curve example. Reprinted by permission of ICF Kaiser Engineers.

P E R F O R M A N C E

JOB #

ALL PHASES - AREA SUMMARY

KEMS 6.2 SAMPLE DATA
COSTS: x1000

03Mar92-13:16
PAGE 1
PERIOD ENDING 31Dec92

AREA DESCRIPTION

AREA DESCRIPTION	PERIOD					AT DATE					AT COMPLETION		
	BUDGET (BCWS)	EARNED (BCWP)	ACTUAL (ACWP)	SCHED. VARIANCE	COST VARIANCE	BUDGET (BCWS)	EARNED (BCWP)	ACTUAL (ACWP)	SCHED. VARIANCE	COST VARIANCE	BUDGET	ESTIMATE	VARIANCE
PROJECT MANAGEMENT	42	38	77	-4	-39	1,784	1,538	1,683	-246	-145	2,017	2,660	-642
GAS COMPRESSION AND METERING	60	111	83	51	28	1,381	1,401	1,404	20	-3	3,887	4,015	-127
GAS WITHDRAWAL AREA	2	253	285	251	-32	2,634	1,998	2,058	-636	-60	2,646	2,737	-91
LIQUID SEPARATION AREA	450	617	582	168	35	3,634	4,036	3,890	403	146	6,642	6,534	108
STATION SUPPORT FACILITIES	644	65	121	-579	-56	1,999	1,643	1,742	-356	-99	2,643	2,778	-134
REPORT TOTALS	1,198	1,084	1,148	-114	-64	11,431	10,616	10,778	-815	-162	17,836	18,723	-886

FIGURE 8.11 Cost and progress report—summary level. Reprinted by permission of ICF Kaiser Engineers.

301

COST AND COMPARISON

KEMS 6.2 SAMPLE DATA
COST

CONSTRUCTION PHASE

DET. ACCOUNT NUMBER	DESCRIPTION	RECORDED PERIOD	RECORDED TO DATE	OPEN COM.	COM.+REC. TO DATE	EST. TO COMP.	ESTIMATE AT COMPLETION CURRENT	ESTIMATE AT COMPLETION PREVIOUS	VARIANCE	CURRENT BUDGET	AT COMP. VARIANCE
AREA:	**PROJECT MANAGEMENT**										
3000.0000	CONSTRUCTION MANAGEMENT	5078	18227	0	18227	52918	71145	71145	0	62775	-8370
AREA:	PROJECT MANAGEMENT	5078	18227	0	18227	52918	71145	71145	0	62775	-8370
AREA:	**GAS COMPRESSION AND METERING**										
3100.1000	COMPRESSION AREA - CLEARING & GRADING	0	76750	0	76750	0	76750	76750	0	80000	3250
3100.2000	COMPRESSION AREA - FOUNDATIONS & STRUCTURAL STEEL	0	0	108000	108000	1094000	1202000	1202000	0	1168000	-34000
3100.4000	COMPRESSION AREA - FINISHING	0	0	18000	18000	126500	144500	144500	0	138000	-6500
3100.5500	COMPRESSION AREA - H.V.A.C.	0	0	184500	184500	102850	287350	287350	0	281500	-5850
3100.6000	COMPRESSION AREA - PIPING	0	0	0	0	221000	221000	221000	0	210000	-11000
3100.7000	COMPRESSION AREA - ELECTRICAL	0	0	0	0	329500	329500	329500	0	310000	-19500
3100.8000	COMPRESSION AREA - INSTRUMENTATION	0	0	123000	123000	38600	161600	161600	0	159000	-2600
AREA:	GAS COMPRESSION AND METERING	0	76750	433500	510250	1912450	2422700	2422700	0	2346500	-76200
AREA:	**GAS WITHDRAWAL AREA**										
3200.1000	GAS WITHDRAWAL - CLEARING & GRADING	0	61400	0	61400	0	61400	61400	0	64000	2600
3200.2000	GAS WITHDRAWAL - FOUNDATIONS & STRUCTURAL STEEL	20914	975998	0	975998	0	975998	960200	-15798	933000	-42998
3200.4000	GAS WITHDRAWAL - FINISHING	3500	3500	11700	15200	101200	116400	116000	0	111200	-5200
3200.5500	GAS WITHDRAWAL - H.V.A.C.	238543	238543	0	238543	0	238543	229280	-9263	224600	-13943
3200.6000	GAS WITHDRAWAL - PIPING	6950	6950	0	6950	169850	176800	176800	0	168000	-8800
3200.7000	GAS WITHDRAWAL - ELECTRICAL	9856	9856	0	9856	253744	263600	263600	0	248000	-15600
3200.8000	GAS WITHDRAWAL - INSTRUMENTATION	2923	2923	98077	101000	30880	131880	131880	0	129800	-2080
AREA:	GAS WITHDRAWAL AREA	282686	1299170	109777	1408947	555674	1964621	1939560	-25061	1878600	-86021
AREA:	**LIQUID SEPARATION AREA**										
3300.1000	SUPPORT FACILITIES - CLEARING & GRADING	5436	84975	0	84975	0	84975	84975	0	88500	3525
3300.1500	SUPPORT FACILITIES - CONNECT UTILITIES	31522	31522	0	31522	0	31522	29050	-2472	28000	-3522
3300.2000	SUPPORT FACILITIES - PRE-ENGINEERED BUILDING	0	0	96000	96000	206550	302550	302550	0	297000	-5550
3300.4000	SUPPORT FACILITIES - FINISHINGS	0	0	17500	17500	150850	168350	168350	0	157000	-11350
3300.5500	SUPPORT FACILITIES - H.V.A.C.	0	0	126000	126000	123420	249420	249420	0	242400	-7020
3300.6000	SUPPORT FACILITIES - PIPING	0	0	0	0	263300	263300	263300	0	250000	-13300
3300.7000	SUPPORT FACILITIES - ELECTRICAL	0	0	0	0	394450	394450	394450	0	371000	-23450
3300.8000	SUPPORT FACILITIES - INSTRUMENTATION	0	0	87700	87700	46320	134020	134020	0	130900	-3120
AREA:	LIQUID SEPARATION AREA	36958	116497	327200	443697	1184890	1628587	1626115	-2472	1564800	-63787
REPORT TOTALS		324722	1510644	870477	2381120	3705932	6087053	6059520	-27533	5852675	-234378

FIGURE 8.12 Cost and progress report—detail level. Reprinted by permission of ICF Kaiser Engineers.

COST PERFORMANCE REPORT (FORMAT 1)

PAGE 1

1. TITLE	
2. REPORTING PERIOD	01Dec91 TO 31Dec91
3. ID NUMBER	
4. PARTICIPANT NAME AND ADDRESS	
6. START DATE	01Jul91
JOB #	
5. COST PLAN DATE	
7. FINISH DATE	31Dec93
8. NEGOTIATED COST	N/A
9. EST. COST AUTH. UNPRICED WORK	N/A
10. TARGET PROFIT/FEE%	N/A
11. TARGET PRICE	N/A
12. ESTIMATED PRICE	N/A
13. SHARE RATIO (31Jul91)	N/A
14. CONTRACT CEILING	N/A
15. ESTIMATED CEILING	N/A

16. WBS ELEMENT

DISCIPLINE	CURRENT PERIOD — Budgeted Cost Work Scheduled	Budgeted Cost Work Performed	Actual Cost of Work Performed	Variance Schedule	Variance Cost	CUMULATIVE TO DATE — Budgeted Cost Work Scheduled	Budgeted Cost Work Performed	Actual Cost of Work Performed	Variance Schedule	Variance Cost	AT COMPLETION Budgeted	Revised Estimate	Variance
MANAGEMENT	47143	25762	28381	(21381)	(2619)	157332	113526	126994	(43806)	(13467)	881886	994397	(112511)
CIVIL	0	0	0	0	0	152376	152376	155253	0	(2877)	152376	155253	(2877)
GEOTECHNICAL						18414	18414	17839	0	575	18414	17839	575
STRUCTURAL	44269	47301	37519	3032	9782	146012	198181	162213	52169	35968	376336	372193	4143
ARCHITECTURAL	43565	11923	24168	(31642)	(12245)	152208	166785	151430	14577	15354	224651	235469	(10818)
MECHANICAL	0	0	0	0	0	92070	92070	100702	0	(8632)	92070	100702	(8632)
H.V.A.C.	28481	12725	19968	(15756)	(7243)	144687	160331	145819	15643	14512	270686	272700	(2014)
PIPING	137339	11381	126193	(125958)	(114813)	717843	876887	836526	159044	40361	1366319	1395091	(28772)
ELECTRICAL	25410	103233	24053	77824	79180	143436	224421	124208	80985	100213	374034	347910	26125
INSTRUMENTATION	46833	16021	12717	(30812)	3303	214929	256699	233476	41770	23223	523648	519735	3913
GENERAL	1374	0	4885	(1374)	(4885)	4359	4316	7711	(43)	(3395)	43158	47474	(4316)
SPECIFICATIONS	0	0	0	0	0	43158	43158	43158	0	0	43158	43158	0
17. WBS SUBTOTAL	374413	228346	277884	(146067)	(49539)	1986824	2307163	2105327	320339	201836	4366735	4501919	(135184)
18. COST OF MONEY	10699	8473	8337	(2226)	137	59605	53085	54728	(6520)	(1643)	131002	135058	(4056)
19. GENERAL & ADMIN.	28532	22596	22231	(5936)	365	158946	141559	145942	(17387)	(4382)	349339	360154	(10815)
20. UNDISTRIBUTED BUDGET											0	0	0
21. SUBTOTAL	413644	259415	308452	(154229)	(49037)	2205375	2501807	2305997	296433	195810	4847076	4997130	(150054)
22. MANAGEMENT RESERVE											0	0	0
23. TOTAL	413644	259415	308452	(154229)	(49037)	2205375	2501807	2305997	296433	195810	4847076	4997130	(150054)

26. DOLLARS EXPRESSED IN: ONES

27. SIGNATURE OF PARTICIPANT'S PROJECT MANAGER

28. DATE

FIGURE 8.13 Cost performance report. Reprinted by permission of ICF Kaiser Engineers.

compliance with the Department of Energy (DOE) and Department of Defense (DOD) Cost/Schedule Controls Systems Criteria (C/SCSC), and the system has the capability to produce all of the required forms for cost performance reports and cost plans that meet DOE and DOD criteria.

System Requirements

KEMS was developed using the ARTEMIS A2 programming language. The current version of ARTEMIS A2 is ARTEMIS 7000 which operates in the VAX/VMS and UNIX operating environments and ARTEMIS 7000/386 which operates in the DOS operating environment. In the VAX/VMS and UNIX operating environments the system requirements must be analyzed for the particular project and are determined by the number of workstations, the proximity of the workstations to each other, and the size and scope of the project. All of these establish the amount of data to be processed and the size of processor that is required. In the DOS environment the minimum hardware requirements are an IBM compatible microcomputer with an Intel 80386 or compatible processor chip with a clock speed of 16 Mhz (for reasonably rapid processing) and 4 megabytes of RAM. More powerful hardware will enable the system to operate faster and more efficiently, and the preferred configuration would be the fastest and most powerful system available at project initiation. The data storage requirements depend on the amount of data to be processed but a 40-Mb hard drive would be a minimum for efficient operation for any size project. A color monitor is recommended to take advantage of the on-screen graphics capabilities, and a mouse is highly desirable for efficient operation. In addition it is essential to have a high-speed, high-quality printer for producing output—a laser printer is recommended since both reports and plots can be produced. A large plotter (E size) is recommended for plotting schedule networks. A modem may be needed to transfer data and an uninterruptible power supply—if the system is located at the construction site—is highly recommended. If the system will be operated on a network, the configuration of the hardware in relation to the network must be determined on a case-by-case basis, which is dependent on project circumstances.

Organizational Factors

The major effect on the organization of a computerized project management system is a reduction in the number of staff required to effectively manage and control a project. It is important to remember that computerized project management systems are only tools to assist project personnel in processing and reporting project data. Many project managers have the misunderstanding that they do not need personnel trained in project management if they have a project management computer system. Since the computerized systems are essentially reporting and data manipulation systems, they merely reduce the data or present it in a particular way. As a result the user of the system must be fully knowledgeable in project management and control procedures and methodologies to be able to analyze the data that is produced and make intelligent decisions for timely corrective action.

Trained personnel are the key to effective use of project management systems. These personnel must not only be knowledgeable in project management philosophy but must also possess a certain level of computer skill and understanding. Projects up to $50 million in value can typically be implemented with one project controls engineer using a single dedicated computer. Projects that are valued in the hundreds of millions of dollars or higher may require several project controls engineers and a dedicated computer systems analyst. Large projects should be set up to operate in a multiple user environment on a LAN to allow real time access to project data so that project personnel from all project functions, such as engineering, procurement, construction field supervision and others, may both enter and retrieve data.

Other Concerns

When setting up a computerized project management system on a new project, there are many things to consider that will affect the overall success of the project. Size of the project and the scope of services to be controlled must be established to ensure proper development of a staffing plan and subsequently a computer hardware requirements plan. Size and scope of the project will also be a factor in determining the computer configuration to be implemented.

Client requirements may also be a factor in the selection of hardware and software. Many clients have their own management information services departments or, on cost reimbursable contracts, will purchase the equipment to be used and will specify the type and configuration of equipment for project use.

Other factors that will determine the computer hardware configuration are the location of the project and the availability of hardware and hardware support in the local area. If the project is situated at a remote overseas location, then the computer equipment selected for use should be locally available or have local support for maintenance. If the project is spread out over a large area or in several locations, or if data such as accounting data must be transferred from another location, then the communication links become a factor in determining the configuration. It would not be wise to invest in expensive modems and communications software if the phone lines to a remote location were not of sufficient quality and reliability to support data communication. In this case couriers or mail links could be established for transporting diskettes for data transfer.

Orientation and training of project personnel is another important area affecting the success of a project. A good training program should be able to rapidly and effectively acquaint users with the operation of the system. This becomes extremely important at remote locations. There are many levels of training that are required on a large project. Some personnel will be fully computer literate, while others will have to be shown the location of the on–off switch; similarly some will be very knowledgeable in project management techniques, while others will require training. A user who has good computer skills and knowledge of project management techniques can usually be trained in about one week. Others will take longer depending on their skill and experience levels.

8.11 HUMAN FACTORS

Volumes have been written on the human factors that make projects success-ful, and it is not the purpose of this section to offer a treatise on this subject. Some important principles do apply, however, and are increasingly used in projects of all types and sizes regardless of the management style of the PM.

From an esoteric point of view it can be said that skills in team building, conflict resolution, and the building of consensus are particularly helpful. While PMs usually have the authority to operate more autocratically and, for example, to unilaterally remove unsatisfactory project personnel, the autocratic approach should be used sparingly to avoid damaging morale. A skilled PM recognizes there are many successful management styles and subordinates need wide latitude in running their own parts of the project; thus slavish copying of the PMs style—or worse yet, mandating his or her style to subordinates—can seriously impact or even defeat the very goals desired. The project team, if well managed, can become cohesive, efficient, and a real sense of synergism can develop.

To properly discharge his or her duties, the PM should operate in a **pro-active** manner. The PM must be adequately involved in the specifics of the project and its ongoing decisions and cannot remain aloof from project activities. The PM, however, should not interfere in the internal workings of the various project departments and groups. This is certain to cause confusion and error, and to destroy any sense of cooperation. The PM is the project team leader and as such sets the example for the team members. The PM's role is leadership, and the project will largely succeed or fail based on his or her ability to motivate and to instill a true sense of **commitment** in the project team members.

Perhaps the most important factor with today's labor force is the need to build **consensus** among the project team. There must be a common goal that is clearly defined and well understood. Whether it is to produce a project with absolutely the lowest cost, shortest schedule, or highest level of quality, the goal needs to be stated in the simplest language possible so that there can be no misunderstandings. On a recent project where schedule was critical, the PM had several large banners posted in the design office listing the number of schedule days remaining until completion with the number on them changed daily. With this constant reminder, there was no question as to what was the goal of the management, and the group performed accordingly.

It should also be remembered that the goal(s) of the project probably will change depending on the phase of the project. Thus **excellence of design** is usually an early goal, while later, when the design is fixed and detailing is taking place, **schedule performance** becomes important. Obviously it is possi-ble for the personnel to become confused, so management has a responsibility to advise the organization whether the goals have changed.

Where there is potential conflict in setting the goals, such as a concern over compromising quality for schedule, there should be sufficient discussion among the project personnel and its management so that the problem is fully aired and the input of the personnel is clearly understood. Where the decision can truly be a shared one, there will certainly be more commitment on the part of the personnel. This will make the work significantly easier and will encourage the best performance from the personnel. The sharing of a decision—if that is the decision of the management—must be made in advance and adhered to so that there is no question that the involvement of the personnel is real, their opinions are given full weight, and the process is not merely an exercise to patronize them. Personnel today are sophisticated, and nothing will destroy morale quicker than what is perceived as game playing by management, particularly where it involves the ideas or commitment of the rank-and-file personnel.

Team building is the natural outcome of consensus building and the sharing of authority and responsibility with project personnel. Considerable thought and effort should go into this, with some companies using seminars, weekend retreats, and ongoing programs to help the project personnel to work together more effectively. As the autocratic approach toward project management is reduced, the personnel will be willing to assume more individual authority, leading to an increase in both efficiency and morale. Further many of the minor decisions that formerly came to the PM can be made at lower levels in the organization. To facilitate this, the PM must make a conscious effort to delegate both authority and responsibility and to explain to the personnel exactly what is being done, who is responsible for what and what their authority is. Following that he or she must stick with it and permit the system to operate. To minimize risk the PM can approach this in a progressive fashion, gradually increasing the responsibility and authority of the personnel, as they demonstrate competance.

Conflict is an inevitable product of the process of group dynamics. While all of the personnel and each of the groups working on a project may share the same goal, their individual priorities and view of the importance of the contributions of others varies significantly.

This problem arises in part because the project will not be successful without the contribution of each person. It is thus natural for each person to feel that his or her work is vital, as indeed it is. The difficulty is to recognize that not all work is vital and further that priority must be given to certain activities and decisions depending on the status of the project. As an example, the PM will not focus on the concerns of the instrumentation group to establish the I/O list for the group's computer applications, in the early stages of the project, when the basic civil engineering criteria for seismic design is being developed. At a later stage the situation will reverse itself with these priorities shifting. A factor of major importance is the need to develop information for the design of major equipment or materials purchases, particularly if they have long lead times. Establishing the performance requirements and specific features for procurement of, say, a liquid nitrogen refrigerator for a cryogenics plant would take priority over more immediate design matters, since the lead time might be on the order of 12 months and it would cause great upset to change these parameters later.

The PM must set the **priorities** and ensure that the various groups understand the need to raise issues of conflicts and priorities and to resolve them as rapidly as possible. The role of the PM is not to make the work force happy, and splitting the difference or finding the compromise position is rarely the correct solution. The correct solution is the best answer to the problem considering the technological requirements, the factors of cost and schedule, and overall impact on the project. If some team members are disturbed, they must recognize that the work is a team effort, that they are members of the team and are expected to support the overall effort. It rarely comes to that but in some cases where persons are disruptive or unwilling to yield to overall project requirements, they may need to be replaced, for overall project harmony is essential.

This does not mean that project personnel are required to agree with every decision made by the PM, but they have the obligation, as professionals, to carry out the decision to the best of their ability. Whatever differences they have with the PM should be aired within the project environment and resolved there. Under no circumstances should they raise these issues with the client or try to use the client to influence the decision. Where this occurs, the PM should immediately and decisively discipline the individual, if necessary removing the person from the project.

8.12 R&D PROJECT MANAGEMENT

The principles to be followed in R&D project management are the same as previously described, with the caveat that both schedule and cost control will be more difficult. This is particularly the case with the research portion of the work. Given the discovery nature of the work it is much more difficult to even define milestones, much less control the work to achieve them on any really predictible schedule. As a result not only the schedule but the cost of the project will often vary widely from that originally anticipated. To provide a better perspective of this type of work, it should be viewed more as time-and-material work than a normal project. If this approach can be taken (or tolerated), false hopes will not be raised and disappointments avoided.

Development work can be scheduled and forecast, more readily and while not approaching the accuracy of conventional project work, it is more nearly evolutionary than research work. Thus the normal techniques of schedule and cost control can be applied to development work, with the recognition that larger contingencies are necessary. Where specific goals are to be achieved on a given timetable, it may be necessary to establish duplicate or multiple teams to pursue the work. During the work particular care must be given to **configuration control,** that is, the establishing of the status of the work at a given point so that the particular revisions of the drawings, specifications, and test data are all clearly defined and can be replicated at a future date if necessary.

One troublesome area is knowing when to **terminate** the work on a unsuccessful project. All research and development work is not successful. While it can be theoretically viewed as a learning experience, the work should be terminated and the personnel reassigned if some economic benefit, sufficient to pay for the cost of the R&D work, is not apparent. As distasteful as it seems, there is a point at which it is necessary to terminate the work and cut your losses. There are usually signs of the work approaching this phase:

- A lack of fresh ideas.
- Economic analyses that require utilization factors in the range of 80% or better to break even.
- Returns on investment that are at or below the prime interest rate.
- A general feeling that there are no more avenues truly worth pursuing and that significant progress cannot be achieved.

8.13 MEASURING PROJECT SUCCESS

It is always useful to determine whether a project is **successful** and to judge the financial performance of the project team, the performance of the individual personnel, and the degree of client satisfaction. To the greatest extent possible quantitative measures should be applied—the normal measures of success involving the meeting of performance criteria tempered by economic ones such as return on investment, rapidity of payoff of investment, and the ability to produce the required quantity and quality of the product. As an example, if projects can produce a payback of the investment in five years or less, the project can usually be considered a success. Similarly, if the rate of earnings on the investment is on the order of the prime rate (averaged over the last three to five years) plus 5%, or if the utilization rate required to break even is less than 65% (of the normal operating hours), the project is typically considered successful.

For some projects the performance criteria may be rather broad, for example, abating a noise source, cleanup of hazardous or toxic materials, and scenic protection. The evaluation of success for this type of work may be more subjective, but some measurements should be identified that will help to quantify the effort.

For certain projects where the project was undertaken for reasons that are not justified fully in terms of profit or economics, there are other somewhat qualitative measures that are useful ways in gauging success. One example is breaking into a new technological field. Here the concern is the acquisition and familiarization of the personnel with new technological areas. This is

---

TABLE 8.5   Measuring Project Success

Quantifiable

Capital cost—Cost versus budget, capitalized cost, ROI, time to amortize investment

Operating costs—Present worth approach, for differential in operating costs. Consider evaluating as an annuity if continuing or long term

Schedule—Value of differential for schedule improvement
1. Reduces escalation of capital cost
2. Additional product produced

Capacity—Value of additional product produced, evaluate as for operating costs above

Quality—Value of reaching grade earlier. Consider the present worth of the value of improved grade or quality over the life of the improved performance

Profitability—ROI, comparison to alternative investments, also prime plus 5%

Nonquantifiable

Client satisfaction
1. Relations during design, construction, and fabrication
2. Operating ease. Some cases may be quantifiable in numbers or classifications of personnel
   Built in safeguards
   Convenience
   Operator training minimized
   Forgiveness of operator error
   Reliability, backups, redundancy, etc.
3. Environmental and regulatory compliance
4. Degree of automation

Penetration of new markets, capture of or increase in market share

Increased capability in new technologies

Maintain staff

Note: Where changes are adverse (longer schedule, increased cost, etc.), evaluations may have negative values.

---

the classic case facing every firm at some point when they wish to broaden their operating base. In some cases the work is taken on a low, or perhaps no profit, basis—with the expectation that future work will prove profitable and will overcome the lost opportunity cost of the money that might have been earned had the firm utilized the staff on more usual work. Where the firm has plenty of normal work and actively seeks new work on this basis, it is fair to assess this cost against the actual work pursued. In some cases it is possible to consider the lost profit as a sunk cost to be amortized over future work and to factor that into the cost charged for future work in the new field. While seemingly a very simple concept, in application it is quite difficult to apply. Once the decision is made to pursue the new work, the sunk costs are incurred, and whether future bidding can be successful when including a pro-rata share of the sunk costs is very uncertain. Perhaps a better way to approach the question is to consider the share of the market to be won for the new technology and to base the judgment on such an analysis. In any event such evaluations are generalized at best and can only serve to provide some level of assurance or caution about work that may be considered.

Where work is taken to maintain the work force, a different set of concerns arises. Typically these relate to the continuing ability of the firm to compete and to be accepted as a serious bidder capable of performing the work that may be bid. There may be additionally a strong interest in avoiding hiring and firing cycles, with the attendant negative effect on the work force. Whatever the reason, it is important that the work force be motivated to perform at the highest professional level. Some firms are very open about the problem of maintaining the work force, and in general when known by the staff, this can be a powerful incentive for them to perform well.

Broadly speaking then, the judgment of whether a project is successful can be broken into two approaches: those that are quantifiable and those that are not readily quantifiable. Table 8.5 summarizes this material.

CHAPTER 9

INTERNATIONAL PROJECTS

While the essential content of a project is likely to be the same whether located in the United States or abroad, there are many factors that complicate the planning and execution of projects in foreign locations. These factors fall into cultural, operational, technical, governmental, and financial areas and have the effect of cutting across all the activities of the project in such a way that virtually every activity is affected.

9.1 BASIC DIFFERENCES

Governments in many countries are not democracies organized as in the United States. Many are of a parlimentary type, with the powers of the parliament often severely limited; others are authoritarian to a greater or lesser degree. As a result it is important to recognize that there are different ways of governing the country, from requirements for driving licenses (in some countries forbidden for women) to internal work permits, residence permits, and exit visas in certain areas. While it is inevitable that comparison will be drawn between the United States and the host country, it is important to bear in mind that, when working abroad, you are the foreigner, and that these countries are **sovereign** and can impose restrictions of any type they wish. While there are many things that may seem undesirable or even ineffective, it is best to keep opinions on these to yourself. The alternate may be expulsion from the host country and removal from the project. Further one should not be too quick to judge other governments until sufficient experience and knowledge of their culture and history has been obtained.

A major difference between the United States and most other nations is the ownership role of the banks. In much of the world the banks own many of the industrial enterprises and literally, often through interlocking director-ates, control both the day-by-day and the long-term operations of them. On a recent project in Western Europe a major bank owned both the client organization and about 200 other enterprises in the nation, including those in shipping, agriculture, manufacturing, transportation, engineering services, heavy construction, textiles, and so forth. As a result there were cases in which members of this family of enterprises were given favorable treatment or awarded contracts without competitive bidding. While no general rules can be given, it is useful to try to learn of these relationships to understand this linkage and the likely implications for the project. It is not useful and is usually counterproductive to try to change them.

The basic **cultural** differences are perhaps the most important ones to affect the project. A project in a Muslim country such as Kuwait would be in a very different cultural climate than one, say, in Taiwan which would in turn have its differences from one in Nigeria. An awareness to constantly bear in mind is that when working in foreign countries, you are the visitor and the host country has a history of cultural and national values including present political ones which will not be changed to suit you. It is necessary to adapt to the surrounding environment without losing the very things that you were brought to the country to provide. In effect you will be working in a multicultural environment, one culture being your own, another that of the local environment, and perhaps a third of the client if located in a different area. If the country does not permit the consumption of alcohol, or restricts gambling, or limits shop hours, or requires that its citizens operate in a certain way, it is important to try to conform to these requirements and not to violate local laws or customs. Many countries do not want to become ''westernized'' and wish to retain their values and culture, so it is necessary to be sensitive to that and to not attempt to convert national personnel to western ways. When ''in-country'' we are guests and should behave as such, and the old adage ''When in Rome do as the Romans do'' is good advice.

The **value** systems of these nations are certain to be different than our own, and it should be kept in mind that the religious, dietary, and cultural customs differ from our own and will affect not only work schedules but the execution of the work as well. This is true even when working in English-speaking nations such as Australia, New Zealand, Canada, and England. Their values have evolved over centuries because of the people, climate, religious or governmental development. Often geography and trade patterns have had a significant effect on the cultural development and have created an attitude toward other nations that reflects itself in a certain way of doing things. Human life often has a different value in other nations, and the entire social fabric is enormously different than what you may be used to. For these and similar reasons it is important to operate in a low profile manner in-country and to try to the greatest extent possible to adopt to the practices of the host country.

Generally host countries are understanding of this and tend to make allowances for foreign personnel. Infractions of local customs are often overlooked, although violations of laws may not be. Overall then, if the project is to get the best from the local personnel and maintain a harmonious working relationship, it is important to try to adapt to the local customs.

The customs of the host country will have an effect on the way in which you conduct yourself. Business is handled in very different ways in different cultures. Often the standards of courtesy are different and require adaptation to local practices. Some questions that should be answered before the first business trip are the following:

1. What is the purpose of the trip or meeting? Will it be decision making, for the exchange of information, or to become acquainted with each other?

2. How is business conducted in-country? Are agents widely used? Is bargaining the norm? Are the real negotiations conducted outside the meeting(s)? Who will the decision makers be?

3. Who will prepare the agenda? Will documents be exchanged in advance?

4. How are business meetings conducted? Do they start on time? Can you expect to be kept waiting? How much time is spent on pleasantries? What is the typical flow of the meeting?

5. Are meetings likely to be conducted in English? If not, what language? Are interpreters needed?

6. Should you have business cards printed in the local language? What are the common practices regarding the exchange of them?

7. What are the pitfalls in local courtesy? Do people shake hands, bow, etc. Are certain gestures to be avoided? Who should sit first?

8. What sort of entertainment is appropriate? Should spouses be invited?

9. Are gifts normally exchanged? If so, what type, value, when given, etc.?

The term **culture shock** has been used to describe the type of change and adjustment that expatriate personnel encounter when they begin to operate in foreign cultures. It can be illustrated in the form of a curve, which is shown in Fig. 9.1.

Upon assignment to the project there is great expectation peaking upon arrival in the host country. Shortly after arrival in-country, however, there is great disillusionment reaching its maximum in about a month, after which, as adjustments take place and the personnel locate housing, schools, a doctor,

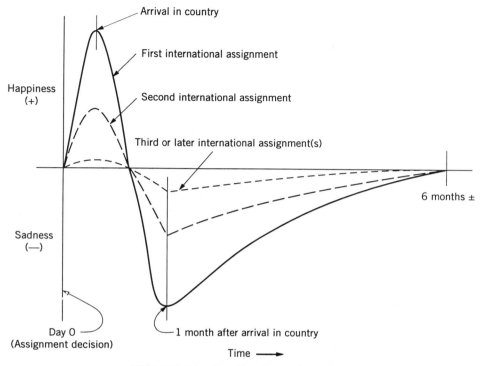

Arrival in country

First international assignment

Second international assignment

Third or later international assignment(s)

Happiness
(+)

6 months ±

Sadness
(—)

Day 0
(Assignment decision)

1 month after arrival in country

Time ⟶

FIGURE 9.1 Emotional reaction curve.

dentist, make friends, etc., they feel progressively better, reaching their preassignment emotional or attitude level in about six months. For a second international assignment the same phenomenon occurs but the degree of enthusiasm or disappointment is only about half as much as the initial assignment, although it still takes about six months to reach a balance. For later assignments the same effects occur, but their amplitude is very small and hardly noticeable.

   Location of international projects is more often than not a problem. Project locations are rarely in or near major cities, and there is usually a lack of facilities for jobsite personnel and their dependents. One example is a project located in the middle of New Guinea several hundred miles from the nearest city in an area where the natives were totally primitive and were rumored to have been headhunters a few generations before. In such a location virtually all services and facilities had to be provided by the project. Even where projects are in more accessible locations, transportation systems are often poor, and a sense of isolation may develop. Where projects are in cities, this problem is reduced, but because of cultural and language differences, some feeling of isolation may still arise, particularly if one is not adventurous.

The convenience and reliability of **communications** systems can vary widely. Usually in-country communication systems are adequate and reasonably reliable, although there are often cases of long waits, as long as two years, for the installation of a residential telephone. In other places the local communication systems are unreliable and fail to operate, for example, when strong winds are blowing. It is wise to set up alternate communication methods, and this may require the establishment of courier systems or amateur radio equipment where permitted by the local authorities. As part of the site survey (discussed later) it is wise to determine the local communication methods, their reliability, compatability with electronic data transmission equipment, and so on.

Communications out of the country may be problematic as well. Many nations today are concerned about information entering and leaving their countries, so "trans-national data flow" is carefully controlled. In these circumstances all communications into and out of the country flow through government facilities and systems, and independent earth satellite or radio systems are either not permitted or severely limited. While this is in itself more of an awkwardness than an outright problem, communications of this sort can be monitored and may not be secure. For these circumstances the PM may occasionally go to an adjoining country to make important confidential telephone calls to headquarters, or a code can be used in conversations. Alternatively, if time is not too critical, mail, telex, or facsimile are rarely interfered with and may be reasonably secure.

Currency differences when working on international projects are always a potential problem. Not only is there a difficulty in bearing in mind the equivalence between the local currency and the dollar, the exchange rate may vary as currencies are allowed to float. If project activities take place in several different countries, there may be three, four, or more currencies at any one time, thus further complicating the conversion question. For this reason, unless the host country currency is extremely strong, it is convenient to express all transactions in one of two currencies, that of the host country for all in-country transactions and dollars for all others.

Most countries do not permit unlimited amounts of their currency to leave the country, and personnel who wish to repatriate their in-country savings may have difficulty in converting the local currency to dollars either directly or by their local bank for electronic transfer to their stateside bank. Host countries understand the need for foreign personnel to be able to repatriate earnings and are usually sympathetic and can provide assistance. In addition the company will normally have some system that makes it possible to convert additional currency. Although it may be tempting, it is not wise to become involved in a black market for money exchange; it is illegal, dangerous, and in the long run merely creates problems for the individual. Perhaps the most important thing to bear in mind is that an international assignment is an opportunity to broaden oneself, and a preoccupation with money and exchange rates tends to distract from the advantages and opportunities of the assignment.

Language differences are always a problem but become particularly troublesome in countries that do not use Latin characters in its written form. With some attention it is possible to learn enough of the spoken language fairly rapidly to be able to ask and understand directions, perform some simple shopping, order meals in a restaurant, and generally get around.

Reading the written language can also be learned fairly readily in most countries, although, as cited above, in places like the Orient, Middle East, or Russia the written language is more difficult to understand. A good way to learn to read the local language is to subscribe to a local language newspaper, which together with local television programming provides excellent practical language usage. With persistance any language can be learned, and the time spent on this will pay handsome dividends both individually and in terms of the effectiveness of one's work. While it may not be possible to truly learn the language, it is essential that enough phrases be learned to say please, thank you, and to handle some of the common courtesies. Even this small attempt to learn and to use the language is appreciated by the local people and usually results in their increased cooperation.

A caveat on the use of the local language pertains to important meetings or correspondence, particularly those involving **contract** matters. For these circumstances, unless you are fully fluent in the language, a translator should be used to ensure that there is no misunderstanding regarding the meaning and intent of the discussion or correspondence.

Most foreign countries have significantly different relative **labor** and **material costs.** In consequence the amount of time spent on design may need to be shifted against the cost of the materials, particularly where materials must be imported and the local currency may not be strong. In these circumstances the design should be refined much more to ensure minimal material cost. By comparison, in the United States, where materials are relatively cheaper and design labor higher, designs often use standard sizes and frequently even approximations to expedite completion of the design. On a project in the Orient it was necessary to move several heavy pressure vessels to a jobsite on a barge, and as part of the design a barge landing area with a sheet piling bulkhead at one end was planned. Personnel in-country determined that rather than build the bulkhead, it would be cheaper to employ local labor to build and later remove a gravel ramp from the barge to the shore for each of the several times it was to be used, since local labor was very cheap and gravel was readily available nearby. Similarly concrete is often transported from the batch plant to the point of placement in 50-lb "head pans" to avoid use of transit mix trucks or conveyors. In both instances the work is performed in a surplus labor area, and it is in the national interest not only to conserve foreign exchange by avoiding the purchase of additional equipment but also to provide employment for the local labor force.

One typical difficulty of working in less-developed countries is the general question of **supply difficulties.** This manifests itself in problems ranging from the availability of plant equipment and engineering supplies to even food or potable water. It is hard to predict in advance what the problems will be, but one can say with certainty that there will be supply difficulties and because of them project personnel will be forced to use less-efficient methods for certain activities. This problem is often exacerbated by restrictions on the importation of certain items that can be produced in-country but that are often either of low quality or limited availability. These difficulties normally do not affect the high-tech items such as instruments or specialized equipment but more common items such as low-pressure valves, core-drilling equipment, and protective coatings (paints). A useful way to identify these problems is to conduct an in-country survey (see Section 9.5), preferably no later than at the contract negotiation or bid stage. Then allowances for these problems can be made either in the contract language or in the project cost estimate and schedule. If it is not possible to perform this survey before the contract is established, it is vital that it be performed immediately thereafter. There will be enough surprises and difficulties with other matters when personnel begin arriving in-country, so additional problems, which could to some extent have been foreseen, are not needed.

Often some percentage of the contract value is required to be furnished in-country. This can often be met with local labor, although targets are sometimes set for the furnishing of materials, equipment, and labor separately. Where the target values cannot be met, early negotiation with the authorities is necessary to avoid significant impact on the project schedule. In some cases it may be necessary to set up fabrication facilities in-country to achieve these goals, and indeed the contract may contemplate such an arrangement.

An associated concept is that of **technology transfer,** whereby nationals of the host country participate at various levels within the project organization. Their purpose is to learn the technology of the project and thus eventually to become sufficiently skilled so that they are able, with a minimum of foreign assistance, to perform virtually all the work themselves on future projects. Often this will take the form of duplicating all the key positions with national personnel who work with the project personnel—seeing all the documents, participating in the meetings, observing the decision-making process, and in general understudying the project personnel. This can often slow down the project, so it is necessary to be alert to its impact to avoid schedule or client relation problems. Many times these personnel will spend much of their time copying all the documents they can lay their hands on whether project related or not. Thus it is useful to segregate the project to prevent nationals from bothering workers on other projects and to protect proprietary information.

Mobilization requires bringing into the host country an ever-increasing number of personnel at the maximum rate, but not so rapidly that their requirements for housing, project working space, and facilities create problems that detract from the critical front-end planning, initial major decisions, and setting up the basic project working procedures. Following contract execution, the PM, together with an accounting manager and perhaps a personnel manager and an administrative assistant, would be the first to arrive in-country. They would be followed, typically within one or two weeks, by a cost/schedule supervisor, a project engineer, and perhaps four to six key project design personnel. In the third group, about a month later, would be about 10 to 20 key personnel to fill out the project team; national personnel are added to the team at this time. Following this mobilization period, more specialized personnel are added on an as-needed basis. The mobilization should take from about three weeks for a small project team to something on the order of three months if the team is a large one. To some extent this schedule is dependent on the availability of housing, which can be hotel, guest house, or permanent quarters. If housing is limited, the mobilization may often be lengthened. Other factors such as the availability of visas, requirements for medical and dental examinations, sequences of inoculations, and background checks of workers could extend the mobilization period, and these should be foreseen during the bidding or negotiation stage.

Letters of credit must often be established before the work can begin. They are in many instances the pacing item in beginning the mobilization itself and can delay mobilization. To minimize the impact on the schedule, some of the early work is done at the home office, but it is limited to only the most pressing activities and rarely includes significant commitments to suppliers or outside parties.

Regardless of location one feature of international work is the fact that there is less help at hand. This requires that the project personnel be more flexible, self-reliant, and able to operate independently. When working at a principal domestic office, it is possible to seek out contemporaries to discuss common problems, possible solutions, and in general to validate courses of action. Overseas this is rarely possible, and despite the best of communications the personnel on international assignments should be selected in part for their ability to operate without a great deal of detailed guidance from the home office. This includes not only the basic content of the work of the project, but the ability also to make the right kinds of decisions. This last item is particularly important, for it is not possible to review all decisions prior to implementation and project personnel with the right feel for the work are essential. In many respects this requires that the people who work on these assignments be more skilled and experienced than their stateside counterparts and that they display considerable **initiative** and willingness to adopt new methods.

Recreation can often be a problem on international assignments. Rarely are the cultural values sufficiently close to those of the United States to permit the same sports or cultural activities. The corollary to this, however, is the opportunity to become involved in new experiences. Local sporting and recreation activities can take the place of those enjoyed stateside and new skills can be learned.

Relations with **consular** authorities is an area completely foreign to stateside personnel, since there is no comparable group in our local governments. Representating the U.S. government, consular personnel will play an important part in most foreign assignments. They are able to provide many suggestions on how to deal with the local personnel. They provide all the typical consular services, including assistance in emergencies, and they can help in developing business contacts in the local community. Often they sponsor chambers of commerce and other types of business groups including the local U.S. business community and can facilitate work in the local economy. They frequently provide libraries of English language books and periodicals, sponsor film series, and similar social and cultural activities.

Consulates are responsible for the general protection and welfare of their citizens living, working, and traveling in the country, and as such they may ask that they be kept advised as to all U.S. citizens in-country, that their passport numbers be registered, and that the normal work location and a contact telephone number and address for each be maintained in their records. This provides significant protection for citizens in the event of political instability or natural disaster. In the case of a medical emergency, or where evacuation of personnel is required, they normally would arrange such matters. Where a death occurs, they would also assist in arrangements for the return of the remains to the United States. While they are not in the business of providing loans to stranded travelers, in certain extreme circumstances they can help make such arrangements. All in all consulates and their personnel play an important part in the in-country activities and can be of considerable help to the project personnel.

9.2   ROLE OF THE PROJECT ENGINEER AND PROJECT MANAGER

The operational roles of the PE and the PM are essentially the same as found on domestic projects. Because of the factors discussed in this chapter, however, their roles will be considerably **expanded,** and they can expect to be given broader authority to make decisions on their own. While representing a challenge, this increases the opportunity for serious error, so more care needs to be shown in the absence of the normal checks and balances found when working in a domestic location.

Relationships with **counterpart** personnel are closer and more continual than on a domestic assignment. Since the in-country firms you will be working with will be attempting to learn the most about the project and the methods used to perform the work, you can expect that there will be many more questions, requests to explain or justify a particular course of action, and studies required. Do not be misled by the apparent lack of ability of the counterpart personnel. They often are lacking in experience, and in many instances are younger than project personnel, but they are almost always the brightest and most promising personnel available and with a modest amount of experience and seasoning become very competent. Often they are the future professional and political leaders of the country, and the contacts formed will be of great value over time. Time spent with them to learn something of the language and culture of the host country will also contribute to the smooth operation of the project, apart from the personal relationships that develop.

A series of relationships will likely develop with host government representatives who will often have a sponsorship, if not an equity, role in the project and its operations. Similar to the counterpart relationships these should be cultivated not only to assist in the operational aspects of the work but also to improve the understanding of the country, its needs, and its concepts. These relationships will likely be more formal, since these officials will often be discharging official duties and will be bound by legal and procedural requirements. Nevertheless, the maintenance of cordial relationships with them is essential if the project is to run smoothly.

9.3 PERSONNEL ADMINISTRATION

Personnel administration is the area that has the greatest difference from practices in the United States. On an international assignment personnel find themselves thrown together more closely and continually. This has the effect of not allowing one to get away from the work, and for some personnel this can become very wearing. Small differences that would never become a problem in the United States may be a source of irritation and frustration, resulting in friction between the expatriates. Where family status is provided and wives and dependents may be present, the problem can become amplified. In this circumstance the employee has his or her project duties, which are similar to those in the states, while the dependents feel thrown into a foreign environment where frequently they cannot speak the language, have little knowledge of the local schooling arrangements for their dependents, don't know where to shop, don't have a dentist or doctor, a beauty parlor, a barber,–the list could go on and on. The problems may become overwhelming for the family, and some percentage of the family status personnel are unable to cope with the environment and return to the United States prematurely. When possible, consideration should be given to the flexibility of personnel and their likelihood of success before assigning them to international projects.

This problem can be ameliorated by creating a system of help, advice, and assistance to families, whereby other spouses and dependents in-country provide advice and guidance to new families for the initial part of their assignment. Typically after two or three months the families are sufficiently oriented to be over the initial effect of the culture shock they experience, and their need for assistance has largely passed.

Being in a foreign environment the expatriate personnel are likely to socialize with each other more extensively and to look at each other as a sort of extended family, and there will arise many more minor problems than in the United States. As a result of this experience, some firms will screen personnel for compatibility before assignment to a foreign project. This has not proved to be too successful, and even in these cases it is a very uncertain evaluation and only where there are clear current indications that the person would have problems should a judgment be made. Naturally, where there are significant health problems, the person should probably not be offered the assignment. But usually health care is not a problem, since medical and dental examinations are mandatory prior to being assigned to an international project. In some cases a certificate or letter provided by the family doctor for each person, including dependents, will meet this requirement.

Personnel assigned to international projects can work in-country on a business trip basis, on a temporary assignment, or as a permanent assignment. **Business trips,** as the term implies, are for short periods typically a week or two. **Temporary status** can be a longer period, often up to six months with special financial incentives such as a temporary increase in salary, a limited shipment of personal effects, and perhaps in-country payments of cost-of-living differentials. Shipments of household effects paid by the company will vary with the length and type of assignment but often 1200 lbs are allowed for temporary assignments of up to six months. Housing is often furnished for temporary assignments as well. Usually temporary status personnel do not bring dependents with them, and if they do, the employee assumes all expenses for them.

Permanent assignments are for a longer period of time, usually two years or for hardship areas often 12 or 18 months. This minimum time is often established to both provide the individual with U.S. tax benefits and to ensure sufficient productive time on the project. Permanent assignments usually carry some form of financial incentive for the overseas assignment, called **uplift.** Uplift is a percentage increase in salary, typically 20% in most locations to perhaps 40 to 50% for remote or hardship areas. For certain difficult assignments a completion bonus may also be offered. In addition more frequent vacations out of country may be available as part of the project conditions. They may include the furnishing of housing or a housing allowance, a cost-of-living differential, allowances for schooling of children, and so forth. Each foreign assignment has its own characteristics and requirements, and the assignment conditions will reflect them. In the end of course it is a question of what does it take to get people to accept the assignment recognizing the factors that make it favorable or unfavorable. Normally an **employment contract** is signed with the employee, which sets forth the salary, working conditions, allowances, work week, duration of the assignment, household effects allowances, and similar conditions. This has the advantage of providing a written record of the agreement with the employee and avoids many potential problems. In some cases for permanent assignments, the company somewhat reduces the weight shipped (often 8000 lbs) but will furnish large appliances (i.e. refrigerator, washing machine, and clothes dryer) in lieu of shipping them from the United States. This has the advantage of compatibility with local electrical power supply and the availability of both service and spare parts.

Married personnel with dependents will cost the project from two to three times as much as single personnel. Thus it is usual to find that there are few married personnel assigned to international projects. Thus assignments providing married status are retained mostly for the senior personnel.

Local hires fall into two categories, **national personnel** and citizens of other countries, commonly called **third country nationals** (TCNs). Local hires, as the term implies, are local citizens hired in-country. Among local citizens hired may be both professional and manual personnel. They are paid in accordance with the prevailing wage scale, although it may be necessary to pay them a bit higher than the local pay rates where there is full employment and the project work is seen as relatively temporary, lasting only a few years. Local personnel should be hired not only for working level positions but, if suitable personnel can be found, for more senior posts as well. In general, the more the local personnel can be integrated into senior project assignments, the smoother the project's operations will be. Some care must be taken, however, in hiring local citizens, since in some cultures, bribes, kickbacks, and similar business practices are common, and the project may need to establish additional controls to ensure proper performance. National personnel can be of great assistance in facilitating approvals by ministries, in obtaining various licenses, and in expediting other similar activities where knowledge of the in-country governmental procedures is important.

Third country nationals are citizens of neither the host country nor the United States. These people are essentially local hires. They may not in fact be resident in the country when hired but may be located elsewhere and willing to travel to the project and work in-country. Usually they are technical personnel not available in-country who are qualified to work at the operational rather than the supervisory level. As a result their employment conditions normally provide fewer benefits, and their pay scales are lower than expatriate personnel. Their pay and employment conditions will vary depending on the market for their services, their skill level, and their personal requirements. A policy for such hires should be developed that recognizes these factors and provides some flexibility to the project administrative personnel. A separate classification system with a typical overlapping set of grades and salaries provides a convenient way to achieve part of this.

Housing on most international projects is a source of continuing concern. Basically there are three arrangements used: project furnished housing, either in a compound or in the general community; open market housing which the project pays for; and open market housing which the employee arranges and for which the employee receives a preestablished housing allowance. These are discussed in more detail in Section 9.10.

Cost-of-living differentials are often provided in the employment contract for international projects as a way to compensate personnel for the added cost to maintain a roughly equivalent standard of living in the host country. Although it is recognized that because of unavailability of commodities and food it is not possible to maintain the same standard of living, an equivalent one can usually be arranged. One device used is a comparable market basket of food, clothing, and similar items, but omitting housing costs. The market basket is costed at the home location in the United States and also at the in-country location, and the difference converted to local currency which comprises the allowance. Periodically, often semiannually, the market basket is reevaluated and the allowance adjusted.

Education of dependents is often a problem on foreign assignments. It is somewhat traditional with British citizens while on foreign assignments, to send their dependents to boarding schools either in Britain or third countries. While not often practiced by American citizens, it has the advantage of ensuring that the quality of the education they receive is predictible. Typically on foreign assignments the dependents attend private schools, since public school education in most countries varies widely or is taught in only the local language, often unknown to the expatriate child. The quality of these schools varies widely, and it is uncertain if the student will receive the level of education desired. On the positive side, the advantages of living in a foreign culture and learning a new language are a real advantage and will often outweigh any shortfall in the schooling. In addition the enrollment at these schools is small, often less than fifty students, and thus more personal instruction is received and close friendships are formed, often lasting a lifetime.

For personnel located at **remote locations** where private or company schools are not found, teaching by the parents using correspondence course materials is often used. The correspondence course materials, including texts and supplementary material, are available from several schools specializing in this in the United States. This system has the advantage of letting the students proceed at their own pace but does not offer the breadth of curriculum found in the more formal school settings. Some have found that this tends to bond the family closer together and thus feel that this system has significant merit, although after three or four years they believe that the student should attend a more traditional school if only to fill in the gaps that may have developed.

Salary payments are typically made both in-country and to a designated bank account in the United States. One system widely used is to pay one-half the employee's pay in dollars into a stateside bank account and one-half in-country in local currency. This has the advantages of providing local currency and avoiding the need to bring additional dollars into the country. Many expatriates have found that among the half-salary paid in local currency, the cost-of-living differential, and the housing allowance, they have sufficient funds to live quite comfortably and that the portion paid in dollars to their stateside account represents a clear savings to them. This, together with the large tax deduction received by overseas workers, provides a significant financial incentive for personnel to work abroad. Some companies have even gone so far as to indemnify overseas personnel against paying more in total taxes than if they had worked in the United States and also against unanticipated changes in the tax rates while on foreign assignments.

In some countries, where it is permitted, the entire salary is paid into a U.S. bank account in dollars, and the employee receives advances in local currency, at the current exchange rate, to cover in-country expenses.

Because the amounts of money saved by an employee and available for investment are often large, it is useful to engage a financial consultant to advise on methods and types of investments to ensure conservation and growth.

Rest and recreation is only a problem on assignments where the in-country cultural conditions are extremely restrictive or where the work location is relatively isolated. Where this is a problem, it is normal to provide periodic vacation time out of country at fairly frequent intervals. This can be as short as every three months to as long as six or nine months and is a function of the employee status, whether married or single, the work week, the number of days and the hours per day, and other working conditions. The time taken for rest and recreation is usually counted as time worked in-country for purposes of the employment contract.

Most employment contracts provide for **home leave** with transportation, including expenses, to the point of origin and return to the foreign assignment being paid by the project. This provides an opportunity for personnel to renew their stateside relationships, look after personal and financial matters, and perhaps obtain medical or dental care that has been deferred. Usually the home leave is granted after one or two years and provides an incentive for personnel to return to their project assignment. The period of home leave is normally two weeks plus travel time.

Passports are the principal identification recognized within the host country and during the early stages of a foreign assignment may often be required on a daily basis. In some countries where exit visas are required to leave the country, it is common for the project managers office to hold all the passports for project personnel. American passports have a high value on the black market in most countries and are often stolen. It is important to guard against loss of a passport as replacement is inconvenient and can be difficult. For this reason it is also wise to register passport numbers with the American consulate to permit their ready replacement if lost. After all the admonitions against never giving up your passport, it is always a surprise the first time one checks into a foreign hotel to be asked to surrender your passport, though they are routinely returned the next morning.

In some countries it is a distinct advantage to maintain one's **tourist** status.

To do this requires that you leave the country typically before 60 or 90 days have expired and to have the passport stamped to demonstrate this. While inconvenient, it avoids the complication of obtaining a local driving license, registering with the police, permits one to drive with tourist license plates (usually less expensive than local ones), and similar restrictions. The advantages and disadvantages for each location vary, and often special governmental waivers are granted to project personnel, so each case must be judged separately. Regardless of the situation, an international driver's license should be obtained prior to leaving for an international assignment.

Personnel **briefings** are an effective way to indoctrinate potential project personnel with the requirements of a particular foreign assignment. They should be held long enough in advance of the assignment to replace personnel who will withdraw when they learn of the specifics of the assignment. The briefings should paint as realistic a picture of the assignment as possible including the negative aspects so that personnel assigned will have no illusions and will be less likely to drop out after arriving at the project location. Where personnel are identified for married status assignments, their spouses should also attend the briefings.

9.4 BUSINESS DEVELOPMENT

Business development activities for international work follow the same methodology as domestic work with two more significant differences, namely the more extensive use of **agents** or **partners** in-country and the likelihood that it will be necessary to either participate in or arrange financing for part or all of the project. Some countries will not permit an extranational firm to operate in-country unless they are in a partnership with a local company, and they may also limit the percentage equity held by the extranational firm. Further, to arrange for introductions to the critical decision makers in-country, it is virtually always necessary to work with an agent who is usually a prominent local businessman. The agent should be utilized not only for the development of business but for the formation of the contract and to assist in solving significant contract problems if they arise in the course of the work. The agent, if well chosen, will be able to identify other projects being considered and can be a great help over the long term. If a local joint company is set up, the agent may become a senior member of the firm, often serving as a director or senior consultant. Usually the agent will wish to remain independent and will rarely become an employee of the joint company. The agent can be paid either a finders fee or a percentage of the contract value for the services.

The importance of maintaining contacts in-country cannot be over stated. If these contacts are at the senior governmental levels, it may benefit future work of the firm. Hence choosing the right agent has vital long term effects.

If it is not possible to find a suitable agent, it is often better to join with a local firm and use their existing business development network.

Similarly it is important to hire both local law and accounting firms to provide ongoing legal advice, particularly with respect to local laws and requirements. Their role is not to take the place of the usual legal and accounting organizations but rather to counsel the project personnel on matters of local importance.

9.5 CONTRACT DEVELOPMENT AND ADMINISTRATION

Contract development would follow the usual steps found in the United States, with the exception that terms that relate to a national participation in furnishing either materials or services should be carefully checked out. An extensive in-country survey is particularly important before the contract is signed to ensure that there are the stated engineering, supplier/fabricator, and construction capability to support the contract requirements. One example was a contract in a third world country that required that all small valves be purchased in-country, a survey indicated that this was possible but likely there would be both delivery and quality problems. As was expected, the valves when delivered were often late, and then only about 60% of them would pass special receiving testing set up to detect quality problems. By knowing of the potential problem in advance, the contract limited the valves to nonspecialty types and thus excluded critical valves such as instrument and safety system components. It also warned the project to order sufficient quantities of valves so that, despite a high rejection rate upon receiving and late deliveries, there was sufficient material to avoid schedule delays.

The in-country survey should also consider the **normal practices.** While not contract items, the use of mechanization, local labor practices, and similar considerations should be incorporated into the project execution planning. In particular, conformance with normal pay scales should be followed to avoid a serious distortion of the local economy.

The **contract** should be certain to state the governing law and language. Often the contract written in the local language will govern, and for this reason it is absolutely essential to have an accurate translation for reference. Similarly a clause on dispute resolution will be invaluable, since there always will be items that require resolution and that can cause, not only great friction, but distraction from the day-to-day operations of the project. A clause setting forth a mediation or arbitration procedure is an excellent conflict resolution device.

Often foreign countries will require that a contract be registered with the local authorities and in certain circumstances with the American consulate. The local agent or law firm can advise about the requirements and proper procedures.

To prevent exploitation of the host country, and to force the company into a greater involvement with the host country economy on a longer-term continuing basis, **repatriation of profits** from a busines in a foreign country is often legally limited. Typically these limit the amount of profit that can be repatriated in any one year and place similar restrictions on the amount of **equity** that can be repatriated in the form of profits or foreign exchange. Usually this is a value on the order of 15% per year. In general the weaker the currency of the host country, the more restrictive their policies will be. However, where a project is sufficiently important to the national interest, the host country may be willing to negotiate a larger amount.

Administration of the contract will follow the normal practices found in the United States excepting that, due to the differences in business customs, administration may be more **rigorous.** Some countries are totally literal, and contract language is interpreted in its narrowest meaning. Under no circumstances should the administration of the contract be less rigorous than would be the case in the United States, since ultimately the client will, at least as a point of departure, judge performance by domestic U.S. standards.

9.6   ENGINEERING

Local practices vary widely around the world, and the execution of the project engineering work should incorporate these practices to the maximum extent possible. One example mentioned earlier was greater refinement in the use of materials, and another the more widespread use of hand labor in lieu of machines. Additionally the availability of indigenous materials will affect design activities. If, for example, cement is produced locally, there will be an incentive to design concrete structures in lieu of using imported structural steel. Further, since most economies also have capability to produce rebar locally, at least in the smaller sizes, this encourages a reinforced concrete structural design. Taking this example further, there will be a significant capability in brick-and-mortar design in a developing country where most structural steel, aluminium, pipe, wire and cable, instruments, and many other processed materials and equipment must be imported.

In addition to the design and importation questions, the local labor force will be more skilled in practices utilizing locally produced materials, so the project may not need to specially train workers. A word of caution, however: despite local labor's experience in using indigenous materials, their practices are often marginal, and there must be consideration given to ensuring that the proper levels of quality are achieved.

To determine any adjustments necessary to the design, it is important that an in-country capability survey be performed. The survey should focus on these factors:

1. *Local practices.* Design methods, use of materials, standard construction techniques, and so on.

2. *Capability of in-country design firms.* Size, sophistications, and scope.
3. *Capability of in-country consultants.*

When assigning work to **local firms** whose capabilities are not well known, it is wise to assign limited, closed-end tasks that permit early evaluation of capability, schedule responsiveness, and cost. Based on the local firm's performance, assignments can be expanded as confidence in their work develops. Wherever possible assignments should be of the block type so that, if difficulties develop, the work can be reassigned to another local firm or sent back to the United States. Where local firms are integrated into the design effort, it is critical that the responsibility for the design not be confused and that the project design work be arranged to be independent of the work of firms not directly under the control of the principal design group. Where major design decisions are dependent on the progress of outside design groups, invariably the design will either be late or of questionable accuracy or quality, and the principal design work will suffer accordingly.

9.7 PROCUREMENT

With the **procurement** area, as with engineering, an in-country survey should be run to establish capability to fabricate or supply materials and equipment likely to be needed for the project. Usually even newly emerging countries will have some capability to fabricate simple pressure vessels, pipe, tankage, rebar, modest weldments, electrical bus bar, and similar noncomplex items. For a major project it may be useful to construct a special fabrication facility. One example is a plant built to produce spiral weld pipe for an oil pipeline in the Middle East, with the skelp being imported but the pipe fabrication (including labor) being supplied locally. At the conclusion of the project the pipe production facility remained in-country and became part of the manufacturing infrastructure.

While these facilities cannot always be justified by the economics on a particular project, often the host government will assist with the financing, since in the long term these facilities are in the national interest.

In some cases it may be possible to import the item(s) into the country in kit form and perform the final assembly and testing using local labor to save cost and increase the national content of the project. This can also form the basis for a future capability if there is a possibility of such work.

Since **shipping** and **transportation** are often considered part of the procurement or material management function. The procurement survey should consider the in-country's ability to receive cargo of large size or heavy weight. The survey should include not only the normal tonnage capacity of the ports, roads, railroads, and airfields but their ability to handle heavy lifts, and extra long or wide items of equipment as well.

The requirements of the local customs authorities should be investigated. Typical concerns are the following:

- What are the documentation requirements?
- What are the import duties by classes of materials?
- What are the normal times to clear various different types of cargo?
- Are there temporary storage and weather protection facilities?
- What are the materials handling capability?
- Are there limitations on the importation of special cargo such as explosives?
- Are there special requirements, such as posting a bond, for materials to be brought into the country, used on project, and then reexported?
- What is the size and makeup of the local labor force? Is it properly equipped to move the tonnage anticipated at the rate required, and so on?

Often the receiving function and the freight clearance operations are performed by a shipping agent. If this is the case, then the capability, experience, and reputation of that organization should be ascertained prior to engaging it. In-country transportation capability should be a part of the survey and cover the areas discussed earlier.

9.8   CONSTRUCTION AND FABRICATION

The **construction** and **fabrication** capabilities in-country should be a part of the survey. Local practices should be incorporated into the design to the maximum extent possible for reasons of cost and familiarity. If extensive use will be made of local contractors and subcontractors, their capability, experience, and reputation must be determined fairly early in the project. In some cases the organization of the project and the physical layout of the work may change because of a local contractor's requirements or work methods.

Heavy lifts are always a concern, whether in the United States or abroad. To avoid difficulty the planned rigging for any heavy lift of, say, 50 tons or larger, should be carefully checked by the project organization. Often such a check will reveal potential problems that require changes in the rigging or in the procedure for the lift itself. For particularly critical operations it may be necessary to engage a rigging consultant to review the proposed methodology.

Vendors' representatives are usually integrated into the construction operation for final assembly and testing work. While few problems result from their work, it is necessary to provide them with the required work crews and to anticipate their arrival so that the project is ready for them when they arrive and that all necessary materials are on hand.

9.9 COST AND SCHEDULE FACTORS

The basic requirements for cost and schedule control remain the same as for work in the United States. But increased emphasis is necessary in the **financial management** area, since international projects are almost always under continuing and often severe financial constraints.

The methods of project financing vary widely depending on the economy of the host country, the type of project, and the general world economy. The most convenient financing arrangement is where the host country has sufficient economic resources to finance the project without outside funding arrangements. Unfortunately, this is becoming increasingly rare. More and more financing arrangements rely on outside lenders or consortia to fund the work.

Where the work is funded in-country, the financial arrangements are similar to those in the United States, and often no unique procedures are required. Where, however, the work is financed by a major developmental bank, such as the International Bank for Reconstruction and Development (World Bank), the Export Import Bank of the United States (Exim), the Asian Development Bank, the African Development Bank, and so on, the recipient usually is asked to provide some portion of the funding from its own foreign exchange. This counterpart funding creates an immediate equity position in the project and requires that the host country (borrower) pay a portion of the project costs, typically 10%, in the currency of the lender. Since the lenders' currencies are "hard," this contribution normally comes from the foreign exchange earned by the host country, and it encourages a deep commitment to the project. Another important incentive is the favorable interest rate that these loans carry, often 2 or 3% below prime, with delays in beginning payments on principal sometimes until after the project is in service a year or two.

The **letters of credit** through which payment to the contractor, designer, or supplier are made typically require that for each and every billing the counterpart payment (often 10%) be made before the balance of the billing (i.e., 90%) is paid (drawn down against the letter of credit). Thus in the administration of the in-country contract there must be strict attention paid to the timing and processing of the invoices and a system developed to minimize delays in the payment of the counterpart portion, so as not to hold up payment of the principal amount of the invoice(s).

Financing of projects by a bank or consortia of banks is becoming increasingly common as the developmental agencies reach their lending limits. These commercial loans have the advantage of less-complicated procedural arrangements and can be arranged more rapidly. Their principal disadvantage is that the interest rates applied to the loans are usually a bit higher, around 2% more than the rates from developmental agencies. In addition these lenders prefer to fund only those proposed projects that have more favorable economic potential, and thus many infrastructure-type projects may not be candidates for such loans.

Of increasing interest are developmental loans for **turnkey** projects in which the project designer/constructor arranges the financing. These loans typically link the project payment or price to performance so that the turnkey contractor assumes some risk for performance. The interest rate these project loans carry is on the order of 2% higher than bank consortia financing and thus are limited to projects with favorable economic performance potential. Many of these projects are planned to pay back the loan over a period of perhaps 20 years. As such, the question of the **stability** of the host country government often is a concern. To alleviate this, insurance is available from the U.S. government to protect domestic companies against forfeiture. The insurance is modestly priced at 1% or less of the project cost.

When working on international projects, the same arrangements for financial services in-country are required as for domestic projects. In addition are the requirements to bring foreign currency into the country. This is normally desired by the host country, since its currency is usually softer than the U.S. or similar currencies. Typically these countries have restrictions on conversion and repatriation of currency. Thus it is important not to convert and bring more money into the country than necessary. A considerable effort is normally expended to minimize this problem, including the practice of advance billings, the widespread use of local currency wherever possible, and very careful cash management.

Where currencies, either hard or soft, are free **floating,** an unfavorable change in the exchange rate can have a serious effect on the financial health of the project. To circumvent this, many companies have adopted the policy to never establish fixed exchange rates either in contracts or in other commitment documents but rather to convert from one currency to the other, for billings, purchases, advances, and all other financial transactions at the exchange rate in effect either at the time of the transaction or the time of the billing.

With increasing world trade with countries that have blocked or extremely weak currencies, the use of **barter** is resurging as a way to finance projects. This is, however, a complex and specialized business unto itself. Unless there is complete familiarity and experience with the barter product, it is invariably better to engage an **agent** to handle the barter activities. Typically involved are supplemental contractural documents, customs and transportation arrangements, importation procedures, conversion to the currency of the contractor (dollars, sterling, etc.), and finally payment to the contractor. The normal commission paid to these barter agents is 5%.

9.10 HOUSING AND CAMP OPERATIONS

In Section 9.1 the subject of **housing** was briefly discussed and three arrangements of housing listed; project furnished housing, open market housing paid for by the project, and open market housing with the employees receiving a housing allowance. In general, the first is the most expensive and carries the most risk for the project. The project that purchases housing units must at the conclusion of the project sell them, taking the economic risk of the rise or fall of their value. From the standpoint of the employees this is the most desirable. The company is taking the entire risk, and the employee is relieved of the problem of arranging for housing. The disadvantage to the project, in addition to the economic risk, is the difficulty in matching the size and availability of this type of housing with the constantly changing needs of the project personnel. It also involves the project in ongoing repairs and maintenance of the housing units—another whole set of problems to deal with.

In nonwestern countries where the cultures are very different, employees are housed in **compounds.** This may be the most convenient way to develop the housing where recreation facilities are required. The compounds can in the extreme case become almost self-contained by including utilities, recreation, dining and social facilities, schools, and even shops. More typical are compounds with housing, recreation facilities, and often schools for dependents. The more elaborate the compound facilities, the wider the scope of problems that must be dealt with. Some basic responsibilities include the construction or contracting for the housing units, their furnishing and maintenance, establishing recreation areas often including swimming pools, restaurants, or snack bars, operating car pools or bus transportation for shopping trips, and security.

Open market housing is a preferred option. It removes the burden of the administrative complications from the project management personnel. Although the apparent first cost may be larger, in the long run this arrangement is usually more practical than providing housing. It permits the personnel to obtain housing suited to their needs as to size and features, and they can also choose where they wish to live. Open market housing can only be implemented in an area where sufficient housing units are available for rent or lease within a reasonable distance, with nearby transportation, and so on. The disadvantage, if the company pays the full rental price for the units, is lack of incentive for employees to exercise restraint in the size and type units, and hence the rent incurred.

A better system that is a variation on the foregoing is for the company to provide a specific housing allowance. This has the further advantage of not tending to inflate rental values as is often the case when the company is paying the entire cost of the rental. The housing allowance is calculated upon the difference for comparable units in the United States and those available on the local market, converted into local currency. Periodically, often at six-month intervals, the allowance is adjusted to reflect changes in conditions.

Camps are operated where housing is not available and where the work location is relatively remote. While most often used at construction jobsites, camps have also been used at locations where largely design and administrative functions are performed. Most of the camp functions are the same as operating a compound, excepting that feeding, security, and medical services assume greater importance. The quality and quantity of food and its service is essential to the work force. If food is of poor quality, there no doubt will be increased turnover of personnel, and many may leave the project prior to completion of their contracts. Companies develop reputations for their camps, and it is a factor in the ability to recruit and maintain a skilled work force.

Where the work force consists of nationalities having significantly different cultural, religious, and dietary practices, it may be advantageous to arrange the camp to provide **separate** areas for the housing, feeding, and recreation of these groups. This will avoid considerable friction and permit each group to continue its practices without interference. This method has been followed successfully in the Middle East and has proved to be useful in other areas as well.

Recreation on site is always a concern because it is necessary to seek the proper balance between the cost of permanent facilities like swimming pools and ensuring that the facilities provided will be utilized. There are wide differences between the preferences of jobsite personnel, and the facilities need to be selected to appeal to the widest group.

On some sites where there are a large number of dependents, it may be necessary to establish a **school.** It is far better to hire teachers rather than to attempt to use family members. The teachers who normally interview for such assignments are usually well qualified and provide excellent instruction. Housing must be provided for them, but if a camp or remote housing operation is established, it is not a significant problem.

9.11 COMPLETION AND CLOSEOUT

Completion and **closeout** of international projects follows the same practices as domestic work with added emphasis on start-up and operations. Normally the contract will require some longer-term involvement to assure the client that the project is functioning properly. This often takes the form of a supervisor resident at the project for up to one year, although six months is more typical. Retention of the final payment until successful performance testing is completed plus a suitable period of satisfactory operation is often used.

Operator selection and **training** is frequently made a part of international project contracts. Typically the client nominates candidates for the key operational positions. These candidates are then further screened by the builder. Those hired usually travel to the design offices for further training in the design features and operational parameters of the project. Where a large labor force will be required, some personnel may be sent to specialized schools at local universities, and/or with manufacturers, to learn the operations and maintenance of particular equipment or systems. The training can last as long as one year if the project is complex and the systems sophisticated. Following that, the employees return to the project in the host country and are integrated into the start-up and operations groups for the completion of the project and its initial operations.

A considerable effort is required to **manage** the training program, not to mention the administration of the client personnel who may be located in half a dozen different locations. These personnel will inevitably have diverse and strange problems, always seemingly at the most awkward times. The best practice is to have the client provide an **administrator** for these employees or, failing that, to establish a senior member of the client personnel who will act in this role. The client representative can handle all the personnel problems and thus permit the project representative to concentrate on the content and progress of the training effort.

PART IV

PROFESSIONAL PRACTICE

CHAPTER 10

PROPOSALS

Proposals represent the principal way in which work is acquired. They are prepared for the total range of projects, from the smallest to the largest, and run the complete spectrum, from the simplest to the most complex. The proposal often represents the first opportunity to present a company in depth to a prospective client and a well-prepared, professionally presented proposal document can present the company in a favorable light.

10.1 BIDDING STRATEGY

The reasons for **bidding** work normally fall into one of three categories, each of which require a somewhat different approach. The most common case is the situation where the work is desired to permit the organization to generate a **normal profit.** In this circumstance the bid preparation follows its normal course and the costing of the proposal including the margins for contingency and profit are the usual ones.

For the second category, where it may be necessary to win the award to maintain all or certain parts of the organization, or to accelerate its growth, the approach must be a bit more adventurous. **Margins** can be reduced, and the work taken as a low or nonprofit project. When making the analysis to proceed, the lost opportunity cost of the funds sunk in the work must be considered as a real cost to the company. Offsetting this is the value to the firm of maintaining the organization, including the skilled personnel and specialists who might otherwise leave the firm. In addition the work may provide an opportunity to expand the experience and data base on certain types of work which may have considerable future value to the firm. One way to evaluate this possibility is to perform a present worth analysis of the value of the profit or revenue generated, over say three years, by the part of the organization that would be terminated, were the work not pursued. Also the ability or difficulty in booking future work with the smaller organization needs to be factored into the decision. In general, these factors are not sufficiently persuasive, and a proposal is offered at very low profit levels only under very severe circumstances. The example below illustrates this situation.

Suppose that a five million dollar job is bid on the basis of only 3% profit before taxes as against a normal 12% profit, the differential of 9% or $450,000 represents lost revenue. If evaluated on a present worth basis of three-year duration, with an 11% discount factor, the present worth of the lost revenue is $329,044—still a considerable amount. If compared to losing say 12 people from the work force and later having to hire and indoctrinate new personnel at a typical cost of $4,000 each, it is apparent that the $48,000 saving in future personnel costs which themselves have a present worth of $35,040 is not sufficient to justify keeping the personnel. In this instance, which is fairly typical, bidding the job on the reduced profit basis is not economically justifiable.

The final case is where it is considered important to enter into an area of **new technology.** This can be accomplished in several ways: by affiliating with or buying a firm already established in the field and thus acquiring the technology, by gradually developing the skills required by evolution, or by hiring experts to provide the kinds of skills desired. The entire purpose of this approach is to position the firm for the long term, and short-term considerations are less important than usual. The offering of bids at low or no-profit levels to win the work is rarely successful in this circumstance, since the client views technical competence as the prime consideration and low price as a secondary factor.

A major problem that arises is the difficulty in most competitive markets of setting future fees high enough to recover the **lost opportunity costs** of the funds not earned by virtue of the low profit approach. One method often taken in this instance is to develop the skills by taking jobs where the new technologies are a smaller portion of the project content and thus gradually developing the expertise desired. As a part of this approach selective hiring of experts in the desired field is often performed. As in the previous case it is necessary to calculate the overall effect on the firm financially so that the true cost of positioning the firm is known.

If the **affiliation** approach is used, there is a hazard that there will be a loss of control, since the firm does not have sufficient technological depth in the field. If another company is purchased, the acquiring firm is often reluctant to allow the acquired firm to continue to operate as it has and may thus damage the ability to operate competitively. While a bit esoteric and difficult to evaluate, these nevertheless are very real problems that must be considered when judging these alternatives.

If the **buy-in** approach is used, it is important to remember that from the standpoint of profitability and productivity of the organization, normal standards cannot be used to judge the project in terms of cost or revenue and that the long-term goals of the decision are the main consideration.

Regardless of the purpose behind the bid, the key element when preparing a proposal is **responsiveness.** The most cleverly crafted proposal with the finest art work and figures, the best presentation, and so forth, will not get serious consideration if it is not responsive to the needs of the client. Regardless of the form and content of the Request For Proposal (RFP), the response must speak to the requirements stated and their particulars. One very effective way to ensure responsiveness is to play back the **words** of the RFP wherever possible. Where differences with the RFP requirements or approach is necessary, it is useful to structure the proposal to respond specifically to the questions and format of the request in its early sections and in later sections to amplify, suggest alternates, and provide reasons and data to support modification of the RFP requirements. Often proposals are evaluated by different groups of client personnel, and it is important that they be able to see **direct** evidence of the compliance on the part of the bidder. Lacking that, the proposal may be given only a cursory review and discarded.

Other factors to be borne in mind are the **effort** to prepare the proposal and the effort the client expends to request and evaluate the proposals. Both activities are significant and consume valuable time and money. It is foolish to waste effort on a poorly prepared proposal, and it is insulting to the client to receive such a document. An early and timely decision on whether a proposal will be prepared must be made and if the decision is no, a courteous regrets letter should be sent to the client immediately. In some cases the client especially wants a proposal from you and may modify the RFP or the due date, and other aspects, to encourage your response. This can be important intelligence. A proposal that is particularly requested will be given extra attention and review and, even if ultimately turned down, improves your position for the next offering.

Where an RFP is general, and wide latitude is provided, detailed responsiveness is not a concern, and there is no problem in offering initially the approach deemed most effective. The concept of providing **value added** to the product, process, or project is often important to the owner and can be used to strengthen the proposal.

In some circumstances the client will not issue a formal request for proposal but will select a bidder and negotiate a contract. Since no scope document, such as an RFP, exists, it is important that the negotiations include the development of sufficient **definition** to ensure a clear understanding of the work and its features, performance requirements, and similar technical details that may not be appropriate to include in the contract itself but that are often made a part of the contract by reference.

This situation is the best possible arrangement for the bidder, but some restraint must be used when developing pricing to avoid the temptation to significantly inflate the profit. Clients have long memories and, once taken advantage of, will be unlikely to contract with the offending firm.

10.2 TYPES

Proposals normally fall into one of four categories:

1. Those that are solicited.
2. Those that are unsolicited.
3. Proposals for government entities.
4. Developmental proposals.

Of these types the **solicited** proposal is relatively easy to prepare, since it has adequate definition to clearly scope the project and its major parameters. RFPs are normally issued by the owner and specify in detail what is wanted and when. Usually they do not specify how the work is to be done, although restraints such as maintenance of operations, noise levels, discharges from the site, and many other similar requirements may be stated. RFPs in the public sector frequently, and in the private sector occasionally, include the evaluation factors or methods to be used in judging the proposals.

Unsolicited proposals, while the most uncertain, are usually the easiest to prepare since there are no stated criteria or requirements to be met other than your own assessment of the need. They must be carefully structured to first establish the need in the client's mind and then to demonstrate how it can be met and why the proposer is the best qualified firm to offer the services described. Unsolicited proposals run the risk of offending the recipient unless a fair amount is known about the business needs of the company. Thus some care must be exercised to ensure that the proposal is specific and truly represents an opportunity for the client. Such proposals also run the risk of being rejected out of hand without any real consideration if they are not fairly precise in their focus.

Proposals for **governmental** entities are highly structured and are discussed in Section 10.6.

Developmental proposals resemble unsolicited ones, since there is usually much less structure to the proposal, and these proposals become generalized project management overview documents as well. In this case the qualifications of the project team members and the experience of the bidder in successfully managing the type of work are most important.

10.3 CONTENT

The **content** of the proposal will be the final factor in determining to whom the work is awarded, and no amount of attractive presentation can overcome basic defects in content. Some RFPs will ask for separate portions of the proposal to cover the technical, management, and commercial aspects of the work, with perhaps a management summary for overall use. This is fairly unusual and normally limited to very large proposals. For the more usual RFP a composite document covering all of these features is satisfactory.

Proper **scope definition** is the most important of all the factors in the proposal. While it may seem fundamental, more proposals have been lost because of this factor than any other. For fixed price proposals it is imperative that the scope be well defined and stated clearly, and precisely to avoid misunderstandings and, more important, to limit the liability of the firm for work that was not originally included. For reimbursable work or development projects, a carefully defined scope permits price(ing) to be evaluated with respect to the type and quantity of work. Where the work is quoted on an hourly basis for different personnel classifications, it still is important to define the scope so as to establish the plan for personnel staffing.

Apart from costing the proposal discussed in Section 10.4, the **technical** and **qualifications** section is basic to the offering. The client will want to know the technical experience and qualifications of the firm. To this end there is normally a section that describes the type and level of technical work done by the firm. This includes particular emphasis upon prior successful work or work currently underway that is identical or similar to that contemplated. The work should be described in sufficient detail to demonstrate this experience. Generally a table listing prior similar projects with brief but specific data on size, schedule problems overcome, cost limitations met, unique technical problems solved, climatic or location problems overcome, and similar information to indicate the experience and skill of the organization is included. The listing must be as direct and relevant as possible, but lacking that, similar or indicative experience can be used. Data on the number of years the firm has been in business and the overall number of projects undertaken is useful, particularly if it has been in business for an extended period of time and large numbers of projects have been completed. Normally the names of references are not provided unless requested by the RFP. Where work in new technologies is to be undertaken, the section on staff qualifications will be of increased importance.

The section on **staff qualifications** and **size** should cover the size of the staff, educational levels, years of work experience, honors and publications, and similar qualifications all with emphasis on the type of work in the proposal. Some care must be taken that when this emphasis is added, it does not mislead the client. The information should be factual and arranged to put the personnel and the organization in the best possible light. With high-tech work it is often useful to identify consultants who will join the project on at least a part-time basis and to use their expertise and qualifications to support the project team. Often they would sit on a Board of Review that oversees the development of the project. Where the work is more conventional, they may be included with the project team.

One concern often evaluated is whether the organization has enough qualified personnel to discharge its present work load and to take on the additional work required. The normal way to resolve this is to indicate the present size of the organization and the personnel required for the proposed work, thus providing a measure of the percentage of the company's work that the new project will represent. For projects running for an extended time, it may be necessary to present this information in the form of a graph or a tabulation running over the life of the proposed project. Staffing concerns can be a bit troublesome. Some clients may want the project to represent the major part of the firms effort to ensure the project will receive the proper attention, while other clients may be uneasy if the project is too large a portion of the work of the firm and there may be insufficient resources to cover unexpected developments. Little can be done about this except to try to make the point elsewhere in the proposal that the firm is prepared to give the project its top priority in personnel and management attention and that, if needed, additional resources will be provided to ensure proper quality and on time performance.

For proposals where **shop capacity** and the availability of specialized equipment are important, the demonstration of these follows the same approach. Tabular or graphical data indicating the present and future shop load with and without the proposed project work will provide the information required. For long projects the table can list total shop capacity and firm work off by quarter, while for shorter projects the tabulation can list the information on a monthly basis. Where the capacity does not appear adequate to handle the additional work, it is useful to point out that additional shifts would be used to produce the work to assure satisfactory schedule deliveries. When using this approach, it is necessary to describe how the required numbers of qualified shop personnel, including supervision and management, for the additional shifts will be provided. There is an underlying concern that the personnel, if newly hired, will be inexperienced, so the work will neither be of acceptable quality nor completed in a timely fashion. Similarly the experience with specialized shop processes and the equipment to provide them may be an important concern. This can be dealt with in the same manner as the question of equipment generally.

Key personnel are always an important subject. The client will want many personnel classified as key so that, as is often the case in the contract, they cannot be reassigned without client permission. At the same time the bidder typically wishes to minimize the number to permit the greatest flexibility in their assignments. If the RFP does not identify the category of key personnel and which project positions they hold, the bidder can preempt this by identifying the key personnel in the proposal and thus hopefully limit their number. Where this issue has been raised in the RFP, a statement to the effect that the firm will work with the client to minimize disruptions in those rare cases where it may be necessary to reassign key personnel can be helpful. Normally this issue is raised in the contract negotiations rather than in the RFP.

Often the **resumes** of key project personnel are included in the proposal. Except for the most extensive proposals it is useful to limit the resumes to one page for each key person. As indicated previously, it is useful to tailor the resumes to the project, emphasizing related experience that will be of interest to the client.

The **stability** of the firm and its ability to provide **financial resources** to ensure satisfactory completion are usually a requested item in RFPs. For many RFPs this is handled by requiring that the bidder furnish both bid and performance **bonds.** The bid bond is a warranty by a third party (surety), usually an insurance or casualty company, that ensures that if the bidder is awarded the work it will be able to furnish a (contract) performance bond in the amount of the bid. The premium for a bid bond is usually fairly small, around $50.00 for a single bid or $200.00 to cover all bids for one year.

Performance bonds are similar to bid bonds but are a financial guarantee of project completion (i.e., discharge of the contract) in event that the bidder fails to perform. Again offered by a surety, the premium for this bond is usually between 3% of the first $100,000 of contract and 2% for excess value if no materials are supplied. Where materials are supplied, the premium is typically 1.5% with a minimum premium of $300.00. Well-managed companies having a good performance record on the type of work being bid are likely to be able to negotiate somewhat lower rates than these. Further, as the size of the project grows larger, the sureties are typically willing to negotiate the cost for their bonds, and the percentage paid could drop to one-half or less than that paid on a smaller project. Most of the major insurance companies offer this type of bonding service.

To qualify for either type bond requires a progressive experience record upon which the bonding companies base their premium rate. If a firm has only been doing half-million dollar projects, it will have difficulty obtaining bonding to bid a twenty-million dollar job. On the other hand, if its experience record is poor, even if it has done the same size project, it will likely be forced to pay a higher than average premium were it to obtain bonding at all. Qualification data similar to that in a proposal is furnished to the bonding company to support a bond application.

Other forms of financial information include requests to furnish **credit** reports such as Dun & Bradstreet reports or **certified financial** statements. Of the two, the credit reports are more desirable, since they do not disclose confidential information and utilize to some extent data furnished by the firm itself. They also tend to present the firm in a more favorable light and are less likely to require explanations and discussion. Where credit or financial statements are required, they should be furnished as separate confidential supplements to the proposal.

Schedules included in the proposal should be brief and not overly detailed. Generally a 20-line overall bar chart, or something shorter, should be used to outline the project schedule. Major milestones should be noted along with important restraints, but only major events should be indicated, for much of the schedule development will take place during the initial phases of the work. Not all critical or long lead items should be shown, merely the longest few. Where uncertainty exists, it is useful to discuss it, not from the standpoint of uncertainty, but from the standpoint that following contract award, the initial steps will include "a detailed evaluation and resolution of . . . ." In some cases it may be necessary to include more detailed schedules of certain critical activities to support the overall project. Where this was not called for in the RFP, it can be used to strengthen the proposal if attention is called, for example, "to the importance of an early decision . . . as indicated by the brief schedule of . . . activities enclosed."

With increasing concern about the **quality** of the work product, it is wise—whether asked for in the RFP or not—to include in the proposal some discussion and information on the quality assurance and quality control programs and methods employed. Chapter 2 included information on this. If it can be shown that a concern for quality is an integral part of the way the company performs its work and that it costs the client nothing extra, this can be an added feature of the proposal. Where the RFP has asked for specific information on these programs, a full and specific response should be included in the proposal.

10.4 COSTING THE PROPOSAL

The process of **costing** the proposal will follow the same steps as developing the project cost estimates described in Chapter 6. For convenience, a brief overview is given below. It emphasizes the differences that affect proposal pricing.

Bids should only be prepared to the level of detail necessary to support the requirements. To the extent practical, they should be of a relatively uniform level of accuracy throughout, although it is widely recognized that variance of accuracy will occur in the different bid items. More important, they must be timely. In some proposals it may be necessary to present the estimate of costs by category. When this is required, the costs are often grouped as shown in Table 10.1.

Direct costs are those related directly to performance of the work. They cannot be general business costs, for these are allocated to the activities of the firm as overhead. Examples in direct support of the work are labor, payroll additives, permanent equipment cost, rented equipment, transportation of materials, office rent, project insurance, bonding, and outside services.

Indirect costs are those general costs necessary for the firm to carry on its business. These costs represent support costs of a more general nature. Typical are support staff salaries including payroll additives, rent, depreciation, utilities, insurance, security, legal services, interest, loan payments, sales costs, public relations expenses, and professional society participation. Broadly speaking, it is possible to have the same type of costs appear in both the direct and indirect categories, the difference being whether the costs are expended in direct support of the project or for the overall benefit of the

TABLE 10.1 Proposal cost categories
Direct costs
Office labor
Field labor
Permanent plant equipment
Equipment rental
Indirect costs
Office labor
Field labor
Consumables
Tooling
Permits
Construction/fabrication support
Subcontracts
Overhead[a]
Taxes
Escalation
Contingency
Bonding
Fee or profit

[a] Can be grouped into one item or more commonly shown against the particular cost category. The term *general and administrative* (G&A) is often used to include groupings of overhead and administrative costs, particularly for governmental work.

organization, in which case only a pro-rata share (the overhead rate) is chargeable to the project.

Certain overhead costs should not be borne by projects and should be in still a separate category. Entertainment is an example of this and should be recovered from overall earnings.

Where the contract is a **reimbursable** one, the makeup of reimbursable costs is usually a point of extensive negotiation during the contract formation phase. There should be a carefully considered position regarding the inclusion of these items in the proposal to avoid possibly losing the work because of a perception of high cost while at the same time preserving a position that permits the negotiators to make some concessions during contract discussions.

When preparing the cost estimate for the proposal, it is important to follow the principle mentioned earlier—which is to have all the **contingency** in a single place. The estimate should be prepared based on the most realistic estimate of the individual items in the scope with the idea that the work will proceed as planned. As a separate element the overall amount of contingency desired for the entire project should be established by senior management based on their view of the work, its risks, and the management's objectives. If the contingency is spread throughout the estimate, it will distort it to the point that the estimate soars, and more important, there will be no way of knowing where the contingency is added and how much is present in the individual items. As a result it may be virtually impossible to correct the estimate, and likely the job will be lost because of a high bid.

Bid quantities may be requested in certain RFPs and should be stated in the terms requested. While it is normal to assign higher costs to early bid items and lower costs to later ones, to improve the front-end cash flow from the project, if carried to an extreme this will generate suspicion and may cause loss of the work. The best advice that can be given is to maintain a degree of reasonableness to the cost figures assigned. It is well if they can be documented by work papers that support the values assigned and that can be utilized if the question arises. Thus the pro-rata distribution of, say, mobilization costs, where they are not a separate line item, would be more heavily weighted against early units of work.

Escalation, if a part of the costing of the proposal, should be considered a separate item. When necessary, these provisions can be estimated and included in fixed price or unit priced proposals if there is no provision for handling these costs separately. Where the RFP is silent on this issue and the project will run over a considerable time, the bidder should consider stating escalation separately with a suitable explanation. If it is possible to clear this with the client prior to submitting the bid, it often is helpful and avoids a possible adverse reaction.

Frequently the cost of **bonding** is requested as a separate bid item. This is the normal case with governmental proposals. Some care needs to be exercised to ensure that overhead percentages, escalation, and profit are not applied to the bonding costs and thus inflate them.

Entertainment was spoken of earlier. Entertainment costs should not be in the estimate if the project is a reimbursable one. If it is a lump sum project, the inclusion of such costs is not a problem. But entertainment allowances do tend to increase the cost of the bid.

Fee or **profit** is of course the payoff for a successful project and as such should be carefully evaluated. It should be included in the bid cost and pro-rated to the major bid items if a bid or line item breakdown is required. It is unwise to disclose the amount of profit calculated into the bid unless the bid form requires that it be stated, and even in this case it should be avoided if possible. If profit must be differentiated, it is useful to somewhat understate the amount estimated to avoid an adverse reaction from the client, while recognizing that the client has access to bids and other data that would indicate normal ranges of profit for the work contemplated.

One hybrid form of fixed price arrangement is a **unit priced** contract. In this document firm unit prices are required for work where it is not possible to accurately determine the quantities, and the number of units can only be estimated. A common example has to do with concealed or underground work such as earthmoving. In this case various categories of work might be described and their unit prices requested in the bid request. An RFP might contain the following categories of work with estimated quantities and require the bidder to fill in the unit price bid: Drill, shoot and excavate rock $8.04/yd; rip and excavate rock $4.62/yd; and general excavation $2.52/yd.

Where these items represent a significant cost factor, and since large volumes of work can be handled more efficiently, the party requesting the bid will often establish a **tiered** pricing structure with larger quantities paid at lower unit prices. It is better to establish these larger volume (and lower unit) prices as a part of the bid while competition exists than to try to negotiate them later, when the owner has little negotiating leverage. In the example that follows the first quantity level values (i.e., 3000, 10,000, and 20,000 cu yds) are slightly over the estimated quantity, with the larger values for each category included to provide a suitable margin. If the actual quantities are significantly larger, a lower unit price will apply.

Class I Excavation

Drill, shoot, and excavate rock	0–3000 cu yd	$8.77/cu yd
	0–4000 cu yd	$8.04/cu yd
	0–5000 cu yd	$7.32/cu yd

Class II Excavation

Rip and excavate rock	0–10,000 cu yd	$5.46/cu yd
	0–15,000 cu yd	$4.62/cu yd
	0–20,000 cu yd	$3.78/cu yd

Class III General Excavation

	0–20,000 cu yd	$2.85/cu yd
	0–30,000 cu yd	$2.52/cu yd
	0–40,000 cu yd	$2.20/cu yd

A similar approach is often used where it may not be possible to accurately estimate the effort required in service or developmental proposals and to price the services of personnel, equipment, and facilities on a unit price basis. Thus a proposal might include unit rates for personnel by different classifications, equipment by type and capacity, and facilities based on size (square footage) or some similar measure. It is also common for certain building trades to bid work on a unit priced basis ($x per electrical outlet, $y per square for roofing, etc.).

10.5 REVENUE EVALUATION

The basic criteria to judge whether or not a proposal will be offered and at what pricing level must include the question of expected **revenue.** In normal circumstances the pre-tax revenue to be earned is a function of the type of work, its complexity, duration of the project, and the prospect for follow-on or continuing work.

Each firm will have some **historical** data on which to base its revenue and profit estimates. These values modified by the circumstances above can be used for analysis. Where historical data do not exist or vary widely, and for single projects where the prospects for follow-on or continuing work are slight, the normal percentages of profit will vary significantly. For a good overall guide to these percentages and other important statistics, the Dun & Bradstreet Information Services publication *Industry Norms & Key Business Ratios* contains useful statistical data on over 800 lines of businesses, updated annually.

Where additional units are likely to be sold, it is often possible to reduce the offered price slightly by assuming that some of the first-time sunk costs will be spread over several units. As a result estimates of the revenue and profit to be earned on the first project, or unit, will be somewhat inflated. Unless there is some form of written option document, or better yet a letter of intent, the basis for such pricing is tenuous. If only one project is in fact performed, the reduction in price that was predicated upon additional units all becomes chargeable to the profit earned on the single unit and can significantly change the economics of the project. To ensure an understanding of the financial risk involved, this possibility should be evaluated as the potential worst case.

A prime concern in operating an engineering practice is the basis on which the engineering services will be rendered. Generally the contractual bases available are lump sum, cost plus fee (either fixed or percentage), and unit priced.

Lump sum contractual arrangements are preferred by clients because the client assumes no risks for overruns by the engineer. Since the engineer assumes all risk, he or she must take great pains to carefully describe the **scope** of activities such that the lump sum offering can be properly defined and **limited.** With a lump sum arrangement, **scope changes** are the basis for "extras," which require negotiation and can be acrimonious. Nevertheless, where indicated, these scope changes should be pursued, since their cost is directly a "bottom-line" cost to the engineer. Particular care should be given to time delays, as these can have serious effects on lump sum work, usually causing cost overruns.

Cost-plus-fee arrangements are preferred by the engineer when scope cannot be well defined or is subject to changes, when schedules are indeterminant, or when new technologies are being utilized for the development of the engineering work. In a cost-plus-fee arrangement the client assumes the risk for overruns but gains the benefit of underruns. Since it is a separate item, the fee represents a secure form of profit for the engineer. However, the fee may not be totally net profit, since entertainment and other costs may be charged against it. Obviously, for those cases where some form of reimbursable contract is entered into, the client will expect to be involved to a much greater extent in engineering decisions made on a day-by-day basis. This has the potential disadvantage of affecting the engineer's freedom. Client involvement can also adversely affect schedule. The client may have difficulty understanding the effects of even small changes that may be requested. In the lump sum contract the engineer can ask for total freedom with no restraints or involvement by the client until the finished product is delivered, because the engineer is furnishing the entire package and assuming the financial risk if the lump sum work overruns its contractual amount. It is not possible to take this position with cost-plus-fee work.

The fee determination for either **fixed** or **percentage fee** can be arranged as necessary to suit the preferences of the owner. Typically fixed fees are used for cases where it is possible to reasonably establish the scope in a very broad way, and the fixed fee is not normally subject to renegotiation except in cases where the scope changes have been significant, say, on the order of 20% or more. The use of percentage fee is not usually favored because of the possibility of expanding the basic scope to automatically earn the additional fee.

For some work where the scope and the extent of the engineering is unpredictable or difficult to anticipate, unit pricing may instead be used. Typically the unit pricing includes both **direct** and **indirect payroll costs, overhead (burden), and fee,** all structured on the basis of dollars per worker-hour. **Unit pricing** will usually establish different unit prices for different categories of work. Thus the unit price in dollars per hour for a drafter will differ from that for a checker, a design engineer, a supervising design engineer, and so on.

Pricing for engineering services typically includes (1) direct payroll (gross amounts paid employees); (2) an allowance for indirect payroll costs, which is a direct multiplier on payroll costs, ranging from 25 to 40% to include the costs for employer payroll taxes, vacation, sick leave, retirement, and so on; (3) an overhead or burden percentage on the order of 75 to 100% to cover a large variety of office expenses (e.g., office rent, furniture and fixtures, insurance, personnel recruiting, training, development and management, legal services, technical society participation, and sales costs including proposal preparation); and (4) an allowance for fee (profit) often arbitrarily set at 10 or 15%.

Typically the overall multiplier of direct payroll (including indirect costs) will range from 2 1/2 to 3 1/2; this includes fee (profit) plus overhead costs. In providing engineering services, there are frequently costs incurred for laboratory testing, transportation and travel, consultants' fees, and so on—all accrued for the benefit of the client. These costs are often passed along to the client with no or a modest 10 to 15% markup. Normal engineering practice does not include the addition of fees to these types of costs.

The ASCE has published fee curves that, although intended for civil engineers, are widely used as guides in engineering consulting practices more generally. The fee curves represent total remuneration, as in the case of lump sum work, and not merely that portion of fee (profit) described earlier. These curves are shown in Fig. 10.1 and Fig. 10.2.

When pricing equipment that consumes large quantities of **spare** or **wearing** parts over its life, some firms offer low or no profit prices with a view to making the profit from the spare parts that will be sold. A large manufacturer of a safety item regularly prices a basic equipment item slightly below cost to ensure that the item is priced sufficiently low to be selected. The manufacturer makes the profit from the replacement parts for the equipment that can only be purchased from its firm. Over the life of the equipment, this results in a very high profit as the volume of the replacement parts is high. Such pricing is a fundamental strategy of some companies, and an individual project is unlikely to initiate a basic change in the way the strategy is implemented. There may be instances where it is appropriate to consider such an approach.

Operating contracts are an effective way to not only generate continuing profits but also permit the firm to position itself for significant future projects. With personnel working closely with client personnel, much early information on long range plans and prospects for projects becomes known, and it is possible to either offer unsolicited proposals or to be in a better position to respond when an RFP is issued. When taking on operating services work, however, the work must bear all its own costs and should not be subsidized. Usually these services are priced on a unit rate basis (i.e., dollars per hour for each classification of personnel employed). Lump sum or fixed price arrangements are only used where it is possible to exactly fix the hours worked, classifications used, supplies needed, support required from the parent organization, and so on. But this is rarely the case. The more usual arrangement is the unit price approach. The support or overhead costs (both direct and indirect) will not normally be as high as with conventional work, since office space and much other support are provided by the client. Off-setting these factors is the ability to charge fairly high **unit rates** to reflect the fact that the personnel provided tend to be more highly graded than the average employee, since they will fill the senior positions and will be required to operate more independently. If they remained in the home office, they could direct groups of personnel in the customary work of the firm and as a result could generate significant revenue. Hence their assignment to operations represents a lost opportunity cost for the revenue they would otherwise generate. In addition they carry more responsibility for proper operations to avoid damage or downtime at the facility. The question of damage to the facility can be guarded against by suitable insurance, but the exposure to downtime is a significant concern and can be used effectively when justifying these senior positions and their accompanying rates.

Operating services are often of an interim nature, lasting only until the client can recruit, train, and indoctrinate its own personnel to operate the facility. As a result the profit earned from this work and its likely continuation should be heavily discounted for future years. A typical arrangement has the owner initially providing at least 50% of the positions, with this growing

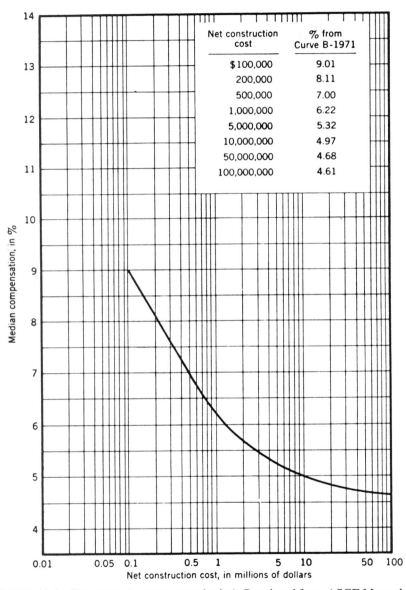

Net construction cost	% from Curve B-1971
$100,000	9.01
200,000	8.11
500,000	7.00
1,000,000	6.22
5,000,000	5.32
10,000,000	4.97
50,000,000	4.68
100,000,000	4.61

FIGURE 10.1 Fee curve (average complexity). Reprinted from ASCE Manual #45, *A Guide for the Engagement of Engineering Services*. Copyright © 1981. Reprinted by permission.

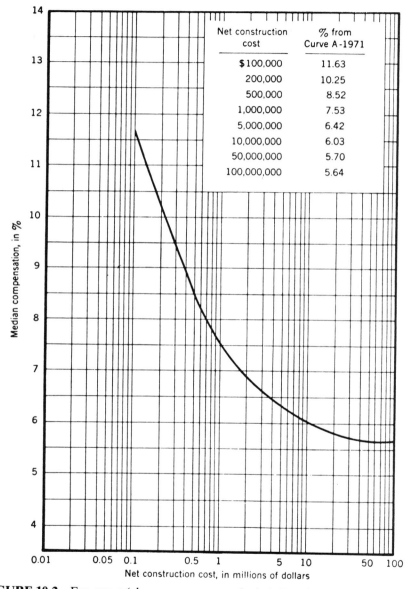

Net construction cost	% from Curve A-1971
$100,000	11.63
200,000	10.25
500,000	8.52
1,000,000	7.53
5,000,000	6.42
10,000,000	6.03
50,000,000	5.70
100,000,000	5.64

FIGURE 10.2 Fee curve (above-average complexity). Reprinted from ASCE Manual #45, #1 *A Guide for the Engagement of Engineering Services*. Copyright © 1981. Reprinted by permission.

to about 80% after a one-year period and to 95% or more after the second year. As a result after the second year only a handful of personnel will be required, and the projections of future revenues should reflect this.

Where operating services are of a more **permanent** nature, it is advantageous to gradually replace the operating personnel with local hires. They can generally be obtained for all but the most complex positions and thus over a relatively short time, say, three years it is possible to reduce the required personnel to the percentages cited above. The advantages in hiring local personnel are several, including the reduction in personnel costs, generating goodwill with local governmental authorities as well as the public, reduction in personnel administrative problems regarding housing, schooling, transportation, and so forth. On the negative side, the local personnel may need training in the plant and its systems, although this can be reduced by their participation in plant start-up and initial operations. The savings in personnel costs can be significant, on the order of 40% per person, and this is a real incentive for adopting such a program. The savings in the other areas permit a reduction in support personnel, and gradually the nonlocal staff at the facility can be scaled down. It may still be necessary to provide a few key people at the facility but they can also be reduced in time so that after perhaps three to five years the facility can be totally run by locally hired personnel. Pricing this into a proposal is not difficult. It is merely necessary to establish a staffing plan of the number of personnel by grade level—broken down by home office type or local hires, extended by the applicable rates including additives and other costs—the entire number extrapolated over the time at the facility and then to perform the various summaries for the different conditions. This staffing plan becomes the scope for the personnel portion of the work, which combined with other costs, such as supplies and outside services, becomes the basis for the proposal. If the work is to be offered on a lump sum basis, suitable contingency (profit, etc.) must be added. If the work is or can be made reimbursable, the staffing plan can be made the scope document, thereby reducing the size of the contingency required.

Costing of **warranty** considerations is a bit more complex than the examples cited above. By far the best approach, if there are historical data available, is to base an assumed repair, replacement, or service involvement on these data, factored for the differences between the data and the facility or process. If a worst-case analysis is prepared, and all the possible scenarios of failure and consequential damage are considered, it will convince one that the work should not be undertaken. Rather it is wise to only consider those events that are credible, remembering that a suitable force majure or liability limitation clause (see Chapter 11) will provide protection. In effect the analysis should be based on what can reasonably be expected to happen. The analysis can use the more sophisticated failure mode and effects analysis method, with probabilities assigned to the various failure modes, but considerable care should be taken to not overstate these probabilities. In many respects this is analogous to the problem of assigning contingency, care must be taken to not overestimate the likelihood of difficulty.

The **cost** for this possibility can be considered a potential charge against future earnings, or it can be covered by actually setting aside a reserve. Of the two, the reserve approach is less desirable because it ties up funds better used for other things and increases the actual cost due to the lost opportunity value of the funds tied up. This approach may in some cases be necessary where there is no wish to encumber future earnings.

10.6 GOVERNMENTAL PROPOSALS

Proposals to **governmental** entities normally respond to detailed RFPs in which the content, format, and exhibits are carefully defined. Often there is little room for a creative approach because of the need to ensure that all the bidders respond uniformly. Nevertheless, for large or complex projects even these proposals can include some more unique ways to approach the work, or to present the qualifications of the bidder. Primary among these is the qualifications section, which asks for information of previous projects together with their size, complexity, and relevance to the specific work in the RFP.

It is often possible to suggest improved or innovative ways to perform the work to reduce cost. Reductions in schedule are also of interest but tend to be less important then **cost** reductions. It is often useful to point out that "The shortened schedule yields resultant reductions in cost of . . . ."

Competitive bidding is usually not employed for Architect-Engineer (A-E) services. Rather the preferred design organization is selected based on qualifications and performance data, and within cost guidelines a contract is then **negotiated** with that company. Where adequate definition of the project is available, firm, fixed price contracts are negotiated, but if there is advanced technical development required or if the scope can not be well defined, other reimbursable or time and material contracts may be used.

Because of the broad nature of public bidding on government work so that the work can be equitably distributed among many firms, there are often more bidders than in private sector work where the bid list can be more restricted. As a consequence it is less likely for one firm to be easily awarded successive work. For cases where competitive bids are required, the price competition will often be severe, and the achievable **profit margins** may be lower than in the private sector.

Contracts are usually classified by dollar size, with smaller contracts having less formal procedures than larger ones. An often used break point is $10,000. For large contracts that may include construction or procurement, typically over $500,000, the firm is required to have an acceptable plan for subcontracting to small, minority- or women-owned busines firms.

Federal government work normally requires that Standard Forms

254, "Architect-Engineer and Related Services Questionnaire," and 255, "Architect-Engineer and Related Services Questionnaire for Specific Project," be submitted. Copies of these forms are found in Appendixes C and D, respectively.

Selection of the A-E firm is based on the following considerations:

1. Professional qualifications.
2. Specialized experience and technical competence in the type of work required.
3. Capacity to accomplish the work in the required time.
4. Past performance on government and private contracts in terms of quality control, quality of work, and compliance with performance schedules.
5. Location in the general geographic area of the project and knowledge of the locality of the project.
6. Volume of work previously awarded to the firm, with the view of equitably distributing the work among qualified firms.

Most government agencies maintain a list of interested bidders who are notified when work is to be advertised for bid. These lists are maintained by each agency **independently,** and application must be made to the agency in which the firm is interested.

For projects that are subject to competitive bidding, more extensive **bonding** is required in public sector work. Fairly typical are the following bonds required for large projects in the $1,000,000 or higher range: bid bond in the amount of 20%, performance bond in the amount of 100% of the bid, and a payment bond in the amount of 50%. These values often decline with larger dollar amount bids. Bid openings normally utilizing sealed bids are open to the public, and the results published shortly after the bid opening.

10.7 PREPARING THE PROPOSAL

Once it has been decided to prepare the proposal, it is critical to immediately advise the client that in response to their RFP a proposal will be submitted. This will not only indicate your interest, but it will ensure that you will receive addenda if they are issued. If there is a standing proposal preparation group in the company, the next step is identifying who will head the team and what are the important due dates. If a new group is to handle the work, a proposal manager must be named immediately and a team of personnel identified to support the proposal work. For large proposals separate technical sections are assigned to different persons, each having technical expertise in the particular section for which he or she is responsible. For smaller proposals one person may be responsible for more than one portion of the proposal with assistance as required from other members of the organization. The persons or group preparing the proposal must be given sufficient authority and support to be able to draw on the entire resources and experience of the organization. Preferably they are housed together in a convenient space, where they have access to historical records, data processing and communications equipment, administrative support, and other services to facilitate their work.

A critical step at this point is to establish a fairly detailed **outline** of the proposal, noting the personnel responsible for the preparation of the work and the resources needed. It may be necessary to establish a special working area and laboratory facilities and to obtain engineering, technical, and administrative support, scheduling and estimating personnel, and perhaps even technical writers and illustrators if the work is extensive or complex.

Most proposals are prepared over a short time. Often this means establishing a two- or three-week schedule and a task list with the persons identified who will be responsible for particular portions. Essentially this should be done on the first or second day of the effort. Milestone dates should be established for the issuance of drafts of material, and in the schedule provision should be made for the required review time by senior management. The review must be sufficiently in advance of the end of the schedule to permit modest revision to the material if required. To provide control for the development of proposal material, certain dates need to be established as cutoff dates. After these dates no further input will be accepted.

A simple 20-line bar schedule of the proposal effort with the principal milestones and restraints shown should be prepared as the first task of the proposal leader. Figure 10.3 shows a typical proposal schedule.

When scheduling the activities, it is normal to assume that the work will take place on an **accelerated** basis with the workweek extending through Saturdays and Sundays if necessary and that the normal workday will be longer than eight hours. Some float should be provided to permit rewriting and final polishing of the proposal. Requests for extensions of the proposal's due date are almost never granted and should not even be contemplated. The effort of the entire team should be geared to the required submission date which must be considered **unchangeable.**

As mentioned earlier, each item in the proposal should be assigned to a specific person. To ensure control, the proposal manager should review the progress of each member of the team no later than every other day, thus permitting changes in staffing or assignments when unexpected problems develop. Unexpected problems will be certain to arise, so flexibility is an essential attribute of the team members.

Some activities have significant **lead times.** For example, for the proposal volume(s), the printing of covers, if not very simple, and the printing and laminating of index tabs can sometimes take a week or more. For a typical proposal it is essential that their preparation be decided upon promptly.

The **supporting activities** of artwork, graph preparation, drawings, and

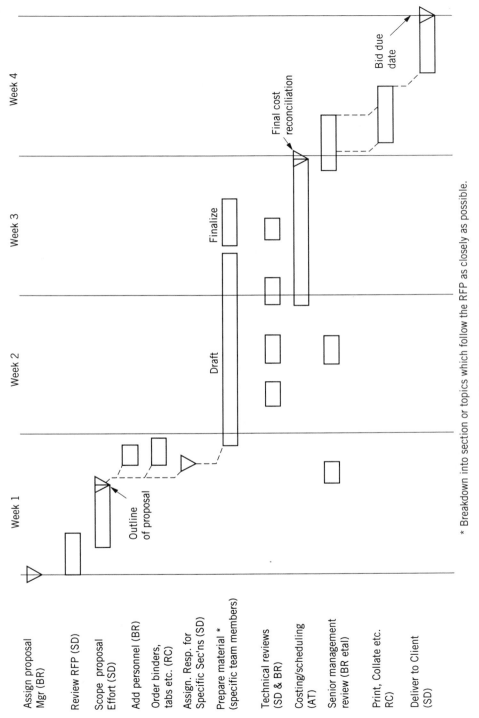

FIGURE 10.3 Proposal schedule.

* Breakdown into section or topics which follow the RFP as closely as possible.

similar items must also begin at the earliest date so that there are enough data to present realistic information. Regardless of whether the data are firm or not, it is important to indicate in the proposal that all drawings and art work are conceptual and subject to change as the design develops.

Hard data such as economic models, cost studies, and estimates should only be qualified if absolutely required. In all cases the criteria used for the study and any limitations on its accuracy or applicability should be stated so that the client will not be misled. This can be handled as footnotes or commentary so as not to diminish the value of the work.

Similarly **client-furnished information,** such as economic factors, flow-sheets, and land boundries, should be clearly identified as such. Any questions on them can be verified when the project commences; questioning client information can be offensive if not indicated tactfully. The legal question of the ability to rely on the data will be handled in the contract and does not need to be considered in the proposal unless a draft contract is a part of the documentation.

Because of the developmental nature of a proposal, many firms will prepare it in a three-ring or comb-type loose leaf binder. This provides maximum flexibility, for it permits the last minute addition or removal of material. Where possible discrete topics should be concluded on a page and the next topic begun on the following page. Doing so tends to slightly increase the size of the proposal but makes its reading, evaluation by the client, and, perhaps as important, its preparation more convenient. Page numbering should be done as the last step. If there is concern that pages are becoming lost or misordered, the final proposal could be assembled with plastic comb ring type or some more elaborate form of binding. In lengthy proposals backup material such as resumes and costs are presented in appendices. This often improves the proposal's overall appearance. For added impact presentation paper with a ruled box margin on all four sides can be used to make the proposal seem more formal and finished. Should an even more formal bound proposal be desired, time must be added to the preparation schedule for the printing, collating, and binding activities, which are usually performed by outside firms.

10.8 WRITING AND STYLE

All proposals should be written in **direct, clear, unambiguous** language. Sentences should be kept short and written in the third person using language that is plain and unadorned, while at the same time being positive and confident. The use of complex or obscure language should be avoided. The goal of the proposal is to communicate to the client why your firm should be given the work, and all of the written material should be slanted in that direction. Technical terms should be used as necessary to improve precision and to demonstrate a general knowledge of the subject but should not be used to try to impress the client with the technical depth of the organization. This is demonstrated in the qualifications section, using previous experience, staff resumes, and such.

From a legal point of view it is important to avoid **puffing,** that is, describing the firm as being particularly expert or possessing unusually unique knowledge or experience. Statements of this type have been interpreted by the courts as requiring that the firm display more than the normal amount of skill and judgment for the industry and type of work, and they increase the exposure of the firm to litigation. The best way to promote your firm is to indicate that it has a large amount of related experience and is extremely well qualified for this work as indicated by the data attached. In the qualifications data the type, size, and number of similar projects, staff size, experience, and so forth, can be used to make the same point without stating it as a fact.

Wherever possible, the words of the RFP should be played back to the client to ensure that the proposal will be construed as responsive. When in doubt, use the words in the RFP, qualifying them only sparingly. Too much equivocation can lose the work.

When the client organization is a small one, care needs to be taken with the proposal to avoid the impression that they might be overwhelmed either by your technical competence or the size and complexity of your organization. The proposal should demonstrate that the project team and organization offered will fit with both the client organization and the project needs.

10.9 PRESENTATION OF PROPOSAL

For most proposals no formal **presentations** are required. However, a form of presentation can occur when the bidder is asked to clarify certain items on the proposal or when meetings are held. Each of these contacts with the owner is an opportunity for presenting the firm's qualifications and the advantages to the owner of using the firm for the work. This opportunity should not be lost, and the wise firm will take advantage of it—in some cases requesting an opportunity to formally review its proposal to ensure that the owner fully understands it and there are no questions or unexplained items.

Where there is a requirement for a formal presentation, as is often the case with large or complex projects, suitable preparation is essential. Dry runs should be held to present the material in the time permitted and organized by subject as the owner may direct. In the absence of instructions it is prudent to develop an outline of the presentation, including the time allocated for the presentation, and to review it with the owner to ensure that it is suitable. When presenting the proposal material, it is far better to use the actual personnel who will direct the project work rather than relying on others. In some cases firms have used professional actors to make such presentations only to find that when questioned by the client, they are unable to answer them because this material was not part of the script. The reaction of the owner in these cases is less than enthusiastic. So, despite the fact that the project personnel may not be polished in their delivery style, they will know their material well and can respond to questions or offer clarifications where needed. Naturally it is wise to rehearse them, using someone who plays the role of the "devil's advocate" to ask the hard and sometimes embarrassing questions. Further when the actual presentation is made, the Project Engineer and Project Manager must be prepared to step in and support their personnel when awkward or difficult moments arise.

If after the proposal has been evaluated the work was lost to another firm, it is appropriate to determine from the owner the reason the work was lost. Much valuable **intelligence** can be gained and used for future offerings. In particular, it is important to identify whether the pricing was competitive and whether the overall qualifications of the firm were adequate. Using this information, future pricing and other proposal data can be appropriately adjusted.

CHAPTER 11

CONTRACTS AND LEGAL CONSIDERATIONS

The contract is the basic document in the family of project documents developed. It represents the statement of intent including the purpose and scope of the work contemplated. As the binding legal document between the parties, the contract sets forth the prime requirements for performance of both parties including their rights and responsibilities. It also provides for the financial arrangement between the parties, so it is essential that its language be clear and unequivocal, and not require interpretation. The contract must effectively be a stand-alone document. Contracts are of many types and forms. The purpose of the material that follows is to provide a frame of reference for the Project Engineer or Project Manager charged with responsibility for contract formation or implementation.

11.1 PROJECT TYPES

The **type** of project will often dictate the type of contract used for the work. The type and length of the contract should reflect the dollar value of the work, its importance, and the reputation and experience of the parties to the contract. In general, small projects will use simple forms of contracts, while complex work more detailed and comprehensive forms. Projects whose scope can be well defined will utilize firm price or lump sum type contracts, while those whose development or scope cannot be well defined will utilize some form of reimbursable arrangement. To permit firm price bidding, contracts will often be written **separately** for design and for construction or fabrication. This has the advantage of enabling definition, and hence contracts can be firm priced, but it suffers from the delays inherent in not being able to award the construction or fabrication contract until substantially after the completion of the design work. Clearly such projects take much longer to complete. But, because of legislative requirements, this practice is almost universally used in public sector work.

On the other hand, minimum **schedules** are important for work in the private sector, and design may be incorporated with construction and/or fabrication in what is often called a **turnkey** project, where the entire project responsibility is assigned to one firm. Typical design projects in the private sector will use a variety of contracts from one-page letter types to the full formal contracts, often running 20 or 30 pages. Heavy emphasis is normally placed on the accuracy of the design effort and its timely execution. If there is a prior satisfactory working relationship between the parties, it may not be necessary to tie down all the loose ends and a brief letter of authorization can get the work underway while some form of simple agreement, often the same one used previously between the parties, is worked out.

Government or municipal work will typically use a more formal and extensive document. Usually these are standardized, although they may vary significantly from agency to agency even within the same governmental entity.

Work that includes extensive fabrication or construction activity will usually require additional clauses to cover the added activities. Similarly turnkey projects, where one firm has the entire responsibility for design, procurement, fabrication and/or construction, and start-up services, will require more extensive contract provisions, with emphasis on liability and responsibility for proper operation.

Research and **development** projects are difficult to scope and hence will typically utilize contracts written on a unit price or reimbursable basis with emphasis on such matters as ownership of patents and process developments that may evolve, laboratory facilities, and so forth. Greater involvement with the client is also a typical requirement. Because of the close interaction with the client and the potential impact on the work, the contractor often becomes an extension of the client organization, and thus internal operations such as overhead rates and definition of reimbursable items become important.

11.2 CONTRACT TYPES

While there are many contract types, most fall into either the fixed price, often called **lump sum,** or the **reimbursable** types. Numerous variations are used to address the limitations of each and to improve on their applicability to different project situations. In the material that follows, the terms "contractor" and "client" (or "owner") will be used to designate the parties to the contract. The contractor may be a design firm, service organization, contractor or fabricator, while the client is the owner, municipality, or other entity with whom the contract is drawn.

Lump sum contracts are by far the most widely used. They may not be easier to formulate and administer than other types, but they avoid drawing the client into the **operations** of the contractor. As mentioned elsewhere, they require that the scope of the work be well defined, for changes in scope during the work places the client at a disadvantage in negotiating changes. With suitable clauses they can be used where the scope is somewhat less certain, although some risk ensues. In these situations the contract normally allocates the **risk** to the party in control. Virtually all contracts for materials and equipment are of the fixed price type. They are particularly useful where the funds for the project are restricted, since they permit the client to predict the total cost for the project with a high degree of confidence. If the fabrication and delivery schedule is a lengthy one, clauses providing for escalation of labor and/or material costs may be required if the duration of the contract exceeds a 12-month period.

Work in the **public sector** will almost entirely be of the fixed price type even if the project is broken into phases that are bid or negotiated separately. The same holds true in the private sector, although some form of reimbursable contract is often used where schedule concerns are important so that the contract can be developed and executed even before the work is fully defined.

Where it is not possible to clearly define the scope, some form of reimbursable contract is used. These contracts normally enable the contractor to be reimbursed for both direct and indirect costs and to obtain a fee or profit for performing the work.

Of all the contract types the **cost plus percentage fee (CPPF)** type is the **least** desirable to the owner because there is no incentive for the contractor to minimize project costs. Indeed, since the fee (profit) earned is a percentage of the cost of the work, there is actually an incentive to spend more than necessary, escalating not only the cost of the project but often lengthening the schedule as well. The only justification for this type of contract arrangement would be where there are literally no bidders, and it is necessary to provide significant encouragement to a contractor. Even in this case there are incentive-type contracts that can be used instead.

A variation on the CPPF type is the **cost plus fixed fee (CPFF)** type. In this form, the costs are reimbursable to the contractor, but the fee earned is fixed. The fee is established at the beginning of the work and remains unchanged unless there are major changes in the scope of the work. When major changes occur, the fixed fee is renegotiated based on the change in scope. While this may sound much like the CPPF, in actual operation fee renegotiation occurs infrequently. As a result the CPFF contract has proved to be equitable, and it is widely used.

Clients often will want to limit their cost exposure for reimbursable work and will ask for a **guaranteed maximum** price. This in effect caps the price paid to the contractor while permitting the client to expand the scope. The risk in this arrangement is borne entirely by the contractor. To protect the contractor, it is necessary to carefully define the work, in effect converting the reimbursable work to lump sum and providing enough **margin** for contingency between the expected cost and the target price. In this way the contractor will have incentive to accept the change. When applied to elements of a larger reimbursable project, it is possible to utilize the same arrangement, and it is often agreed to by the contractor merely to maintain the goodwill of the client.

Unit-priced work is used where it may not be possible to fix the quantities of the work initially but where the elements of the work can be well defined in advance, and where the elements are repetitious. This sort of arrangement is often used, for example, where an electrical contractor will bid a firm price per outlet, lighting panel, light fixture, and so forth, in advance of the actual drawings of the work being available. The client gets the benefit of a firm price for each unit of work but retains flexibility if the overall scope (or quantities) vary significantly. In these cases it is common for the bidder to establish limits of the minimum number of units to which these unit prices apply. It is uncommon for the bidder to establish maximum quantities. Doing so generally leads to increased efficiency with lowered unit costs (to the contractor) and thus higher profit. To compensate for this, the client when negotiating the contract will usually set upper limits, with perhaps a breakpoint in the unit pricing for units in excess of these limits.

A similar form of contract is the **time and material** type, which is often used where the number of units cannot be easily defined. In this arrangement the contractor performs the work required and is paid for his costs of labor and materials including agreed additives for overhead and other indirect costs. An allowance for profit is added to the costs. Usually the profit allowance is a percentage of the dollar value of the total of the other costs and in some respects duplicates the CPPF arrangement. Time and material (T&M) contracts are usually only used to perform disputed or additional work, where it is not possible to conveniently establish a price in advance. As a result the scope of the work is much smaller, and the CPPF disadvantages are not substantial. To implement such an arrangement, it is preferable to establish beforehand the percentages of additives and other indirect charges which are appropriate. Following that it is merely necessary to authorize the work on a T&M basis. The important factor is to establish exactly what the scope of the T&M work is to be to assure no later misunderstandings result.

There are many kinds of **incentives** that can be written into contracts. Among the most common is the **bonus penalty** clause used to encourage improved performance of cost, schedule, or quality. Another increases the contractor's profit or, in the case of a fixed price contract, provides a bonus for improved performance. In either case the incentive requires a clear and precise definition of the event or standard of performance that triggers payment of the bonus or incentive. Often, where defined by schedule, beneficial occupancy enables the facility or project to begin operation, although all portions may not be completed. Process plant completions can be expedited using this approach, often benefiting the client who can use the facility earlier than scheduled. A good example of this was a two-month early completion of a paper mill which permitted shipments of the product in advance of projections, yielding 20% increased revenues for the first year with a 10% savings in the cost of the project.

The bonus penalty clause in many contracts serves as an incentive for the contractor because of the **increase** in contract value for early completion and the **penalty** for completion after the contract completion date. Usually the bonus or penalty amounts are equal and are stated on a per day basis. To protect both parties, a limit is placed on the number of days of penalty or bonus, with 90 days often being used for large complex projects having overall project schedules of one to two years. For reimbursable work a similar incentive is included when cost overruns or underruns are shared by the client and contractor. This permits both parties to benefit from reduced cost and has proved to be very effective.

For work in the public sector it is common for the design and construction work to be contracted **separately,** since often the design firm is prohibited by law from bidding on the construction work. Generally the contracts are of the fixed price type, since a complete design allows for lump sum bidding of the construction work. While simplifying the administration of the contracts for the client, this practice considerably extends the project schedule. Since design and construction cannot overlap, the time to advertise, take bids, analyze them, and award the construction contract can be significant. For most pubic sector work the advantage to the client of using lump sum bidding outweighs the disadvantage of the additional time required.

In the construction industry the term **subcontract** is used in a particular way. It covers the delegation of a contractor's duties to perform part of the work to another contractor but without relinquishing the original contractor's obligations to the owner. Normally subcontract work is performed on the jobsite and includes the application of labor. If the contractor were to perform work off site, even using a large labor force, the work would come under a supply contract or a purchase order rather than a subcontract. Subcontracts contain most of the features of the main contract between the client (owner) and the contractor, usually called the "prime contract," and will take one of the forms discussed above.

11.3 IMPORTANT CONTRACT PROVISIONS

There are literally hundreds of different provisions found in complex contracts today. Many are written for unique circumstances and do not appear in most contract situations. What follows is a review and commentary on the contract provisions most likely to be encountered in normal circumstances. The information is intended to give background and some guidelines on general use. When legal issues arise, advice of suitable legal counsel should be obtained.

Fundamental to the contract is the principle that it is the **binding legal document,** that all activities by the parties must be carried out. It is not possible to dismiss terms or requirements merely because they are onerous or overlooked. Requirements in contracts cannot be unilaterally waived or avoided. If there is to be a change, it must be presented to the other party in good faith and negotiated in accordance with the provisions of the contract.

Basic to any contract is a meeting of the minds or **mutual assent.** By signing a contract there is an implied meeting of the minds, and this issue is only challenged if there is indication that the challenging party was misled or did not understand the terms and requirements of the contract because of an ambiguity. This challenge is only open to the party signing the contract, and not the one who drew up the document, since it is assumed that the contract's writer is the more knowledgeable of the two parties.

While **verbal** contracts are widely thought to be binding, an oral promise is not enforceable in most jurisdictions, and a written statement is required to document the transaction. Thus for any work a **written** contract should be employed. The contract can be as short as a few paragraphs or a lengthy document including appendixes and supplementary material totaling several volumes. The contract should be written in simple, direct language, avoiding complex legal phrases. A good rule to follow is that if there is any question about the meaning or possible interpretation of some part of the contract, it should be rewritten to be unequivocal and not subject to misinterpretation. As mentioned above, the **writer** of the contract bears a larger burden than the signatory, and in a dispute the writer will be expected to have been more knowledgeable and expert and thus to have the advantage in the drawing of the contract. Still it is always good practice, unless the contract is a familiar standard type previously used satisfactorily, to have the contract reviewed by legal counsel prior to its signing.

Scope definition is central to the contract. It is the legal requirement for the work to be performed. The basic principle to bear in mind is that the definition of scope for lump sum work needs to be as precise as possible. Any vagueness in this definition leads almost certainly to a dispute, and thus to ill will and difficulties between the parties, and damages the trust and cooperation that may have been painfully developed. Both parties suffer when the scope is poorly defined, and in the long run no one benefits.

Apart from the details of the definition of scope, there is the broader question of overall responsibility. Does the agreement require that the contractor perform procurement services, training, personnel selection, permit application, additional special studies, start-up services, financing assistance, testimony at public hearings, geotechnical exploration, and so forth? This consideration will affect not only the cost of the work but also the ability of the contractor to perform because of personnel, equipment, laboratory, or other limitations.

Most contracts require **timely** performance and will include a clause stating that time is of the essence in the performance of the work. While seemingly routine, this clause can become vexing if some of the work is disputed or interrupted. In such circumstances it is wise to obtain legal guidance before stopping any of the work.

Another important issue is whether the client will furnish any data, equipment, and materials. The contract should be clear on the degree to which the contractor can **rely** on the accuracy of client-furnished data, whether the data needs to be confirmed by testing or other means, and what the responsibility of the contractor will be in using the data. Among these data would be flow sheet information, topographic details, process information, chemical or physical data, materials properties, and so on. In addition the extent of responsibility of the contractor for furnished materials and equipment should be spelled out. "Is the contractor responsible for performing special tests or examinations of this material or can the contractor merely rely upon the material as delivered and utilize it directly?"

The contract should always establish the applicable **codes and standards.** Usually a clause is added stating that the work will be conducted in accordance with "all applicable codes and standards," and this places the burden of definition upon the contractor. Although normally sufficient, in some cases the client may wish to point out special requirements to ensure that the contractor does not overlook them. Then the language can be changed to read "all applicable codes and standards including but not limited to . . . ".

Where a client wants to include certain standards not legally required, mention should be made in this portion of the contract. In the absence of a statement regarding codes and standards, all legal requirements of the jurisdiction would govern the work and compliance with them is required.

Licenses and **permits** can affect project schedules and in some cases even the economic feasibility of projects. For this reason it is wise to indicate the permits and licenses known to be needed and who is responsible for obtaining them. It is also useful to include a disclaimer regarding the schedule and cost impact if the licenses or permits are delayed. The disclaimer should be careful not to limit the licenses to those stated, since this might relieve the contractor of significant responsibility. Usually the principal licenses are well known, so there is little difficulty in planning for them. The problem arises with the secondary permits, licenses, and studies required. To avoid consequent delays, contracts will often establish an effective start date for schedule considerations, such as "after receipt of the last governing permit or license."

Environmental protection is becoming an integral part of the activities defined in the contract. Control of discharges of materials off site, maintaining a safe and acceptable environment on site, and obtaining the necessary agreements of environmental authorities need to be defined in the contract.

Some care must be taken, however, not to limit these concerns to a few, for then the contractor's responsibility would cover only those listed. The best way to handle environmental matters is to name those that are known and add a general statement that other applicable environmental regulations must be complied with as well.

In lump sum work the client rarely will require that **key** personnel be named or committed to the work. In reimbursable or developmental work this is common practice, since the client wants to be assured that the key members of the project team will not be changed without client approval. The rationale for this is that the personnel who know the project have learned it at client's expense, and changing personnel exposes the project to delays and errors while the new personnel are learning their jobs. As a result there is often a clause in the contract requiring that changes or substitutions of personnel for certain key positions not be made without the **prior** agreement of the client. To limit this and provide the maximum flexibility, it is usual practice to reduce the number of positions defined as "key" to as few as possible, typically no more than four or five, even on large complex projects. Including more positions than this, severely limits the flexibility of the contractor and increases the administrative complexity of the project. It may also work to the disadvantage of the project, since the contractor would be unlikely to offer substitutions until severe problems were apparent, possibly too late to avoid major adverse affects. Further, the personnel involved may feel that their professional careers are "on hold." It is often possible to get a client to accept personnel substitutions if the new candidate has been on the project for some time, has performed well, and the proposed substitution is a promotion for the individual.

From a contractural point of view the way to handle this is to incorporate language that approval of substitutions will not be needlessly withheld if the substitute candidate is suitably qualified. Even with this clause difficulties can arise, and the ultimate assurance that can be given the client is to develop and maintain cordial working relationships and to only make those changes that are truly necessary.

Quality control and **quality assurance** programs and provisions are often included in the contract. They are usually general, stipulating the overall program that must be complied with, but there are some contracts in which the provisions are very specific and go far beyond merely programmatic requirements. It is better to keep these requirements at the program level to provide the maximum flexibility in the implementation of them. Wherever possible, the contract should be arranged on this basis. Frequently contracts will require that certain **tests** and **inspections** be carried out by **third parties.** This is extremely undesirable. Third parties have no interest in the progress of the project and may in fact create problems merely to enhance their own work. A far better approach is to be sure the contract language retains responsibility and control in the hands of the contractor and utilizes these parties only as advisers to the project. Further it is wise to try to establish an agreement between the contractor and the client on the exact scope and responsibilities of these independent quality assurance/quality control organizations. It is also useful to identify the quality assurance/quality control organization to ensure they are knowledgeable, practical, and will not unnecessarily affect the progress of the work.

Chapter 10 discussed the method of costing unit-priced work and gave an example of the way in which quantities may be offered in a proposal or cost report. In the contract it is important to define the method of **payment,** its basis, and whether **retentions** are to be made. Retentions represent an incentive for the contractor to perform and complete the work, and many arguments over minor matters can be avoided if retentions are utilized. Usually the retentions are 10% of the individual invoice rendered. On large multimillion dollar contracts the cumulative 10% can become very large and thus in the later stages of these projects, when the risk of poor performance has passed, they may be reduced to 5% or even a lower value. To maintain flexibility, from the client's point of view, it is wise not to include this reduction in the contract but to use it for negotiations during the late stages of the contract. Where bonds have been provided for the work and the retention is reduced, the bonding company must be notified that its financial interest in the project has been reduced. The reason for this is that if the contractor defaults, the surety is entitled to the retained funds to complete the work.

The frequency of **payment,** currency and method of determining exchange rate to be used (if the work is international), and method of measurement for payment are all important questions that should be part of the contract's language. Often advanced payments will be requested by the contractor to which the client may object. Similarly the method of measurement for payment, its review, and certification is important. For lump sum, unit-priced, and reimbursable work, independent confirmation of the quantities billed will usually be undertaken by the client. For many contracts it is common to include a method of calculating interest payments due the contractor for **late** payment of invoices by the client. Typically they provide for payment no later than 20 days after receipt of a correct and complete invoice, with interest accruing, often at prime plus 1 or 2%, on the unpaid amount. Disputed amounts in invoices do not usually bear interest, although, if abused, this can become a contract administration problem.

To encourage contractor performance, many contracts will include bonus/penalty clauses. The bonus or penalty that accrues to the contractor can be stated in the contract or negotiated later. In its most common form the bonus or penalty hinges on a single **milestone,** such as project completion, delivery of the first article, or completion of a performance test. In other cases, where the project is large and takes place over an extended period, intermediate milestones may be used as continuous incentives. A portion of the fee may be distributed or a bonus established to be earned by meeting intermediate milestones such as submittal of the Environmental Impact Report, award of the data-processing contract, or issuance of the first 200 drawings to the piping fabricator. Typically the milestones are established at about 90-day intervals with, say, four or five milestones determining the amount of the fee risked or bonus earned at each interval. This has proved to be an effective system, although only the larger projects seem to be able to utilize the frequent milestone approach because of the administrative work involved.

One important principle to bear in mind is that courts will not enforce a contract with a penalty clause alone. It is possible, however, to include a bonus clause without a penalty. Normally the amounts of the bonus and penalty are equal, and thus the contract is neutral as regards the parties. It is common to place a limit on the amount of the bonus or penalty, and if time is the determinant, it is frequently limited to 90 days (early or late), with a specific dollar amount per day being stated.

Escalation is often a factor in contracts that run for a significant period of time. In these cases increases due to escalation are provided for by either specific percentage changes in the values of the contract amounts or in establishing a formula by which these values can be calculated. The easier method to establish and administer is the fixed percentage approach where, for example, "all labor costs including additives billed for and after the month of January 1994 would increase 4%." The actual percentages and timing of the change (in this case an increase) would depend upon the negotiated renewals of the labor agreements. One variation on this method is to apply different percentages to different crafts or classifications of labor where the labor force is large and increasing. If a labor agreement of one or more of the crafts is up for renewal during the contract period, the contract needs to reflect this fact.

The more complex escalation arrangements occur where both labor and materials are subject to escalation and where the rates vary independently. A common approach is to index the costs to Bureau of Labor Statistics indices by type of craft (ironworkers, carpenters, etc.) and by type of material (forest products, iron and steel, etc.). This is normally used for the longer-term contracts running more than one year where it is not possible to reasonably predict material cost escalation and wage rate labor settlements. With these arrangements the adjustments are made only a few times per year to avoid the administrative complications of more frequent calculations.

Perhaps the most disputed and argumentative part of the contract is the portion that deals with **liability.** The general rule is that when the contract is silent on this subject, there is **no limitation** on the liability of the signatories. For this reason, and as a matter of good business practice, it is essential from the contractor's point of view to establish some limit on the liability stated in the contract. It is normal to limit the liability to only those acts of the contractor's own organization and employees when performing work for the benefit of the client. This excludes all other forms of damage and liability such as accidents off site and failure to perform. Consequential damages and implied warranties would fall into this category and are normally not included. Only under the rarest of circumstances would the contractor accept exposure to loss of profit, failure or incapacity of others to perform either as a result of actions or lack of actions by the contractor, and so forth.

The usual practice is to negotiate a liability limit that is **not greater** than the amount covered in the contractor's comprehensive general liability insurance policy. This cap provides overall protection such that, regardless of the event, there will be adequate insurance coverage. Usually the contract also provides that the deficiencies that require correction under this provision must be discovered within one year of the completion of the project and reported to the contractor in writing within ten days of discovery. The normal remedy is that the contractor would repair or replace the deficiencies at his or her cost up to the liability limit of its liability coverage if it is available.

One arrangement often used in design-type service contracts is to limit liability to no more than one-half the fee to be earned so that some profit is made on the work even under the worst circumstances. This same approach is sometimes used for more complex projects, particularly those of a reimbursable type. Again the remedy usually stated is that the contractor (design firm) would correct the design at its cost but not to exceed a contracturally specified liability limit.

The requirement for **insurance** coverage is almost always a provision of contracts. This takes two forms: insurance for the work on the project, and general business insurance for the contractor. The later is discussed in Chapter 13. The minimum insurance dollar amounts and types are almost always established by the client and include insurance for workers compensation and employers liability, general liability for accidents, improper actions by employees of the contractor, and owned or leased automobiles or other vehicles. Certificates of Insurance should be furnished by the contractor to the client before on-site work is begun. In some cases **special coverage** is added to the insurance by the contractor for the project because of unique operations such as tunneling, or because of hazardous materials in use such as radioactive tracers. Care must be taken to ensure that the insurance remains in force for the duration of the work. A current insurance certificate will reflect this.

Virtually all contracts will include a **force majeure** clause. This is a protection for both parties in the event of external events over which neither has control, such as fire, flood, and other acts of God or strikes or insurrection. The clause would also operate where some outside event has delayed the delivery to the site of equipment furnished by the owner to the contractor. In general, these clauses provide that the contract will be automatically extended by the amount of time delay caused by the event and that neither party will be penalized for the effect. Since recoverable costs and fees will continue during the period of force majeure, reimbursable contracts provide for appropriate contractual adjustments. For lump sum contracts, some overhead charges continue, and these also require that the contract amount be adjusted. Timely notification to the client together with an appropriate explanation are required by the contract.

Acceptance of the work on behalf of the client is an important contract provision, for it provides a mechanism for the formal conclusion of the work. This clause normally requires a written request for acceptance by the client, with time limitations for either acceptance or notification to the contractor of deficiencies. On most sizable projects the client and contractor will have close and continuing contact during the late phases of the work, and deficiencies will be promptly brought to the contractor's attention. As a result notification for acceptance of the work will usually occur only after a substantial amount of informal discussion and the cleaning up of the punch list or completion list developed by the client. Nevertheless, it is important to follow the contract provisions and present a written request for acceptance to the client. This establishes a formal date for completion under the contract as well as for contractural scheduling and payment and bonuses or penalties. For the bonding companies it establishes a limit on the duration of their involvement in the project and their exposure.

Deficiencies are normally discussed in a paragraph of the contract that requires that the contractor be notified in writing of the deficiency and that the notification be made within 10 days of its discovery. From the client's point of view it is important to establish a system that provides for prompt written notification to the contractor when deficiencies are discovered. If the notification is delayed beyond the duration provided for in the contract, the contractor can plead that the contract was not complied with and is not obligated to take corrective action. While job diaries and similar internal records may be helpful in the event of dispute, a proper system for prompt written notification will go a long way toward avoiding this type problem.

Warranties are usually provided by the design firm and run for a specific period, typically one year after completion or acceptance, whichever is sooner. Any deficiencies discovered after that period are not bound to be corrected under the contract. Further design firms almost always limit their liability to correcting the design at their expense and do not accept liability for labor and materials necessary to correct their errors.

Warranties for materials purchased by the contractor as agent for the client are normally passed directly along to the client. Often the client will want extended warranties which the supplier is unwilling to provide and will request them from the contractor. This is extremely hazardous and is rarely agreed to by the contractor. The usual language in the contract is that the contractor "will use his or her best efforts to obtain the warranties from the suppliers . . . ".

Title to work and **ownership** of plans is normally covered in the contract, which usually provides that they pass to the client on completion of the work. The client's title to these is usually limited to work in connection with the specific project, for example, for future modifications of the project, and they cannot be used for other projects or purposes without the written consent of the contractor. In addition the contract usually provides that the contractor can hold copies of the design and use it for other purposes in the future. This is not a problem except where proprietary information is involved, in which case the contract would include suitable language.

For purchase orders the **title** and **risk of loss** relates primarily to the point at which legal ownership transfers from the supplier to the purchaser, normally at the FOB or FAS point. This has little effect except that the purchaser should ensure that insurance coverage exists from that point. Since the party having title will have to pursue insurance claims in the event of loss or damage, and since most claims concern damage or loss in transit, it is advantageous to the purchaser to have the FOB point established as the destination rather than the supplier's plant. This places the burden of any claims administrative work on the supplier.

Purchase orders can carry a clause that gives the purchaser the right to **reject** goods for failure to meet the terms of the pruchase order. Usually this covers inadequate quality, furnishing of different material than called for, excessive quantities, and so forth. Only in the case of extreme delays in shipment where substitute material may have been purchased is it used in connection with late receipt of material.

Details on **termination** of the work can be provided in the contract. Termination takes place at the convenience of the client,—and less often at the convenience of the contractor—with 30 days written notice often being required. On smaller projects a shorter time may be adequate, and the contract may call for 10 days. This interval provides for an orderly conclusion to the activities, with protection for sensitive materials or processes. After the termination and the payment of outstanding invoices and other remaining activities, title to the design, work, and materials will pass to the client.

Termination is a very serious matter and only in the most unusual case should the contractor be surprised by a termination. A large part of the work of contract administration is to maintain a close and continuing relationship with the client, and termination normally arises out of fundamental differences that have not been resolved despite extensive, ongoing discussion and negotiations.

The naming of representatives and an individual to receive **notices** for both parties is usually included. This not only identifies the individual who is named, normally by project title, but also states that he or she is empowered to act on behalf of the contractor (or client). Contracts for work in foreign countries where English is not the legal language will usually state which is the controlling language for questions of contract interpretation. The responsibility for obtaining bonds, permits, and other special documents is often noted in the contract, particularly where they are unusual or require a significant effort.

A more recent development in contracts is the addition of **dispute resolution clauses.** These clauses take several forms, depending on the type of dispute resolution preferred. The principle types are **mediation, arbitration,** and **mini-trials.** In mediation the parties present their dispute to a mediator who hears the issues and the arguments of the parties and attempts to guide the parties toward a mutually agreeable settlement. There is no judgment handed down by the mediator, and the sole purpose is to bring the parties to an agreement.

Arbitration is a process where the dispute is heard by an arbitrator or, in the case of larger disputes, a panel of three arbitrators, who will reach a decision based on the material presented. The arbitrator(s) are attornies, judges, and trained executives who are experienced in the industry. Their decision is binding and will be enforced by the courts. This requirement is incorporated in the contract by the clause that includes the recommended language of the American Arbitration Association:

> Any controversy or claim arising out of or relating to this contract, or the breach thereof, shall be settled by arbitration in accordance with the Rules of the American Arbitration Association, and judgment upon the award rendered by the arbitrator(s) may be entered in any court having jurisdiction thereof.*

Similar clauses are available for incorporating the requirements for mediation in the contract. One variation on the mediation process involves a mini-trial in which the parties conduct a brief simulated trial before a judge and in the presence of their senior management. This permits the arguments to be presented in a nonbinding way but is usually sufficient to cause the senior representatives to reach an agreement, avoiding further litigation.

One other approach to resolve disputes is the establishment of **dispute resolution boards** at the start of the project. The board meets over the life of the project, whenever a dispute arises that requires their attention. This permits disputes to be resolved when they are small and current, before positions harden and issues become emotional. This method has proved to be particularly effective on large complex projects.

Contracts can also contain a clause that states that the written contract is the **entire** agreement between the parties and supercedes any previous understandings and agreements.

Contracts will often have appendices, which can be quite extensive, that provide details essential to the contract. Typical is the definition of various categories of cost for reimbursable contracts. Examples of these are the distinction between recoverable costs, other direct costs, standard rates, overhead rates, and fee. Other listings include schedules of rates, and charges, personnel classifications, and so forth. These listings can often run to several pages if the contract is extensive and complex.

---

* Reprinted by permission of the American Arbitration Association

11.4 PITFALLS

Because of the importance of the contract, mistakes made during negotiations or administration can have a far-reaching effect. Certainly the most important of these are the very contract negotiations themselves. First, and foremost, is the need for developing a written record of the agreements reached in the negotiations. There should be no interest in developing side letters or agreements, the contract should include all of these and become a **stand-alone** document. As mentioned earlier, some details may be placed in appendices, but they are still part of the main contract. Any other agreements or understandings incorporated in the contract by reference become part of the contract as well. The difficulty arises when side letters and other such matter are not incorporated into the contract and thus have questionable authority. With the entire agreement clause described previously, the status and enforceability of other side agreements is clouded and may create added difficulties.

Contracts are often broken into **packages** to facilitate the work. As a result there is opportunity for error. Conflict between the packages, duplication of scope, or omissions can create severe difficulties in both administration and execution of the contract. Duplication is the easiest of these to deal with, since scope reductions are easier to resolve than conflicts or omissions. Conflicts can be resolved usually with an engineering review to ensure that the intent is properly met, with the excess work being treated as a duplication. It is wise to limit the number of packages into which the work is broken to avoid these potential problems. This is also consistant with the owner's point of view, which is to make the packages as broad as possible to pick up any areas that would have been omitted.

Omissions are easily dealt with by adding the required work to one of the existing contracts. The difficulty which arises is one of schedule as the existence of an omission is not usually apparent until schedule slip has taken place.

Where the client furnishes data to the contractor, it is important that the **precedence** of documents mentioned in Chapter 7 be established and, where possible, resolved as a part of the contract negotiations to avoid a potential conflict at a later date. Where the client does not furnish data, this requirement can be included in subcontracts and purchase orders.

The need for **specificity** cannot be overemphasized. The contract, above all other documents, needs to be precise and unequivocal in its language and requirements. Against this, standard **all** contract clauses and requirements should be tested. There is absolutely no worse situation than at a later date to have a dispute arise, where each of the parties, in good faith, believes that the language of the contract means something different to them. If there is any doubt or ambiguity, the contract's wording should be recast.

An important concept in many contracts is the requirement that the plans and specifications be completed by the contractor and that the data provided by the client merely establishes overall guidance or requirements. A further potential problem arises with this type of open-end requirement when the client wishes to approve some if not most of the design documents as they are finalized. For reasons cited earlier, it is vital to define the degree of completion of the data provided, whether it is to be **controlling** or whether the design is to be finalized by the contractor. Similarly the requirement for client approvals must be clearly defined, including the type of data to be **approved,** the responsibility assumed by the client, and the time permitted in the schedule for the approval cycle. Without these criteria it will be impossible to control the development of the design, and the project schedule will be meaningless.

Literal interpretations of the contract can often become unrealistic and work to the disadvantage of the client. On a project to build a power plant, the contract specified that the velocity of the cooling water through the condenser tubing was to be 7 ft per second. Since the pumps providing the cooling water used ocean water, the pumping system was designed to deliver water at 7 ft per second at low tide. When the tide flooded, the pumps had a slightly reduced lift, and the flow would increase to 7.4 ft per second, still well below the range of erosion problems which would not start until velocities of 9 to 10 ft per second. This increased flow reduced the condenser's back pressure somewhat and permitted the generation of about 3% more power. When the client became aware of the velocity increase, he insisted that at the higher tidal condition, the velocity be no greater than the 7 ft per second cited in the contract. This required that a bypass be built around the condenser to operate during this high-water condition. Not only did the bypass cost the contractor about $500,000 to build, it reduced the output of the plant at the high-water condition and thus cost the client significant revenue over the life of the plant. In this instance the client personnel were not knowledgeable about the technology of the problem. By acting both bureaucratic and suspicious, they in the end injured not only themselves but the contractor as well. In retrospect, had the contract specified a velocity between 7 and 8 ft per second, or had the client been less literal in the interpretation, the problem would not have arisen.

In a similar vein **strained interpretations** of the contract can be difficult to deal with. The best advice that can be offered is to be patient and try to explain why the interpretation is incorrect and to convince the client that the work will be performed with his or her best interests in mind. While legal counsel can assist in developing the arguments why the interpretation is strained, in general, it is best not to confront the client with counsel until all other means of discussion and resolution have been exhausted.

A broad **distribution** of the contract to all the project team members will materially assist in the efficient execution of the work. If there is a concern regarding some terms, fee structure, or other financial data, that information can be omitted from the copies distributed. In general, it is better to distribute the contract to too many personnel than too few.

By any standard **litigation** is to be avoided if possible. To this end, prompt notification to the client or contractor of changed conditions, errors, omissions in the work, or other problems is essential. An efficient, rapid method of identifying changes requested or needed and obtaining approval, if the contract requires it, will greatly assist in reducing tension between the parties and permit resolution of problems before positions become polarized and issues become enlarged. The maintenance of diaries, daybooks, and similar documents is extremely helpful but does not take the place of the formal notification that may be required by the contract. It is essential to adhere to the word and letter of the contract but to so arrange the administration that **another level** of documentation exists to support the position taken. In the event that it is not possible to resolve the question, it is wise to notify the client that the work will be conducted as requested, but under protest, and that resolution will take place at a future date.

Many contracts require that **bonds** of one form or another be furnished. Bonds, while a form of insurance to the client, are not an insurance policy for the contractor. Rather they represent a financial guarantee. In the event of **failure to perform,** the contractor has financial exposure for the full recovery of any monies spent by the surety to complete the work. Thus any actions that breech the contract and cause the surety to act to complete the work can become a direct liability of the contractor.

With the low cost and ease of performing microfilming, and computer data duplication for records retention, it is wise to retain more documents for longer periods than to prematurely discard them. Legal counsel should provide data on the recommended **retention time** for different types of contractural documents. Records retention was discussed in Chapter 2; the retention periods appropriate for design documents may not be appropriate for contractural documents, which are normally retained for longer periods. When in doubt, the following guidelines have proved helpful:

1. The contract should be retained permanently together with significant working papers used in its development.
2. Significant correspondence relating to contract execution, performance, close out, and so forth, should be retained either permanently or for ten years.
3. Documentation on changes should be retained from five to ten years, until it is reasonably certain that further issues will not arise.

11.5 SAMPLE CONTRACTS

The American Institute of Architects (AIA) has developed sample contracts, two of which are applicable to a wide variety of situations. Copies of these, AIA Document B141, Standard Form of Agreement between Owner and Architect, and AIA Document A201, General Conditions of the Contract for Construction, are included as Appendixes A and B.

CHAPTER 12

HUMAN RESOURCES

Of all the areas in which the engineer or manager works, human resources is the one that has the greatest potential for improvement and that can yield enormous benefits for relatively modest effort. The need of the enterprise to obtain superior performance from its personnel, given the constant change found today, makes the requirement for employee involvement and recognition more essential than ever. The competitive market requires that more be done with fewer resources—financial, physical, and human. As a result the continual need for ever more efficiency places ongoing demands on the population of the firm to continually perform at higher levels and to work smarter not harder.

The enlightened firm will integrate these needs into a program in which a "win-win" situation results. Despite the high turnover of today's work force and the perceived lack of long-term stability and employment with a particular firm, the goals of the company and the employees can be made consistent and compatible. If this is achieved, the company will prosper and the employees will remain productive. Employees are sophisticated and will respond to the type of leadership and management provided. If it is consistent, fair, and humanistic, the employees will respond accordingly. If it is capricious, uncaring, and indifferent, the employee turnover will be high, morale poor, and, most important, the performance of the organization will be substandard.

12.1 THE WORK FORCE

Employees today require a share in the **decision making** and "ownership" of their own work, which in the past was retained almost entirely by management. If made a partner in these activities, superior employee performance will result. The performance may be based more on their responsibility to the profession than necessarily to the firm, but the firm will nevertheless reap the benefit of the quality of their work. The employees should be involved as well with innovation and improvement of the work processes and product quality at every step of the way.

Employees today also expect to share in the financial benefits of the company, and their performance will be influenced by the type and immediacy of the financial incentives offered.

The sum of these factors presents management with its greatest challenge: "How do I manage this very diverse, everchanging work force who have little company loyalty to ensure the highest quality product produced with the minimal cost?" "How do I plan for the future to ensure the correct set of skills and experience in the work force to properly position the firm for the longer term?"

As a result of these problems, the role of the Human Resources (HR) group has assumed increasing importance in today's organization. Most HR groups fail because they do not understand the business and do not participate in the major **strategic** decisions of the firm. It is important to integrate them into the mainstream of the business and to make a full business partner of them. They should be fully aware of the operating plan for the company, both the near term and the longer term. In many firms they are an active participant in and contribute to the writing of the operating plan. Only with this type of involvement is it possible to support the objectives of the firm and develop the necessary plans to support the financial objectives, timing, and number and type of personnel required. The responsibility of senior management, in addition to directing the day-by-day operations of the firm, is to set the broad direction for the firm. An important element of this is to develop a clear **vision** of what the firm is and what it wants to be. The setting of clear goals, publishing of mission statements, and objectives and similar writings are important ways to disseminate this information to the organization. This problem is made more acute by the dynamics of business today, and thus flexibility in the development, implementation, and revision of these writings and in the methods of their execution is essential.

It is necessary to develop and instill in the firm's members a concern regarding the interests and requirements of the **customer** as well as a commitment to the product itself in terms not only of its **excellence,** but it's maintainability and overall performance. Similarly it is essential to develop in the minds of the staff a concern for the **financial** performance of the firm.

12.2 STAFFING

Acquiring and maintaining the correct **staff** to perform the work is a continual process of change and balance with the workload and its requirements. Rarely is the work so consistent that changes in the number and composition of the staff are not required. With the dynamic and rapidly changing business climate, changes in personnel are constant, and only modest stability is found in most firms today. As a result little employee loyalty is developed, and personnel with 20 or 30 years of service are rarely found. Further the rapid market changes that some firms face require severe changes in the numbers and types of personnel. In this circumstance many firms are using **lateral** moves within the company as a way to avoid layoffs, provide job enrichment, and to cross-train personnel to develop a more flexible work force. Where this is not done and a hire-fire policy is used, desirable employees often will not apply for work at the company, given its employment reputation. Further, whether formally or informally, new employees must be oriented to the firm. This is both costly and time-consuming. For these reasons it is better, wherever possible, to maintain the work force through cross-training, rotation, and similar techniques.

The usual situation that arises is that the work is similar to that previously performed but the specifics require more engineers of a particular discipline than are available, and a surplus of others exists. The redeeming feature in this circumstance is that most personnel are **flexible** and, with proper coaching and suitable organization of their activities, can be given work that they will be able to perform satisfactorily even though it is outside their design specialty. It should be remembered that engineers are trained, at least in part, in all of the major disciplines, and while a mechanical engineer could not alone handle all the work of a complicated structural analysis, he or she under the guidance of an experienced structural engineer could perform many of the routine calculations, sizing of structural members, and materials selections. The engineer, while working at a somewhat lower level than a true journeyman, is nevertheless making a significant contribution to the overall goals of the company, and replacement with a structural "specialist" may not be necessary. The corollary to this is that the employee must be flexible, and a part of the orientation of the new employees is to stress the importance of flexibility and their willingness to work in a variety of roles. The role of the HR group is to ensure that there is a suitable assessment of the attitudes, skills, and experience of the personnel so that supervisors and managers can have these data available when considering other assignments.

For the longer run the firm must make staffing decisions not only based on present needs but with an eye to the **future.** It is absolutely essential that the work in hand be covered, and this needs to be the primary thrust of any staffing program. Still it is important not to forget the future, and where possible, the staff should be recruited and developed with that in mind. The usual problem is to determine even in a general manner what the future staffing distribution should be. There is no easy answer to this, although one approach is to assume that the ratio of sales dollars per employee will improve from year to year. A 5 to 10% improvement (increase) is often used. Since the cost per employee will increase slightly from year to year, 3 to 5% is a commonly applied factor. Another more qualitative approach often used is to assume that the basic work of the firm will continue with the same product line or project mix as experience has shown, but that it will likely develop in the direction of the greatest **interest** of the management. Defining the direction of interest of the management is not too difficult if one evaluates whether its fundamental interest lies in maximizing profits, expanding its market by moving into associated technical fields, being on the cutting edge of technology, working only in known and comfortable technical areas, or perhaps moving into a completely different technological area over time.

The answer to this question will largely dictate whether to operate with a primarily specialist organization with the potential for significant personnel turnover, whether to try to staff the organization with a larger proportion of generalist engineers, or whether to maintain a relatively small staff and utilize significant numbers of temporary personnel. In general, while temporary personnel may be less expert in the fine points of a technology, they usually have broad and varied experience, are adaptable, and can do the basic work required with relatively fewer specialists providing the technical guidance for their work. This decision wil also be influenced by the type of organization utilized, whether functional, project, or matrix type. Each have advantages and disadvantages, as discussed in Chapter 3.

At the same time there must be a consideration of the **balance** of the organization in terms of the experience level of the personnel. For convenience engineers are often spoken of as being either **entry**-level, **journeyman,** or **supervisory** personnel. The distribution is important to the organization. If the organization is top heavy with the more senior personnel, there will be many instances where the personnel work is below their level of capability, and this may lead to some minor dissatisfaction. But the major problem will be the cost of performing the work, since senior personnel are usually more highly paid. A more severe problem presents itself if the organization has too many junior personnel. In this case, while the cost to perform the work will be minimized, the ability to perform adequately may be jeopardized. The preferred balance is at least one journeyman or supervising engineer for every two or three entry-level ones. Again the availability of a skills assessment program will permit fitting personnel into assignments most closely suited to their skills and experience and will tend to reduce or minimize this problem.

Recruitement of personnel is always a concern. Unless there is a major event such as a contract award with significant press and television coverage, the public is generally not aware that a particular firm needs personnel. Further there is a policy in some companies that they need to recruit for the future, and they may maintain ongoing college recruitment programs to identify promising personnel with hope for their long-term growth within the company. There may even be a need to recruit a few very senior personnel from time to time, for which executive recruiters, "headhunters" may be utilized. Regardless of which method of recruitment is utilized, it is important to work out a recruiting program that covers the geographic areas of interest.

A large national firm would have a three-part program consisting of local, regional, and national recruiting, while a smaller firm might only have a local and regional program.

The best source of recruitment is from personnel already working within the firm. While many firms rely upon word of mouth or recommendations of the existing staff, more are coming to rely on a formal **position posting** system, which permits a person in the firm to apply for the position and be properly considered. For many firms the posting system has several advantages, it forces the originator to define the position and its responsibilities very precisely, it avoids overlooking qualified personnel elsewhere in the organization, and it opens the application process to all persons in the organization who feel they meet the job requirements. From the legal standpoint this last factor is extremely important, since it reduces the potential for litigation. If the person is selected for a position at a higher grade and receives a **promotion** as a result (i.e., promotion from within), the effect on morale is also beneficial. In addition no advertising is required, and since the personnel being considered are in the same location, relocation is avoided. Following receipt of applications under this system, the candidates are interviewed, and the system follows normal recruitment processes.

An excellent source of recruitment for the firm is **referrals** by existing employees. An employee will not make a referral if he or she is not satisfied with the firm and at the outset a favorable climate exists. There is no better reference than one's own employees, and in many cases from 30 to 40% of openings can be filled by references from employees. In addition, since the values of the employee and the candidate are likely to be similar, there will be a better chance of easily fitting into the firm. What is more, there is no cost to the firm for referrals. While some companies will pay a bonus to employees who successfully refer candidates who are later hired, many other firms view that as a part of the responsibility of working for the company.

Some firms have a specific policy prohibiting **nepotism,** that is, the employing of close relatives. In general, this is not the concern that it formerly was and the employment of related personnel has many advantages. In particular, if the personnel are treated equitably, it reinforces the loyalty to the company and tends to build a more tight knit group, since the family members tend to identify more closely with the goals of the company. It also provides a ready pool of candidates for recruitment where their skills fit the overall requirements of the firm. Some care needs to be taken, however, to ensure that salary increases, promotions, and desirable work assignments are not influenced by family relationships. There are certain minimal safeguards, as dictated by good business management, that should also be in place to prevent initiation or approval of purchase orders, checks for goods or services, expense account reports, and similar financial activities by relatives.

Normal staff recruitment uses **advertising** in daily **newspapers** of large urban centers where there is a significant available work force with the necessary technical skills for the work. Typically newspaper advertising fills only 5 to 10% of the vacancies, mostly with candidates for positions that are not too specialized. To develop a larger pool of applicants and thus more selectivity in hiring, advertising is run for a series of days or weeks, rather than merely for a few days. Ads should be fairly large, one column inch or larger to attract attention, even if the position is relatively junior. The objective is to catch the eye of the reader, and a large ad does get attention. The difference in cost for the large size ad is not really significant when considering the importance of acquiring better personnel and the long-term benefits that result from improved staffing. It is not necessary to have illustrations in the newspaper ads, for they merely take space without delivering information on the specifics of the position(s). While the newspaper staff can provide some assistance in the writing of the advertisement, the basic information on the position, including the qualifications and responsibilities, need to be specific and brief. Newspaper ads often generate a large response of both qualified and unqualified applicants, so additional time for applicant review needs to be provided.

Advertising in **technical publications** follows the same guidelines outlined above, but it must be modified to apply to the less frequent, typically monthly, publication. It is common for these advertisements to deal with the core positions, namely long-term, permanent positions where potential for growth into senior management levels is a major consideration.

If recruitment is for positions in the general area of the firm's home office, it is better to advertise in local papers no more than one and one-half hour commute away. If the positions to be filled are more senior and require special skills, it may be necessary to widen the advertised area.

The use of **search firms** is normally limited to perhaps 5% of the positions needed, since their work is very costly (up to 40% of the annual salary of the person) and often time-consuming (usually requiring months to fill the position). As a result they are only consulted for senior or highly specialized positions.

Another source of personnel is agencies that provide **temporary employees.** These personnel do not expect permanent employment and can be released back to their parent firm on short notice, typically a week or less. As mentioned earlier, these personnel are often highly experienced and are usually able to make a contribution to the work almost immediately.

College recruiting is favored where it is anticipated that there will be a continuing need for new personnel for a number of years. College recruiting programs are ongoing long-term commitments by the firm. Their success can be almost directly measured by the degree to which the firm establishes good working relationships with the directors of placement at colleges. Since there are many firms recruiting and usually a limited availability of space, favorable recommendations to the students by the faculty and the treatment received by previous graduates recruited will affect not only the number but the quality of students requesting interviews. The most effective programs are those where technical managers perform recruitment interviews, since they can answer the students' questions directly and often informally counsel the students as well. While costly, this approach has proved to be very useful.

The use of **co-op programs,** where the students work six months and go to school six months, provides the students with income and direct practical experience with the firm. It permits the firm to determine the abilities of the student and whether career growth to senior management level is likely. This last point is important, since many firms use the career path potential as an evaluation criteria for co-op students. If the co-op student has a potential to grow professionally, there is likely to be continuation in the company program for several summers and hiring upon graduation. If the student does not show this potential, release from the program is made, and this should be done as early in the program as possible. Such release does not necessarily create a problem for the former student, since co-op or work experience is considered an advantage by potential employers.

For both college recruits and co-op students it is useful to periodically review their work history to see if they are staying with the firm long enough to justify the effort for their recruitment. Some firms have found that this is not justified and prefer to hire engineers two or three years out of school who have been given their early training by another firm, saving the expense and the lower productivity of their initial years.

Job fairs are another source of candidates for recruitment. One advantage, if they are held locally, is that the personnel hired will not need to be relocated. The principle disadvantage is that they often only provide a yield of about 5% for the open positions. Job fairs, while not the cheapest way to recruit personnel, can produce satisfactory results however. Highly experienced personnel will generally be reluctant to attend these fairs, so the fairs will attract mostly junior or less-skilled personnel. They are an excellent way to see a large number of people in a short time and are reasonably cost effective.

Increased active involvement with **minority groups** in the community provides a way to improve the recruitment of these personnel through referrals and recommendations. An ongoing program of cooperation and work with minority groups in both the business and residential communities is very effective and can provide not only assistance in the personnel area but community support as well. This furthers the goal of creating a work force, that is representative of the community and has ethnic diversity. To implement this, additional recruiting in areas of minority representation may be required. There is also the reassurance that this effort meets the guidelines set forth by the Equal Employment Opportunity Commission.

If **relocation** of new employees is provided by the company, it is practical to recruit from virtually any region of the country, although beyond, say, 1000 miles the cost to transport the household goods of the employee rises significantly. Many firms have greatly reduced their long-distance recruiting because of this high cost. For example, the cost to relocate a family of four from Boston to Portland, Oregon, is on the order of $60,000 after moving, house hunting, settling in, and other costs are tabulated. In addition some companies become involved in the buying and selling of housing, and this can be both costly and time-consuming. For the most part firms will recruit over long distances only for **key positions,** which comprise about 3 to 5% of the work force. Some firms avoid this problem by providing a cap on allowances for long-distance relocation moves; for example, for any move over 1000 miles, the employee bears a portion of the cost. Normally the relocation provides settling-in allowances for lodging, meals, and incidental expenses for from two weeks to one month, in which time the employee should find permanent housing. To avoid excessive spending by the employees, these amounts exclude personal expenses. In addition, to ensure that the employee does not relocate at the firm's expense and then change jobs at the new location, the employee may be liable for the expenses spent on his or her behalf until some period of employment has lapsed, usually a year or two, with the liability declining on a monthly basis. In some cases the firm will finance an advance house-hunting trip for the employee and spouse, which, if successful, can avoid the need for the settling-in period and bring a great deal of peace of mind to the employee and family. This also cuts down on absences from the office for house hunting once the employee arrives and helps integrate the employee into both the firm and the community.

Interviews of candidates should be conducted in a professional nonthreatening manner with the emphasis on determining whether the candidate possesses the skills and attitudes to perform the work and whether the employee can or has demonstrated reliability and dependability. An interview by the prospective supervisor is essential, and his or her recommendation should be given the most weight in the hiring decision. In some cases multiple interviews may be appropriate and useful. While some firms utilize extensive testing, most professional positions are filled based on credentials and prior experience. As a result it is normal to peform a **check** on the educational credentials stated in the employment application as well as the duration of employment and salary levels paid by previous employers. Because of time this checking is often not completed until the employee has been hired. As a result candidates should be advised that falsification of any data on the employment application is a serious matter and may be cause for termination. In general, previous employers, because of their concern for liability, will not disclose any performance data on the candidate apart from the dates of employment and initial and final classification of the person. Some former employers will provide information on first and last salary paid but not in most cases. It is rare that you will be able to obtain an **evaluation** of the quality of the work of the individual, although the interviewer by asking questions during the interview such as "Why did you leave the positions noted on your application (resume)?" can often get an indication of whether the person was a satisfactory employee. Evidence of past accomplishments and results achieved can often be very useful. Together with the length of employment this may provide some indication of the stability and, in the case of long service and promotions, the previous level of performance of the person.

While it is important that employment interviews be conducted in an in-depth manner, it is equally important to bear in mind that items that are not job related are illegal and should not be brought up in the interview. Questions such as the marital situation of the candidate, religious or cultural affiliations, and other personal matters should not be raised. Further stress-type interviews in which the candidate is required to make decisions or recommendations while under induced stress are undesirable, and in general inappropriate, since they do not represent the working conditions that the candidate will encounter. Wherever possible the interviewers should reduce their observations to writing to facilitate the hiring decision and to avoid potential future litigation by candidates passed over. The written material should be objective and deal with the performance requirements of the position and the perceived ability of the candidate to perform the work required.

Orientation of new employees is often a hit or miss effort with little or no formal orientation occurring in many firms. While it is true that in small firms the orientation can be quite brief, if the firm is of any size, it is wise to provide an in-depth orientation for the new employee so that the policies of the company and its principles are understood. If a statement of company policy exists, it is a good starting point, together with some discussion of the corporate culture in the firm. Where a policy manual exists, it should be reviewed with the employee, together with the usual routine material regarding time records, expense accounts (if provided), operational responsibilities, supervisory ladder, material on performance evaluations, salary administration, bonus system, employee benefits, vacation accrual and scheduling, sick leave, normal work schedule, and other similar administrative matters.

Many companies have established a system of **mentors** where a new employee is assigned a senior member, usually in the same or a similar work unit, to act as a mentor to answer most of the normal questions that arise and in general to guide the employee through the first few months with the firm. Often called the "buddy system," this has proved to be effective and provides the employee with someone to turn to in an informal way without involving the immediate supervisor who is likely to be very busy. In addition some supervisors ask different staff members to take the new employee to lunch during the first few weeks to answer other questions and to more rapidly integrate the employee into the organization. The overall cost of these programs is not very large, and firms that use them believe that significant dividends result. They are enthusiastic about these programs and recommend their use to others.

When compared to the cost of recruitment, relocation, and hiring, orientation costs are small and represent yet another way to integrate new employees into the firm. An orientation program helps avoid costly initial mistakes and also avoids premature resignation of some employees who may otherwise feel isolated and lose interest in the position. Some firms go so far as to involve the family in the orientation and provide assistance to the spouse and family in finding not only housing, but doctors, dentists, and similar community services.

Of great importance to many employees is the question of whether a dual (promotion) ladder exists, whereby personnel who prefer to remain in the technical areas of the work can advance to the same levels of salary and grade as persons who move into the areas of management. In many firms a **technical specialist** can make as much money as a manager, although the technical ladder often stops one or two levels below top management. This is normally sufficiently high that the technical staff feel there is enough room for growth and recognition. Some firms have established special categories of these personnel who are called by titles such as senior advisors, principal engineer, fellows, or technical speciality vice president. The difficulty arises where the technical personnel feel that there are perhaps only one or two levels available to them and that they are regarded as merely a skilled work force.

When recruiting personnel, the concept of using **agency,** part-time, temporary, or contract workers is one of the fastest growing practices today. In general, there is no need for extensive interviews of agency personnel because they can be released immediately if found to be inadequate. For part-time or contract personnel the normal interview process should be used because they will either join the firm as an employee (part time) or will work under a special contract. In either case they should be fully interviewed and their qualifications determined as a part of the interview and post interview process. During the staffing interviews some candidates will have a record of several jobs lasting perhaps a year or two, which may indicate a tendency to be a job hopper. Most employers do not wish to hire such personnel. Thus it is incumbent on the interviewer to carefully determine the reasons for these job changes. They may all be legitimate and caused by factors outside the control of the employee, such as mergers, contract completions, and contract cancellations, and the employee may work out very well in the proposed position.

12.3 COMPENSATION AND BENEFITS

Compensation and **benefits** is an area of fundamental interest to the employee and is subject to both rational understanding and emotional reaction. There are always employees who feel they are underpaid, not only with regard to their peers but also in the absolute sense in that they feel they are not earning what they are worth. In some cases this is true, but overall it is the effect of supply and demand and the conditions of the marketplace. In other words, the availability of personnel and the number of vacancies actually set the prevailing wage and benefits levels.

Compensation programs should consider three forms of reward for the individual, **base salary** level, **bonuses** based on performance, and some form of **equity participation** in the firm. Salary is based on the value of the position to the firm and is often the only form of compensation considered.

Most salary programs are based on the concept of **merit,** and the term merit pay is widely used. In this approach the various levels of personnel classification, and the salary ranges within them, provide a way in which personnel can be rewarded based on their merit and performance. The actual salary levels for the various personnel classifications are set by a comparison with primarily local competition and to a lesser extent industry in general. From this a set of salary levels is established that may be then modified to reflect local factors such as cost of living, desirability of location, and so forth.

Normal compensation planning utilizes a series of salary grades that **overlap** and thus provide room for growth within a particular grade. For personnel whose careers are proceeding normally, the 75th percentile of the salary range is often considered the point at which promotion is made. For personnel whose development has leveled out and who are not progressing, the entire range is available. This is particularly advantageous for personnel who may be making a contribution to the firm but who have limited potential and are not likely to advance significantly. By utilizing the entire range available prior to promotion, it is possible to provide for reasonable salary growth while avoiding the need to promote personnel who are not qualified. The ranges of salary available for each grade level are adjusted as necessary to reflect the effect of supply and demand and other economic factors such as inflation and cost-of-living adjustments. An example of a salary schedule is presented in Table 12.1.

A weakness of highly structured salary programs is that in the administration of the program the goal of rewarding **performance** and excellence is often lost, and all employees tend to receive increases that are similar. This occurs for several reasons, perhaps the most important of which is the effect of **inflation** and increases in the cost of living. If the cost of living is increasing at say 4% per year, any salary increase less than that does not theoretically keep pace, and the employees are actually receiving a cut in their pay. This tends to place a floor under the lowest salary increases of about 5%, in this case, and in effect diverts this amount from the pool of money allocated for merit salary increases. As a result raises to cover the cost of living become a form of entitlement, and the entire level of salaries is escalated.

A second cause is the concern to keep the employees happy so that they will not **resign.** While low turnover is desirable, it should not be purchased at the cost of badly skewing salary administration. In fact, if an employee is merely doing an adequate job, it may be more appropriate to not provide a raise and encourage either improved performance or a resignation to an-

TABLE 12.1 Salary schedule: Monthly salary ranges for number grades

Salary grade	1st quartile	2nd quartile	3rd quartile	4th quartile	
	Minimum	25%	Midpoint	75%	Maximum
11	$1570	$1755	$1935	2115	2300
12	1755	1960	2160	2360	2565
13	1975	2205	2430	2655	2885
14	2245	2505	2760	3015	3275
15	2570	2865	3160	3455	3750
16	2940	3280	3615	3950	4290
17	3345	3735	4120	4505	4895
18	3725	4155	4585	5015	5445
19	49,400	55,100	60,800	66,500	72,200
	(4117)	(4592)	(5067)	(5542)	(6017)
20	55,300	61,600	68,000	74,900	80,700
	(4609)	(5134)	(5667)	(6200)	(6725)
21	62,000	69,100	76,300	83,500	90,600
	(5167)	(5759)	(6359)	(6959)	(7551)
22	69,200	77,100	85,000	92,900	100,800
	(5767)	(6425)	(7092)	(7759)	(8417)

Note: Monthly equivalents in parentheses for salary grades 19–22.

other firm where the employee's talents better fit the work. Turnover rates of between 5 and 10% per year are no cause for alarm, and in some instances even higher rates are healthy. Conversely, an extremely low turnover rate may indicate a stagnant firm. Moderate turnover becomes a particularly important factor if there is an ongoing program to hire new graduates and provide for new blood in the firm.

Another cause often cited for not providing a wide range of increases to employees is the question that "If I give Fred a 15% increase this year, what will he expect next year?—I certainly don't want to set a precedent that will be hard to correct at a later date." The corollary to this question is the concern that "If I provide too large an increase, a grade change will be necessary, and I don't think Fred is ready for that yet." Both of these questions can be readily handled if the purpose of the salary program is kept in mind. Each year (or salary review period) stands on its own. The criteria against which salary is measured is performance, and the performance of the individual is evaluated for that period. Outside economic factors need to be considered because often the profitability of the firm will affect the ability to pay increased salaries. Performance changes from year to year and the administration of the salary program should reflect these factors. If a promotion is indicated by the performance but for some reason not desired by the supervisor, virtually all programs provide for exceptions to be made.

Similarly, if the salary increase would exceed the range for the grade in question, an exception can again be made.

While the foregoing apparently makes the administration of a salary program simple, there are at least two complicating factors: first, the effect on established salaries of the salaries paid to new graduates and, second, the concern that some valuable employees may be in positions where they do not truly have an opportunity to perform and make a significant contribution, even though they have done so in the past and are still capable of doing so.

The effect of inflation has created a seemingly constant escalation in the starting salaries paid to **new** college **graduates** and is a continual source of stress on the established salaries for the lower-grade personnel in any firm. Over the past decade starting salaries have risen at a rate of about 4% per year, placing upward pressure on existing salaries to avoid the condition where a new graduate earns as much as one who has been with the firm for a year or slightly longer. As a result there is a tendency to provide increases of at least this amount to keep the previous new hires on a par, causing lower-grade salary growth which is largely inflation driven and not representative of the merit or performance of the personnel.

While quite different in cause, the situation where a proven employee is assigned by the firm to work that does not provide an **opportunity** for (or visibility of) significant performance creates a situation where some departure from the strict "pay for performance" standard may be necessary. If the employee is truly a valued one whom the firm wishes to retain, it may be useful to provide salary adjustments approximating the average to provide for "normal" salary growth and keep the person in line with peers. This is predicated on the belief that the person will perform well when given the opportunity on a near-term future assignment. If there is any doubt of this, or if the person has peaked out and is likely not to be promoted in the future, consideration should be given to no salary action or to extending the duration until the next salary action. Extending the time until the salary action can often be used when personnel are nearing the top of their salary range or when they are in the late stages of their careers and their performance has leveled off.

An alternate to providing salary increases when economic conditions are adverse is to consider **job enrichment** by way of **lateral assignments.** This has the effect of providing new and different experiences and challenges for the employee, while from the firm's point of view the employee is broadened and becomes more valuable. In the long run this may be the most important action that can be taken to ensure employment stability.

Implementation of the salary program for a specific group normally occurs through a **salary plan.** This plan is drawn up periodically, usually annually, and based on the grade level of the person, the current salary, and current performance, and it establishes a planned salary adjustment for the individual. This plan often uses a dollar pool concept to provide for the compensation of the entire group, with the manager allocating the pool money among the personnel based on the factors noted above. For a rough first cut, the manager sets aside a specific amount on the order of 2% of the pool for outstanding performance and, using the remainder, calculates the average change in the pool amount as a percentage of the total payroll for the group, establishing an overall average increase. Following that, adjustments are made to increase the amounts for above-average personnel and decrease the amounts for others. Poor performers often are given no increase in the plan on the assumption that their performance will not improve. For outstanding personnel the managers' retention fund is used to provide significant adjustments perhaps as large as 15 or 20%. In some cases promotions are appropriate, and the increase in salary accompanying them can be accommodated from the managers' retention fund as well. A commonly used percentage is 5% minimum for the promotion plus whatever percentage of increase was due the employee for the duration and level of performance displayed.

The actual execution of the salary plan usually is staggered through the year with a **performance evaluation** of the employee taking place roughly on the anniversary date of their employment and a salary action shortly thereafter. Some care must be taken in administering the plan, not to follow it blindly but rather to make adjustments upward or downward in the dollar amounts provided depending on the performance of the employee. Unless some care is taken, recent performance may be given more weight than performance earlier in the year, and thus the evaluation may be distorted. In some firms the salary adjustments are made on the same day for the entire staff. This simplifies the administrative work and tends to yield the same effect as the staggered plan described above. A portion of a staggered salary plan is shown in Fig. 12.1.

In increasing use are **bonuses** of one form or another that link compensation to performance of the firm as a whole, the work of the individual, or work unit, or a combination of both. Linking the bonus to the performance of both the firm and the individual has the advantage of providing an incentive for the employee to look beyond merely his or her own activities and consider the benefit of overall improvements. It also provides a way to increase the interest of the employee in the welfare and financial well-being of the company. However, bonus plans must be **simple** enough so that employees can readily calculate their own bonuses. This is a function of the elements of the plan and may not be appropriate to all circumstances.

Bonuses are often used in lieu of salary increases to reward performance. A major advantage is that the bonus is tied to and relates specifically to performance, and an added benefit is that this arrangement avoids the need for increasing the salary level of the individual and tends to avoid salary escalation. Under this concept the performance for each period stands alone and offers the major advantage that, when poor business conditions are present, salaries have not been inflated and bonuses can be omitted.

A pitfall to the bonus system is a tendency for it to become **routine** and for bonuses to be expected by the personnel. This is particularly the case if

Name	Last Action						Performance[a]	%	Date	Planned		Comments
	Grade	Salary	Quartile	%	Date	Amount				Salary	Change[b]	
Josh	11	1825	II	8	6/93	135	1.05	7	8/94	1950	125	
Zach	13	2450	III	6	4/93	140	1.0	5	4/94	2570	120	
Rachael	13	2495	III	6	12/93	145	.95	4	2/94	2595	100	
George	14	3150	IV	3	5/93	90	.75	3	11/94	3240	90	
Stephanie	15	3475	IV	4	10/92	135	.85	4	4/94	3615	140	
Jason	18	4200	II	9	8/93	350	1.10	7	6/94	4495	295	
		$17,595/mo.								$18,465/mo.		

[a] Where 1.00 is expected normal performance.
[b] Net change $870/mo. = 4.94%

FIGURE 12.1 Salary plan.

bonuses are granted based primarily on salary, grade, or position rather than specific performance. As with basic salary administration it is important to guard against them becoming a form of entitlement and thus destroying their effectiveness.

Equity participation is available in a myriad of forms, from options to purchase stock at discounted values to ESOP (Employee Stock Ownership Plans, see Chapter 13), where in some cases the employees themselves actually are the owners and stockholders of the firm. They have a significant advantage that when used in lieu of bonuses, they usually do not require cash disbursements and have little or no effect on current earnings or distributions. As with bonuses there is an increasing interest and use of equity arrangements to encourage increased employee concern for the welfare and performance of the company. These plans usually have significant **tax** implications, and careful legal advice should be obtained prior to initiating them.

Employee **performance reviews** are the basis for an effective and equitable salary administration program, and they are an extremely valuable tool to improve personnel performance. They must be constructive, accurate, and represent an honest evaluation of the person including strengths and weaknesses. In particular, they should form a basis for a way in which the employees can be encouraged to improve their performance. This improvement can be brought about by the supervisor performing coaching, consciously providing a variety of work assignments to broaden the person, including formal schooling, specialized seminars, technical society involvement, writing papers and technical articles, and participation in community or civic groups. With this wide variety of alternatives there is no reason that some specific steps to produce a higher level of performance cannot be pursued if the employee is interested.

When an employee enters a new classification, a **work-planning** session must be held to establish the requirements of the position. This normally uses a description of the job and a careful review of its characteristics. The performance requirements of the position are reviewed against the standard review form, and each of them identified as being a key category, an applicable category, or one that is not applicable. Not all characteristics are applicable to a position, although higher-graded positions will utilize more of them. When a performance review is subsequently held, typically annually, the basis for performance has been established, and the review relates directly to performance against those characteristics.

Reviews are most effective if they encourage an open dialogue between the supervisor and the employee. The forms used for this are normally developed with that objective, and while they vary widely, most tend to consider the following categories:

Productivity	Job knowledge
	Quality of work
	Quantity of work

	Meeting schedules
	Resource utilization
Work habits	Dependability
	Flexibility
Relationships	Communications
	Cooperation
	Client relations
	Response to supervision
Responsibility	Self-development
	Initiative
	Problem analysis
Supervision/management*	Decision making and execution
	Delegation
	Leadership
	Planning and controlling
	Performance reviews
	Employee development
	Handling employee problems
	Policies and procedures

When actually conducting a performance review, each characteristic is graded on a scale of meets, exceeds, or needs improvement. The overall performance review can then be summarized by the key categories: for example, nine meets, six exceeds, and zero does not meet. This provides a quick overview and permits ready comparison between personnel.

An important part of the review is the development of recommendations and action items for the employee. Often these involve such comments as "More attention to detail is required," or "John should take a course in statistics at the local junior college." Sometimes the action items involve the supervisor, who may commit to providing a greater variety of work for the employee, permitting more initiation of activities, and so forth.

It is tempting, when performing employee reviews, to **skew** the rating upward, and as a result many employees appear to be better than they truly are. Where they are used for salary administration, this has the effect of levelizing the entire rating and salary system and damages the basic reason for the salary program. It is extremely important to guard against this, and some firms have even gone so far as to limit the percentage of employees who can be classified as "meets requirements" and "exceeds requirements" to, for example, 60 and 25%, respectively. Another way to deal with this, as used by a senior executive whose line managers consistently rated large numbers of their personnel as "exceeds," was to warn them that all personnel so rated would be made available for transfer to other divisions to try to

* Unless supervision is performed the supervision/management categories are considered "not applicable."

help the company solve some of its overall problems. As expected, the ratings soon came into line. Still another approach uses a table in which the rater indicates what percentage of the employees rated fell into each category (i.e., meets, exceeds, or needs improvement), thus permitting evaluation of the rater.

Employees should be reviewed at least annually. For new employees semiannual reviews are widely used. Reviews should also be prepared when there is a change in supervisors, assuming that the employee has worked for the supervisor long enough for a valid assessment. If there is a significant change in the work assignment, a review should also be prepared. Following the preparation, it is reviewed with the rater's immediate supervisor, and adjustments made as appropriate. After this the review should be reviewed carefully with the employee. There should be sufficient time allowed, a half-hour as a bare minimum and longer as necessary. The discussion should not be interrupted, and if necessary a separate room or office should be used. The employee should be encouraged to discuss his or her performance, interests, areas of concern, and areas where personnel development are appropriate. The reviewer should place emphasis on the positive aspects of the review, stressing areas of improvement or further development.

A difficult problem arises when it is necessary to **reduce staff** due to project completion, reduction in work load, or other similar factors. The problem is more complicated than the hiring decision, since personnel may be engaged on work from which they cannot be easily removed and replaced. One way to handle the problem is to review the personnel, **forced ranking** them on a point basis in terms of ability within classifications. Considerable care should be taken to ensure that the forced ranking categories and, within them, the basis for awarding points are directly related to the performance requirements of the position and the business of the firm.

Starting from the bottom (i.e., the lowest ranked individual), identify those persons working on critical work who cannot be readily replaced, indicating the earliest date when replacement could be made. Using these data, determine who will be laid off, starting from the lowest-ranked person and working upward.

When doing this it may be possible to take special measures to substitute higher-ranked personnel for lower-ranked ones, and thus retain them in the organization, for example, the case where a lower-ranked person on an active project has only four more months of work and a person ranked significantly higher has no work left at all. The possibility of replacing a lower-ranked person with the higher-ranked one should be considered even though the person would be working below their capacity. It would protect the higher-ranked person for a few additional months during which time new work might be acquired. Alternatively, the higher-ranked person could be given other work in the firm or take a leave or vacation. If the project is reimbursable, the client may permit the substitution if the firm absorbs part of the cost of the orientation to the project or the salary differential.

Whatever the method used, it is important to follow the principle that the most qualified personnel be retained and that the decision to reduce staffing be in writing, including the reasons and the logic behind the decision. Almost certainly there will be dissatisfied personnel who will attempt to bring suit for **wrongful discharge,** and such **documentation** is invaluable for establishing, often years later, the reasons for the action. To further reduce this exposure, some large firms employ additional internal review procedures, normally requiring review of the decision of the operating unit, by both the personnel group and senior management prior to employee notification. This provides assurance that candidates for layoff have received proper consideration and that the system for selection for layoff has been **equitably** implemented. In the example of a retention analysis form shown in Fig. 12.2, the forced ranking used to classify personnel for retention has four categories. The categories are 40% for present and sustained performance with a range of 1–16 points, 20% for attitude and technical speciality with a range of 1–8 points, 10% for willingness to relocate with a range of 1–4 points, and 30% for potential for advancement with a range of 1–12 points. Hard criteria must be used to establish the basis for the point score and remove concern that there is excessive subjectivity in the ranking. For example, where potential relocation is important, award one point for willingness to relocate within a state, two points for willingness to locate anywhere in the United States, and four points for willingness to relocate overseas.

Many firms today will make significant attempts to find jobs for staff being released. This **outplacement** activity ranges from assistance in writing resumes and providing minimal secretarial assistance and temporary office space, to hiring outside firms who specialize in placement of personnel. In many cases these services are quite expensive, and where offered, the general pattern is that the separating firm pays for the services provided. In the case of mergers affecting employees of long or significant service, severance payment packages are often available.

Benefit plans are frequently the deciding factor in a decision by an employee who holds more than one employment offer. They are also of great importance to the existing staff, and with rising costs, they are of increasing concern to management as well. Benefit plans include the common concerns of health and perhaps dental insurance, life insurance, vacation and sick leave policies, and tax-deferred savings plans which in many cases include matching funds by the employer. Of increasing interest are plans for child care and longer-term maternity leave as the work force includes larger numbers of women of child-bearing age. Many firms have sponsored employee credit unions to provide attractive rates for loans for automobiles and other large purchases and slightly higher interest rates on deposits.

While these benefits are often taken for granted, their cost is significant, with the employer's share costing on average about 40 to 45% of the cost of direct payroll. As a way to contain these costs and avoid their continuing increase, many firms offer a **cafeteria** approach where the employee is given

RETENTION ANALYSIS

(based on forced ranking of 9/08/94)

Name	Grade	Performance Evaluation M / E / Ni	Forced Ranking Performance	Attitude & Technical Specialty	Relocation	Potential	Total	Assignment Completes
Williams	14	8 / 5 / 0	12	6	2	9	29	3/95
Jones	14	9 / 4 / 0	14	4	0	8	26	11/15
Swift	15	8 / 4 / 1	10	7	4	10	31	11/6
Johnson	15	7 / 6 / 0	11	8	2	7	28	5/95
Adams	15	7 / 5 / 1	9	6	9	6	25	12/31

Recommendation: Unless additional work is booked prior to 11/15, Swift to be placed on overhead work, Jones to be released 11/15, and Adams to be held through the holidays and released in early January. Four weeks advanced notice to be given to Jones and Adams.

Signed: _____	Date: _____	Dept. Head
Concur: _____	Date: _____	Human Relations
Approved: _____	Date: _____	Manager

FIGURE 12.2 Retention analysis.

in effect a dollar allowance to be used for the benefits of most importance to the individual. A young healthy woman, with a child might elect to carry major medical insurance with coverage only for serious illness but might wish to also receive child-care benefits, whereas a married male employee age 50, whose children are fully grown, might wish medical insurance with a minimum deductible together with maximum contribution to a tax-deferred retirement fund. Desired coverage that exceeds the company allowance is then paid by the employee. Basically then, the cafeteria plan permits each employee to select the benefits package that most closely fits his or her needs.

An area of recent interest is preventive medical care to contain rising medical costs. Screenings are provided to detect the early onset of heart, pulmonary, and similar diseases, and physical fitness programs are often offered as well.

Other **benefits** usually not considered by the employee are company-paid travel insurance, memberships in technical societies or professional groups, company sponsorship of professional registration and attendance at technical society meetings, and company sponsored employee clubs and interest groups.

The task facing the company is to contain the rising costs of these benefit programs while proving a sufficient range to remain attractive to present and potential employees. To this end, the greater variety of benefits available, including variations on individual components such as variable deductibles for medical insurance, the more attractive the program becomes. The difficulty is to provide essentially error-free, efficient administration of programs having such a wide range that virtually each employee has a different plan. One partial solution is to group as much of the plan with a single carrier as possible, thus shifting much of the administrative burden to that organization.

Retirement, while widely professed to be a career goal of many, often creates psychological problems for the individual. As a result many companies provide pre-retirement counseling and seminars to older members of the work force. Such programs are highly desirable, but they are usually only offered in the last year or so before retirement. To be truly effective, planning for retirement should begin in **midcareer.** While the planning may not be firm, consideration should be given to starting savings plans for retirement, developing outside interests, and so forth. On the other hand, the trend toward early retirement has created a situation where many employees are reluctant to disclose to their employers that they are contemplating early retirement, so they do not take advantage of company-sponsored courses. They believe that disclosure will remove them from salary increases and promotions, and leave them with less-challenging work assignments. They must therefore depend on outside courses, seminars, lectures, and so forth. Fortunately there is an abundance of this information available, and it is relatively easy to educate oneself on the significant concerns in retirement and, as important, the transition during the initial phase of retirement.

Among the most important choices to be made is whether the employee will truly retire or whether a second career in perhaps an entirely different field is what is actually desired. With improved health and the lengthening of life expectancy, retirees can look forward to perhaps 10 or 20 active and productive years before they are unable to work, and a second career is becoming more common and more attractive. Most employees favor some form of second career, even if it is a part-time or volunteer community service-type activity. If there are sufficient financial resources available, there may be no need to supplement income, and the money earned may be a very secondary consideration. This opens up a wide range of possibilities and permits the retiree to work at activities that were merely earlier dreams.

Financial planning for retirement should start as early as possible. Many persons start when they are 35 to 40 years old. The importance of financial planning of course increases as the employee ages. The use of **tax-deferred** savings plans offered by the company or taken out privately by the employee are highly advantageous, and with disciplined, regular additions can grow substantially over time. Portability of retirement plans is of increased interest as duration of employment with a firm decreases. Development of alternate sources of income to augment Social Security payments is very important. Many retirees have derived additional income from speaking, consulting, teaching, writing, and hobbies.

Broadly speaking, it is necessary to provide from 60 to 75% of normal income in retirement to maintain the same standard of living and activities as before retirement. Of this Social Security will normally provide about one-third. Clearly it is necessary to plan so that there will be sufficient financial resources to make up the difference. Since monies invested will earn some return prior to being drawn down, it is possible to estimate approximately how long retirement savings will last. Table 12.2 indicates this general relationship.

One useful way to begin planning for retirement is to establish three files—retirement financial, retirement occupational, and retirement recreational—and to begin clipping articles and filing data in each. This has the advantage of gathering information without necessarily being too restrictive and provides for **self-education** long before retirement is anticipated. An added advantage is that a wide range of material can be accumulated, and possible options are less likely to be overlooked. Close to or even after retirement, when goals are a bit better established the files can be pruned and inappropriate material discarded.

12.4 OPERATIONS

To properly operate a human resources function, it is necessary to establish a **policy** and **procedure** manual. The manual should include the items discussed in this chapter as well as a section on the legal considerations and

TABLE 12.2   Retirement fund table: Number of years money will last

	Annual % Rate of Return																		
Annual % Rate of Withdrawal	1	2	3	4	5	6	7	8	9	10	11	12	13	14	15	16	17	18	19
20	5	5	5	6	6	6	6	7	7	7	7	8	9	9	10	11	12	14	17
19	5	6	6	6	6	7	7	7	7	8	8	9	9	10	11	12	14	18	
18	6	6	6	6	7	7	7	8	8	9	9	10	10	11	13	15	18		
17	6	6	7	7	7	7	8	8	9	9	10	11	12	13	15	19			
16	6	7	7	7	7	8	9	9	10	10	11	12	14	16	20				
15	7	7	8	8	8	9	9	10	11	12	13	14	16	21					
14	7	8	8	9	9	10	11	11	12	13	15	17	22						
13	8	8	9	9	10	11	11	12	14	15	18	23							
12	9	9	10	10	11	12	13	14	16	19	24								
11	10	10	11	12	12	14	15	17	20	25									
10	11	11	12	13	14	16	18	21	27										
9	12	13	14	15	17	19	22	29											
8	13	15	16	18	20	24	31												
7	15	17	19	22	26	33													
6	18	20	23	28	37														
5	22	26	31	41															
4	29	35	47																
3	41	55																	
2	70																		

requirements of employment. The manual should be widely distributed and all personnel should be encouraged to read it.

As an adjunct to the formal personnel policies and procedures, frequent **communication** between management and the work force is essential. This is particularly important in a technical organization where the personnel often apply their own interpretations and extrapolations to events and may distort their intention. When rumors start, much energy is lost to these nonproductive activities. Frequent communication will help significantly to reduce the spread of rumors and will provide an opportunity not only to distribute positive information about the firm but, if performed well, to enhance the reputation of the firm in the eyes of the employees. An important principle needs to be borne in mind, however. Communication should be timely, honest, and neither patronizing nor alarmist. Today's work force is better educated and more sophisticated than in previous years. Similarly, as the work force becomes more of a partner in the work of the firm, it is important to treat personnel as such and to provide as much information to them as possible.

A suitable structuring of the work force is necessary with the development of **job descriptions** and **work planning** for each position in the organization. Job descriptions outline the requirements and functions of the various positions in the organization. Work planning carries this one step further and sets forth the function of the positions and the interfaces with other persons and groups, and the responsibilities and authority of the person holding each position. Thus a job description might set forth the qualifications and experience requirements for a position as senior engineer, while senior engineers might hold any of several different positions as group leader, technical specialist, or start-up supervisor, each of which would have significantly different operational responsibilities.

While the principal effort in developing **motivation** in the work force may be expended by management, professional employees carry part of the responsibility themselves. It is incumbent upon the professional to give the employer his or her best performance and not to merely go through the motions or put forward half-efforts. If one feels that he or she cannot do this, that individual should leave the firm and seek other work. Where the firm is working on projects or in areas that differ fundamentally with the views of an individual, that person should make his or her concerns known to the manager and request reassignment to other work that is not offensive. If this is not possible, then the person should resign from the firm, with the understanding that reasonable people can disagree and still recognize each other's position.

Similarly a firm should develop a feeling of **trust** among the employees by treating them evenhandedly and equitably and by sharing information on the well-being of the firm, including finances and work load. It goes without saying that the firm will try to motivate its employees to improve their performance and to continually improve the product. This attitude should be established early in the career of an individual, for with professional growth the individual will be given responsibility for motivating others. To this end it is useful to recognize that motivation results from an acceptance of common goals. To get the best from the work force, more authority and responsibility should be given to the employee. The days of the supervisor making all decisions are long gone. More companies are vesting increased authority with working level personnel and raising their commitment to company goals. **Ownership** of the production process (be it engineering, fabrication, or service) offers the best way for employees to be brought into the company's effort to efficiently produce a successful product. The establishment of small work groups with **authority** to make changes to improve the efficiency and quality of their work has proved to be an extremely effective method. The results have indicated large gains in productivity and quality, as well as reductions in errors, scrap and rework, employee turnover, and absenteeism. Of course as employee morale improves, a whole group of secondary benefits result as well.

These programs have largely become the way in which the quality assurance and improvement programs are now implemented.

It is fundamental that the employee provide input and participate actively in these **continuous** improvement programs. The same attitude affects the career of an individual, for self-improvement programs are essential to continued career growth. This is discussed in more detail in Section 12.5.

Stress is always present in the workplace and can encourage the best performance from each of us. The difficulty that many people have is to distinguish between stress, which is not necessarily unhealthy, and distress, which can create emotional and health problems. While many of our reactions to situations at work are a result of our upbringing, as adults we are responsible for our own actions and can modify our responses to improve both our effectiveness and the results of our work. It is important that we deal with those things that are within our ability to change and not needlessly fret over those things outside our ability to change. Often a word to one's supervisor will save much agonizing over what to do.

There is no job that does not have some **conflict** inherent in it. Goals of individuals in different groups are not the same, and yet they are called on to work together toward a common end. As a result each must compromise and yield a bit to achieve the end desired. The important thing is to establish clear goals, to recognize what the end desired is, and to work toward that end, even if a bit different then what a more narrow view would define.

The idea behind personnel policies and programs is **equity** and **nondiscrimination.** Apart from any moral questions, the federal and state legislation on this subject is very clear. Equal Employment Opportunity and Affirmitive Action requirements are a reality of the operations of the firm. Surely it does not make good business sense to overlook likely sources of qualified and committed personnel. Despite this, and even with the best run programs, there will be times when employees feel that they have been discriminated against in terms of hiring, promotion, retention, or job assignments. To guard against such action, it is essential to have a written policy and procedures that follow the legal requirements and establish how the various aspects of the program are to be implemented. To ensure that the policy and procedures are adequate, it is wise to have them reviewed by legal counsel prior to implementation. It is also necessary to maintain adequate records in each of these areas to demonstrate that the procedures were followed. With this sort of documentation, few if any cases will go to litigation in the courts, and those that do will usually be decided in favor of the employer.

Despite the best efforts of both the management and the staff, there will inevitably arise **grievances** that, whether real or perceived, seem real to the employee. A formal procedure to appeal this within the company will permit them to be resolved more readily. At a minimum, a documented, well-structured, and administered program will provide a significant defense if litigation is forced on the firm. One practice widely used is to have a system of grievance review that includes two levels of review (above the working level). An important element in the grievance procedure is to ensure an impartial hearing. The review system should provide for this and permit timely correction of the problem, including possible reinstatement, back pay, or transfer to another unit.

While sometimes called *professional responsibility,* **ethics** in the discharge of engineering or other professional responsibilities is a paramount concern. It establishes a standard against which to judge the basis for decisions made. The engineering profession, in particular, can place large numbers of the public at risk if its work is not performed correctly. Thus the fundamental principle upon which proper ethical performance rests is protection of the health and safety of the public. This concern transcends all considerations of cost and schedule.

This principle does not mean that designs are to be **overdesigned** to eliminate risk. Rather it means that normal engineering practice must follow the use of suitable conservative factors of safety and the proper application of the codes and standards that have been developed specifically for the protection of the public. Where developmental activities require that they be infringed upon, it should be done in a limited and controlled way to avoid risk to the general public, and only after performance has been proved can the design be widely implemented.

Over the years the engineering profession has developed a code of ethics intended to ensure integrity and competence in the conduct of its work. A copy of this standard and its detailed guidelines are shown in Fig. 12.3.

Effective **supervision** of the work force is a primary responsibility of management. The concept of ownership of certain functions by the staff (also called *empowerment*) has been previously discussed and can represent a major challenge to the first line supervisor. This sharing of authority (and responsibility) for decision making, as well as other aspects of the work, is where most supervisors have difficulty. At the extremes we have supervisors

Accreditation Board for Engineering and Technology*

CODE OF ETHICS OF ENGINEERS

THE FUNDAMENTAL PRINCIPLES

Engineers uphold and advance the integrity, honor and dignity of the engineering profession by:

I.   using their knowledge and skill for the enhancement of human welfare;

II. being honest and impartial, and serving with fidelity the public, their employers and clients;

III. striving to increase the competence and prestige of the engineering profession; and

IV. supporting the professional and technical societies of their disciplines.

THE FUNDAMENTAL CANONS

1. Engineers shall hold paramount the safety, health and welfare of the public in the performance of their professional duties.

2. Engineers shall perform services only in the areas of their competence.

3. Engineers shall issue public statements only in an objective and truthful manner.

4. Engineers shall act in professional matters for each employer or client as faithful agents or trustees, and shall avoid conflicts of interest.

5. Engineers shall build their professional reputation on the merit of their services and shall not compete unfairly with others.

6. Engineers shall act in such a manner as to uphold and enhance the honor, integrity and dignity of the profession.

7. Engineers shall continue their professional development throughout their careers and shall provide opportunities for the professional development of those engineers under their supervision.

345 East 47th Street New York, NY 10017

*Formerly Engineers' Council for Professional Development. (Approved by the ECPD Board of Directors, October 5, 1977)

AB-54 2/85

FIGURE 12.3 Code of ethics of engineers. Reprinted by permission of the Accreditation Board for Engineering and Technology, Inc.

Accreditation Board for Engineering and Technology*

SUGGESTED
GUIDELINES FOR USE WITH
THE FUNDAMENTAL CANONS OF ETHICS

1. Engineers shall hold paramount the safety, health and welfare of the public in the performance of their professional duties.

   a. Engineers shall recognize that the lives, safety, health and welfare of the general public are dependent upon engineering judgments, decisions and practices incorporated into structures, machines, products, processes and devices.

   b. Engineers shall not approve nor seal plans and/or specifications that are not of a design safe to the public health and welfare and in conformity with accepted engineering standards.

   c. Should the Engineers' professional judgment be overruled under circumstances where the safety, health, and welfare of the public is endangered, the Engineers shall inform their clients or employers of the possible consequences and notify other proper authority of the situation, as may be appropriate.

      (c.1) Engineers shall do whatever possible to provide published standards, test codes and quality control procedures that will enable the public to understand the degree of safety or life expectancy associated with the use of the design, products and systems for which they are responsible.

      (c.2) Engineers will conduct reviews of the safety and reliability of the design, products or systems for which they are responsible before giving their approval to the plans for the design.

      (c.3) Should Engineers observe conditions which they believe will endanger public safety or health, they shall inform the proper authority of the situation.

   d. Should Engineers have knowledge or reason to believe that another person or firm may be in violation of any of the provisions of these Guidelines, they shall present such information to the proper authority in writing and shall cooperate with the proper authority in furnishing such further information or assistance as may be required.

      (d.1) They shall advise proper authority if an adequate review of the safety and reliability of the products or systems has not been made or when the design imposes hazards to the public through its use.

      (d.2) They shall withhold approval of products or systems when changes or modifications are made which would affect adversely its performance insofar as safety and reliability are concerned.

   e. Engineers should seek opportunities to be of constructive service in civic affairs and work for the advancement of the safety, health and well-being of their communities.

   f. Engineers should be commited to improving the environment to enhance the quality of life.

2. Engineers shall perform services only in areas of their competence.

   a. Engineers shall undertake to perform engineering assignments only when qualified by education or experience in the specific technical field of engineering involved.

   b. Engineers may accept an assignment requiring education or experience outside of their own fields of competence, but only to the extent that their services are restricted to those phases of the project in which they are qualified. All other phases of such project shall be performed by qualified associates, consultants, or employees.

   c. Engineers shall not affix their signatures and/or seals to any engineering plan or document dealing with subject matter in which they lack competence by virtue of education or experience, nor to any such plan or document not prepared under their direct supervisory control.

3. Engineers shall issue public statements only in an objective and truthful manner.

   a. Engineers shall endeavor to extend public knowledge, and to prevent misunderstandings of the achievements of engineering.

   b. Engineers shall be completely objective and truthful in all professional reports, statements, or testimony. They shall include all relevant and pertinent information in such reports, statements, or testimony.

   c. Engineers, when serving as expert or technical witnesses before any court, commission, or other tribunal, shall express an engineering opinion only when it is founded upon adequate knowledge of the facts in issue, upon a background of technical competence in the subject matter, and upon honest conviction of the accuracy and propriety of their testimony.

   d. Engineers shall issue no statements, criticisms, nor arguments on engineering matters which are inspired or paid for by an interested party, or parties, unless they have prefaced their comments by explicitly identifying themselves, by disclosing the identities of the party or parties on whose behalf they are speaking, and by revealing the existence of any pecuniary interest they may have in the instant matters.

   e. Engineers shall be dignified and modest in explaining their work and merit, and will avoid any act tending to promote their own interests at the expense of the integrity, honor and dignity of the profession.

4. Engineers shall act in professional matters for each employer or client as faithful agents or trustees, and

*Formerly Engineers' Council for Professional Development.

FIGURE 12.3 (continued).   Code of ethics of engineers (Page 2 of 5).

shall avoid conflicts of interest.

a. Engineers shall avoid all known conflicts of interest with their employers or clients and shall promptly inform their employers or clients of any business association, interests, or circumstances which could influence their judgment or the quality of their services.

b. Engineers shall not knowingly undertake any assignments which would knowingly create a potential conflict of interest between themselves and their clients or their employers.

c. Engineers shall not accept compensation, financial or otherwise, from more than one party for services on the same project, nor for services pertaining to the same project, unless the circumstances are fully disclosed to, and agreed to, by all interested parties.

d. Engineers shall not solicit nor accept financial or other valuable considerations, including free engineering designs, from material or equipment suppliers for specifying their products.

e. Engineers shall not solicit nor accept gratuities, directly or indirectly, from contractors, their agents, or other parties dealing with their clients or employers in connection with work for which they are responsible.

f. When in public service as members, advisors, or employees of a governmental body or department, Engineers shall not participate in considerations or actions with respect to services provided by them or their organization in private or product engineering practice.

g. Engineers shall not solicit nor accept an engineering contract from a governmental body on which a principal, officer or employee of their organization serves as a member.

h. When, as a result of their studies, Engineers believe a project will not be successful, they shall so advise their employer or client.

i. Engineers shall treat information coming to them in the course of their assignments as confidential, and shall not use such information as a means of making personal profit if such action is adverse to the interests of their clients, their employers, or the public.

(i.1) They will not disclose confidential information concerning the business affairs or technical processes of any present or former employer or client or bidder under evaluation, without his consent.

(i.2) They shall not reveal confidential information nor findings of any commission or board of which they are members.

(i.3) When they use designs supplied to them by clients, these designs shall not be duplicated by the Engineers for others without express permission.

(i.4) While in the employ of others, Engineers will not enter promotional efforts or negotiations for work or make arrangements for other employment as principals or to practice in connection with specific projects for which they have gained particular and specialized knowledge without the consent of all interested parties.

j. The Engineer shall act with fairness and justice to all parties when administering a construction (or other) contract.

k. Before undertaking work for others in which Engineers may make improvements, plans, designs, inventions, or other records which may justify copyrights or patents, they shall enter into a positive agreement regarding ownership.

l. Engineers shall admit and accept their own errors when proven wrong and refrain from distorting or altering the facts to justify their decisions.

m. Engineers shall not accept professional employment outside of their regular work or interest without the knowledge of their employers.

n. Engineers shall not attempt to attract an employee from another employer by false or misleading representations.

o. Engineers shall not review the work of other Engineers except with the knowledge of such Engineers, or unless the assignments/or contractual agreements for the work have been terminated.

(o.1) Engineers in governmental, industrial or educational employment are entitled to review and evaluate the work of other engineers when so required by their duties.

(o.2) Engineers in sales or industrial employment are entitled to make engineering comparisons of their products with products of other suppliers.

(o.3) Engineers in sales employment shall not offer nor give engineering consultation or designs or advice other than specifically applying to equipment, materials or systems being sold or offered for sale by them.

5. Engineers shall build their professional reputation on the merit of their services and shall not compete unfairly with others.

a. Engineers shall not pay nor offer to pay, either directly or indirectly, any commission, political contribution, or a gift, or other consideration in order to secure work, exclusive of securing salaried positions through employment agencies.

b. Engineers should negotiate contracts for professional services fairly and only on the basis of demonstrated competence and qualifications for the type of professional service required.

c. Engineers should negotiate a method and rate of compensation commensurate with the agreed upon scope of services. A meeting of the minds of the parties to the contract is essential to mutual confidence. The public interest requires that the cost of engineering services be fair and reasonable, but not the controlling consideration in selection of individuals or firms to provide these services.

(c.1) These principles shall be applied by Engineers

FIGURE 12.3 (continued). Code of ethics of engineers (Page 3 of 5).

in obtaining the services of other professionals.

d. Engineers shall not attempt to supplant other Engineers in a particular employment after becoming aware that definite steps have been taken toward the others' employment or after they have been employed.

(d.1) They shall not solicit employment from clients who already have Engineers under contract for the same work.

(d.2) They shall not accept employment from clients who already have Engineers for the same work not yet completed or not yet paid for unless the performance or payment requirements in the contract are being litigated or the contracted Engineers' services have been terminated in writing by either party.

(d.3) In case of termination of litigation, the prospective Engineers before accepting the assignment shall advise the Engineers being terminated or involved in litigation.

e. Engineers shall not request, propose nor accept professional commissions on a contingent basis under circumstances under which their professional judgments may be compromised, or when a contingency provision is used as a device for promoting or securing a professional commission.

f. Engineers shall not falsify nor permit misrepresentation of their, or their associates', academic or professional qualifications. They shall not misrepresent nor exaggerate their degree of responsibility in or for the subject matter of prior assignments. Brochures or other presentations incident to the solicitation of employment shall not misrepresent pertinent facts concerning employers, employees, associates, joint ventures, or their past accomplishments with the intent and purpose of enhancing their qualifications and work.

g. Engineers may advertise professional services only as a means of identification and limited to the following:

(g.1) Professional cards and listings in recognized and dignified publications, provided they are consistent in size and are in a section of the publication regularly devoted to such professional cards and listings. The information displayed must be restricted to firm name, address, telephone number, appropriate symbol, names of principal participants and the fields of practice in which the firm is qualified.

(g.2) Signs on equipment, offices and at the site of projects for which they render services, limited to firm name, address, telephone number and type of services, as appropriate.

(g.3) Brochures, business cards, letterheads and other factual representations of experience, facilities, personnel and capacity to render service, providing the same are not misleading relative to the extent of participation in the projects cited and are not indiscriminately distributed.

(g.4) Listings in the classified section of telephone directories, limited to name, address, telephone number and specialties in which the firm is qualified without resorting to special or bold type.

h. Engineers may use display advertising in recognized dignified business and professional publications, providing it is factual, and relates only to engineering, is free from ostentation, contains no laudatory expressions or implication, is not misleading with respect to the Engineers' extent of participation in the services or projects described.

i. Engineers may prepare articles for the lay or technical press which are factual, dignified and free from ostentations or laudatory implications. Such articles shall not imply other than their direct participation in the work described unless credit is given to others for their share of the work.

j. Engineers may extend permission for their names to be used in commercial advertisements, such as may be published by manufacturers, contractors, material suppliers, etc., only by means of a modest dignified notation acknowledging their participation and the scope thereof in the project or product described. Such permission shall not include public endorsement of proprietary products.

k. Engineers may advertise for recruitment of personnel in appropriate publications or by special distribution. The information presented must be displayed in a dignified manner, restricted to firm name, address, telephone number, appropriate symbol, names of principal participants, the fields of practice in which the firm is qualified and factual descriptions of positions available, qualifications required and benefits available.

l. Engineers shall not enter competitions for designs for the purpose of obtaining commissions for specific projects, unless provision is made for reasonable compensation for all designs submitted.

m. Engineers shall not maliciously or falsely, directly or indirectly, injure the professional reputation, prospects, practice or employment of another engineer, nor shall they indiscriminately criticize another's work.

n. Engineers shall not undertake nor agree to perform any engineering service on a free basis, except professional services which are advisory in nature for civic, charitable, religious or non-profit organizations. When serving as members of such organizations, engineers are entitled to utilize their personal engineering knowledge in the service of these organizations.

o. Engineers shall not use equipment, supplies, laboratory nor office facilities of their employers to carry on outside private practice without consent.

p. In case of tax-free or tax-aided facilities, engineers should not use student services at less than rates of other employees of comparable competence, including fringe benefits.

FIGURE 12.3 (continued). Code of ethics of engineers (Page 4 of 5).

6. Engineers shall act in such a manner as to uphold and enhance the honor, integrity and dignity of the profession.

   a. Engineers shall not knowingly associate with nor permit the use of their names nor firm names in business ventures by any person or firm which they know, or have reason to believe, are engaging in business or professional practices of a fraudulent or dishonest nature.

   b. Engineers shall not use association with non-engineers, corporations, nor partnerships as 'cloaks' for unethical acts.

7. Engineers shall continue their professional development throughout their careers, and shall provide opportunities for the professional development of those engineers under their supervision.

   a. Engineers shall encourage their engineering employees to further their education.

   b. Engineers should encourage their engineering employees to become registered at the earliest possible date.

   c. Engineers should encourage engineering employees to attend and present papers at professional and technical society meetings.

d. Engineers should support the professional and technical societies of their disciplines.

e. Engineers shall give proper credit for engineering work to those to whom credit is due, and recognize the proprietary interests of others. Whenever possible, they shall name the person or persons who may be responsible for designs, inventions, writings or other accomplishments.

f. Engineers shall endeavor to extend the public knowledge of engineering, and shall not participate in the dissemination of untrue, unfair or exaggerated statements regarding engineering.

g. Engineers shall uphold the principle of appropriate and adequate compensation for those engaged in engineering work.

h. Engineers should assign professional engineers duties of a nature which will utilize their full training and experience insofar as possible, and delegate lesser functions to subprofessionals or to technicians.

i. Engineers shall provide prospective engineering employees with complete information on working conditions and their proposed status of employment, and after employment shall keep them informed of any changes.

Accreditation Board for Engineering and Technology
345 East 47th Street
New York, NY 10017

AB-59 2/85

FIGURE 12.3 (continued). Code of ethics of engineers (Page 5 of 5).

who are totally autocratic and require that all matters be referred to them and those who have no real interest in the operational side of the work and abdicate their responsibility. The proper balance is achieved between these extremes based on the ability and willingness of the supervisor and employee to share and accept authority. Depending on their respective levels of comfort (or security), they will achieve an operational balance. The balance will not be static but will shift toward the employee as more experience is gained and confidence developed. As a result the supervisor will find that less formal supervision is necessary, and the working relationship will evolve into that between peers. If we remember that another prime responsibility of a supervisor is to **develop personnel,** this evolution serves that purpose as well. On balance then, rather than weakening the supervisory position, it simplifies the job of the supervisor and provides for growth of the employee. A further benefit arises in that, as employee satisfaction is increased, work quality improves and fewer discipline problems tend to arise.

Rewards for outstanding work range from an informal compliment to formal systems, even cash awards. Whatever system is used, a word of appreciation from the immediate supervisor is an essential first step and is often more important to the employee than any other recognition that may occur. In most cases more elaborate systems are not justified, and the reward often loses its significance when handled in a routine manner. In particular, systems that seemingly reward people for position rather than contribution are seen through by the staff and rapidly lose their value. An elaborate system of awards may not be necessary, and much of the cost and effort to administer such a program may not be well spent. Timeliness is a critical element in effective recognition and appreciation.

One system widely used is to give the employee a cash award for a night on the town, a weekend away, or a similar activity with family or spouse. Less popular are a luncheon with the employee's supervisor and his or her supervisor as well. Normally there is in addition the presentation of a momento, followed by a brief write-up naming the employee, the work product, the ideas, and so on, in a company newsletter.

While not a pleasant subject, it is nevertheless essential that a system of **discipline** be defined in the manual on personnel policies. The system should be based on an initial orientation that sets forth standards of acceptable performance and conduct. From this baseline discipline is imposed depending on the severity of the deviation. Informal approaches such as advice and consuling should always be used first, with additional actions such as verbal and written warnings if improvement does not result. For severe violations such as fighting, drunkenness, or possession of weapons, suspension or termination may be immediately imposed. The system should include the complaint procedure described earlier as well as various levels of disciplinary action that can be taken for offenses of varying severity. It is wise to establish the levels of discipline in advance of the offense to avoid an improvised disciplinary action. Although disciplinary levels and consequences may not need to be advertised to the work force, they should not be hidden from them either and should be made available upon request. A typical system of discipline might have the levels of classification as shown below:

1. Verbal warning.
2. Written warning.
3. Final warning.
4. Suspension with pay.
5. Suspension without pay.
6. Dismissal.

All disciplinary actions should be **documented** to avoid misunderstandings and reduce exposure to litigation. Normally a handwritten note to the personnel file of the individual is sufficient, with a more formal written memo to the employee for written warnings, suspension, and dismissal.

Promotion, while a welcome event to the employee, is also a cause for some concern to the management. The unspoken question is whether the employee will perform well in the new higher position. In general, the answer is yes. However, in a sensitive position or where the demands of the job are unusual, it is not always possible to predict future performance. Where possible, some firms try out the employee at the higher level on work of a less-sensitive nature until the employee demonstrates ability to handle the higher position. Where this can be done without being an obvious trial assignment, it is a useful way to limit exposure. Still it does lengthen the time required to demonstrate employee capability, so some firms prefer to promote employees and let them sink or swim. The advantage to the employee of not being in a sheltered position is that he or she must perform from day one at the new level. It also utilizes in a positive way the high motivation that occurs immediately after a promotion and encourages the employee to fully apply him or herself to the new position.

In general, it is a poor idea to try out an employee at a new level and promote the employee later if performance is satisfactory. Such a tentative arrangement is damaging to the morale of the individual, and subordinates may not give their full support to the individual whom they feel may not be permanent in the position. It is far better to carefully evaluate and promote the employee. It is a positive sign of the confidence of the management in the individual and provides assurance at the very time needed, when the person is new on the job and needs maximum support.

Time management is becoming ever more important in the work of the individual. With the ever-increasing flow of data it is vital that the trivial be separated from the important and that time be conserved so that it can be expended on priority items. One widely used technique is to establish, at the start of the day, a "to do" list. This list sets forth what is to be accomplished that day. The basic idea is that the most important or unpleasant tasks should be listed and dealt with first.

Initially only one or two items on a list of perhaps 10 or 12 items will be achieved. But after a few days it will become apparent that more care must be exercised in the way one's time is structured. As a result a larger number of items will be completed on the following days. Since many items that require action depend on others and are beyond the control of the individual, it is necessary to establish some flexibility in the listed items. If this approach is followed, items that are carried over to later days will not be critical ones.

The use of a calendar to **schedule** the work day can be helpful, but it must use realistic time estimates. One approach is to allow one-third more time than expected when scheduling work activities and, then based on the results, to use those values in the future. Before scheduling an activity or meeting, the need for it should be clearly established. Unless this sort of discipline is exerted, the scheduling will be ineffective, and one will constantly wonder where the time went and what was accomplished. Another principle widely followed is to try to handle a piece of paper only once. Once picked up, an action is taken and the task is disposed of. The exception to this are items that are not timely or require action by others. For these a **follow-up file** can be established. In its manual form the follow-up file uses a 30-day file in which items for future action are kept until the day for their action arises. Each item for follow-up is assigned a date as it is placed in the file. Each day the file, or pocket, for that day is emptied, and whatever is in it dealt with. There are many more aspects to effective time management. Short courses or seminars can assist in developing techniques if time still seems to be a problem.

The widespread use of **electronic** or **e-mail** is taking the place of much paperwork. While extremely convenient, e-mail has the disadvantage of facilitating wide distribution of information that often is of no particular interest to the addressees. Because of the large amount of extraneous matter, some discipline must be exercised by system users. Fortunately it is relatively easy to scroll through a large number of items and only read those of interest, discarding the rest. To assist in managing the messages, it is useful for the senior personnel who normally receive the most e-mail to impress on their subordinates the importance of limiting their messages to matters of real interest and not to routinely provide copies of their messages.

Meetings, though often useful and essential, can just as often not be necessary and create an enormous waste of time, not only for the individual but for everyone in attendance. Meetings must have a purpose—an agenda—a specific time for adjournment, and as few attendees as possible. It is not necessary and highly undesirable to have extra people at the meeting. The cost to the firm is high, and often very little gets decided that could not have been done by circulating a note to the individuals soliciting their views. Very often meetings become a way to share the responsibility for a decision, rather than discharging the responsibility of the supervisor for obtaining the information and making the decision.

The first decision to make is, "Is the meeting really necessary?" Unless the answer is a strong yes, the meeting should not take place. It may be possible to make the decision without the input of the others if the risk is low, or perhaps the attendees can be contacted by telephone. In any event all reasonable steps, and perhaps a few extreme ones, should be taken to avoid any but absolutely essential meetings.

The use of **newsletters** as a form of internal communication is extremely valuable and should be pursued in all firms larger than 30 employees, they are even useful in smaller firms as well. In general, the newsletter should contain information on work outlook for the company, personnel promotions, awards for performance, new products or developments, background information on the industry or its technology, developments or changes in personnel policies, and employee-related human interest items. The newsletter should be published regularly and on a firm schedule. Where possible comments or questions from the readers should be solicited, with questions and answers published in subsequent issues. The wide availability of desktop publishing capability permits the work to be done in-house at low cost. But they are always somewhat dated because of the time it takes to write the material, publish it, and distribute it to the employees. Many companies in addition use one-on-one meetings or staff meetings as a way to disseminate information on a faster basis. This permits the supervisor to meet more frequently with his or her personnel and takes place in smaller groups which promotes more communication. One practice that seems to have caught on is to hold an informal meeting every Friday afternoon to bring everyone up to date on current matters. These meetings can be attended by a senior officer and are structured so that questions are welcomed. They are usually limited in size to maintain informality and openness.

Company-sponsored **picnics** and **open houses** provide a way to involve and support the families of employees, although these practices have been diminishing in popularity. In some companies bowling leagues and other sport or recreational activities provide this type linkage. With the increasing concern for physical fitness and an interest in providing shower-and-change facilities for the employees who wish to exercise during the work day, more forward-looking companies are installing exercise facilities or providing membership assistance to local facilities, such as the Y, where the employees can join fitness programs. A corollary to the physical fitness question is the overall question of executive health. All personnel should be encouraged to take annual vacations of at least a week at a time to provide for their mental health. Similarly routine physical examinations oriented to maintenance of health should be encouraged. Some firms will pay for a part of the cost of the physical examinations.

From the company's point of view participation in professional **technical society** activities is useful, not only as a form of advertising but for personnel recruitment as well. It also provides a way that the firm can support the technical society. Sponsorship of technical meetings, field trips, providing speakers and technical materials, allowing use of company space for both committee and technical meetings, advertising in technical organs, and support of society activities by employees are all ways in which the company can further the goals of the society. To best carry this out, it is useful to establish an overall coordinator for technical activities. This individual establishes a budget for the activities and allocates funds to the technical societies involved based on importance and usefulness. Afterward, the various participants are assigned a budget against which their activities, including travel and seminar tuition, are charged. Periodically the budget is reviewed, and adjustments are made depending on its status and expenditures.

A similar situation exists where a firm sponsors industry-type **seminars** for its clients, potential clients or suppliers. The budget is handled somewhat like that for a technical society, with a coordinator taking overall charge of the program in all its aspects. Ideas on attendance for such meetings should be obtained from the sales and operations departments who usually have definite information on the clients, firms, levels of personnel, and specific individuals to be invited. These industry-type meetings are generally widely attended. They are almost invariably well worth the funds spent on them but attendees must be well prepared and smoothly present their material, with handouts for others to study at the conclusion of the meetings. Recreation is often provided at these meetings including popular sports such as golf and tennis. But the most important factor is that there be specific content to the meeting. It should not be merely an excuse for recreation. In effect elaborate meetings are less successful and tend to project the wrong image as compared to more austere ones. Where the choice exists, it is better to be a bit understated than to be too lavish.

Normally from one-third to over half of the staff are graduates of local educational institutions. Thus relations with **educational institutions** are always an important concern of the senior management and the human resources staff. It is wise to cultivate good relations not only with the senior college or university deans and administrators but with the heads of placement and with professors in related technical disciplines. While financial aid such as grants, endowments, matching of employee gifts, and such are desired, much can be done without any significant financial impact. The furnishing of guest lecturers, arranging of field trips to both the firm's offices and nearby projects, are useful ways to reach the student population and to obtain the goodwill and favorable recommendations of the professors. One large company sponsors an occasional conference at its offices for the professors of the local university and presents data on its recent major projects, special studies, and economic outlook. The meeting is always well received by the faculty and generates much goodwill. Participation on advisory panels is helpful and provides the university with the point of view of industry, which they always find useful.

12.5 TRAINING AND DEVELOPMENT

Informal **training** and **development** will continue throughout one's professional career. This "on-the-job" training often is hit or miss, depending on the timing and type of assignments that arise. A professional career is too important to leave professional development to chance, and some form of plan for self-development is essential. This need not be too formalized, but a clear recognition of career goals, both near and long term, is an essential first step. After identifying the goals, it is useful to consider the body of knowledge required at each level and to outline a course of study—reading, classes, seminars, professional society involvement, and similar activities—that will meet those requirements. It is important to note that included in these requirements should be a **network** of friends and acquaintances. The self-development plan should not limit itself to merely technical areas but should include other areas and activities as well. The important principle to bear in mind is that these improvement programs are long-term ongoing aspects of the profession and that the learning process will continue throughout one's professional career.

Two areas of self-development that should be pursued by all persons, particularly early in their careers, are **time management** and **memory improvement.** There are numerous courses available for each of these areas. Because of the broad application of such skills, they can over one's career be more beneficial than any specific technical material. In particular, the ability to organize and effectively utilize time, as previously mentioned, is an important management skill that can improve one's effectiveness. The ability to remember names and other data can be improved through simple readily learned techniques. An improved memory will materially benefit and simplify one's work as well as help to assist the development of both business and personal relationships.

Some firms have established training programs for new hires and junior personnel, and these programs should be utilized where possible. Firms will often participate in a **tuition refund** plan where all, but more usually part, of the tuition paid for courses satisfactorily completed is refunded to the employee. The courses must have some fairly direct relationship to the work of the employee, and approval prior to registration is usually a requirement of the reimbursement program. Some firms will sponsor an employee in a program for an advanced degree, usually in a management field, to compliment their technical skills and prepare them for more significant future assignments. These courses tend to be intensive, with the degree often earned in a year or fifteen months.

Rotation to assignments in other parts of the company is often included in development programs. These assignments can provide the opportunity for an employee to broaden practical knowledge and obtain more on the job experience. Some care must be taken, however, to ensure that the assignments are truly rotations and not merely a way to move someone out of a department. The best way to do this is to obtain some understanding that sponsorship of the individual and his or her career development will remain with the parent organization.

Registration of professional engineers should be an integral part of any plan for personnel development. While multiple registrations are helpful, after the second registration they are less important than a broad range of skills and experience. Registrations in multiple states should also be considered for those personnel who prefer to stay in the technical side of the operation and wish to become engineering specialists.

CHAPTER 13

BUSINESS OPERATIONS

Business operations involve the entire spectrum of activities including organizational factors, financial management, advertising and promotion, administration, and planning for the long-term operations of the firm. Inherent in these are the responsibilities and involvements of the staff of the firm.

13.1 ORGANIZATION

The most usual form of organization for medium- and large-size firms is the **corporation.** The corporation is an artificial fictional entity that is distinct from its owners, who are known as shareholders (stockholders). It is organized to permit the entity to operate without shareholders assuming individual liability and to permit the operations to continue after the death of one or more of the shareholders. The liability of the corporation is limited to its own net worth plus any insurance it may carry. As a result an individual investor can only lose his or her own investment in the corporation, since it is part of the net worth of the corporation.

The corporation is incorporated and **chartered** in and by a state. The charter defines what type of business the corporation may engage in. Some states, notably Delaware, have encouraged incorporation in their state and have established statutes that are advantageous to corporations and simplify their operations. Even though they may actually transact no business within Delaware itself, many U.S. corporations are chartered in that state.

The corporation can own property and can sue or be sued as would a real person. It is a legal entity with an **indefinite** life and is not affected by the death, illness, or bankruptcy of its shareholder(s). In general terms a corporation is considered the same as a person except for the specific legal status that it enjoys.

The corporation activities are normally directed by a **board of directors** who meet several times a year and are elected by the shareholders. The board in turn elects or appoints an executive to perform the full-time activities connected with managing the operations of the corporation. The current trend is to weaken the liability protection of directors of the corporation, however, and to expose them to litigation. Much of this is civil interest litigation, where directors have allowed activities of the corporation that appear to be violations against the public interest, such as illegal dumping. This does not affect the liability of the shareholders who are not directors, although litigation may affect the assets of the corporation and thus indirectly the shareholders.

Ownership of the corporation is vested in the **shareholders** who through their stock purchases provide the initial funds (capitalization) for the activities of the corporation. Additional issues of stock may be made when the shareholders authorize them. The **stock** types commonly utilized are either **common** or **preferred,** although there are a large number of more limited types available in certain circumstances. Preferred stock has the first call upon the earnings of the corporation for dividends and is normally ensured a specific dividend that may vary for different issues (series). Common stock is paid dividends depending on the earnings of the corporation available after payment of the preferred dividends. Earnings of the corporation are taxed as corporate profits prior to distribution to the shareholders and are later taxed again as income when received by the shareholders. This double taxation of earnings is felt by many to be unfair, and frequent efforts to revise this provision of the internal revenue code have been made. In some tax years dividends have been exempt from taxation either by virtue of the type of investment or by virtue of the dollar amount received.

Some corporations whose articles of incorporation include a special provision limiting the number of shareholders are classified as **close** (closely held) corporations. The limitation on the number of shareholders will vary from state to state and from corporation to corporation, thirty-five being a typical value. This arrangement permits the corporation to operate in a simplified manner, not only structurally but managerially as well. In particular, decision making is more convenient with fewer, more involved shareholders. This corporate form requires written shareholder agreements among all the shareholders, allocating authority to control corporate activities. By this device the corporation, while retaining liability protection for the shareholders, can structure its organization and operations as if it were a partnership. This provides great flexibility in its operations and significantly simplifies decision making and the conduct of day-by-day operations. An additional advantage is that there can be less pressure on the management to show short-range benefits—for example, earnings' improvements from quarter to quarter—because long-range considerations are deemed more important. Usually the close corporation operates more informally and includes most or all shareholders in daily decisions. In addition to the legal limits, close corporations will often establish their own limitations on the transfer or sale of the shareholders' interest to avoid increasing the number of shareholders and thus dilution of control.

A special form of corporation, called an **S corporation,** is a regular corporation that has elected to be taxed as if it were a partnership. Current law limits the number of shareholders to thirty-five. There are in addition many specific provisions that define and limit these corporations, but for many small- to moderate-size enterprises this form or organization may be the most advantageous. The advantage is that both the profits and losses of the corporation are passed directly to the shareholders, on a proportional basis, and are reflected directly as income or losses on their personal income tax returns. Where income is retained in the business and not distributed to the shareholders, it is taxed at the S corporation level, and when later distributed to the shareholders, is not further taxable.

Employee Stock Ownership Plans, (ESOPs) are being used in increasing numbers to provide a way for employees to become more highly motivated by participating in the ownership of the firm. Under this arrangement the employees actually own all or a very large share of the company through ownership of the stock of the firm. While not a cure all for the problems of a company, experience has demonstrated that the productivity of the employees rises, absenteeism falls, and on an overall basis the quality of the work product improves. An increase in the price of the stock of the firm raises the value of the employees' holdings, which provides incentive for further improvements. The results tend to build upon themselves and usually lead to a substantial improvement in the performance and value of the firm.

ESOPs can be valuable tools for corporate **financing,** and the cash generated from them used to provide working capital, retire outstanding debt, buy out an existing owner, and so forth. Because of the vested interest this type ownership creates, the plans provide protection against hostile takeovers. In addition there are significant tax advantages to both the company and the employee owners. ESOPs are applicable to any type of firm that uses a stock form of capitalization, although certain tax regulations affect the details and specificity of the way in which they can be arranged.

Many sizable firms, and in particular those dealing in services, are organized as **partnerships.** The essential elements in the partnership are that the parties will place their efforts, capital, labor, and skill at risk in performance of the business and that they will **share** in the profits and losses of the business operations. While many of these groupings are informal, partnership agreements should be carefully prepared and reduced to writing. It is important to ensure that the degree of participation of the individuals is defined and to establish the basis for both profits from and liability for the operations of the partnership. In the absence of a specific statement of interest in the partnership agreement, each partner shares equally in the profits and losses of the enterprise regardless of that partner's contributions financially or otherwise. Many partners do not make any direct financial investment in the partnership but may provide the use of their name, proprietary ownership of a process or patent, their services, and so forth. Upon dissolution of the partnership, however, such partners would not be entitled to a return of capital unless agreed to by all the partners or stated in the agreement.

The partnership is a legal entity and can own, buy, sell, transfer, exchange, and handle property as if it were an individual. Unlike the corporation, however, the partnership **does not** shield the individual members from liability for the activities or debts of the partnership. In the event of dissolution, the partners may be personally liable, and if they become insolvent, the balance of the debts may become the responsibility of the remaining partners. A corporation may be a member of a partnership, and indeed such happens with joint ventures when the members are corporations.

Since the partnership does not have the indefinite life of a corporation, the death or withdrawal of a partner is a common problem. Most partnership agreements have provisions covering this possibility that enable the remaining partner(s) to purchase the interest of the deceased or withdrawn partner. To facilitate this, it is important that the partnership agreement cover this possibility and that terms of and methods for the valuation for sale or transfer be defined. It is also useful to forestall problems with the decedent's heirs, to establish the timing and manner of payment. In partnerships with only a few members, the partners will often hold life insurance on each other so that there will be sufficient liquidity to pay off the interest of the decedent, and normally the partnership itself will pay the premiums on such insurance, the premiums being treated as a nontax deductible business expense by the partnership.

The Uniform Partnership Act is widely recognized and provides guidance as to the recommended provisions of the agreement. As a minimum it provides a useful checklist of the normal provisions found in an agreement.

Sole proprietorships have few of the complications listed above, although the necessary business licenses must be obtained as for the other forms of organization. As in the case of a partnership there is no limitation on the liability of the individual, and all the assets of the individual may be used to satisfy debts of the business. The principal operating advantage is the ability to reach rapid decisions without requiring the agreement of other parties. The typical disadvantage is the difficulty occurring with the growth of the firm, in that after it reaches a modest size, the type and volume of decisions to be made require more complex internal organization and delegation of authority. Many sole proprietors are unable to change their management style to accommodate this, and this often represents a growth limit. The hiring of a professional manager(s) to control specific areas of the day-by-day operations normally solves this problem.

Joint ventures are becoming more widely used as firms seek to take on work that requires a wider range of experience or resources than they can provide alone. Joint ventures are really a form of partnership, with a limited life and the general form of organization that follows accordingly. Of particular importance is a clear statement of the responsibility of each of the venturers for furnishing funds, personnel, materials, equipment, proprietary information, and so on, together with definition of the interest each has in the profits or losses of the operations. Responsibility for decisions in specific areas and direction of the work must also be defined, along with responsibility for the furnishing of personnel for specific assignments within the joint venture. Often special joint boards of control will be established to manage the work, and frequently one of the firms will be identified as the lead joint venturer having overall management responsibility for the work.

The life of the agreement must be carefully defined with provision for protection against delayed litigation, for often litigation does not arise until years after completion of the work and dissolution of the venture. One way to reduce this exposure is to obtain from the client a suitable statement of completion, including a release from future liability.

Where any of the above forms of organization are considered, competent legal counsel should be used to draw up the necessary documents and provide advice regarding the applicability of the provisions and special requirements of the jurisdiction where the enterprise will operate.

As an alternative to joint ventures, it is often possible to **subcontract** work to another firm. This is particularly the case with professional services, such as engineering design. Some care should be taken to ensure that the subcontract agreement holds the subcontractor fully and solely responsible for its own work and that the purchaser of the services is fully indemnified against errors or omissions. Again legal counsel should be used to draw the agreement to ensure proper protection.

13.2 FINANCIAL MANAGEMENT

The heart of any enterprise is **financial management.** While a primary consideration in the private sector, where profit is a fundamental goal of the enterprise, in the public sector it often assumes an even larger role. Part of this reason is the often absolute limitation on funds and the purpose for which they can be utilized. Frequently there are strict prohibitions on transfers of funds, and they must be used only for specific purposes. Every business must establish suitable banking and financing arrangements to support its operations. These include not only the normal arrangements of checking accounts, credit lines, and so forth, but also provisions for initial capitalization and for increases in capitalization as circumstances may later require. Of all causes of business failure, **undercapitalization** and the inability to raise capital lead the list by far. For established firms, sale of additional stock provides a convenient way to increase capitalization, although some care needs to be taken as this can often dilute the value of the existing stock. In some cases capitalization can be increased by paying dividends in stock so that in effect the capital is increased by the dividends retained in the business.

For newly organized firms **venture capital** can be obtained from financial companies specializing in this, as well as through the venture capital departments of large brokerage houses. As with all other financial items, rates and services offered are competitive, and by shopping around, it is possible to obtain rates that may be more favorable. The Small Business Administration of the U.S. government provides assistance to small businesses and can assist in obtaining financing under certain conditions.

Measuring the financial health of a company on an ongoing basis is essential to its management. A series of **ratios** are widely used to provide knowledge not only of the financial condition of the company but also of its operations. For many projects these same ratios, with only slight modification, can also be used. These ratios fall into three general categories: measures of **solvency, liquidity,** and **profitability.** For any particular analysis some of these are of more interest than others, but all provide information from which an analysis can be made. It is useful to note also that special circumstances will cause some of the ratios to change widely, and they must be interpreted with a knowledge of the normal ratios for the industry as well as the recent operations of the particular business.

While each business will use different measures to determine its own performance, the most widely used ratios are those shown in Table 13.1.

An effective way to manage a large, complex enterprise is to periodically, often monthly, review the ratios for an overview of the operations. Setting goals for each of them, together with the reviews mentioned above, will add precision and provide measurable goals to the overall management of the company. Ratios that measure the activity and health of a project can be developed and utilized in the same manner. While each project has different criteria, it is possible to use specific operational measures to an advantage, for example, accounts-receivable turnover, or average days to collect receivables (billings), inventory turnover, or average number of times the inventory is turned in a year.

Astute financial management can yield significant dividends to the firm and provide a source of funds in and of itself apart from the actual operations of the firm. When interest rates a few years ago were in double digits, a number of firms with large cash reserves were earning as much from the conservative investment of these funds as from the basic operations of the companies themselves. While this was somewhat unique, accurate billings, careful handling of funds, speeding up of cash flows, reducing billing time, rapid resolution of disputed amounts of billings, and similar measures, all go a long way to improving the financial performance of a project.

Of all the areas of financial management, perhaps the one most poorly performed is the area of **cash flow.** Some smaller firms seem to be unable to ever fully control cash flow and are continually forced to delay payments to

TABLE 13.1 Financial Ratios	
Ratio	Normal Range
Solvency	
Debt to equity	Varies with Industry[a]
Long-term debt to fixed assets	Varies with Industry[a]
Times interest earned per year	Varies with Industry[a]
Operating income as % of total assets	Varies with Industry[a]
Liquidity	
Current ratio	> 1.50 : 1
Quick assets/current liabilities (acid test ratio)	> 1 : 1
Inventory turns per year	3 to 5
Receivables average time	< 35 days
Profitability	
Gross profit as % of sales	> 20%
Operating profit as % of sales	> 4 to 15%
Net income as % of sales	> 2 to 8%
Return on assets used	> Prime plus 3 to 6%
Return on stockholders equity	> Prime plus 1 to 3%

[a] These ratios vary widely with the type of industry—whether capital intensive, high technology, or long-established technology fields. For guidance, typical values by industry, updated annually, can be found in most major libraries in several publications such as *Industry Norms and Key Business Ratios* published by Dun & Bradstreet Information Services, Murray Hill, NJ, and *RMA Annual Statement Studies* published by Robert Morris Associates, Philadelphia, PA.

suppliers, borrow to meet payroll and governmental tax payments, and in some cases delay payroll payments to employees. Any of these shortcomings are in themselves a problem, but the inability to pay suppliers and to lose the advantage of trade discounts is tantamount to throwing money away. There are available today numerous accounting programs suitable for use on personal computers that can assist in cash flow management and permit accurate day-by-day control of **minimum** or **zero-balance** accounts together with all manner of analysis and prediction of cash requirements. Often this service is available through a local bank which relieves the company of the work.

Some partial answers to the problem include the use of **advance payments** (normally requiring a contract provision), **prompt billing** to clients with appropriate follow-up to expedite payment, the use of **discounts** to encourage prompt payment, and the prompt paying of billings where discounts can be taken. Use of a bank line of credit also provides a way to obtain cash and can be used in conjunction with a zero-balance account to minimize banking balances not earning interest. In extreme cases the **factoring** of receivables can provide for immediate cash, although many consider this an extreme step to be taken only in highly unusual circumstances.

The thrust of the financial management program for a project is to ensure that billings for work completed are rendered accurately and promptly. It is important that the contract include provisions for **interest** to accrue on unpaid invoices and that the rate of interest be specified (12% annual interest compounded monthly, an annual interest rate of 1% over prime compounded weekly, etc.). Typical provisions state that interest will start to accrue either 10 or 20 days after receipt by the client.

Wherever possible the contract should also provide that advanced billings be permitted, for this represents a significant savings in the cost of funds committed to the work. Usually this provides for billing at midmonth for work contemplated through the entire month. With the billing presented about four weeks early, cash flow is speeded up by about a month as well.

Accuracy is absolutely essential for invoices to be accepted without being returned for correction. If consistently well performed, attention to accuracy will build confidence that the invoices are correct. Over the long run this will tend to speed up their payment, since it will reduce the scrutiny they receive and facilitate their acceptance without long explanations and in some cases negotiations. Where invoices include **contested** items, and this occurs primarily in reimbursable work, the client should be encouraged to pay the uncontested portion and to temporarily set the other aside in a suspense account for payment at the earliest possible date, when the problem is resolved. If the contract is properly written, these delayed amounts would bear interest if found to be correct, thus discouraging the client from contesting items merely for the purpose of delaying their payment.

All **billings** should be forwarded to the client via the Project Manager, who will make the actual presentation of them to the client. Since the Project Manager is familiar with the provisions of the contract, this permits a review, independent of the accounting department, to ensure that there are no major conflicts between the billing and the contract. It also provides a buffer between the accounting personnel and the client and permits them to be brought in to assist the Project Manager in the case of disputed items of an accounting nature. In addition, since the Project Manager will follow up on the billings to ensure prompt payment, this places him or her directly in contact with the client to record and monitor both presentation and payment of the invoice(s). It also provides an incentive to the client to pay invoices promptly, since the Project Manager is usually in continual touch with the client, indeed often sharing offices with the client's counterpart personnel. As a result the PM is constantly at hand to remind the client of the status of invoices rendered for payment but for which payment has not been received.

It is essential to establish suitable **banking** arrangements that can provide several types of services. Generally these include separating the payroll accounts from other accounts and operating all the non-interest-bearing accounts on a zero-balance basis. Often an interest-bearing treasury account is established, which is the basic account in which the project funds are held and from which transfers to the payroll and other accounts are made as needed. The banking arrangements should also provide a line of credit to cover periods when the several operating accounts may be temporarily exhausted. This is of particular importance when working overseas. Overseas it is normally undesirable to convert any more dollars to local currency than absolutely necessary, and the line of credit can often be used to bridge the gap between a local payroll and receipt of a pending payment in local currency.

In addition, deposits to the treasury account can serve as a **compensating balance** when it is necessary to establish these with the bank utilized. While banks will normally try to establish a minimum compensating balance, the amount of the balance, the interest charged on the line of credit, the amount of the line of credit, and the monthly cost to maintain other commercial accounts, such as a payroll account, are all negotiable items. With some modest shopping it is possible to reduce banking costs markedly below those initially offered. Where banks ask for guarantees, it is wise to immediately obtain experienced advice, since the wrong guarantees, such as turning over receivables or pledging all inventory, are traps that can have severely negative effects on the future profitability and often the survivability of the company.

While it is true that **taxes** are, like death, unavoidable, there are numerous strategies to defer and reduce them. Through careful review of quarterly earnings, for example, it is possible to only pay estimated taxes for a low-earnings period rather than one-quarter of the annual taxes due and conserve funds. Because of the complexity of the tax laws, their constant changing nature, and the severe penalties for noncompliance, competent tax planning advice should be obtained and overall strategies developed to minimize taxes and delay their payment until absolutely required.

An **independent** overview of the operations of the firm is vital to ensure that the content of financial statements is accurate and that the operations conform to company policies and procedures. These normally take the form of **audits**, both **financial** and **operational.** Financial audits are normally performed by outside firms who specialize in audits and who are highly experienced in their conduct. The usual concern of such audits is to certify that the financial statements are accurate and that the results of operations are fairly stated. They may also determine whether suitable internal controls are in place and being followed. In many cases the audit firm is also the outside accounting firm who routinely works with the company, since it's generally familiar with the company's operations including special financial or operational arrangements.

Some large firms have internal audit departments who perform many of the same functions noted above but who have the advantage of being members of the firm, and thus being totally familiar with all the operational details of the departments being audited. This is an economic way to operate in some circumstances and provides an added training opportunity to senior personnel who make up the internal audit teams by exposing them to other projects and operating departments.

Operational audits are of two types, both normally being performed by personnel in-house but from departments different than those being audited. The normal operational audit is a nontechnical one and reviews conformance to policy and procedures. Examples of these are audits for Equal Employment Opportunity Commission (EEOC) compliance, conformance to design control procedures, and records storage. While these are valuable, they merely confirm conformance and that the proper procedural steps have been taken. They do not in themselves deal with the technical content of the work.

The question of whether to employ only outside **legal counsel** or to establish an in-house legal department arises when the amount of annual fees paid to outside firms for legal support to ongoing operations reaches the equivalent of three persons, one of whom would be paid at the level of a senior manager. At about this point it seems financially attractive to perform this work with one's own employees. There are, however, some pitfalls to be considered. First among these is the requirement that there will be enough continuing legal work to justify the associated costs. When the outside firm is used, despite what may seem high hourly billing rates, their work is or can be limited, for when business conditions do not require it, they perform no work. The mere existence of an in-house legal group will attract some inquiries that would not otherwise have taken place, and there may be a tendency to overkill minor problems. In addition, and perhaps more important, a small corporate legal department will not have the breadth of expertise of most outside law firms and may really only be useful for conventional run-of-the-mill problems. Because of the high billing rates and the rapid rate at which costs for outside counsel can mount up, in any circumstance it is wise to establish one person as the contact with the outside law firm to authorize any work. All requests for assistance should flow through this person, and no other requests should be honored unless unusual circumstances dictate. It is useful for that same person to receive and review all the invoices from outside counsel to ensure that they are in response to valid requests and in conformance with the agreed billing rates.

Insurance, while a necessary adjunct to business, is often neither obtained nor administered in a businesslike way. It is useful to remember that the insurance industry is a competitive business and that rates are negotiable. As a result it is common for the same coverage to be quoted premiums that vary widely depending on the carrier, its knowledge of your business, and, most important, your detailed method of operation and control. To the extent possible all insurance should be handled by a single agent or firm who will come to understand in detail the operations of the company. This will tend to avoid duplicate coverage, which can occur when different agents each provide a portion of the insurance coverage. The agent can, as a result, represent the company to the insurance carrier(s) in the most effective way to yield the greatest benefit for the lowest premium.

Most insurance carriers classify companies into **risk categories** depending on their industry and often look no further into the performance of an individual company. As a result the premiums quoted are for average or normal risks, even though they could be lower for the same deductibles and coverage, where a firm's operations are more conservative and careful and thus safer for the carrier. If premiums are higher than desired, look into reducing or modifying the deductibles offered to reduce the exposure of the carrier. This is really a form of **self-insurance,** which in certain instances may be completely appropriate to your operations. Small companies doing business with very large ones do not always carry insurance for routine, low-risk operations but rather assume the risk for themselves. As an extreme case there are several small firms with a low net worth doing environmental hazard remedial design work that carry no insurance but carefully pick their clients. They only work for large clients which they feel will not, because of their low net worth, sue them in the event of a problem. While an unusual approach, with the high cost of insurance for certain types of operations, more firms are looking into this arrangement.

Apart from the more usual types of insurance, including those that may be mandated by the state, such as Workman's Compensation, it is important to protect the firm against **liabilities** from the activities of the employees. Product liability coverage is also necessary for firms in the manufacturing sector, as is some form of business interruption coverage. The liability coverage can take several forms depending on the firm, its size, and experience. For medium to large size firms the deductibles may be fairly large, but often the insurance carrier provides reimbursement for the cost-of-litigation defense.

With the increasing number of stockholder suits or personal liability actions brought against corporate officers, officer's and director's insurance is becoming necessary and should be considered by medium and large size firms. Likewise **error-and-omission** coverage for design or service-type firms is considered essential. Newly established firms will normally find the initial premiums high, but they tend to fall off rapidly after three years of coverage as a favorable claims record is built up. Premiums for this type of insurance can run from 0.5 to 1% of the value of the work performed. With good claims experience and negotiation with the carriers, these values can be significantly reduced.

Business interruption insurance is intended to provide coverage to permit the business to continue operating in the event of an unforeseen event such as a fire or flood. While important, if the business can readily and rapidly resume operations with only a minimum of interruption, the cost of the premiums should be carefully evaluated against the risks covered. In many cases higher deductibles can be utilized to provide catastrophy coverage, while maintaining the cost of the premiums at a low level. This implies of course that minor problems are borne by the firm, again as a form of self-insurance.

Employees who occasionally use their own vehicles for company business expose the company to liability. The company **automobile** or **liability** policies should be reviewed to ensure that there is coverage for this (nonowned automobile) or hired automobile coverage for vehicles rented by the company and used for business activities.

There are a myriad of other insurance coverages available that may be useful depending on the business, including key man, disability buyout, salary continuation, credit for receivables, leased property, and of course medical and life insurance on employees. All of these fall into the same categories as listed above and can be effectively negotiated with the carriers. The best way to evaluate this is to do a realistic cost analysis for each, considering the premiums required for various levels of deductibles versus the risk involved. Having done this it is then important to shop around, as the premiums will certainly vary widely from carrier to carrier. If however, all the insurance is given to one carrier, significant additional reductions, as described earlier, may be possible. It is necessary to continually compare the premiums paid by splitting the coverage among the carriers as against placing all the insurance with a single carrier.

Patent rights for new discoveries made in connection with work at the firm should be obtained from all employees. Company ownership of patents is normally covered in the employment agreement at the time of hire. The assignment of patent rights to the company for discoveries while working on company projects or during off hours, utilizing techniques or information developed at work are typical definitions. The company rather than the individual would in these circumstances pursue a patent application and obtain one if it is felt warranted. In many circumstances the company has no interest in the patentable item, and the employee may ask and should be granted a **release** to obtain a patent in his or her own name. Where this can be granted, it will be advantageous to employee morale. Upon separation from the company, many firms will require that they remain the assignee for patentable ideas originating for typically two years after separation, if the ideas apply to the technology used or learned while with the firm. Rather than have this sort of item negotiable at the time of separation, it should be a condition of the employment contract mentioned above.

In a similar vein is the obtaining of **copyrights** for writings prepared for the company. As with patents, copyright rests with the company, and permission to publish must be obtained where appropriate. It is usually not difficult to obtain permission, since companies normally are willing to permit publication of their employees' writings with suitable citations or copyright notices. This is usually done with no fee, or only a modest one, and is readily arranged with a simple release form. Normally, in contrast to patents, companies do not give up ownership of the copyright. In some cases sharing of publication royalties for employee-written work using copyrighted material has been arranged.

The extent to which a firm is willing to share its **profits** with its employees is often critical in attracting and retaining highly qualified personnel. Most firms today have some form of pension plan where deposits are made to an employee's account. Typically these can be as much as 5 to 7% of annual salary and are held in a tax-deferred fund, so pension earnings can grow rapidly. Often this amount is built into the payroll additive rate so that earning of this amount is automatic.

In addition some firms have a profit-sharing plan where a percentage of annual profit, often as high as 25%, is distributed among the employees as a bonus. The distribution is often pro-rated among the employees depending on their total annual earnings. The distribution can either be a current or a deferred one. Where a distribution is made, it is often performed before the end of the calendar year to avoid earnings' carryover.

13.3 ADVERTISING AND PROMOTION

Since every firm is in the process of working off backlog, it is essential to have some organized way in which new business is developed. This includes not only the promotion of new work from present and former clients or customers but also the development of new clients or customers. Generally the field in which the firm works defines the market that it seeks, and thus the population from which the work will emerge. From this definition of the market it serves, the firm should develop a plan with quantified values for its various components. A business development plan would carefully define the market, identify the firms or groups that are potential clients, establish a broad schedule of contacts to be made with each group, estimate the number and value of the work or contracts that would result, estimate the cost of the effort, and, finally, evaluate the revenue to the firm. When broken down on a quarterly or monthly basis, it becomes an overall plan for implementation by the sales group. When dealing with professional services, contacts within the industry and reputation are primary factors in obtaining new work. As a result the business development plan should include these elements as well.

Marketing of the firm and its products is normally handled directly by representatives who are employees of the firm, representatives who are independent business people and who are paid on a commission basis, or through dealerships or franchises. The use of representatives, whether direct or independent, seems to be the most effective for offering professional services. While marketing involves an extensive amount of personal contact and involvement with potential clients, direct mail advertising and inquiries are inexpensive and can often identify promising areas. As a result a balanced program of personal contact and suitable direct mailing is often used.

Many large firms and public entities have rather standardized methods of soliciting bids for work through their purchasing or contracts department. In addition, since the bids are usually evaluated for technical adequacy by the engineering or operations group, it is wise to develop contacts within these groups. Merely registering with these groups will not in itself generate any significant interest, and it is necessary to actively and periodically call on them to remind them of your continued interest and to expand upon the firm's capability as it grows.

For very large firms, formal programs with publications and the use of advertising agencies with more sophisticated approaches are common but are expensive and only justified for perhaps 1% of the firms in the field.

13.4 ADMINISTRATION

Administrative work cuts across the entire range of activities of the firm and can affect not only operational aspects of the work but the profitability of the company as well. Some of the more important ones are discussed in the material that follows.

To provide a framework for the operations of the firm and as a source of reference for employees working in unfamiliar areas, a **procedure** manual should be prepared. This manual does not need to be ponderous. It should state briefly the proper methods for the operations of the organization. Usually the operating departments will each prepare their own manuals, often in the form of the desk book described earlier. For work in large firms, or where more formal programs are required either by clients or regulatory bodies, the procedure manual is treated as a more formal document, with audits often run to ensure compliance. Where formal quality programs exist, the procedure manual is in itself a controlled document with revisions and changes subject to a rigorous system of review and approval.

Flexible **working hours** are of increasing importance to the work force. With the rising concerns over the congestion in cities where most offices seem to be located, the use of **flex time,** with a specific set of core hours, is becoming widespread. During the core hours, typically five or six hours of the day, all employees are present. This has the advantage of spreading out the commute rush and often avoiding traffic congestion, while permitting the employees to choose working hours more suited to their personal needs. It has the disadvantage of somewhat complicating any employee transportation pooling arrangements, although in practice these are readily resolved. The flex time approach usually adds a band of two hours at each end of the day, thus utilizing the offices four more hours a day. While some extension of security and janitorial services may be made, the effect is often minimal.

A variation on flex time that is becoming more widespread is the **flexible workweek.** In this arrangement the employees work a longer day, often an additional hour, and receive, for example, every other Friday off, creating a three-day weekend every other week. This arrangement normally requires that employees sign up for an A or B group so that half the work force is available on any particular Friday. The obvious disadvantage that the employee needed is off on a particular Friday has not proved to be a major problem, especially when the system is formally established within the firm and the work scheduled accordingly.

Telecommuting is becoming a more accepted practice as electronic transfer of data becomes easier. Many employees can perform the bulk of their work at a computer and do not need to be continually in the company offices. Indeed some companies are providing the computer to the employee, together with the supporting furniture, to permit this type of work at home. The difficulty that sometimes arises is the feeling of separation from the company offices and the need to occasionally visit the offices to meet with other employees. This can be readily achieved through scheduled visits to, or days of work, at the office, perhaps once a week. Another more subtle disadvantage is the lack of informal contact with other employees. The opportunity to have lunch together, meet accidently in the halls, have the seemingly unrelated question arise, and so forth, are not available to the telecommuter in the same way. This does not seem to be a major disadvantage, and on balance the advantages seem to outweigh the disadvantages.

An adjunct to telecommuting is the establishment of **satellite** offices. More and more operating units can function remotely from the main business location and can be housed in branch locations. This usually affords significant savings in the cost of office space and avoids the congestion and commuting difficulties with offices in urban locations. Wage levels in such locations tend to be lower, and a further savings is made not only in direct payroll costs but in associated payroll costs as well.

Electronic Data Interchange (EDI) is rapidly becoming a standard way to exchange information with purchasers and suppliers and to improve operations in other ways as well. In this mode computers at one location connect directly to computers at another location, and data are transmitted to be processed, and used without human intervention. A common example is the electronic transfer of funds used in banking transactions. Expansions of this concept now permit many forms of documents and data to be transferred and many operations to be performed automatically. While this technique originally developed between parties that have a form of partnering, as described in Chapter 7, it is rapidly expanding to other less rigorous relationships.

EDI requires that the computers use a standard protocol for communication, and ANSI Standard X 12 has developed uniform transaction sets that define the data and its formatting for electronic handling. EDI permits the exchange of technical data in the form of drawings as well as general business data. Thus it is possible to exchange purchase orders, invoices, shipping documents, credit memos, and similar financial data, including electronic transfer of funds for completed transactions. The transfer of funds largely removes the **float** of two or more days, which typically existed with checks and mailing, and may affect cash management requirements. With the possibility of diversion of funds or data, some care must be taken to ensure that suitable system **security** measures exist.

While the basic advantages of EDI—namely improved accuracy, reduced inventories, and more rapid billing and payments—are valuable, a much larger savings is possible if internal systems are improved. This is achieved by using these electronic data directly, without additional human intervention. Thus the use of EDI for order entry should include, as its next logical step, direct entry into production planning and releases and ultimately into Computer-Integrated Manufacturing (CIM). With a modest review of the operations of most companies, the potential benefits of EDI will rapidly become apparent.

Guidelines for **travel** by employees is most important to service-type organizations whose personnel are frequently called upon to travel. Some criteria that have proved to be both economic and useful include the use of coach class air travel for all trips of three hours or less travel time, with business class used for longer trips. For "red-eye" (overnight) flights an upgrade to the next level is usually permitted, since the red-eye fares are reduced fares. For executive travel, first-class should be used on a limited basis, and a good rule is that no more than three senior executives or members of the same project team should travel on the same aircraft. Where charter aircraft are employed, this rule should be enforced with particular care.

Firms will normally enter into both **car rental** and **hotel agreements** to provide discounted services. With their large number available, it is wise to enter into multiple agreements to permit some choice to be made and, more important, to develop some competition in the costs for the services. Where a project will generate significant local business, it may be feasible to enter into a special agreement with one or more local hotels or car rental agencies.

While the general rule is that all necessary and reasonable travel expenses are reimbursed, it is wise to establish **maximum** allowances for meals and incidental expenses, apart from transportation and hotels, and to not, without special approval, reimburse employees for expenses when these are exceeded. Normally all expenditures greater than, say, $25 should be supported by receipts, when travel expense reports are submitted. Differing travel policies for reimbursement are used by various companies regarding such things as laundry expenses, personal telephone calls, and entertainment. In general, laundry is reimbursed when a trip exceeds four or five days, and a personal telephone call home every other day from domestic locations is appropriate. Usually entertainment is considered a personal expense and not reimbursed.

The physical demands of travel are sufficiently rigorous that the benefit an employee gains by frequent traveler awards is truly earned. As a consequence travel awards for frequent flyers and hotel users are usually left in the name of the traveler and not claimed by the company. Travel should only be taken when truly necessary, and itineraries should utilize the shortest, most economic routing, and discourage the tendancy of some employees to arrange their travel to maximize their frequent traveler awards. Prior approval of the need for the travel and the use of an outside travel agency for travel arrangements should be required. Only in the most unusual cases should the employees be permitted to arrange their own travel. An independent audit by the accounting group can assist in ensuring compliance by the operating departments. Of increasing feasibility is telephone and **electronic conferencing,** which should be used wherever possible permitting a reduction in travel with commensurate time and cost savings.

Security concerns divide themselves into two groups: security of the work and security of the workplace (i.e., physical security and security of senior members of the firm—or personal security). Because of its cost and impact on operations, before a security program is established, there should be a clearly demonstrated need for it.

Physical security is fairly straightforward and primarily involves access control. The use of guards and barriers to control access, the use of visitors' logs and badges, and the requirement for escorts to and from the workplace are all useful ways to control access and limit the ability of a visitor to roam about. Where work of a secure nature is being performed, badges worn in plain sight are a very effective way to control access and permit instant identification of unauthorized personnel. Special care should be taken to ensure that delivery personnel pass a security control point. Truck entrances, garages, and similar locations should have security check points. Clients will often have work that is proprietary or classified, and some work areas may need to be segregated and employ additional security measures. This may require that the work be located in a separate building away from the normal work area.

Physical security usually requires the presence of security personnel any time workers may need to be in the facility, and this often includes nights and weekends. As a result it may be more cost effective to use alarm systems for off-hours security, with periodic checks to ensure its effectiveness.

Personal security is normally related to prevention of kidnapping for financial, or in some cases political, motives. It is usually a concern only for the senior executives of large firms. Primary among the precautions is an effective system of physical security with limits on access. In addition directories should not list addresses or telephone numbers of senior personnel, and this information should only be available on a need-to-know basis. Travel between residences and the office should be varied, using different routes and times so that a pattern cannot be readily established. In some cases defensive driving courses should be taken to permit evasive action to be taken. The use of portable (cellular) telephones in cars should be encouraged to provide instant communication in the event of emergency.

For travel it is important that **itineraries** not be published. Rather this information should be restricted to the office of the executive, again on a need-to-know basis. Where particular concern exists, although rarely used, it may even be useful to consider hotel registrations under assumed names. When traveling, a low profile is useful, and attention should not unnecessarily be called to oneself. For travel abroad, it is wise to check with the U.S. State Department for traveler advisories that may be in effect. While it may not be as easy to maintain a low profile abroad because of physical characteristics, it is wise to not travel in-country alone, and someone should at all times know where the traveler can be reached. The local U.S. embassy or consulate can provide some assistance and advice in these situations, and it is wise to let them know you are in-country if the visit is longer than a few days.

Some form of **public relations** activities occur in all firms, with more formal programs used in the larger ones. There will be continual requests to join organizations and to contribute to charities, which, if not carefully controlled, can create a problem with the more vocal groups getting support while other meritorious ones receive nothing. The best approach is to set up a **budget** of the funds or time involvement that the firm will provide. If done on an annual basis, it provides a way to budget for these activities and permits an organization to be turned down without ill feelings if the reasons are explained to them.

To qualify for inclusion on the list, the firm needs to establish appropriate **criteria.** One usual item is that the particular group would serve the self-interests of the business and contribute to its growth and profitability. Other criteria are an interest in furthering art, music, child care, recreation, and creating a favorable public image. Thus in some cases it may be necessary to support organizations whose goals have little to do with the work of the company. Normally businesses will not wish to associate or support militant organizations, since they tend to create a poor public image. As with all other administrative items a single person should be identified as the administrator for such activities, with all requests being forwarded to him or her. It is wise to provide the administrator with some training in dealing with the media so that the firm can be presented in the best possible light.

For smaller firms it is usually easier to furnish **nonfinancial** aid such as luncheon speakers, judges at science fairs, or advisors to Junior Achievement groups. Often such more personalized involvement is more effective than direct financial support.

For all but the smallest firms it is inevitable that there will be occasional **media** requests for information on work currently underway or other items of interest. It is useful to establish favorable contacts with media personnel, both print and broadcast. This will often permit a story to be checked with the firm before its broadcast or distribution and should, as a minimum, allow the company point of view to be presented as well. Media personnel want to present accurate information, and the firm will be better served if its representative answers inquiries fully and promptly. Cooperation with the media is not only useful but highly desirable.

13.5 THE BUSINESS PLAN

A **business plan** is an essential management tool. Its purpose is to set the goals for the firm for the next year, three years, and perhaps five years. While it is true that the growth of a business is affected by many external factors outside the direct control of the firm, there are many things that in fact can be controlled. To proceed without a plan is to give up control over many of the factors that affect the future of the company. It is also true that the further in time one looks ahead, the less certain the data become. Despite that, the mere act of developing a business plan forces attention to a variety of activities and factors that have a major impact on business. Although the thrust and direction of the questions and analyses are similar, business plans for new businesses are somewhat different from those of established ones.

The plan should contain an overview of the present status of the company including financial information, data on the product line, clients, market share, and personnel resources. This information should be presented in numerical form wherever possible to permit specific goal setting and measurement. From these data the plan develops goals and plans for the specific time periods of interest.

Perhaps the most important part of the plan is the review of the business the company is engaged in. This should be a broad overview of the industry, with a consideration of parallel and complimentary activities that fit the basic approach of the company. There should be considerable flexibility in defining the business. For example, an airline would not define itself as merely an airline but rather as in the transportation business. When defined in this manner, a whole range of allied activities is opened up. In this case transportation of freight, delivery systems, aircraft servicing, aircraft maintenance and overhaul, and pilot and personnel training are all other areas where the business could operate without getting too far from the basic activities that it now performs. Thus, in broadly defining the company business, a group of opportunities for future activities can be developed. These opportunities then are evaluated in terms of the company, the industry, the overall market, and the preferences of the management. In particular a detailed analysis of the specific market segments considered should be performed. This is important because most activities develop in niche businesses, which can become significant in themselves.

New business plans place greater stress on development of either clients or business sectors and necessarily require significant attention to the initial financial operations. An outline of the key questions that should be considered in such a plan is presented in Table 13.2. The plan is for a service company establishing itself for the first time, but with minor adjustment it applies equally well to older established businesses and those that produce products for sale.

Financial and Human resources are an important part of a business strategy, as is the time required to **penetrate** new markets. These are further considerations of investment in new plant or equipment, rental of offices, hiring of personnel, and training. To provide a balanced approach, it is useful to prepare a high, low, and likely scenario for each of these items.

The likely revenue and benefits available for each of these activities, including their risks, can then be estimated. In estimating risk, one computes not only the direct risk by virtue of financial loss but the near-term effect

TABLE 13.2   Outline of a New Business Plan

I. Development of sales/client base
   A. Where will the first client come from?
      1. Geographical area
      2. Specialized services or niche markets
      3. Current contacts
   B. Why should they come to us instead of another?
   C. Marketing Strategies
      1. Trade publications
      2. Trade shows & conventions
      3. News letters
      4. Personal contacts
II. Financial plan
   A. Pre-operating expenses
      1. Deposits and down payments
         Rent & utilities
         Communications services
         Furniture and equipment
      2. Professional services, legal, accounting, insurance
      3. Stationary and supplies
      4. Business promotion
      5. Business licenses and permits
   B. Staying power
      1. Funds to pay for operating expenses until sufficient cash is generated from billings to pay operating expenses
      2. Cash needed by owners for living expenses
   C. Budgets and projections
      1. Estimates of A and B, and when cash will be needed
      2. Estimates of billings, when generated and timing of collections
   D. Source, timing and amount of financing
      1. Investment by owners/shareholders
      2. Financial institutions
      3. Venture capital
      4. Other lenders
   E. Funding provisions if projections are not met
   F. Pricing of product/services
      1. How to establish hourly rates
      2. Calculation of overhead factor to be applied to various types of work
      3. Use different rates for different types of work or use standard rate for each level of employee? (If standard rate, should there be any adjustment for extraordinary knowledge and/or proficiency?)

TABLE 13.2   *(Continued)*

  G. Billing and collection
     1. Frequency
     2. Use of advance payments
     3. Delays in payments
     4. Disputed billings
     5. Allowances for uncollectibles
  H. Business continuity
     1. Disability insurance for key personnel
     2. Business continuation insurance related to loss of productivity of key personnel and cost of replacement of them
     3. "Key personnel" life insurance
     4. Disability buyout insurance and life insurance on owners

on earnings, alternate return on investment, and so forth. Following that, there can be a rational decision made on the relative importance of the various alternates. After ranking them, the company now has in hand an overall plan for review and approval by top management.

The plan needs to have a mechanism for **feedback** on the performance against the plan. This should include specific and measurable goals. The reporting against these goals should be more frequent than merely annually. Often it is done on a quarterly basis. The reporting goals need to be kept simple and fairly limited. It is not necessary to measure everything, merely those items that indicate overall performance. Finally, based on the actual performance, the plan should be revised periodically, often annually.

APPENDIX A

STANDARD FORM OF AGREEMENT BETWEEN OWNER AND ARCHITECT, AIA DOCUMENT B 141

T H E   A M E R I C A N   I N S T I T U T E   O F   A R C H I T E C T S

AIA Document B141

Standard Form of Agreement Between Owner and Architect

1987 EDITION

THIS DOCUMENT HAS IMPORTANT LEGAL CONSEQUENCES; CONSULTATION WITH AN ATTORNEY IS ENCOURAGED WITH RESPECT TO ITS COMPLETION OR MODIFICATION.

AGREEMENT

made as of the
Nineteen Hundred and
day of
in the year of

BETWEEN the Owner:
(Name and address)

and the Architect:
(Name and address)

For the following Project:
(Include detailed description of Project, location, address and scope.)

The Owner and Architect agree as set forth below.

TERMS AND CONDITIONS OF AGREEMENT BETWEEN OWNER AND ARCHITECT

ARTICLE 1
ARCHITECT'S RESPONSIBILITIES

1.1 ARCHITECT'S SERVICES

1.1.1 The Architect's services consist of those services performed by the Architect, Architect's employees and Architect's consultants as enumerated in Articles 2 and 3 of this Agreement and any other services included in Article 12.

1.1.2 The Architect's services shall be performed as expeditiously as is consistent with professional skill and care and the orderly progress of the Work. Upon request of the Owner, the Architect shall submit for the Owner's approval a schedule for the performance of the Architect's services which may be adjusted as the Project proceeds, and shall include allowances for periods of time required for the Owner's review and for approval of submissions by authorities having jurisdiction over the Project. Time limits established by this schedule approved by the Owner shall not, except for reasonable cause, be exceeded by the Architect or Owner.

1.1.3 The services covered by this Agreement are subject to the time limitations contained in Subparagraph 11.5.1.

ARTICLE 2
SCOPE OF ARCHITECT'S BASIC SERVICES

2.1 DEFINITION

2.1.1 The Architect's Basic Services consist of those described in Paragraphs 2.2 through 2.6 and any other services identified in Article 12 as part of Basic Services, and include normal structural, mechanical and electrical engineering services.

2.2 SCHEMATIC DESIGN PHASE

2.2.1 The Architect shall review the program furnished by the Owner to ascertain the requirements of the Project and shall arrive at a mutual understanding of such requirements with the Owner.

2.2.2 The Architect shall provide a preliminary evaluation of the Owner's program, schedule and construction budget requirements, each in terms of the other, subject to the limitations set forth in Subparagraph 5.2.1.

2.2.3 The Architect shall review with the Owner alternative approaches to design and construction of the Project.

2.2.4 Based on the mutually agreed-upon program, schedule and construction budget requirements, the Architect shall prepare, for approval by the Owner, Schematic Design Documents consisting of drawings and other documents illustrating the scale and relationship of Project components.

2.2.5 The Architect shall submit to the Owner a preliminary estimate of Construction Cost based on current area, volume or other unit costs.

2.3 DESIGN DEVELOPMENT PHASE

2.3.1 Based on the approved Schematic Design Documents and any adjustments authorized by the Owner in the program, schedule or construction budget, the Architect shall prepare, for approval by the Owner, Design Development Documents consisting of drawings and other documents to fix and describe the size and character of the Project as to architectural, structural, mechanical and electrical systems, materials and such other elements as may be appropriate.

2.3.2 The Architect shall advise the Owner of any adjustments to the preliminary estimate of Construction Cost.

2.4 CONSTRUCTION DOCUMENTS PHASE

2.4.1 Based on the approved Design Development Documents and any further adjustments in the scope or quality of the Project or in the construction budget authorized by the Owner, the Architect shall prepare, for approval by the Owner, Construction Documents consisting of Drawings and Specifications setting forth in detail the requirements for the construction of the Project.

2.4.2 The Architect shall assist the Owner in the preparation of the necessary bidding information, bidding forms, the Conditions of the Contract, and the form of Agreement between the Owner and Contractor.

2.4.3 The Architect shall advise the Owner of any adjustments to previous preliminary estimates of Construction Cost indicated by changes in requirements or general market conditions.

2.4.4 The Architect shall assist the Owner in connection with the Owner's responsibility for filing documents required for the approval of governmental authorities having jurisdiction over the Project.

2.5 BIDDING OR NEGOTIATION PHASE

2.5.1 The Architect, following the Owner's approval of the Construction Documents and of the latest preliminary estimate of Construction Cost, shall assist the Owner in obtaining bids or negotiated proposals and assist in awarding and preparing contracts for construction.

2.6 CONSTRUCTION PHASE—ADMINISTRATION OF THE CONSTRUCTION CONTRACT

2.6.1 The Architect's responsibility to provide Basic Services for the Construction Phase under this Agreement commences with the award of the Contract for Construction and terminates at the earlier of the issuance to the Owner of the final Certificate for Payment or 60 days after the date of Substantial Completion of the Work, unless extended under the terms of Subparagraph 10.3.3.

2.6.2 The Architect shall provide administration of the Contract for Construction as set forth below and in the edition of AIA Document A201, General Conditions of the Contract for Construction, current as of the date of this Agreement, unless otherwise provided in this Agreement.

2.6.3 Duties, responsibilities and limitations of authority of the Architect shall not be restricted, modified or extended without written agreement of the Owner and Architect with consent of the Contractor, which consent shall not be unreasonably withheld.

2.6.4 The Architect shall be a representative of and shall advise and consult with the Owner (1) during construction until final payment to the Contractor is due, and (2) as an Additional Service at the Owner's direction from time to time during the correction period described in the Contract for Construction. The Architect shall have authority to act on behalf of the Owner only to the extent provided in this Agreement unless otherwise modified by written instrument.

2.6.5 The Architect shall visit the site at intervals appropriate to the stage of construction or as otherwise agreed by the Owner and Architect in writing to become generally familiar with the progress and quality of the Work completed and to determine in general if the Work is being performed in a manner indicating that the Work when completed will be in accordance with the Contract Documents. However, the Architect shall not be required to make exhaustive or continuous on-site inspections to check the quality or quantity of the Work. On the basis of on-site observations as an architect, the Architect shall keep the Owner informed of the progress and quality of the Work, and shall endeavor to guard the Owner against defects and deficiencies in the Work. *(More extensive site representation may be agreed to as an Additional Service, as described in Paragraph 3.2.)*

2.6.6 The Architect shall not have control over or charge of and shall not be responsible for construction means, methods, techniques, sequences or procedures, or for safety precautions and programs in connection with the Work, since these are solely the Contractor's responsibility under the Contract for Construction. The Architect shall not be responsible for the Contractor's schedules or failure to carry out the Work in accordance with the Contract Documents. The Architect shall not have control over or charge of acts or omissions of the Contractor, Subcontractors, or their agents or employees, or of any other persons performing portions of the Work.

2.6.7 The Architect shall at all times have access to the Work wherever it is in preparation or progress.

2.6.8 Except as may otherwise be provided in the Contract Documents or when direct communications have been specially authorized, the Owner and Contractor shall communicate through the Architect. Communications by and with the Architect's consultants shall be through the Architect.

2.6.9 Based on the Architect's observations and evaluations of the Contractor's Applications for Payment, the Architect shall review and certify the amounts due the Contractor.

2.6.10 The Architect's certification for payment shall constitute a representation to the Owner, based on the Architect's observations at the site as provided in Subparagraph 2.6.5 and on the data comprising the Contractor's Application for Payment, that the Work has progressed to the point indicated and that, to the best of the Architect's knowledge, information and belief, quality of the Work is in accordance with the Contract Documents. The foregoing representations are subject to an evaluation of the Work for conformance with the Contract Documents upon Substantial Completion, to results of subsequent tests and inspections, to minor deviations from the Contract Documents correctable prior to completion and to specific qualifications expressed by the Architect. The issuance of a Certificate for Payment shall further constitute a representation that the Contractor is entitled to payment in the amount certified. However, the issuance of a Certificate for Payment shall not be a representation that the Architect has (1) made exhaustive or continuous on-site inspections to check the quality or

quantity of the Work, (2) reviewed construction means, methods, techniques, sequences or procedures, (3) reviewed copies of requisitions received from Subcontractors and material suppliers and other data requested by the Owner to substantiate the Contractor's right to payment or (4) ascertained how or for what purpose the Contractor has used money previously paid on account of the Contract Sum.

2.6.11 The Architect shall have authority to reject Work which does not conform to the Contract Documents. Whenever the Architect considers it necessary or advisable for implementation of the intent of the Contract Documents, the Architect will have authority to require additional inspection or testing of the Work in accordance with the provisions of the Contract Documents, whether or not such Work is fabricated, installed or completed. However, neither this authority of the Architect nor a decision made in good faith either to exercise or not to exercise such authority shall give rise to a duty or responsibility of the Architect to the Contractor, Subcontractors, material and equipment suppliers, their agents or employees or other persons performing portions of the Work.

2.6.12 The Architect shall review and approve or take other appropriate action upon Contractor's submittals such as Shop Drawings, Product Data and Samples, but only for the limited purpose of checking for conformance with information given and the design concept expressed in the Contract Documents. The Architect's action shall be taken with such reasonable promptness as to cause no delay in the Work or in the construction of the Owner or of separate contractors, while allowing sufficient time in the Architect's professional judgment to permit adequate review. Review of such submittals is not conducted for the purpose of determining the accuracy and completeness of other details such as dimensions and quantities or for substantiating instructions for installation or performance of equipment or systems designed by the Contractor, all of which remain the responsibility of the Contractor to the extent required by the Contract Documents. The Architect's review shall not constitute approval of safety precautions or, unless otherwise specifically stated by the Architect, of construction means, methods, techniques, sequences or procedures. The Architect's approval of a specific item shall not indicate approval of an assembly of which the item is a component. When professional certification of performance characteristics of materials, systems or equipment is required by the Contract Documents, the Architect shall be entitled to rely upon such certification to establish that the materials, systems or equipment will meet the performance criteria required by the Contract Documents.

2.6.13 The Architect shall prepare Change Orders and Construction Change Directives, with supporting documentation and data if deemed necessary by the Architect as provided in Subparagraphs 3.1.1 and 3.3.3, for the Owner's approval and execution in accordance with the Contract Documents, and may authorize minor changes in the Work not involving an adjustment in the Contract Sum or an extension of the Contract Time which are not inconsistent with the intent of the Contract Documents.

2.6.14 The Architect shall conduct inspections to determine the date or dates of Substantial Completion and the date of final completion, shall receive and forward to the Owner for the Owner's review and records written warranties and related documents required by the Contract Documents and assembled by the Contractor, and shall issue a final Certificate for Payment upon compliance with the requirements of the Contract Documents.

2.6.15 The Architect shall interpret and decide matters concerning performance of the Owner and Contractor under the requirements of the Contract Documents on written request of either the Owner or Contractor. The Architect's response to such requests shall be made with reasonable promptness and within any time limits agreed upon.

2.6.16 Interpretations and decisions of the Architect shall be consistent with the intent of and reasonably inferable from the Contract Documents and shall be in writing or in the form of drawings. When making such interpretations and initial decisions, the Architect shall endeavor to secure faithful performance by both Owner and Contractor, shall not show partiality to either, and shall not be liable for results of interpretations or decisions so rendered in good faith.

2.6.17 The Architect's decisions on matters relating to aesthetic effect shall be final if consistent with the intent expressed in the Contract Documents.

2.6.18 The Architect shall render written decisions within a reasonable time on all claims, disputes or other matters in question between the Owner and Contractor relating to the execution or progress of the Work as provided in the Contract Documents.

2.6.19 The Architect's decisions on claims, disputes or other matters, including those in question between the Owner and Contractor, except for those relating to aesthetic effect as provided in Subparagraph 2.6.17, shall be subject to arbitration as provided in this Agreement and in the Contract Documents.

ARTICLE 3
ADDITIONAL SERVICES

3.1   GENERAL

3.1.1 The services described in this Article 3 are not included in Basic Services unless so identified in Article 2, and they shall be paid for by the Owner as provided in this Agreement, in addition to the compensation for Basic Services. The services described under Paragraphs 3.2 and 3.4 shall only be provided if authorized or confirmed in writing by the Owner. If services described under Contingent Additional Services in Paragraph 3.3 are required due to circumstances beyond the Architect's control, the Architect shall notify the Owner prior to commencing such services. If the Owner deems that such services described under Paragraph 3.3 are not required, the Owner shall give prompt written notice to the Architect. If the Owner indicates in writing that all or part of such Contingent Additional Services are not required, the Architect shall have no obligation to provide those services.

3.2   PROJECT REPRESENTATION BEYOND BASIC SERVICES

3.2.1 If more extensive representation at the site than is described in Subparagraph 2.6.5 is required, the Architect shall provide one or more Project Representatives to assist in carrying out such additional on-site responsibilities.

3.2.2 Project Representatives shall be selected, employed and directed by the Architect, and the Architect shall be compensated therefor as agreed by the Owner and Architect. The duties, responsibilities and limitations of authority of Project Representatives shall be as described in the edition of AIA Document B352 current as of the date of this Agreement, unless otherwise agreed.

3.2.3 Through the observations by such Project Representatives, the Architect shall endeavor to provide further protection for the Owner against defects and deficiencies in the Work, but the furnishing of such project representation shall not modify the rights, responsibilities or obligations of the Architect as described elsewhere in this Agreement.

3.3   CONTINGENT ADDITIONAL SERVICES

3.3.1 Making revisions in Drawings, Specifications or other documents when such revisions are:

  .1 inconsistent with approvals or instructions previously given by the Owner, including revisions made necessary by adjustments in the Owner's program or Project budget;

  .2 required by the enactment or revision of codes, laws or regulations subsequent to the preparation of such documents; or

  .3 due to changes required as a result of the Owner's failure to render decisions in a timely manner.

3.3.2 Providing services required because of significant changes in the Project including, but not limited to, size, quality, complexity, the Owner's schedule, or the method of bidding or negotiating and contracting for construction, except for services required under subparagraph 5.2.5.

3.3.3 Preparing Drawings, Specifications and other documentation and supporting data, evaluating Contractor's proposals, and providing other services in connection with Change Orders and Construction Change Directives.

3.3.4 Providing services in connection with evaluating substitutions proposed by the Contractor and making subsequent revisions to Drawings, Specifications and other documentation resulting therefrom.

3.3.5 Providing consultation concerning replacement of Work damaged by fire or other cause during construction, and furnishing services required in connection with the replacement of such Work.

3.3.6 Providing services made necessary by the default of the Contractor, by major defects or deficiencies in the Work of the Contractor, or by failure of performance of either the Owner or Contractor under the Contract for Construction.

3.3.7 Providing services in evaluating an extensive number of claims submitted by the Contractor or others in connection with the Work.

3.3.8 Providing services in connection with a public hearing, arbitration proceeding or legal proceeding except where the Architect is party thereto.

3.3.9 Preparing documents for alternate, separate or sequential bids or providing services in connection with bidding, negotiation or construction prior to the completion of the Construction Documents Phase.

3.4   OPTIONAL ADDITIONAL SERVICES

3.4.1 Providing analyses of the Owner's needs and programming the requirements of the Project.

3.4.2 Providing financial feasibility or other special studies.

3.4.3 Providing planning surveys, site evaluations or comparative studies of prospective sites.

3.4.4 Providing special surveys, environmental studies and submissions required for approvals of governmental authorities or others having jurisdiction over the Project.

3.4.5 Providing services relative to future facilities, systems and equipment.

3.4.6 Providing services to investigate existing conditions or facilities or to make measured drawings thereof.

3.4.7 Providing services to verify the accuracy of drawings or other information furnished by the Owner.

3.4.8 Providing coordination of construction performed by separate contractors or by the Owner's own forces and coordination of services required in connection with construction performed and equipment supplied by the Owner.

3.4.9 Providing services in connection with the work of a construction manager or separate consultants retained by the Owner.

3.4.10 Providing detailed estimates of Construction Cost.

3.4.11 Providing detailed quantity surveys or inventories of material, equipment and labor.

3.4.12 Providing analyses of owning and operating costs.

3.4.13 Providing interior design and other similar services required for or in connection with the selection, procurement or installation of furniture, furnishings and related equipment.

3.4.14 Providing services for planning tenant or rental spaces.

3.4.15 Making investigations, inventories of materials or equipment, or valuations and detailed appraisals of existing facilities.

3.4.16 Preparing a set of reproducible record drawings showing significant changes in the Work made during construction based on marked-up prints, drawings and other data furnished by the Contractor to the Architect.

3.4.17 Providing assistance in the utilization of equipment or systems such as testing, adjusting and balancing, preparation of operation and maintenance manuals, training personnel for operation and maintenance, and consultation during operation.

3.4.18 Providing services after issuance to the Owner of the final Certificate for Payment, or in the absence of a final Certificate for Payment, more than 60 days after the date of Substantial Completion of the Work.

3.4.19 Providing services of consultants for other than architectural, structural, mechanical and electrical engineering portions of the Project provided as a part of Basic Services.

3.4.20 Providing any other services not otherwise included in this Agreement or not customarily furnished in accordance with generally accepted architectural practice.

ARTICLE 4
OWNER'S RESPONSIBILITIES

4.1 The Owner shall provide full information regarding requirements for the Project, including a program which shall set forth the Owner's objectives, schedule, constraints and criteria, including space requirements and relationships, flexibility, expandability, special equipment, systems and site requirements.

4.2 The Owner shall establish and update an overall budget for the Project, including the Construction Cost, the Owner's other costs and reasonable contingencies related to all of these costs.

4.3 If requested by the Architect, the Owner shall furnish evidence that financial arrangements have been made to fulfill the Owner's obligations under this Agreement.

4.4 The Owner shall designate a representative authorized to act on the Owner's behalf with respect to the Project. The Owner or such authorized representative shall render decisions in a timely manner pertaining to documents submitted by the Architect in order to avoid unreasonable delay in the orderly and sequential progress of the Architect's services.

4.5 The Owner shall furnish surveys describing physical characteristics, legal limitations and utility locations for the site of the Project, and a written legal description of the site. The surveys and legal information shall include, as applicable, grades and lines of streets, alleys, pavements and adjoining property and structures; adjacent drainage; rights-of-way, restrictions, easements, encroachments, zoning, deed restrictions, boundaries and contours of the site; locations, dimensions and necessary data pertaining to existing buildings, other improvements and trees; and information concerning available utility services and lines, both public and private, above and below grade, including inverts and depths. All the information on the survey shall be referenced to a project benchmark.

4.6 The Owner shall furnish the services of geotechnical engineers when such services are requested by the Architect. Such services may include but are not limited to test borings, test pits, determinations of soil bearing values, percolation tests, evaluations of hazardous materials, ground corrosion and resistivity tests, including necessary operations for anticipating subsoil conditions, with reports and appropriate professional recommendations.

4.6.1 The Owner shall furnish the services of other consultants when such services are reasonably required by the scope of the Project and are requested by the Architect.

4.7 The Owner shall furnish structural, mechanical, chemical, air and water pollution tests, tests for hazardous materials, and other laboratory and environmental tests, inspections and reports required by law or the Contract Documents.

4.8 The Owner shall furnish all legal, accounting and insurance counseling services as may be necessary at any time for the Project, including auditing services the Owner may require to verify the Contractor's Applications for Payment or to ascertain how or for what purposes the Contractor has used the money paid by or on behalf of the Owner.

4.9 The services, information, surveys and reports required by Paragraphs 4.5 through 4.8 shall be furnished at the Owner's expense, and the Architect shall be entitled to rely upon the accuracy and completeness thereof.

4.10 Prompt written notice shall be given by the Owner to the Architect if the Owner becomes aware of any fault or defect in the Project or nonconformance with the Contract Documents.

4.11 The proposed language of certificates or certifications requested of the Architect or Architect's consultants shall be submitted to the Architect for review and approval at least 14 days prior to execution. The Owner shall not request certifications that would require knowledge or services beyond the scope of this Agreement.

AIA DOCUMENT B141 • OWNER-ARCHITECT AGREEMENT • FOURTEENTH EDITION • AIA® • ©1987
THE AMERICAN INSTITUTE OF ARCHITECTS, 1735 NEW YORK AVENUE, N.W., WASHINGTON, D.C. 20006

ARTICLE 5
CONSTRUCTION·COST

5.1   DEFINITION

5.1.1 The Construction Cost shall be the total cost or estimated cost to the Owner of all elements of the Project designed or specified by the Architect.

5.1.2 The Construction Cost shall include the cost at current market rates of labor and materials furnished by the Owner and equipment designed, specified, selected or specially provided for by the Architect, plus a reasonable allowance for the Contractor's overhead and profit. In addition, a reasonable allowance for contingencies shall be included for market conditions at the time of bidding and for changes in the Work during construction.

5.1.3 Construction Cost does not include the compensation of the Architect and Architect's consultants, the costs of the land, rights-of-way, financing or other costs which are the responsibility of the Owner as provided in Article 4.

5.2   RESPONSIBILITY FOR CONSTRUCTION COST

5.2.1 Evaluations of the Owner's Project budget, preliminary estimates of Construction Cost and detailed estimates of Construction Cost, if any, prepared by the Architect, represent the Architect's best judgment as a design professional familiar with the construction industry. It is recognized, however, that neither the Architect nor the Owner has control over the cost of labor, materials or equipment, over the Contractor's methods of determining bid prices, or over competitive bidding, market or negotiating conditions. Accordingly, the Architect cannot and does not warrant or represent that bids or negotiated prices will not vary from the Owner's Project budget or from any estimate of Construction Cost or evaluation prepared or agreed to by the Architect.

5.2.2 No fixed limit of Construction Cost shall be established as a condition of this Agreement by the furnishing, proposal or establishment of a Project budget, unless such fixed limit has been agreed upon in writing and signed by the parties hereto. If such a fixed limit has been established, the Architect shall be permitted to include contingencies for design, bidding and price escalation, to determine what materials, equipment, component systems and types of construction are to be included in the Contract Documents, to make reasonable adjustments in the scope of the Project and to include in the Contract Documents alternate bids to adjust the Construction Cost to the fixed limit. Fixed limits, if any, shall be increased in the amount of an increase in the Contract Sum occurring after execution of the Contract for Construction.

5.2.3 If the Bidding or Negotiation Phase has not commenced within 90 days after the Architect submits the Construction Documents to the Owner, any Project budget or fixed limit of Construction Cost shall be adjusted to reflect changes in the general level of prices in the construction industry between the date of submission of the Construction Documents to the Owner and the date on which proposals are sought.

5.2.4 If a fixed limit of Construction Cost (adjusted as provided in Subparagraph 5.2.3) is exceeded by the lowest bona fide bid or negotiated proposal, the Owner shall:

.1   give written approval of an increase in such fixed limit;

.2   authorize rebidding or renegotiating of the Project within a reasonable time;

.3   if the Project is abandoned, terminate in accordance with Paragraph 8.3; or

.4   cooperate in revising the Project scope and quality as required to reduce the Construction Cost.

5.2.5 If the Owner chooses to proceed under Clause 5.2.4.4, the Architect, without additional charge, shall modify the Contract Documents as necessary to comply with the fixed limit, if established as a condition of this Agreement. The modification of Contract Documents shall be the limit of the Architect's responsibility arising out of the establishment of a fixed limit. The Architect shall be entitled to compensation in accordance with this Agreement for all services performed whether or not the Construction Phase is commenced.

ARTICLE 6
USE OF ARCHITECT'S DRAWINGS, SPECIFICATIONS AND OTHER DOCUMENTS

6.1 The Drawings, Specifications and other documents prepared by the Architect for this Project are instruments of the Architect's service for use solely with respect to this Project and, unless otherwise provided, the Architect shall be deemed the author of these documents and shall retain all common law, statutory and other reserved rights, including the copyright. The Owner shall be permitted to retain copies, including reproducible copies, of the Architect's Drawings, Specifications and other documents for information and reference in connection with the Owner's use and occupancy of the Project. The Architect's Drawings, Specifications or other documents shall not be used by the Owner or others on other projects, for additions to this Project or for completion of this Project by others, unless the Architect is adjudged to be in default under this Agreement, except by agreement in writing and with appropriate compensation to the Architect.

6.2 Submission or distribution of documents to meet official regulatory requirements or for similar purposes in connection with the Project is not to be construed as publication in derogation of the Architect's reserved rights.

ARTICLE 7
ARBITRATION

7.1 Claims, disputes or other matters in question between the parties to this Agreement arising out of or relating to this Agreement or breach thereof shall be subject to and decided by arbitration in accordance with the Construction Industry Arbitration Rules of the American Arbitration Association currently in effect unless the parties mutually agree otherwise.

7.2 Demand for arbitration shall be filed in writing with the other party to this Agreement and with the American Arbitration Association. A demand for arbitration shall be made within a reasonable time after the claim, dispute or other matter in question has arisen. In no event shall the demand for arbitration be made after the date when institution of legal or equitable proceedings based on such claim, dispute or other matter in question would be barred by the applicable statutes of limitations.

7.3 No arbitration arising out of or relating to this Agreement shall include, by consolidation, joinder or in any other manner, an additional person or entity not a party to this Agreement, except by written consent containing a specific reference to this Agreement signed by the Owner, Architect, and any other person or entity sought to be joined. Consent to arbitration involving an additional person or entity shall not constitute consent to arbitration of any claim, dispute or other matter in question not described in the written consent or with a person or entity not named or described therein. The foregoing agreement to arbitrate and other agreements to arbitrate with an additional person or entity duly consented to by the parties to this Agreement shall be specifically enforceable in accordance with applicable law in any court having jurisdiction thereof.

7.4 The award rendered by the arbitrator or arbitrators shall be final, and judgment may be entered upon it in accordance with applicable law in any court having jurisdiction thereof.

ARTICLE 8
TERMINATION, SUSPENSION OR ABANDONMENT

8.1 This Agreement may be terminated by either party upon not less than seven days' written notice should the other party fail substantially to perform in accordance with the terms of this Agreement through no fault of the party initiating the termination.

8.2 If the Project is suspended by the Owner for more than 30 consecutive days, the Architect shall be compensated for services performed prior to notice of such suspension. When the Project is resumed, the Architect's compensation shall be equitably adjusted to provide for expenses incurred in the interruption and resumption of the Architect's services.

8.3 This Agreement may be terminated by the Owner upon not less than seven days' written notice to the Architect in the event that the Project is permanently abandoned. If the Project is abandoned by the Owner for more than 90 consecutive days, the Architect may terminate this Agreement by giving written notice.

8.4 Failure of the Owner to make payment to the Architect in accordance with this Agreement shall be considered substantial nonperformance and cause for termination.

8.5 If the Owner fails to make payment when due the Architect for services and expenses, the Architect may, upon seven days' written notice to the Owner, suspend performance of services under this Agreement. Unless payment in full is received by the Architect within seven days of the date of the notice, the suspension shall take effect without further notice. In the event of a suspension of services, the Architect shall have no liability to the Owner for delay or damage caused the Owner because of such suspension of services.

8.6 In the event of termination not the fault of the Architect, the Architect shall be compensated for services performed prior to termination, together with Reimbursable Expenses then due and all Termination Expenses as defined in Paragraph 8.7.

8.7 Termination Expenses are in addition to compensation for Basic and Additional Services, and include expenses which are directly attributable to termination. Termination Expenses shall be computed as a percentage of the total compensation for Basic Services and Additional Services earned to the time of termination, as follows:

   .1 Twenty percent of the total compensation for Basic and Additional Services earned to date if termination occurs before or during the predesign, site analysis, or Schematic Design Phases; or

   .2 Ten percent of the total compensation for Basic and Additional Services earned to date if termination occurs during the Design Development Phase; or

   .3 Five percent of the total compensation for Basic and Additional Services earned to date if termination occurs during any subsequent phase.

ARTICLE 9
MISCELLANEOUS PROVISIONS

9.1 Unless otherwise provided, this Agreement shall be governed by the law of the principal place of business of the Architect.

9.2 Terms in this Agreement shall have the same meaning as those in AIA Document A201, General Conditions of the Contract for Construction, current as of the date of this Agreement.

9.3 Causes of action between the parties to this Agreement pertaining to acts or failures to act shall be deemed to have accrued and the applicable statutes of limitations shall commence to run not later than either the date of Substantial Completion for acts or failures to act occurring prior to Substantial Completion, or the date of issuance of the final Certificate for Payment for acts or failures to act occurring after Substantial Completion.

9.4 The Owner and Architect waive all rights against each other and against the contractors, consultants, agents and employees of the other for damages, but only to the extent covered by property insurance during construction, except such rights as they may have to the proceeds of such insurance as set forth in the edition of AIA Document A201, General Conditions of the Contract for Construction, current as of the date of this Agreement. The Owner and Architect each shall require similar waivers from their contractors, consultants and agents.

9.5 The Owner and Architect, respectively, bind themselves, their partners, successors, assigns and legal representatives to the other party to this Agreement and to the partners, successors, assigns and legal representatives of such other party with respect to all covenants of this Agreement. Neither Owner nor Architect shall assign this Agreement without the written consent of the other.

9.6 This Agreement represents the entire and integrated agreement between the Owner and Architect and supersedes all prior negotiations, representations or agreements, either written or oral. This Agreement may be amended only by written instrument signed by both Owner and Architect.

9.7 Nothing contained in this Agreement shall create a contractual relationship with or a cause of action in favor of a third party against either the Owner or Architect.

9.8 Unless otherwise provided in this Agreement, the Architect and Architect's consultants shall have no responsibility for the discovery, presence, handling, removal or disposal of or exposure of persons to hazardous materials in any form at the Project site, including but not limited to asbestos, asbestos products, polychlorinated biphenyl (PCB) or other toxic substances.

9.9 The Architect shall have the right to include representations of the design of the Project, including photographs of the exterior and interior, among the Architect's promotional and professional materials. The Architect's materials shall not include the Owner's confidential or proprietary information if the Owner has previously advised the Architect in writing of the specific information considered by the Owner to be confidential or proprietary. The Owner shall provide professional credit for the Architect on the construction sign and in the promotional materials for the Project.

ARTICLE 10
PAYMENTS TO THE ARCHITECT

10.1 DIRECT PERSONNEL EXPENSE

10.1.1 Direct Personnel Expense is defined as the direct salaries of the Architect's personnel engaged on the Project and the portion of the cost of their mandatory and customary contributions and benefits related thereto, such as employment taxes and other statutory employee benefits, insurance, sick leave, holidays, vacations, pensions and similar contributions and benefits.

10.2 REIMBURSABLE EXPENSES

10.2.1 Reimbursable Expenses are in addition to compensation for Basic and Additional Services and include expenses incurred by the Architect and Architect's employees and consultants in the interest of the Project, as identified in the following Clauses.

10.2.1.1 Expense of transportation in connection with the Project; expenses in connection with authorized out-of-town travel; long-distance communications; and fees paid for securing approval of authorities having jurisdiction over the Project.

10.2.1.2 Expense of reproductions, postage and handling of Drawings, Specifications and other documents.

10.2.1.3 If authorized in advance by the Owner, expense of overtime work requiring higher than regular rates.

10.2.1.4 Expense of renderings, models and mock-ups requested by the Owner.

10.2.1.5 Expense of additional insurance coverage or limits, including professional liability insurance, requested by the Owner in excess of that normally carried by the Architect and Architect's consultants.

10.2.1.6 Expense of computer-aided design and drafting equipment time when used in connection with the Project.

10.3 PAYMENTS ON ACCOUNT OF BASIC SERVICES

10.3.1 An initial payment as set forth in Paragraph 11.1 is the minimum payment under this Agreement.

10.3.2 Subsequent payments for Basic Services shall be made monthly and, where applicable, shall be in proportion to services performed within each phase of service, on the basis set forth in Subparagraph 11.2.2.

10.3.3 If and to the extent that the time initially established in Subparagraph 11.5.1 of this Agreement is exceeded or extended through no fault of the Architect, compensation for any services rendered during the additional period of time shall be computed in the manner set forth in Subparagraph 11.3.2.

10.3.4 When compensation is based on a percentage of Construction Cost and any portions of the Project are deleted or otherwise not constructed, compensation for those portions of the Project shall be payable to the extent services are performed on those portions, in accordance with the schedule set forth in Subparagraph 11.2.2, based on (1) the lowest bona fide bid or negotiated proposal, or (2) if no such bid or proposal is received, the most recent preliminary estimate of Construction Cost or detailed estimate of Construction Cost for such portions of the Project.

10.4 PAYMENTS ON ACCOUNT OF ADDITIONAL SERVICES

10.4.1 Payments on account of the Architect's Additional Services and for Reimbursable Expenses shall be made monthly upon presentation of the Architect's statement of services rendered or expenses incurred.

10.5 PAYMENTS WITHHELD

10.5.1 No deductions shall be made from the Architect's compensation on account of penalty, liquidated damages or other sums withheld from payments to contractors, or on account of the cost of changes in the Work other than those for which the Architect has been found to be liable.

10.6 ARCHITECT'S ACCOUNTING RECORDS

10.6.1 Records of Reimbursable Expenses and expenses pertaining to Additional Services and services performed on the basis of a multiple of Direct Personnel Expense shall be available to the Owner or the Owner's authorized representative at mutually convenient times.

ARTICLE 11
BASIS OF COMPENSATION

The Owner shall compensate the Architect as follows:

11.1 AN INITIAL PAYMENT of   Dollars (**$**   )
shall be made upon execution of this Agreement and credited to the Owner's account at final payment.

11.2 BASIC COMPENSATION

11.2.1 FOR BASIC SERVICES, as described in Article 2, and any other services included in Article 12 as part of Basic Services, Basic Compensation shall be computed as follows:

(Insert basis of compensation, including stipulated sums, multiples or percentages, and identify phases to which particular methods of compensation apply, if necessary.)

11.2.2 Where compensation is based on a stipulated sum or percentage of Construction Cost, progress payments for Basic Services in each phase shall total the following percentages of the total Basic Compensation payable:
(Insert additional phases as appropriate.)

Schematic Design Phase:	percent (%)
Design Development Phase:	percent (%)
Construction Documents Phase:	percent (%)
Bidding or Negotiation Phase:	percent (%)
Construction Phase:	percent (%)
Total Basic Compensation:	one hundred percent (100%)

11.3 COMPENSATION FOR ADDITIONAL SERVICES

11.3.1 FOR PROJECT REPRESENTATION BEYOND BASIC SERVICES, as described in Paragraph 3.2, compensation shall be computed as follows:

11.3.2 FOR ADDITIONAL SERVICES OF THE ARCHITECT, as described in Articles 3 and 12, other than (1) Additional Project Representation, as described in Paragraph 3.2, and (2) services included in Article 12 as part of Additional Services, but excluding services of consultants, compensation shall be computed as follows:

(Insert basis of compensation, including rates and/or multiples of Direct Personnel Expense for Principals and employees, and identify Principals and classify employees, if required. Identify specific services to which particular methods of compensation apply, if necessary.)

11.3.3 FOR ADDITIONAL SERVICES OF CONSULTANTS, including additional structural, mechanical and electrical engineering services and those provided under Subparagraph 3.4.19 or identified in Article 12 as part of Additional Services, a multiple of ( ) times the amounts billed to the Architect for such services.
(Identify specific types of consultants in Article 12, if required.)

11.4 REIMBURSABLE EXPENSES

11.4.1 FOR REIMBURSABLE EXPENSES, as described in Paragraph 10.2, and any other items included in Article 12 as Reimbursable Expenses, a multiple of ( ) times the expenses incurred by the Architect, the Architect's employees and consultants in the interest of the Project.

11.5 ADDITIONAL PROVISIONS

11.5.1 IF THE BASIC SERVICES covered by this Agreement have not been completed within ( ) months of the date hereof, through no fault of the Architect, extension of the Architect's services beyond that time shall be compensated as provided in Subparagraphs 10.3.3 and 11.3.2.

11.5.2 Payments are due and payable ( ) days from the date of the Architect's invoice. Amounts unpaid ( ) days after the invoice date shall bear interest at the rate entered below, or in the absence thereof at the legal rate prevailing from time to time at the principal place of business of the Architect.
(Insert rate of interest agreed upon.)

(Usury laws and requirements under the Federal Truth in Lending Act, similar state and local consumer credit laws and other regulations at the Owner's and Architect's principal places of business, the location of the Project and elsewhere may affect the validity of this provision. Specific legal advice should be obtained with respect to deletions or modifications, and also regarding requirements such as written disclosures or waivers.)

AIA DOCUMENT B141 • OWNER-ARCHITECT AGREEMENT • FOURTEENTH EDITION • AIA® • ©1987
THE AMERICAN INSTITUTE OF ARCHITECTS, 1735 NEW YORK AVENUE, N.W., WASHINGTON, D.C. 20006

11.5.3 The rates and multiples set forth for Additional Services shall be annually adjusted in accordance with normal salary review practices of the Architect.

<u>ARTICLE 12</u>
OTHER CONDITIONS OR SERVICES

(Insert descriptions of other services, identify Additional Services included within Basic Compensation and modifications to the payment and compensation terms included in this Agreement.)

This Agreement entered into as of the day and year first written above.

OWNER   ARCHITECT

_____   _____
(Signature)   *(Signature)*

_____   _____
(Printed name and title)   *(Printed name and title)*

APPENDIX B

GENERAL CONDITIONS OF THE CONTRACT FOR CONSTRUCTION, AIA DOCUMENT A201

T H E   A M E R I C A N   I N S T I T U T E   O F   A R C H I T E C T S

Because AIA Documents are revised from time to time, users should ascertain from the AIA the current edition of this document. Copies of the current edition of this AIA document may be purchased from The American Institute of Architects or its local distributors.

The text of this document is not "model language" (language taken from an existing document and incorporated, without attribution or permission, into a newly-created document). Rather, it is a standard form which is intended to be modified by appending separate amendment sheets and/or filling in provided blank spaces.

AIA Document A201

General Conditions of the Contract for Construction

THIS DOCUMENT HAS IMPORTANT LEGAL CONSEQUENCES; CONSULTATION WITH AN ATTORNEY IS ENCOURAGED WITH RESPECT TO ITS MODIFICATION

1987 EDITION
TABLE OF ARTICLES

This document has been approved and endorsed by the Associated General Contractors of America.

AIA   CAUTION: You should use an original AIA document which has this caution printed in red. An original assures that changes will not be obscured as may occur when documents are reproduced.

INDEX

AIA DOCUMENT A201 • GENERAL CONDITIONS OF THE CONTRACT FOR CONSTRUCTION • FOURTEENTH EDITION
AIA® • ©1987 THE AMERICAN INSTITUTE OF ARCHITECTS, 1735 NEW YORK AVENUE, N.W., WASHINGTON, D.C. 20006

GENERAL CONDITIONS OF THE CONTRACT FOR CONSTRUCTION

ARTICLE 1

GENERAL PROVISIONS

1.1   BASIC DEFINITIONS

1.1.1   THE CONTRACT DOCUMENTS

The Contract Documents consist of the Agreement between Owner and Contractor (hereinafter the Agreement), Conditions of the Contract (General, Supplementary and other Conditions), Drawings, Specifications, addenda issued prior to execution of the Contract, other documents listed in the Agreement and Modifications issued after execution of the Contract. A Modification is (1) a written amendment to the Contract signed by both parties, (2) a Change Order, (3) a Construction Change Directive or (4) a written order for a minor change in the Work issued by the Architect. Unless specifically enumerated in the Agreement, the Contract Documents do not include other documents such as bidding requirements (advertisement or invitation to bid, Instructions to Bidders, sample forms, the Contractor's bid or portions of addenda relating to bidding requirements).

1.1.2   THE CONTRACT

The Contract Documents form the Contract for Construction. The Contract represents the entire and integrated agreement between the parties hereto and supersedes prior negotiations, representations or agreements, either written or oral. The Contract may be amended or modified only by a Modification. The Contract Documents shall not be construed to create a contractual relationship of any kind (1) between the Architect and Contractor, (2) between the Owner and a Subcontractor or Subsubcontractor or (3) between any persons or entities other than the Owner and Contractor. The Architect shall, however, be entitled to performance and enforcement of obligations under the Contract intended to facilitate performance of the Architect's duties.

1.1.3   THE WORK

The term "Work" means the construction and services required by the Contract Documents, whether completed or partially completed, and includes all other labor, materials, equipment and services provided or to be provided by the Contractor to fulfill the Contractor's obligations. The Work may constitute the whole or a part of the Project.

1.1.4   THE PROJECT

The Project is the total construction of which the Work performed under the Contract Documents may be the whole or a part and which may include construction by the Owner or by separate contractors.

1.1.5   THE DRAWINGS

The Drawings are the graphic and pictorial portions of the Contract Documents, wherever located and whenever issued, showing the design, location and dimensions of the Work, generally including plans, elevations, sections, details, schedules and diagrams.

1.1.6   THE SPECIFICATIONS

The Specifications are that portion of the Contract Documents consisting of the written requirements for materials, equip-

ment, construction systems, standards and workmanship for the Work, and performance of related services.

1.1.7   THE PROJECT MANUAL

The Project Manual is the volume usually assembled for the Work which may include the bidding requirements, sample forms, Conditions of the Contract and Specifications.

1.2   EXECUTION, CORRELATION AND INTENT

1.2.1 The Contract Documents shall be signed by the Owner and Contractor as provided in the Agreement. If either the Owner or Contractor or both do not sign all the Contract Documents, the Architect shall identify such unsigned Documents upon request.

1.2.2 Execution of the Contract by the Contractor is a representation that the Contractor has visited the site, become familiar with local conditions under which the Work is to be performed and correlated personal observations with requirements of the Contract Documents.

1.2.3 The intent of the Contract Documents is to include all items necessary for the proper execution and completion of the Work by the Contractor. The Contract Documents are complementary, and what is required by one shall be as binding as if required by all; performance by the Contractor shall be required only to the extent consistent with the Contract Documents and reasonably inferable from them as being necessary to produce the intended results.

1.2.4 Organization of the Specifications into divisions, sections and articles, and arrangement of Drawings shall not control the Contractor in dividing the Work among Subcontractors or in establishing the extent of Work to be performed by any trade.

1.2.5 Unless otherwise stated in the Contract Documents, words which have well-known technical or construction industry meanings are used in the Contract Documents in accordance with such recognized meanings.

1.3   OWNERSHIP AND USE OF ARCHITECT'S DRAWINGS, SPECIFICATIONS AND OTHER DOCUMENTS

1.3.1 The Drawings, Specifications and other documents prepared by the Architect are instruments of the Architect's service through which the Work is to be executed by the Contractor is described. The Contractor may retain one contract record set. Neither the Contractor nor any Subcontractor, Subsubcontractor or material or equipment supplier shall own or claim a copyright in the Drawings, Specifications and other documents prepared by the Architect, and unless otherwise indicated the Architect shall be deemed the author of them and will retain all common law, statutory and other reserved rights, in addition to the copyright. All copies of them, except the Contractor's record set, shall be returned or suitably accounted for to the Architect, on request, upon completion of the Work. The Drawings, Specifications and other documents prepared by the Architect, and copies thereof furnished to the Contractor, are for use solely with respect to this Project. They are not to be used by the Contractor or any Subcontractor, Subsubcontractor or material or equipment supplier on other projects or for additions to this Project outside the scope of the

Work without the specific written consent of the Owner and Architect. The Contractor, Subcontractors, Sub-subcontractors and material or equipment suppliers are granted a limited license to use and reproduce applicable portions of the Drawings, Specifications and other documents prepared by the Architect appropriate to and for use in the execution of their Work under the Contract Documents. All copies made under this license shall bear the statutory copyright notice, if any, shown on the Drawings, Specifications and other documents prepared by the Architect. Submittal or distribution to meet official regulatory requirements or for other purposes in connection with this Project is not to be construed as publication in derogation of the Architect's copyright or other reserved rights.

1.4   CAPITALIZATION

1.4.1 Terms capitalized in these General Conditions include those which are (1) specifically defined, (2) the titles of numbered articles and identified references to Paragraphs, Subparagraphs and Clauses in the document or (3) the titles of other documents published by the American Institute of Architects.

1.5   INTERPRETATION

1.5.1 In the interest of brevity the Contract Documents frequently omit modifying words such as "all" and "any" and articles such as "the" and "an," but the fact that a modifier or an article is absent from one statement and appears in another is not intended to affect the interpretation of either statement.

ARTICLE 2

OWNER

2.1   DEFINITION

2.1.1 The Owner is the person or entity identified as such in the Agreement and is referred to throughout the Contract Documents as if singular in number. The term "Owner" means the Owner or the Owner's authorized representative.

2.1.2 The Owner upon reasonable written request shall furnish to the Contractor in writing information which is necessary and relevant for the Contractor to evaluate, give notice of or enforce mechanic's lien rights. Such information shall include a correct statement of the record legal title to the property on which the Project is located, usually referred to as the site, and the Owner's interest therein at the time of execution of the Agreement and, within five days after any change, information of such change in title, recorded or unrecorded.

2.2   INFORMATION AND SERVICES REQUIRED OF THE OWNER

2.2.1 The Owner shall, at the request of the Contractor, prior to execution of the Agreement and promptly from time to time thereafter, furnish to the Contractor reasonable evidence that financial arrangements have been made to fulfill the Owner's obligations under the Contract. *[Note: Unless such reasonable evidence were furnished on request prior to the execution of the Agreement, the prospective contractor would not be required to execute the Agreement or to commence the Work.]*

2.2.2 The Owner shall furnish surveys describing physical characteristics, legal limitations and utility locations for the site of the Project, and a legal description of the site.

2.2.3 Except for permits and fees which are the responsibility of the Contractor under the Contract Documents, the Owner shall secure and pay for necessary approvals, easements, assess-

ments and charges required for construction, use or occupancy of permanent structures or for permanent changes in existing facilities.

2.2.4 Information or services under the Owner's control shall be furnished by the Owner with reasonable promptness to avoid delay in orderly progress of the Work.

2.2.5 Unless otherwise provided in the Contract Documents, the Contractor will be furnished, free of charge, such copies of Drawings and Project Manuals as are reasonably necessary for execution of the Work.

2.2.6 The foregoing are in addition to other duties and responsibilities of the Owner enumerated herein and especially those in respect to Article 6 (Construction by Owner or by Separate Contractors), Article 9 (Payments and Completion) and Article 11 (Insurance and Bonds).

2.3   OWNER'S RIGHT TO STOP THE WORK

2.3.1 If the Contractor fails to correct Work which is not in accordance with the requirements of the Contract Documents as required by Paragraph 12.2 or persistently fails to carry out Work in accordance with the Contract Documents, the Owner, by written Order signed personally or by an agent specifically so empowered by the Owner in writing, may order the Contractor to stop the Work, or any portion thereof, until the cause for such order has been eliminated; however, the right of the Owner to stop the Work shall not give rise to a duty on the part of the Owner to exercise this right for the benefit of the Contractor or any other person or entity, except to the extent required by Subparagraph 6.1.3.

2.4   OWNER'S RIGHT TO CARRY OUT THE WORK

2.4.1 If the Contractor defaults or neglects to carry out the Work in accordance with the Contract Documents and fails within a seven-day period after receipt of written notice from the Owner to commence and continue correction of such default or neglect with diligence and promptness, the Owner may after such seven-day period give the Contractor a second written notice to correct such deficiencies within a second seven-day period. If the Contractor within such second seven-day period after receipt of such second notice fails to commence and continue to correct any deficiencies, the Owner may, without prejudice to other remedies the Owner may have, correct such deficiencies. In such case an appropriate Change Order shall be issued deducting from payments then or thereafter due the Contractor the cost of correcting such deficiencies, including compensation for the Architect's additional services and expenses made necessary by such default, neglect or failure. Such action by the Owner and amounts charged to the Contractor are both subject to prior approval of the Architect. If payments then or thereafter due the Contractor are not sufficient to cover such amounts, the Contractor shall pay the difference to the Owner.

ARTICLE 3

CONTRACTOR

3.1   DEFINITION

3.1.1 The Contractor is the person or entity identified as such in the Agreement and is referred to throughout the Contract Documents as if singular in number. The term "Contractor" means the Contractor or the Contractor's authorized representative.

3.2 REVIEW OF CONTRACT DOCUMENTS AND FIELD CONDITIONS BY CONTRACTOR

3.2.1 The Contractor shall carefully study and compare the Contract Documents with each other and with information furnished by the Owner pursuant to Subparagraph 2.2.2 and shall at once report to the Architect errors, inconsistencies or omissions discovered. The Contractor shall not be liable to the Owner or Architect for damage resulting from errors, inconsistencies or omissions in the Contract Documents unless the Contractor recognized such error, inconsistency or omission and knowingly failed to report it to the Architect. If the Contractor performs any construction activity knowing it involves a recognized error, inconsistency or omission in the Contract Documents without such notice to the Architect, the Contractor shall assume appropriate responsibility for such performance and shall bear an appropriate amount of the attributable costs for correction.

3.2.2 The Contractor shall take field measurements and verify field conditions and shall carefully compare such field measurements and conditions and other information known to the Contractor with the Contract Documents before commencing activities. Errors, inconsistencies or omissions discovered shall be reported to the Architect at once.

3.2.3 The Contractor shall perform the Work in accordance with the Contract Documents and submittals approved pursuant to Paragraph 3.12.

3.3 SUPERVISION AND CONSTRUCTION PROCEDURES

3.3.1 The Contractor shall supervise and direct the Work, using the Contractor's best skill and attention. The Contractor shall be solely responsible for and have control over construction means, methods, techniques, sequences and procedures and for coordinating all portions of the Work under the Contract, unless Contract Documents give other specific instructions concerning these matters.

3.3.2 The Contractor shall be responsible to the Owner for acts and omissions of the Contractor's employees, Subcontractors and their agents and employees, and other persons performing portions of the Work under a contract with the Contractor.

3.3.3 The Contractor shall not be relieved of obligations to perform the Work in accordance with the Contract Documents either by activities or duties of the Architect in the Architect's administration of the Contract, or by tests, inspections or approvals required or performed by persons other than the Contractor.

3.3.4 The Contractor shall be responsible for inspection of portions of Work already performed under this Contract to determine that such portions are in proper condition to receive subsequent Work.

3.4 LABOR AND MATERIALS

3.4.1 Unless otherwise provided in the Contract Documents, the Contractor shall provide and pay for labor, materials, equipment, tools, construction equipment and machinery, water, heat, utilities, transportation, and other facilities and services necessary for proper execution and completion of the Work, whether temporary or permanent and whether or not incorporated or to be incorporated in the Work.

3.4.2 The Contractor shall enforce strict discipline and good order among the Contractor's employees and other persons carrying out the Contract. The Contractor shall not permit employment of unfit persons or persons not skilled in tasks assigned to them.

3.5 WARRANTY

3.5.1 The Contractor warrants to the Owner and Architect that materials and equipment furnished under the Contract will be of good quality and new unless otherwise required or permitted by the Contract Documents, that the Work will be free from defects not inherent in the quality required or permitted, and that the Work will conform with the requirements of the Contract Documents. Work not conforming to these requirements, including substitutions not properly approved and authorized, may be considered defective. The Contractor's warranty excludes remedy for damage or defect caused by abuse, modifications not executed by the Contractor, improper or insufficient maintenance, improper operation, or normal wear and tear under normal usage. If required by the Architect, the Contractor shall furnish satisfactory evidence as to the kind and quality of materials and equipment.

3.6 TAXES

3.6.1 The Contractor shall pay sales, consumer, use and similar taxes for the Work or portions thereof provided by the Contractor which are legally enacted when bids are received or negotiations concluded, whether or not yet effective or merely scheduled to go into effect.

3.7 PERMITS, FEES AND NOTICES

3.7.1 Unless otherwise provided in the Contract Documents, the Contractor shall secure and pay for the building permit and other permits and governmental fees, licenses and inspections necessary for proper execution and completion of the Work which are customarily secured after execution of the Contract and which are legally required when bids are received or negotiations concluded.

3.7.2 The Contractor shall comply with and give notices required by laws, ordinances, rules, regulations and lawful orders of public authorities bearing on performance of the Work.

3.7.3 It is not the Contractor's responsibility to ascertain that the Contract Documents are in accordance with applicable laws, statutes, ordinances, building codes, and rules and regulations. However, if the Contractor observes that portions of the Contract Documents are at variance therewith, the Contractor shall promptly notify the Architect and Owner in writing, and necessary changes shall be accomplished by appropriate Modification.

3.7.4 If the Contractor performs Work knowing it to be contrary to laws, statutes, ordinances, building codes, and rules and regulations without such notice to the Architect and Owner, the Contractor shall assume full responsibility for such Work and shall bear the attributable costs.

3.8 ALLOWANCES

3.8.1 The Contractor shall include in the Contract Sum all allowances stated in the Contract Documents. Items covered by allowances shall be supplied for such amounts and by such persons or entities as the Owner may direct, but the Contractor shall not be required to employ persons or entities against which the Contractor makes reasonable objection.

3.8.2 Unless otherwise provided in the Contract Documents:

    .1 materials and equipment under an allowance shall be selected promptly by the Owner to avoid delay in the Work;

    .2 allowances shall cover the cost to the Contractor of materials and equipment delivered at the site and all required taxes, less applicable trade discounts;

.3 Contractor's costs for unloading and handling at the site, labor, installation costs, overhead, profit and other expenses contemplated for stated allowance amounts shall be included in the Contract Sum and not in the allowances;

.4 whenever costs are more than or less than allowances, the Contract Sum shall be adjusted accordingly by Change Order. The amount of the Change Order shall reflect (1) the difference between actual costs and the allowances under Clause 3.8.2.2 and (2) changes in Contractor's costs under Clause 3.8.2.3.

3.9 SUPERINTENDENT

3.9.1 The Contractor shall employ a competent superintendent and necessary assistants who shall be in attendance at the Project site during performance of the Work. The superintendent shall represent the Contractor, and communications given to the superintendent shall be as binding as if given to the Contractor. Important communications shall be confirmed in writing. Other communications shall be similarly confirmed on written request in each case.

3.10 CONTRACTOR'S CONSTRUCTION SCHEDULES

3.10.1 The Contractor, promptly after being awarded the Contract, shall prepare and submit for the Owner's and Architect's information a Contractor's construction schedule for the Work. The schedule shall not exceed time limits current under the Contract Documents, shall be revised at appropriate intervals as required by the conditions of the Work and Project, shall be related to the entire Project to the extent required by the Contract Documents, and shall provide for expeditious and practicable execution of the Work.

3.10.2 The Contractor shall prepare and keep current, for the Architect's approval, a schedule of submittals which is coordinated with the Contractor's construction schedule and allows the Architect reasonable time to review submittals.

3.10.3 The Contractor shall conform to the most recent schedules.

3.11 DOCUMENTS AND SAMPLES AT THE SITE

3.11.1 The Contractor shall maintain at the site for the Owner one record copy of the Drawings, Specifications, addenda, Change Orders and other Modifications, in good order and marked currently to record changes and selections made during construction, and in addition approved Shop Drawings, Product Data, Samples and similar required submittals. These shall be available to the Architect and shall be delivered to the Architect for submittal to the Owner upon completion of the Work.

3.12 SHOP DRAWINGS, PRODUCT DATA AND SAMPLES

3.12.1 Shop Drawings are drawings, diagrams, schedules and other data specially prepared for the Work by the Contractor or a Subcontractor, Sub-subcontractor, manufacturer, supplier or distributor to illustrate some portion of the Work.

3.12.2 Product Data are illustrations, standard schedules, performance charts, instructions, brochures, diagrams and other information furnished by the Contractor to illustrate materials or equipment for some portion of the Work.

3.12.3 Samples are physical examples which illustrate materials, equipment or workmanship and establish standards by which the Work will be judged.

3.12.4 Shop Drawings, Product Data, Samples and similar submittals are not Contract Documents. The purpose of their submittal is to demonstrate for those portions of the Work for which submittals are required the way the Contractor proposes to conform to the information given and the design concept expressed in the Contract Documents. Review by the Architect is subject to the limitations of Subparagraph 4.2.7.

3.12.5 The Contractor shall review, approve and submit to the Architect Shop Drawings, Product Data, Samples and similar submittals required by the Contract Documents with reasonable promptness and in such sequence as to cause no delay in the Work or in the activities of the Owner or of separate contractors. Submittals made by the Contractor which are not required by the Contract Documents may be returned without action.

3.12.6 The Contractor shall perform no portion of the Work requiring submittal and review of Shop Drawings, Product Data, Samples or similar submittals until the respective submittal has been approved by the Architect. Such Work shall be in accordance with approved submittals.

3.12.7 By approving and submitting Shop Drawings, Product Data, Samples and similar submittals, the Contractor represents that the Contractor has determined and verified materials, field measurements and field construction criteria related thereto, or will do so, and has checked and coordinated the information contained within such submittals with the requirements of the Work and of the Contract Documents.

3.12.8 The Contractor shall not be relieved of responsibility for deviations from requirements of the Contract Documents by the Architect's approval of Shop Drawings, Product Data, Samples or similar submittals unless the Contractor has specifically informed the Architect in writing of such deviation at the time of submittal and the Architect has given written approval to the specific deviation. The Contractor shall not be relieved of responsibility for errors or omissions in Shop Drawings, Product Data, Samples or similar submittals by the Architect's approval thereof.

3.12.9 The Contractor shall direct specific attention, in writing or on resubmitted Shop Drawings, Product Data, Samples or similar submittals, to revisions other than those requested by the Architect on previous submittals.

3.12.10 Informational submittals upon which the Architect is not expected to take responsive action may be so identified in the Contract Documents.

3.12.11 When professional certification of performance criteria of materials, systems or equipment is required by the Contract Documents, the Architect shall be entitled to rely upon the accuracy and completeness of such calculations and certifications.

3.13 USE OF SITE

3.13.1 The Contractor shall confine operations at the site to areas permitted by law, ordinances, permits and the Contract Documents and shall not unreasonably encumber the site with materials or equipment.

3.14 CUTTING AND PATCHING

3.14.1 The Contractor shall be responsible for cutting, fitting or patching required to complete the Work or to make its parts fit together properly.

3.14.2 The Contractor shall not damage or endanger a portion of the Work or fully or partially completed construction of the Owner or separate contractors by cutting, patching or otherwise altering such construction, or by excavation. The Contractor shall not cut or otherwise alter such construction by the

Owner or a separate contractor except with written consent of the Owner and of such separate contractor; such consent shall not be unreasonably withheld. The Contractor shall not unreasonably withhold from the Owner or a separate contractor the Contractor's consent to cutting or otherwise altering the Work.

3.15 CLEANING UP

3.15.1 The Contractor shall keep the premises and surrounding area free from accumulation of waste materials or rubbish caused by operations under the Contract. At completion of the Work the Contractor shall remove from and about the Project waste materials, rubbish, the Contractor's tools, construction equipment, machinery and surplus materials.

3.15.2 If the Contractor fails to clean up as provided in the Contract Documents, the Owner may do so and the cost thereof shall be charged to the Contractor.

3.16 ACCESS TO WORK

3.16.1 The Contractor shall provide the Owner and Architect access to the Work in preparation and progress wherever located.

3.17 ROYALTIES AND PATENTS

3.17.1 The Contractor shall pay all royalties and license fees. The Contractor shall defend suits or claims for infringement of patent rights and shall hold the Owner and Architect harmless from loss on account thereof, but shall not be responsible for such defense or loss when a particular design, process or product of a particular manufacturer or manufacturers is required by the Contract Documents. However, if the Contractor has reason to believe that the required design, process or product is an infringement of a patent, the Contractor shall be responsible for such loss unless such information is promptly furnished to the Architect.

3.18 INDEMNIFICATION

3.18.1 To the fullest extent permitted by law, the Contractor shall indemnify and hold harmless the Owner, Architect, Architect's consultants, and agents and employees of any of them from and against claims, damages, losses and expenses, including but not limited to attorneys' fees, arising out of or resulting from performance of the Work, provided that such claim, damage, loss or expense is attributable to bodily injury, sickness, disease or death, or to injury to or destruction of tangible property (other than the Work itself) including loss of use resulting therefrom, but only to the extent caused in whole or in part by negligent acts or omissions of the Contractor, a Subcontractor, anyone directly or indirectly employed by them or anyone for whose acts they may be liable, regardless of whether or not such claim, damage, loss or expense is caused in part by a party indemnified hereunder. Such obligation shall not be construed to negate, abridge, or reduce other rights or obligations of indemnity which would otherwise exist as to a party or person described in this Paragraph 3.18.

3.18.2 In claims against any person or entity indemnified under this Paragraph 3.18 by an employee of the Contractor, a Subcontractor, anyone directly or indirectly employed by them or anyone for whose acts they may be liable, the indemnification obligation under this Paragraph 3.18 shall not be limited by a limitation on amount or type of damages, compensation or benefits payable by or for the Contractor or a Subcontractor under workers' or workmen's compensation acts, disability benefit acts or other employee benefit acts.

3.18.3 The obligations of the Contractor under this Paragraph 3.18 shall not extend to the liability of the Architect, the Archi-

tect's consultants, and agents and employees of any of them arising out of (1) the preparation or approval of maps, drawings, opinions, reports, surveys, Change Orders, designs or specifications, or (2) the giving of or the failure to give directions or instructions by the Architect, the Architect's consultants, and agents and employees of any of them provided such giving or failure to give is the primary cause of the injury or damage.

ARTICLE 4

ADMINISTRATION OF THE CONTRACT

4.1 ARCHITECT

4.1.1 The Architect is the person lawfully licensed to practice architecture or an entity lawfully practicing architecture identified as such in the Agreement and is referred to throughout the Contract Documents as if singular in number. The term "Architect" means the Architect or the Architect's authorized representative.

4.1.2 Duties, responsibilities and limitations of authority of the Architect as set forth in the Contract Documents shall not be restricted, modified or extended without written consent of the Owner, Contractor and Architect. Consent shall not be unreasonably withheld.

4.1.3 In case of termination of employment of the Architect, the Owner shall appoint an architect against whom the Contractor makes no reasonable objection and whose status under the Contract Documents shall be that of the former architect.

4.1.4 Disputes arising under Subparagraphs 4.1.2 and 4.1.3 shall be subject to arbitration.

4.2 ARCHITECT'S ADMINISTRATION OF THE CONTRACT

4.2.1 The Architect will provide administration of the Contract as described in the Contract Documents, and will be the Owner's representative (1) during construction, (2) until final payment is due and (3) with the Owner's concurrence, from time to time during the correction period described in Paragraph 12.2. The Architect will advise and consult with the Owner. The Architect will have authority to act on behalf of the Owner only to the extent provided in the Contract Documents, unless otherwise modified by written instrument in accordance with other provisions of the Contract.

4.2.2 The Architect will visit the site at intervals appropriate to the stage of construction to become generally familiar with the progress and quality of the completed Work and to determine in general if the Work is being performed in a manner indicating that the Work, when completed, will be in accordance with the Contract Documents. However, the Architect will not be required to make exhaustive or continuous on-site inspections to check quality or quantity of the Work. On the basis of on-site observations as an architect, the Architect will keep the Owner informed of progress of the Work, and will endeavor to guard the Owner against defects and deficiencies in the Work.

4.2.3 The Architect will not have control over or charge of and will not be responsible for construction means, methods, techniques, sequences or procedures, or for safety precautions and programs in connection with the Work, since these are solely the Contractor's responsibility as provided in Paragraph 3.3. The Architect will not be responsible for the Contractor's failure to carry out the Work in accordance with the Contract Documents. The Architect will not have control over or charge of and will not be responsible for acts or omissions of the Con-

AIA DOCUMENT A201 • GENERAL CONDITIONS OF THE CONTRACT FOR CONSTRUCTION • FOURTEENTH EDITION
AIA® • ©1987 THE AMERICAN INSTITUTE OF ARCHITECTS, 1735 NEW YORK AVENUE, N.W., WASHINGTON, D.C. 20006

tractor, Subcontractors, or their agents or employees, or of any other persons performing portions of the Work.

4.2.4 Communications Facilitating Contract Administration. Except as otherwise provided in the Contract Documents or when direct communications have been specially authorized, the Owner and Contractor shall endeavor to communicate through the Architect. Communications by and with the Architect's consultants shall be through the Architect. Communications by and with Subcontractors and material suppliers shall be through the Contractor. Communications by and with separate contractors shall be through the Owner.

4.2.5 Based on the Architect's observations and evaluations of the Contractor's Applications for Payment, the Architect will review and certify the amounts due the Contractor and will issue Certificates for Payment in such amounts.

4.2.6 The Architect will have authority to reject Work which does not conform to the Contract Documents. Whenever the Architect considers it necessary or advisable for implementation of the intent of the Contract Documents, the Architect will have authority to require additional inspection or testing of the Work in accordance with Subparagraphs 13.5.2 and 13.5.3, whether or not such Work is fabricated, installed or completed. However, neither this authority of the Architect nor a decision made in good faith either to exercise or not to exercise such authority shall give rise to a duty or responsibility of the Architect to the Contractor, Subcontractors, material and equipment suppliers, their agents or employees, or other persons performing portions of the Work.

4.2.7 The Architect will review and approve or take other appropriate action upon the Contractor's submittals such as Shop Drawings, Product Data and Samples, but only for the limited purpose of checking for conformance with information given and the design concept expressed in the Contract Documents. The Architect's action will be taken with such reasonable promptness as to cause no delay in the Work or in the activities of the Owner, Contractor or separate contractors, while allowing sufficient time in the Architect's professional judgment to permit adequate review. Review of such submittals is not conducted for the purpose of determining the accuracy and completeness of other details such as dimensions and quantities, or for substantiating instructions for installation or performance of equipment or systems, all of which remain the responsibility of the Contractor as required by the Contract Documents. The Architect's review of the Contractor's submittals shall not relieve the Contractor of the obligations under Paragraphs 3.3, 3.5 and 3.12. The Architect's review shall not constitute approval of safety precautions or, unless otherwise specifically stated by the Architect, of any construction means, methods, techniques, sequences or procedures. The Architect's approval of a specific item shall not indicate approval of an assembly of which the item is a component.

4.2.8 The Architect will prepare Change Orders and Construction Change Directives, and may authorize minor changes in the Work as provided in Paragraph 7.4.

4.2.9 The Architect will conduct inspections to determine the date or dates of Substantial Completion and the date of final completion, will receive and forward to the Owner for the Owner's review and records written warranties and related documents required by the Contract and assembled by the Contractor, and will issue a final Certificate for Payment upon compliance with the requirements of the Contract Documents.

4.2.10 If the Owner and Architect agree, the Architect will provide one or more project representatives to assist in carrying out the Architect's responsibilities at the site. The duties, responsibilities and limitations of authority of such project representatives shall be as set forth in an exhibit to be incorporated in the Contract Documents.

4.2.11 The Architect will interpret and decide matters concerning performance under and requirements of the Contract Documents on written request of either the Owner or Contractor. The Architect's response to such requests will be made with reasonable promptness and within any time limits agreed upon. If no agreement is made concerning the time within which interpretations required of the Architect shall be furnished in compliance with this Paragraph 4.2, then delay shall not be recognized on account of failure by the Architect to furnish such interpretations until 15 days after written request is made for them.

4.2.12 Interpretations and decisions of the Architect will be consistent with the intent of and reasonably inferable from the Contract Documents and will be in writing or in the form of drawings. When making such interpretations and decisions, the Architect will endeavor to secure faithful performance by both Owner and Contractor, will not show partiality to either and will not be liable for results of interpretations or decisions so rendered in good faith.

4.2.13 The Architect's decisions on matters relating to aesthetic effect will be final if consistent with the intent expressed in the Contract Documents.

4.3   CLAIMS AND DISPUTES

4.3.1 Definition. A Claim is a demand or assertion by one of the parties seeking, as a matter of right, adjustment or interpretation of Contract terms, payment of money, extension of time or other relief with respect to the terms of the Contract. The term "Claim" also includes other disputes and matters in question between the Owner and Contractor arising out of or relating to the Contract. Claims must be made by written notice. The responsibility to substantiate Claims shall rest with the party making the Claim.

4.3.2 Decision of Architect. Claims, including those alleging an error or omission by the Architect, shall be referred initially to the Architect for action as provided in Paragraph 4.4. A decision by the Architect, as provided in Subparagraph 4.4.4, shall be required as a condition precedent to arbitration or litigation of a Claim between the Contractor and Owner as to all such matters arising prior to the date final payment is due, regardless of (1) whether such matters relate to execution and progress of the Work or (2) the extent to which the Work has been completed. The decision by the Architect in response to a Claim shall not be a condition precedent to arbitration or litigation in the event (1) the position of Architect is vacant, (2) the Architect has not received evidence or has failed to render a decision within agreed time limits, (3) the Architect has failed to take action required under Subparagraph 4.4.4 within 30 days after the Claim is made, (4) 45 days have passed after the Claim has been referred to the Architect or (5) the Claim relates to a mechanic's lien.

4.3.3 Time Limits on Claims. Claims by either party must be made within 21 days after occurrence of the event giving rise to such Claim or within 21 days after the claimant first recognizes the condition giving rise to the Claim, whichever is later. Claims must be made by written notice. An additional Claim made after the initial Claim has been implemented by Change Order will not be considered unless submitted in a timely manner.

4.3.4 Continuing Contract Performance. Pending final resolution of a Claim including arbitration, unless otherwise agreed in writing the Contractor shall proceed diligently with performance of the Contract and the Owner shall continue to make payments in accordance with the Contract Documents.

4.3.5 Waiver of Claims: Final Payment. The making of final payment shall constitute a waiver of Claims by the Owner except those arising from:

   .1 liens, Claims, security interests or encumbrances arising out of the Contract and unsettled;

   .2 failure of the Work to comply with the requirements of the Contract Documents; or

   .3 terms of special warranties required by the Contract Documents.

4.3.6 Claims for Concealed or Unknown Conditions. If conditions are encountered at the site which are (1) subsurface or otherwise concealed physical conditions which differ materially from those indicated in the Contract Documents or (2) unknown physical conditions of an unusual nature, which differ materially from those ordinarily found to exist and generally recognized as inherent in construction activities of the character provided for in the Contract Documents, then notice by the observing party shall be given to the other party promptly before conditions are disturbed and in no event later than 21 days after first observance of the conditions. The Architect will promptly investigate such conditions and, if they differ materially and cause an increase or decrease in the Contractor's cost of, or time required for, performance of any part of the Work, will recommend an equitable adjustment in the Contract Sum or Contract Time, or both. If the Architect determines that the conditions at the site are not materially different from those indicated in the Contract Documents and that no change in the terms of the Contract is justified, the Architect shall so notify the Owner and Contractor in writing, stating the reasons. Claims by either party in opposition to such determination must be made within 21 days after the Architect has given notice of the decision. If the Owner and Contractor cannot agree on an adjustment in the Contract Sum or Contract Time, the adjustment shall be referred to the Architect for initial determination, subject to further proceedings pursuant to Paragraph 4.4.

4.3.7 Claims for Additional Cost. If the Contractor wishes to make Claim for an increase in the Contract Sum, written notice as provided herein shall be given before proceeding to execute the Work. Prior notice is not required for Claims relating to an emergency endangering life or property arising under Paragraph 10.3. If the Contractor believes additional cost is involved for reasons including but not limited to (1) a written interpretation from the Architect, (2) an order by the Owner to stop the Work where the Contractor was not at fault, (3) a written order for a minor change in the Work issued by the Architect, (4) failure of payment by the Owner, (5) termination of the Contract by the Owner, (6) Owner's suspension or (7) other reasonable grounds, Claim shall be filed in accordance with the procedure established herein.

4.3.8 Claims for Additional Time

4.3.8.1 If the Contractor wishes to make Claim for an increase in the Contract Time, written notice as provided herein shall be given. The Contractor's Claim shall include an estimate of cost and of probable effect of delay on progress of the Work. In the case of a continuing delay only one Claim is necessary.

4.3.8.2 If adverse weather conditions are the basis for a Claim for additional time, such Claim shall be documented by data substantiating that weather conditions were abnormal for the period of time and could not have been reasonably anticipated, and that weather conditions had an adverse effect on the scheduled construction.

4.3.9 Injury or Damage to Person or Property. If either party to the Contract suffers injury or damage to person or property because of an act or omission of the other party, of any of the other party's employees or agents, or of others for whose acts such party is legally liable, written notice of such injury or damage, whether or not insured, shall be given to the other party within a reasonable time not exceeding 21 days after first observance. The notice shall provide sufficient detail to enable the other party to investigate the matter. If a Claim for additional cost or time related to this Claim is to be asserted, it shall be filed as provided in Subparagraphs 4.3.7 or 4.3.8.

4.4   RESOLUTION OF CLAIMS AND DISPUTES

4.4.1 The Architect will review Claims and take one or more of the following preliminary actions within ten days of receipt of a Claim: (1) request additional supporting data from the claimant, (2) submit a schedule to the parties indicating when the Architect expects to take action, (3) reject the Claim in whole or in part, stating reasons for rejection, (4) recommend approval of the Claim by the other party or (5) suggest a compromise. The Architect may also, but is not obligated to, notify the surety, if any, of the nature and amount of the Claim.

4.4.2 If a Claim has been resolved, the Architect will prepare or obtain appropriate documentation.

4.4.3 If a Claim has not been resolved, the party making the Claim shall, within ten days after the Architect's preliminary response, take one or more of the following actions: (1) submit additional supporting data requested by the Architect, (2) modify the initial Claim or (3) notify the Architect that the initial Claim stands.

4.4.4 If a Claim has not been resolved after consideration of the foregoing and of further evidence presented by the parties or requested by the Architect, the Architect will notify the parties in writing that the Architect's decision will be made within seven days, which decision shall be final and binding on the parties but subject to arbitration. Upon expiration of such time period, the Architect will render to the parties the Architect's written decision relative to the Claim, including any change in the Contract Sum or Contract Time or both. If there is a surety and there appears to be a possibility of a Contractor's default, the Architect may, but is not obligated to, notify the surety and request the surety's assistance in resolving the controversy.

4.5   ARBITRATION

4.5.1 Controversies and Claims Subject to Arbitration. Any controversy or Claim arising out of or related to the Contract, or the breach thereof, shall be settled by arbitration in accordance with the Construction Industry Arbitration Rules of the American Arbitration Association, and judgment upon the award rendered by the arbitrator or arbitrators may be entered in any court having jurisdiction thereof, except controversies or Claims relating to aesthetic effect and except those waived as provided for in Subparagraph 4.3.5. Such controversies or Claims upon which the Architect has given notice and rendered a decision as provided in Subparagraph 4.4.4 shall be subject to arbitration upon written demand of either party. Arbitration may be commenced when 45 days have passed after a Claim has been referred to the Architect as provided in Paragraph 4.3 and no decision has been rendered.

4.5.2 Rules and Notices for Arbitration. Claims between the Owner and Contractor not resolved under Paragraph 4.4 shall, if subject to arbitration under Subparagraph 4.5.1, be decided by arbitration in accordance with the Construction Industry Arbitration Rules of the American Arbitration Association currently in effect, unless the parties mutually agree otherwise. Notice of demand for arbitration shall be filed in writing with the other party to the Agreement between the Owner and Contractor and with the American Arbitration Association, and a copy shall be filed with the Architect.

4.5.3 Contract Performance During Arbitration. During arbitration proceedings, the Owner and Contractor shall comply with Subparagraph 4.3.4.

4.5.4 When Arbitration May Be Demanded. Demand for arbitration of any Claim may not be made until the earlier of (1) the date on which the Architect has rendered a final written decision on the Claim, (2) the tenth day after the parties have presented evidence to the Architect or have been given reasonable opportunity to do so, if the Architect has not rendered a final written decision by that date, or (3) any of the five events described in Subparagraph 4.3.2.

4.5.4.1 When a written decision of the Architect states that (1) the decision is final but subject to arbitration and (2) a demand for arbitration of a Claim covered by such decision must be made within 30 days after the date on which the party making the demand receives the final written decision, then failure to demand arbitration within said 30 days' period shall result in the Architect's decision becoming final and binding upon the Owner and Contractor. If the Architect renders a decision after arbitration proceedings have been initiated, such decision may be entered as evidence, but shall not supersede arbitration proceedings unless the decision is acceptable to all parties concerned.

4.5.4.2 A demand for arbitration shall be made within the time limits specified in Subparagraphs 4.5.1 and 4.5.4 and Clause 4.5.4.1 as applicable, and in other cases within a reasonable time after the Claim has arisen, and in no event shall it be made after the date when institution of legal or equitable proceedings based on such Claim would be barred by the applicable statute of limitations as determined pursuant to Paragraph 13.7.

4.5.5 Limitation on Consolidation or Joinder. No arbitration arising out of or relating to the Contract Documents shall include, by consolidation or joinder or in any other manner, the Architect, the Architect's employees or consultants, except by written consent containing specific reference to the Agreement and signed by the Architect, Owner, Contractor and any other person or entity sought to be joined. No arbitration shall include, by consolidation or joinder or in any other manner, parties other than the Owner, Contractor, a separate contractor as described in Article 6 and other persons substantially involved in a common question of fact or law whose presence is required if complete relief is to be accorded in arbitration. No person or entity other than the Owner, Contractor or a separate contractor as described in Article 6 shall be included as an original third party or additional third party to an arbitration whose interest or responsibility is insubstantial. Consent to arbitration involving an additional person or entity shall not constitute consent to arbitration of a dispute not described therein or with a person or entity not named or described therein. The foregoing agreement to arbitrate and other agreements to arbitrate with an additional person or entity duly consented to by parties to the Agreement shall be specifically enforceable under applicable law in any court having jurisdiction thereof.

4.5.6 Claims and Timely Assertion of Claims. A party who files a notice of demand for arbitration must assert in the demand all Claims then known to that party on which arbitration is permitted to be demanded. When a party fails to include a Claim through oversight, inadvertence or excusable neglect, or when a Claim has matured or been acquired subsequently, the arbitrator or arbitrators may permit amendment.

4.5.7 Judgment on Final Award. The award rendered by the arbitrator or arbitrators shall be final, and judgment may be entered upon it in accordance with applicable law in any court having jurisdiction thereof.

ARTICLE 5

SUBCONTRACTORS

5.1   DEFINITIONS

5.1.1 A Subcontractor is a person or entity who has a direct contract with the Contractor to perform a portion of the Work at the site. The term "Subcontractor" is referred to throughout the Contract Documents as if singular in number and means a subcontractor or an authorized representative of the Subcontractor. The term "Subcontractor" does not include a separate contractor or subcontractors of a separate contractor.

5.1.2 A Sub-subcontractor is a person or entity who has a direct or indirect contract with a Subcontractor to perform a portion of the Work at the site. The term "Sub-subcontractor" is referred to throughout the Contract Documents as if singular in number and means a Sub-subcontractor or an authorized representative of the Sub-subcontractor.

5.2   AWARD OF SUBCONTRACTS AND OTHER CONTRACTS FOR PORTIONS OF THE WORK

5.2.1 Unless otherwise stated in the Contract Documents or the bidding requirements, the Contractor, as soon as practicable after award of the Contract, shall furnish in writing to the Owner through the Architect the names of persons or entities (including those who are to furnish materials or equipment fabricated to a special design) proposed for each principal portion of the Work. The Architect will promptly reply to the Contractor in writing stating whether or not the Owner or the Architect, after due investigation, has reasonable objection to any such proposed person or entity. Failure of the Owner or Architect to reply promptly shall constitute notice of no reasonable objection.

5.2.2 The Contractor shall not contract with a proposed person or entity to whom the Owner or Architect has made reasonable and timely objection. The Contractor shall not be required to contract with anyone to whom the Contractor has made reasonable objection.

5.2.3 If the Owner or Architect has reasonable objection to a person or entity proposed by the Contractor, the Contractor shall propose another to whom the Owner or Architect has no reasonable objection. The Contract Sum shall be increased or decreased by the difference in cost occasioned by such change and an appropriate Change Order shall be issued. However, no increase in the Contract Sum shall be allowed for such change unless the Contractor has acted promptly and responsively in submitting names as required.

5.2.4 The Contractor shall not change a Subcontractor, person or entity previously selected if the Owner or Architect makes reasonable objection to such change.

5.3 SUBCONTRACTUAL RELATIONS

5.3.1 By appropriate agreement, written where legally required for validity, the Contractor shall require each Subcontractor, to the extent of the Work to be performed by the Subcontractor, to be bound to the Contractor by terms of the Contract Documents, and to assume toward the Contractor all the obligations and responsibilities which the Contractor, by these Documents, assumes toward the Owner and Architect. Each subcontract agreement shall preserve and protect the rights of the Owner and Architect under the Contract Documents with respect to the Work to be performed by the Subcontractor so that subcontracting thereof will not prejudice such rights, and shall allow to the Subcontractor, unless specifically provided otherwise in the subcontract agreement, the benefit of all rights, remedies and redress against the Contractor that the Contractor, by the Contract Documents, has against the Owner. Where appropriate, the Contractor shall require each Subcontractor to enter into similar agreements with Sub-subcontractors. The Contractor shall make available to each proposed Subcontractor, prior to the execution of the subcontract agreement, copies of the Contract Documents to which the Subcontractor will be bound, and, upon written request of the Subcontractor, identify to the Subcontractor terms and conditions of the proposed subcontract agreement which may be at variance with the Contract Documents. Subcontractors shall similarly make copies of applicable portions of such documents available to their respective proposed Sub-subcontractors.

5.4 CONTINGENT ASSIGNMENT OF SUBCONTRACTS

5.4.1 Each subcontract agreement for a portion of the Work is assigned by the Contractor to the Owner provided that:

.1 assignment is effective only after termination of the Contract by the Owner for cause pursuant to Paragraph 14.2 and only for those subcontract agreements which the Owner accepts by notifying the Subcontractor in writing; and

.2 assignment is subject to the prior rights of the surety, if any, obligated under bond relating to the Contract.

5.4.2 If the Work has been suspended for more than 30 days, the Subcontractor's compensation shall be equitably adjusted.

ARTICLE 6

CONSTRUCTION BY OWNER OR BY SEPARATE CONTRACTORS

6.1 OWNER'S RIGHT TO PERFORM CONSTRUCTION AND TO AWARD SEPARATE CONTRACTS

6.1.1 The Owner reserves the right to perform construction or operations related to the Project with the Owner's own forces, and to award separate contracts in connection with other portions of the Project or other construction or operations on the site under Conditions of the Contract identical or substantially similar to these including those portions related to insurance and waiver of subrogation. If the Contractor claims that delay or additional cost is involved because of such action by the Owner, the Contractor shall make such Claim as provided elsewhere in the Contract Documents.

6.1.2 When separate contracts are awarded for different portions of the Project or other construction or operations on the site, the term "Contractor" in the Contract Documents in each case shall mean the Contractor who executes each separate Owner-Contractor Agreement.

6.1.3 The Owner shall provide for coordination of the activities of the Owner's own forces and of each separate contractor with the Work of the Contractor, who shall cooperate with them. The Contractor shall participate with other separate contractors and the Owner in reviewing their construction schedules when directed to do so. The Contractor shall make any revisions to the construction schedule and Contract Sum deemed necessary after a joint review and mutual agreement. The construction schedules shall then constitute the schedules to be used by the Contractor, separate contractors and the Owner until subsequently revised.

6.1.4 Unless otherwise provided in the Contract Documents, when the Owner performs construction or operations related to the Project with the Owner's own forces, the Owner shall be deemed to be subject to the same obligations and to have the same rights which apply to the Contractor under the Conditions of the Contract, including, without excluding others, those stated in Article 3, this Article 6 and Articles 10, 11 and 12.

6.2 MUTUAL RESPONSIBILITY

6.2.1 The Contractor shall afford the Owner and separate contractors reasonable opportunity for introduction and storage of their materials and equipment and performance of their activities and shall connect and coordinate the Contractor's construction and operations with theirs as required by the Contract Documents.

6.2.2 If part of the Contractor's Work depends for proper execution or results upon construction or operations by the Owner or a separate contractor, the Contractor shall, prior to proceeding with that portion of the Work, promptly report to the Architect apparent discrepancies or defects in such other construction that would render it unsuitable for such proper execution and results. Failure of the Contractor so to report shall constitute an acknowledgment that the Owner's or separate contractors' completed or partially completed construction is fit and proper to receive the Contractor's Work, except as to defects not then reasonably discoverable.

6.2.3 Costs caused by delays or by improperly timed activities or defective construction shall be borne by the party responsible therefor.

6.2.4 The Contractor shall promptly remedy damage wrongfully caused by the Contractor to completed or partially completed construction or to property of the Owner or separate contractors as provided in Subparagraph 10.2.5.

6.2.5 Claims and other disputes and matters in question between the Contractor and a separate contractor shall be subject to the provisions of Paragraph 4.3 provided the separate contractor has reciprocal obligations.

6.2.6 The Owner and each separate contractor shall have the same responsibilities for cutting and patching as are described for the Contractor in Paragraph 3.14.

6.3 OWNER'S RIGHT TO CLEAN UP

6.3.1 If a dispute arises among the Contractor, separate contractors and the Owner as to the responsibility under their respective contracts for maintaining the premises and surrounding area free from waste materials and rubbish as described in Paragraph 3.15, the Owner may clean up and allocate the cost among those responsible as the Architect determines to be just.

AIA DOCUMENT A201 • GENERAL CONDITIONS OF THE CONTRACT FOR CONSTRUCTION • FOURTEENTH EDITION
AIA® • ©1987 THE AMERICAN INSTITUTE OF ARCHITECTS, 1735 NEW YORK AVENUE, N.W., WASHINGTON, D.C. 20006

ARTICLE 7

CHANGES IN THE WORK

7.1   CHANGES

7.1.1 Changes in the Work may be accomplished after execution of the Contract, and without invalidating the Contract, by Change Order, Construction Change Directive or order for a minor change in the Work, subject to the limitations stated in this Article 7 and elsewhere in the Contract Documents.

7.1.2 A Change Order shall be based upon agreement among the Owner, Contractor and Architect; a Construction Change Directive requires agreement by the Owner and Architect and may or may not be agreed to by the Contractor; an order for a minor change in the Work may be issued by the Architect alone.

7.1.3 Changes in the Work shall be performed under applicable provisions of the Contract Documents, and the Contractor shall proceed promptly, unless otherwise provided in the Change Order, Construction Change Directive or order for a minor change in the Work.

7.1.4 If unit prices are stated in the Contract Documents or subsequently agreed upon, and if quantities originally contemplated are so changed in a proposed Change Order or Construction Change Directive that application of such unit prices to quantities of Work proposed will cause substantial inequity to the Owner or Contractor, the applicable unit prices shall be equitably adjusted.

7.2   CHANGE ORDERS

7.2.1 A Change Order is a written instrument prepared by the Architect and signed by the Owner, Contractor and Architect, stating their agreement upon all of the following:

.1 a change in the Work;

.2 the amount of the adjustment in the Contract Sum, if any; and

.3 the extent of the adjustment in the Contract Time, if any.

7.2.2 Methods used in determining adjustments to the Contract Sum may include those listed in Subparagraph 7.3.3.

7.3   CONSTRUCTION CHANGE DIRECTIVES

7.3.1 A Construction Change Directive is a written order prepared by the Architect and signed by the Owner and Architect, directing a change in the Work and stating a proposed basis for adjustment, if any, in the Contract Sum or Contract Time, or both. The Owner may by Construction Change Directive, without invalidating the Contract, order changes in the Work within the general scope of the Contract consisting of additions, deletions or other revisions, the Contract Sum and Contract Time being adjusted accordingly.

7.3.2 A Construction Change Directive shall be used in the absence of total agreement on the terms of a Change Order.

7.3.3 If the Construction Change Directive provides for an adjustment to the Contract Sum, the adjustment shall be based on one of the following methods:

.1 mutual acceptance of a lump sum properly itemized and supported by sufficient substantiating data to permit evaluation;

.2 unit prices stated in the Contract Documents or subsequently agreed upon;

.3 cost to be determined in a manner agreed upon by the parties and a mutually acceptable fixed or percentage fee; or

.4 as provided in Subparagraph 7.3.6.

7.3.4 Upon receipt of a Construction Change Directive, the Contractor shall promptly proceed with the change in the Work involved and advise the Architect of the Contractor's agreement or disagreement with the method, if any, provided in the Construction Change Directive for determining the proposed adjustment in the Contract Sum or Contract Time.

7.3.5 A Construction Change Directive signed by the Contractor indicates the agreement of the Contractor therewith, including adjustment in Contract Sum and Contract Time or the method for determining them. Such agreement shall be effective immediately and shall be recorded as a Change Order.

7.3.6 If the Contractor does not respond promptly or disagrees with the method for adjustment in the Contract Sum, the method and the adjustment shall be determined by the Architect on the basis of reasonable expenditures and savings of those performing the Work attributable to the change, including, in case of an increase in the Contract Sum, a reasonable allowance for overhead and profit. In such case, and also under Clause 7.3.3.3, the Contractor shall keep and present, in such form as the Architect may prescribe, an itemized accounting together with appropriate supporting data. Unless otherwise provided in the Contract Documents, costs for the purposes of this Subparagraph 7.3.6 shall be limited to the following:

.1 costs of labor, including social security, old age and unemployment insurance, fringe benefits required by agreement or custom, and workers' or workmen's compensation insurance;

.2 costs of materials, supplies and equipment, including cost of transportation, whether incorporated or consumed;

.3 rental costs of machinery and equipment, exclusive of hand tools, whether rented from the Contractor or others;

.4 costs of premiums for all bonds and insurance, permit fees, and sales, use or similar taxes related to the Work; and

.5 additional costs of supervision and field office personnel directly attributable to the change.

7.3.7 Pending final determination of cost to the Owner, amounts not in dispute may be included in Applications for Payment. The amount of credit to be allowed by the Contractor to the Owner for a deletion or change which results in a net decrease in the Contract Sum shall be actual net cost as confirmed by the Architect. When both additions and credits covering related Work or substitutions are involved in a change, the allowance for overhead and profit shall be figured on the basis of net increase, if any, with respect to that change.

7.3.8 If the Owner and Contractor do not agree with the adjustment in Contract Time or the method for determining it, the adjustment or the method shall be referred to the Architect for determination.

7.3.9 When the Owner and Contractor agree with the determination made by the Architect concerning the adjustments in the Contract Sum and Contract Time, or otherwise reach agreement upon the adjustments, such agreement shall be effective immediately and shall be recorded by preparation and execution of an appropriate Change Order.

7.4   MINOR CHANGES IN THE WORK

7.4.1 The Architect will have authority to order minor changes in the Work not involving adjustment in the Contract Sum or extension of the Contract Time and not inconsistent with the intent of the Contract Documents. Such changes shall be effected by written order and shall be binding on the Owner and Contractor. The Contractor shall carry out such written orders promptly.

ARTICLE 8

TIME

8.1   DEFINITIONS

8.1.1 Unless otherwise provided, Contract Time is the period of time, including authorized adjustments, allotted in the Contract Documents for Substantial Completion of the Work.

8.1.2 The date of commencement of the Work is the date established in the Agreement. The date shall not be postponed by the failure to act of the Contractor or of persons or entities for whom the Contractor is responsible.

8.1.3 The date of Substantial Completion is the date certified by the Architect in accordance with Paragraph 9.8.

8.1.4 The term "day" as used in the Contract Documents shall mean calendar day unless otherwise specifically defined.

8.2   PROGRESS AND COMPLETION

8.2.1 Time limits stated in the Contract Documents are of the essence of the Contract. By executing the Agreement the Contractor confirms that the Contract Time is a reasonable period for performing the Work.

8.2.2 The Contractor shall not knowingly, except by agreement or instruction of the Owner in writing, prematurely commence operations on the site or elsewhere prior to the effective date of insurance required by Article 11 to be furnished by the Contractor. The date of commencement of the Work shall not be changed by the effective date of such insurance. Unless the date of commencement is established by a notice to proceed given by the Owner, the Contractor shall notify the Owner in writing not less than five days or other agreed period before commencing the Work to permit the timely filing of mortgages, mechanic's liens and other security interests.

8.2.3 The Contractor shall proceed expeditiously with adequate forces and shall achieve Substantial Completion within the Contract Time.

8.3   DELAYS AND EXTENSIONS OF TIME

8.3.1 If the Contractor is delayed at any time in progress of the Work by an act or neglect of the Owner or Architect, or of an employee of either, or of a separate contractor employed by the Owner, or by changes ordered in the Work, or by labor disputes, fire, unusual delay in deliveries, unavoidable casualties or other causes beyond the Contractor's control, or by delay authorized by the Owner pending arbitration, or by other causes which the Architect determines may justify delay, then the Contract Time shall be extended by Change Order for such reasonable time as the Architect may determine.

8.3.2 Claims relating to time shall be made in accordance with applicable provisions of Paragraph 4.3.

8.3.3 This Paragraph 8.3 does not preclude recovery of damages for delay by either party under other provisions of the Contract Documents.

ARTICLE 9

PAYMENTS AND COMPLETION

9.1   CONTRACT SUM

9.1.1 The Contract Sum is stated in the Agreement and, including authorized adjustments, is the total amount payable by the Owner to the Contractor for performance of the Work under the Contract Documents.

9.2   SCHEDULE OF VALUES

9.2.1 Before the first Application for Payment, the Contractor shall submit to the Architect a schedule of values allocated to various portions of the Work, prepared in such form and supported by such data to substantiate its accuracy as the Architect may require. This schedule, unless objected to by the Architect, shall be used as a basis for reviewing the Contractor's Applications for Payment.

9.3   APPLICATIONS FOR PAYMENT

9.3.1 At least ten days before the date established for each progress payment, the Contractor shall submit to the Architect an itemized Application for Payment for operations completed in accordance with the schedule of values. Such application shall be notarized, if required, and supported by such data substantiating the Contractor's right to payment as the Owner or Architect may require, such as copies of requisitions from Subcontractors and material suppliers, and reflecting retainage if provided for elsewhere in the Contract Documents.

9.3.1.1 Such applications may include requests for payment on account of changes in the Work which have been properly authorized by Construction Change Directives but not yet included in Change Orders.

9.3.1.2 Such applications may not include requests for payment of amounts the Contractor does not intend to pay to a Subcontractor or material supplier because of a dispute or other reason.

9.3.2 Unless otherwise provided in the Contract Documents, payments shall be made on account of materials and equipment delivered and suitably stored at the site for subsequent incorporation in the Work. If approved in advance by the Owner, payment may similarly be made for materials and equipment suitably stored off the site at a location agreed upon in writing. Payment for materials and equipment stored on or off the site shall be conditioned upon compliance by the Contractor with procedures satisfactory to the Owner to establish the Owner's title to such materials and equipment or otherwise protect the Owner's interest, and shall include applicable insurance, storage and transportation to the site for such materials and equipment stored off the site.

9.3.3 The Contractor warrants that title to all Work covered by an Application for Payment will pass to the Owner no later than the time of payment. The Contractor further warrants that upon submittal of an Application for Payment all Work for which Certificates for Payment have been previously issued and payments received from the Owner shall, to the best of the Contractor's knowledge, information and belief, be free and clear of liens, claims, security interests or encumbrances in favor of the Contractor, Subcontractors, material suppliers, or other persons or entities making a claim by reason of having provided labor, materials and equipment relating to the Work.

9.4   CERTIFICATES FOR PAYMENT

9.4.1 The Architect will, within seven days after receipt of the Contractor's Application for Payment, either issue to the

Owner a Certificate for Payment, with a copy to the Contractor, for such amount as the Architect determines is properly due, or notify the Contractor and Owner in writing of the Architect's reasons for withholding certification in whole or in part as provided in Subparagraph 9.5.1.

9.4.2 The issuance of a Certificate for Payment will constitute a representation by the Architect to the Owner, based on the Architect's observations at the site and the data comprising the Application for Payment, that the Work has progressed to the point indicated and that, to the best of the Architect's knowledge, information and belief, quality of the Work is in accordance with the Contract Documents. The foregoing representations are subject to an evaluation of the Work for conformance with the Contract Documents upon Substantial Completion, to results of subsequent tests and inspections, to minor deviations from the Contract Documents correctable prior to completion and to specific qualifications expressed by the Architect. The issuance of a Certificate for Payment will further constitute a representation that the Contractor is entitled to payment in the amount certified. However, the issuance of a Certificate for Payment will not be a representation that the Architect has (1) made exhaustive or continuous on-site inspections to check the quality or quantity of the Work, (2) reviewed construction means, methods, techniques, sequences or procedures, (3) reviewed copies of requisitions received from Subcontractors and material suppliers and other data requested by the Owner to substantiate the Contractor's right to payment or (4) made examination to ascertain how or for what purpose the Contractor has used money previously paid on account of the Contract Sum.

9.5   DECISIONS TO WITHHOLD CERTIFICATION

9.5.1 The Architect may decide not to certify payment and may withhold a Certificate for Payment in whole or in part, to the extent reasonably necessary to protect the Owner, if in the Architect's opinion the representations to the Owner required by Subparagraph 9.4.2 cannot be made. If the Architect is unable to certify payment in the amount of the Application, the Architect will notify the Contractor and Owner as provided in Subparagraph 9.4.1. If the Contractor and Architect cannot agree on a revised amount, the Architect will promptly issue a Certificate for Payment for the amount for which the Architect is able to make such representations to the Owner. The Architect may also decide not to certify payment or, because of subsequently discovered evidence or subsequent observations, may nullify the whole or a part of a Certificate for Payment previously issued, to such extent as may be necessary in the Architect's opinion to protect the Owner from loss because of:

.1   defective Work not remedied;

.2   third party claims filed or reasonable evidence indicating probable filing of such claims;

.3   failure of the Contractor to make payments properly to Subcontractors or for labor, materials or equipment;

.4   reasonable evidence that the Work cannot be completed for the unpaid balance of the Contract Sum;

.5   damage to the Owner or another contractor;

.6   reasonable evidence that the Work will not be completed within the Contract Time, and that the unpaid balance would not be adequate to cover actual or liquidated damages for the anticipated delay; or

.7   persistent failure to carry out the Work in accordance with the Contract Documents.

9.5.2 When the above reasons for withholding certification are removed, certification will be made for amounts previously withheld.

9.6   PROGRESS PAYMENTS

9.6.1 After the Architect has issued a Certificate for Payment, the Owner shall make payment in the manner and within the time provided in the Contract Documents, and shall so notify the Architect.

9.6.2 The Contractor shall promptly pay each Subcontractor, upon receipt of payment from the Owner, out of the amount paid to the Contractor on account of such Subcontractor's portion of the Work, the amount to which said Subcontractor is entitled, reflecting percentages actually retained from payments to the Contractor on account of such Subcontractor's portion of the Work. The Contractor shall, by appropriate agreement with each Subcontractor, require each Subcontractor to make payments to Sub-subcontractors in similar manner.

9.6.3 The Architect will, on request, furnish to a Subcontractor, if practicable, information regarding percentages of completion or amounts applied for by the Contractor and action taken thereon by the Architect and Owner on account of portions of the Work done by such Subcontractor.

9.6.4 Neither the Owner nor Architect shall have an obligation to pay or to see to the payment of money to a Subcontractor except as may otherwise be required by law.

9.6.5 Payment to material suppliers shall be treated in a manner similar to that provided in Subparagraphs 9.6.2, 9.6.3 and 9.6.4.

9.6.6 A Certificate for Payment, a progress payment, or partial or entire use or occupancy of the Project by the Owner shall not constitute acceptance of Work not in accordance with the Contract Documents.

9.7   FAILURE OF PAYMENT

9.7.1 If the Architect does not issue a Certificate for Payment, through no fault of the Contractor, within seven days after receipt of the Contractor's Application for Payment, or if the Owner does not pay the Contractor within seven days after the date established in the Contract Documents the amount certified by the Architect or awarded by arbitration, then the Contractor may, upon seven additional days' written notice to the Owner and Architect, stop the Work until payment of the amount owing has been received. The Contract Time shall be extended appropriately and the Contract Sum shall be increased by the amount of the Contractor's reasonable costs of shut-down, delay and start-up, which shall be accomplished as provided in Article 7.

9.8   SUBSTANTIAL COMPLETION

9.8.1 Substantial Completion is the stage in the progress of the Work when the Work or designated portion thereof is sufficiently complete in accordance with the Contract Documents so the Owner can occupy or utilize the Work for its intended use.

9.8.2 When the Contractor considers that the Work, or a portion thereof which the Owner agrees to accept separately, is substantially complete, the Contractor shall prepare and submit to the Architect a comprehensive list of items to be completed or corrected. The Contractor shall proceed promptly to complete and correct items on the list. Failure to include an item on such list does not alter the responsibility of the Contractor to complete all Work in accordance with the Contract Documents. Upon receipt of the Contractor's list, the Architect will make an inspection to determine whether the Work or designated portion thereof is substantially complete. If the Architect's inspection discloses any item, whether or not included on the Contractor's list, which is not in accordance with the requirements of the Contract Documents, the Contractor shall, before issuance of the Certificate of Substantial Completion, complete or correct such item upon notification by the Architect. The Contractor shall then submit a request for another inspection by the Architect to determine Substantial Completion. When the Work or designated portion thereof is substantially complete, the Architect will prepare a Certificate of Substantial Completion which shall establish the date of Substantial Completion, shall establish responsibilities of the Owner and Contractor for security, maintenance, heat, utilities, damage to the Work and insurance, and shall fix the time within which the Contractor shall finish all items on the list accompanying the Certificate. Warranties required by the Contract Documents shall commence on the date of Substantial Completion of the Work or designated portion thereof unless otherwise provided in the Certificate of Substantial Completion. The Certificate of Substantial Completion shall be submitted to the Owner and Contractor for their written acceptance of responsibilities assigned to them in such Certificate.

9.8.3 Upon Substantial Completion of the Work or designated portion thereof and upon application by the Contractor and certification by the Architect, the Owner shall make payment, reflecting adjustment in retainage, if any, for such Work or portion thereof as provided in the Contract Documents.

9.9   PARTIAL OCCUPANCY OR USE

9.9.1 The Owner may occupy or use any completed or partially completed portion of the Work at any stage when such portion is designated by separate agreement with the Contractor, provided such occupancy or use is consented to by the insurer as required under Subparagraph 11.3.11 and authorized by public authorities having jurisdiction over the Work. Such partial occupancy or use may commence whether or not the portion is substantially complete, provided the Owner and Contractor have accepted in writing the responsibilities assigned to each of them for payments, retainage if any, security, maintenance, heat, utilities, damage to the Work and insurance, and have agreed in writing concerning the period for correction of the Work and commencement of warranties required by the Contract Documents. When the Contractor considers a portion substantially complete, the Contractor shall prepare and submit a list to the Architect as provided under Subparagraph 9.8.2. Consent of the Contractor to partial occupancy or use shall not be unreasonably withheld. The stage of the progress of the Work shall be determined by written agreement between the Owner and Contractor or, if no agreement is reached, by decision of the Architect.

9.9.2 Immediately prior to such partial occupancy or use, the Owner, Contractor and Architect shall jointly inspect the area to be occupied or portion of the Work to be used in order to determine and record the condition of the Work.

9.9.3 Unless otherwise agreed upon, partial occupancy or use of a portion or portions of the Work shall not constitute acceptance of Work not complying with the requirements of the Contract Documents.

9.10   FINAL COMPLETION AND FINAL PAYMENT

9.10.1 Upon receipt of written notice that the Work is ready for final inspection and acceptance and upon receipt of a final Application for Payment, the Architect will promptly make

such inspection and, when the Architect finds the Work acceptable under the Contract Documents and the Contract fully performed, the Architect will promptly issue a final Certificate for Payment stating that to the best of the Architect's knowledge, information and belief, and on the basis of the Architect's observations and inspections, the Work has been completed in accordance with terms and conditions of the Contract Documents and that the entire balance found to be due the Contractor and noted in said final Certificate is due and payable. The Architect's final Certificate for Payment will constitute a further representation that conditions listed in Subparagraph 9.10.2 as precedent to the Contractor's being entitled to final payment have been fulfilled.

9.10.2 Neither final payment nor any remaining retained percentage shall become due until the Contractor submits to the Architect (1) an affidavit that payrolls, bills for materials and equipment, and other indebtedness connected with the Work for which the Owner or the Owner's property might be responsible or encumbered (less amounts withheld by Owner) have been paid or otherwise satisfied, (2) a certificate evidencing that insurance required by the Contract Documents to remain in force after final payment is currently in effect and will not be cancelled or allowed to expire until at least 30 days' prior written notice has been given to the Owner, (3) a written statement that the Contractor knows of no substantial reason that the insurance will not be renewable to cover the period required by the Contract Documents, (4) consent of surety, if any, to final payment and (5), if required by the Owner, other data establishing payment or satisfaction of obligations, such as receipts, releases and waivers of liens, claims, security interests or encumbrances arising out of the Contract, to the extent and in such form as may be designated by the Owner. If a Subcontractor refuses to furnish a release or waiver required by the Owner, the Contractor may furnish a bond satisfactory to the Owner to indemnify the Owner against such lien. If such lien remains unsatisfied after payments are made, the Contractor shall refund to the Owner all money that the Owner may be compelled to pay in discharging such lien, including all costs and reasonable attorneys' fees.

9.10.3 If, after Substantial Completion of the Work, final completion thereof is materially delayed through no fault of the Contractor or by issuance of Change Orders affecting final completion, and the Architect so confirms, the Owner shall, upon application by the Contractor and certification by the Architect, and without terminating the Contract, make payment of the balance due for that portion of the Work fully completed and accepted. If the remaining balance for Work not fully completed or corrected is less than retainage stipulated in the Contract Documents, and if bonds have been furnished, the written consent of surety to payment of the balance due for that portion of the Work fully completed and accepted shall be submitted by the Contractor to the Architect prior to certification of such payment. Such payment shall be made under terms and conditions governing final payment, except that it shall not constitute a waiver of claims. The making of final payment shall constitute a waiver of claims by the Owner as provided in Subparagraph 4.3.5.

9.10.4 Acceptance of final payment by the Contractor, a Subcontractor or material supplier shall constitute a waiver of claims by that payee except those previously made in writing and identified by that payee as unsettled at the time of final Application for Payment. Such waivers shall be in addition to the waiver described in Subparagraph 4.3.5.

ARTICLE 10

PROTECTION OF PERSONS AND PROPERTY

10.1 SAFETY PRECAUTIONS AND PROGRAMS

10.1.1 The Contractor shall be responsible for initiating, maintaining and supervising all safety precautions and programs in connection with the performance of the Contract.

10.1.2 In the event the Contractor encounters on the site material reasonably believed to be asbestos or polychlorinated biphenyl (PCB) which has not been rendered harmless, the Contractor shall immediately stop Work in the area affected and report the condition to the Owner and Architect in writing. The Work in the affected area shall not thereafter be resumed except by written agreement of the Owner and Contractor if in fact the material is asbestos or polychlorinated biphenyl (PCB) and has not been rendered harmless. The Work in the affected area shall be resumed in the absence of asbestos or polychlorinated biphenyl (PCB), or when it has been rendered harmless, by written agreement of the Owner and Contractor, or in accordance with final determination by the Architect on which arbitration has not been demanded, or by arbitration under Article 4.

10.1.3 The Contractor shall not be required pursuant to Article 7 to perform without consent any Work relating to asbestos or polychlorinated biphenyl (PCB).

10.1.4 To the fullest extent permitted by law, the Owner shall indemnify and hold harmless the Contractor, Architect, Architect's consultants and agents and employees of any of them from and against claims, damages, losses and expenses, including but not limited to attorneys' fees, arising out of or resulting from performance of the Work in the affected area if in fact the material is asbestos or polychlorinated biphenyl (PCB) and has not been rendered harmless, provided that such claim, damage, loss or expense is attributable to bodily injury, sickness, disease or death, or to injury to or destruction of tangible property (other than the Work itself) including loss of use resulting therefrom, but only to the extent caused in whole or in part by negligent acts or omissions of the Owner, anyone directly or indirectly employed by the Owner or anyone for whose acts the Owner may be liable, regardless of whether or not such claim, damage, loss or expense is caused in part by a party indemnified hereunder. Such obligation shall not be construed to negate, abridge, or reduce other rights or obligations of indemnity which would otherwise exist as to a party or person described in this Subparagraph 10.1.4.

10.2 SAFETY OF PERSONS AND PROPERTY

10.2.1 The Contractor shall take reasonable precautions for safety of, and shall provide reasonable protection to prevent damage, injury or loss to:

  .1 employees on the Work and other persons who may be affected thereby;

  .2 the Work and materials and equipment to be incorporated therein, whether in storage on or off the site, under care, custody or control of the Contractor or the Contractor's Subcontractors or Sub-subcontractors; and

  .3 other property at the site or adjacent thereto, such as trees, shrubs, lawns, walks, pavements, roadways, structures and utilities not designated for removal, relocation or replacement in the course of construction.

10.2.2 The Contractor shall give notices and comply with applicable laws, ordinances, rules, regulations and lawful orders of public authorities bearing on safety of persons or property or their protection from damage, injury or loss.

10.2.3 The Contractor shall erect and maintain, as required by existing conditions and performance of the Contract, reasonable safeguards for safety and protection, including posting danger signs and other warnings against hazards, promulgating safety regulations and notifying owners and users of adjacent sites and utilities.

10.2.4 When use or storage of explosives or other hazardous materials or equipment or unusual methods are necessary for execution of the Work, the Contractor shall exercise utmost care and carry on such activities under supervision of properly qualified personnel.

10.2.5 The Contractor shall promptly remedy damage and loss (other than damage or loss insured under property insurance required by the Contract Documents) to property referred to in Clauses 10.2.1.2 and 10.2.1.3 caused in whole or in part by the Contractor, a Subcontractor, a Sub-subcontractor, or anyone directly or indirectly employed by any of them, or by anyone for whose acts they may be liable and for which the Contractor is responsible under Clauses 10.2.1.2 and 10.2.1.3, except damage or loss attributable to acts or omissions of the Owner or Architect or anyone directly or indirectly employed by either of them, or by anyone for whose acts either of them may be liable, and not attributable to the fault or negligence of the Contractor. The foregoing obligations of the Contractor are in addition to the Contractor's obligations under Paragraph 3.18.

10.2.6 The Contractor shall designate a responsible member of the Contractor's organization at the site whose duty shall be the prevention of accidents. This person shall be the Contractor's superintendent unless otherwise designated by the Contractor in writing to the Owner and Architect.

10.2.7 The Contractor shall not load or permit any part of the construction or site to be loaded so as to endanger its safety.

10.3 EMERGENCIES

10.3.1 In an emergency affecting safety of persons or property, the Contractor shall act, at the Contractor's discretion, to prevent threatened damage, injury or loss. Additional compensation or extension of time claimed by the Contractor on account of an emergency shall be determined as provided in Paragraph 4.3 and Article 7.

ARTICLE 11

INSURANCE AND BONDS

11.1 CONTRACTOR'S LIABILITY INSURANCE

11.1.1 The Contractor shall purchase from and maintain in a company or companies lawfully authorized to do business in the jurisdiction in which the Project is located such insurance as will protect the Contractor from claims set forth below which may arise out of or result from the Contractor's operations under the Contract and for which the Contractor may be legally liable, whether such operations be by the Contractor or by a Subcontractor or by anyone directly or indirectly employed by any of them, or by anyone for whose acts any of them may be liable:

  .1 claims under workers' or workmen's compensation, disability benefit and other similar employee benefit acts which are applicable to the Work to be performed;

.2 claims for damages because of bodily injury, occupational sickness or disease, or death of the Contractor's employees;

.3 claims for damages because of bodily injury, sickness or disease, or death of any person other than the Contractor's employees;

.4 claims for damages insured by usual personal injury liability coverage which are sustained (1) by a person as a result of an offense directly or indirectly related to employment of such person by the Contractor, or (2) by another person;

.5 claims for damages, other than to the Work itself, because of injury to or destruction of tangible property, including loss of use resulting therefrom;

.6 claims for damages because of bodily injury, death of a person or property damage arising out of ownership, maintenance or use of a motor vehicle; and

.7 claims involving contractual liability insurance applicable to the Contractor's obligations under Paragraph 3.18.

11.1.2 The insurance required by Subparagraph 11.1.1 shall be written for not less than limits of liability specified in the Contract Documents or required by law, whichever coverage is greater. Coverages, whether written on an occurrence or claims-made basis, shall be maintained without interruption from date of commencement of the Work until date of final payment and termination of any coverage required to be maintained after final payment.

11.1.3 Certificates of Insurance acceptable to the Owner shall be filed with the Owner prior to commencement of the Work. These Certificates and the insurance policies required by this Paragraph 11.1 shall contain a provision that coverages afforded under the policies will not be cancelled or allowed to expire until at least 30 days' prior written notice has been given to the Owner. If any of the foregoing insurance coverages are required to remain in force after final payment and are reasonably available, an additional certificate evidencing continuation of such coverage shall be submitted with the final Application for Payment as required by Subparagraph 9.10.2. Information concerning reduction of coverage shall be furnished by the Contractor with reasonable promptness in accordance with the Contractor's information and belief.

11.2 OWNER'S LIABILITY INSURANCE

11.2.1 The Owner shall be responsible for purchasing and maintaining the Owner's usual liability insurance. Optionally, the Owner may purchase and maintain other insurance for self-protection against claims which may arise from operations under the Contract. The Contractor shall not be responsible for purchasing and maintaining this optional Owner's liability insurance unless specifically required by the Contract Documents.

11.3 PROPERTY INSURANCE

11.3.1 Unless otherwise provided, the Owner shall purchase and maintain, in a company or companies lawfully authorized to do business in the jurisdiction in which the Project is located, property insurance in the amount of the initial Contract Sum as well as subsequent modifications thereto for the entire Work at the site on a replacement cost basis without voluntary deductibles. Such property insurance shall be maintained, unless otherwise provided in the Contract Documents or otherwise agreed in writing by all persons and entities who are beneficiaries of such insurance, until final payment has been made as provided in Paragraph 9.10 or until no person or entity

other than the Owner has an insurable interest in the property required by this Paragraph 11.3 to be covered, whichever is earlier. This insurance shall include interests of the Owner, the Contractor, Subcontractors and Sub-subcontractors in the Work.

11.3.1.1 Property insurance shall be on an all-risk policy form and shall insure against the perils of fire and extended coverage and physical loss or damage including, without duplication of coverage, theft, vandalism, malicious mischief, collapse, falsework, temporary buildings and debris removal including demolition occasioned by enforcement of any applicable legal requirements, and shall cover reasonable compensation for Architect's services and expenses required as a result of such insured loss. Coverage for other perils shall not be required unless otherwise provided in the Contract Documents.

11.3.1.2 If the Owner does not intend to purchase such property insurance required by the Contract and with all of the coverages in the amount described above, the Owner shall so inform the Contractor in writing prior to commencement of the Work. The Contractor may then effect insurance which will protect the interests of the Contractor, Subcontractors and Sub-subcontractors in the Work, and by appropriate Change Order the cost thereof shall be charged to the Owner. If the Contractor is damaged by the failure or neglect of the Owner to purchase or maintain insurance as described above, without so notifying the Contractor, then the Owner shall bear all reasonable costs properly attributable thereto.

11.3.1.3 If the property insurance requires minimum deductibles and such deductibles are identified in the Contract Documents, the Contractor shall pay costs not covered because of such deductibles. If the Owner or insurer increases the required minimum deductibles above the amounts so identified or if the Owner elects to purchase this insurance with voluntary deductible amounts, the Owner shall be responsible for payment of the additional costs not covered because of such increased or voluntary deductibles. If deductibles are not identified in the Contract Documents, the Owner shall pay costs not covered because of deductibles.

11.3.1.4 Unless otherwise provided in the Contract Documents, this property insurance shall cover portions of the Work stored off the site after written approval of the Owner at the value established in the approval, and also portions of the Work in transit.

11.3.2 Boiler and Machinery Insurance. The Owner shall purchase and maintain boiler and machinery insurance required by the Contract Documents or by law, which shall specifically cover such insured objects during installation and until final acceptance by the Owner; this insurance shall include interests of the Owner, Contractor, Subcontractors and Sub-subcontractors in the Work, and the Owner and Contractor shall be named insureds.

11.3.3 Loss of Use Insurance. The Owner, at the Owner's option, may purchase and maintain such insurance as will insure the Owner against loss of use of the Owner's property due to fire or other hazards, however caused. The Owner waives all rights of action against the Contractor for loss of use of the Owner's property, including consequential losses due to fire or other hazards however caused.

11.3.4 If the Contractor requests in writing that insurance for risks other than those described herein or for other special hazards be included in the property insurance policy, the Owner shall, if possible, include such insurance, and the cost thereof shall be charged to the Contractor by appropriate Change Order.

AIA DOCUMENT A201 • GENERAL CONDITIONS OF THE CONTRACT FOR CONSTRUCTION • FOURTEENTH EDITION
AIA® • ©1987 THE AMERICAN INSTITUTE OF ARCHITECTS, 1735 NEW YORK AVENUE, N.W., WASHINGTON, D.C. 20006

11.3.5 If during the Project construction period the Owner insures properties, real or personal or both, adjoining or adjacent to the site by property insurance under policies separate from those insuring the Project, or if after final payment property insurance is to be provided on the completed Project through a policy or policies other than those insuring the Project during the construction period, the Owner shall waive all rights in accordance with the terms of Subparagraph 11.3.7 for damages caused by fire or other perils covered by this separate property insurance. All separate policies shall provide this waiver of subrogation by endorsement or otherwise.

11.3.6 Before an exposure to loss may occur, the Owner shall file with the Contractor a copy of each policy that includes insurance coverages required by this Paragraph 11.3. Each policy shall contain all generally applicable conditions, definitions, exclusions and endorsements related to this Project. Each policy shall contain a provision that the policy will not be cancelled or allowed to expire until at least 30 days' prior written notice has been given to the Contractor.

11.3.7 Waivers of Subrogation. The Owner and Contractor waive all rights against (1) each other and any of their subcontractors, sub-subcontractors, agents and employees, each of the other, and (2) the Architect, Architect's consultants, separate contractors described in Article 6, if any, and any of their subcontractors, sub-subcontractors, agents and employees, for damages caused by fire or other perils to the extent covered by property insurance obtained pursuant to this Paragraph 11.3 or other property insurance applicable to the Work, except such rights as they have to proceeds of such insurance held by the Owner as fiduciary. The Owner or Contractor, as appropriate, shall require of the Architect, Architect's consultants, separate contractors described in Article 6, if any, and the subcontractors, sub-subcontractors, agents and employees of any of them, by appropriate agreements, written where legally required for validity, similar waivers each in favor of other parties enumerated herein. The policies shall provide such waivers of subrogation by endorsement or otherwise. A waiver of subrogation shall be effective as to a person or entity even though that person or entity would otherwise have a duty of indemnification, contractual or otherwise, did not pay the insurance premium directly or indirectly, and whether or not the person or entity had an insurable interest in the property damaged.

11.3.8 A loss insured under Owner's property insurance shall be adjusted by the Owner as fiduciary and made payable to the Owner as fiduciary for the insureds, as their interests may appear, subject to requirements of any applicable mortgagee clause and of Subparagraph 11.3.10. The Contractor shall pay Subcontractors their just shares of insurance proceeds received by the Contractor, and by appropriate agreements, written where legally required for validity, shall require Subcontractors to make payments to their Sub-subcontractors in similar manner.

11.3.9 If required in writing by a party in interest, the Owner as fiduciary shall, upon occurrence of an insured loss, give bond for proper performance of the Owner's duties. The cost of required bonds shall be charged against proceeds received as fiduciary. The Owner shall deposit in a separate account proceeds so received, which the Owner shall distribute in accordance with such agreement as the parties in interest may reach, or in accordance with an arbitration award in which case the procedure shall be as provided in Paragraph 4.5. If after such loss no other special agreement is made, replacement of damaged property shall be covered by appropriate Change Order.

11.3.10 The Owner as fiduciary shall have power to adjust and settle a loss with insurers unless one of the parties in interest shall object in writing within five days after occurrence of loss to the Owner's exercise of this power; if such objection be made, arbitrators shall be chosen as provided in Paragraph 4.5. The Owner as fiduciary shall, in that case, make settlement with insurers in accordance with directions of such arbitrators. If distribution of insurance proceeds by arbitration is required, the arbitrators will direct such distribution.

11.3.11 Partial occupancy or use in accordance with Paragraph 9.9 shall not commence until the insurance company or companies providing property insurance have consented to such partial occupancy or use by endorsement or otherwise. The Owner and the Contractor shall take reasonable steps to obtain consent of the insurance company or companies and shall, without mutual written consent, take no action with respect to partial occupancy or use that would cause cancellation, lapse or reduction of insurance.

11.4   PERFORMANCE BOND AND PAYMENT BOND

11.4.1 The Owner shall have the right to require the Contractor to furnish bonds covering faithful performance of the Contract and payment of obligations arising thereunder as stipulated in bidding requirements or specifically required in the Contract Documents on the date of execution of the Contract.

11.4.2 Upon the request of any person or entity appearing to be a potential beneficiary of bonds covering payment of obligations arising under the Contract, the Contractor shall promptly furnish a copy of the bonds or shall permit a copy to be made.

ARTICLE 12

UNCOVERING AND CORRECTION OF WORK

12.1   UNCOVERING OF WORK

12.1.1 If a portion of the Work is covered contrary to the Architect's request or to requirements specifically expressed in the Contract Documents, it must, if required in writing by the Architect, be uncovered for the Architect's observation and be replaced at the Contractor's expense without change in the Contract Time.

12.1.2 If a portion of the Work has been covered which the Architect has not specifically requested to observe prior to its being covered, the Architect may request to see such Work and it shall be uncovered by the Contractor. If such Work is in accordance with the Contract Documents, costs of uncovering and replacement shall, by appropriate Change Order, be charged to the Owner. If such Work is not in accordance with the Contract Documents, the Contractor shall pay such costs unless the condition was caused by the Owner or a separate contractor in which event the Owner shall be responsible for payment of such costs.

12.2   CORRECTION OF WORK

12.2.1 The Contractor shall promptly correct Work rejected by the Architect or failing to conform to the requirements of the Contract Documents, whether observed before or after Substantial Completion and whether or not fabricated, installed or completed. The Contractor shall bear costs of correcting such rejected Work, including additional testing and inspections and compensation for the Architect's services and expenses made necessary thereby.

12.2.2 If, within one year after the date of Substantial Completion of the Work or designated portion thereof, or after the date for commencement of warranties established under Sub-paragraph 9.9.1, or by terms of an applicable special warranty required by the Contract Documents, any of the Work is found to be not in accordance with the requirements of the Contract Documents, the Contractor shall correct it promptly after receipt of written notice from the Owner to do so unless the Owner has previously given the Contractor a written acceptance of such condition. This period of one year shall be extended with respect to portions of Work first performed after Substantial Completion by the period of time between Substantial Completion and the actual performance of the Work. This obligation under this Subparagraph 12.2.2 shall survive acceptance of the Work under the Contract and termination of the Contract. The Owner shall give such notice promptly after discovery of the condition.

12.2.3 The Contractor shall remove from the site portions of the Work which are not in accordance with the requirements of the Contract Documents and are neither corrected by the Contractor nor accepted by the Owner.

12.2.4 If the Contractor fails to correct nonconforming Work within a reasonable time, the Owner may correct it in accordance with Paragraph 2.4. If the Contractor does not proceed with correction of such nonconforming Work within a reasonable time fixed by written notice from the Architect, the Owner may remove it and store the salvable materials or equipment at the Contractor's expense. If the Contractor does not pay costs of such removal and storage within ten days after written notice, the Owner may upon ten additional days written notice sell such materials and equipment at auction or at private sale and shall account for the proceeds thereof, after deducting costs and damages that should have been borne by the Contractor, including compensation for the Architect's services and expenses made necessary thereby. If such proceeds of sale do not cover costs which the Contractor should have borne, the Contract Sum shall be reduced by the deficiency. If payments then or thereafter due the Contractor are not sufficient to cover such amount, the Contractor shall pay the difference to the Owner.

12.2.5 The Contractor shall bear the cost of correcting destroyed or damaged construction, whether completed or partially completed, of the Owner or separate contractors caused by the Contractor's correction or removal of Work which is not in accordance with the requirements of the Contract Documents.

12.2.6 Nothing contained in this Paragraph 12.2 shall be construed to establish a period of limitation with respect to other obligations which the Contractor might have under the Contract Documents. Establishment of the time period of one year as described in Subparagraph 12.2.2 relates only to the specific obligation of the Contractor to correct the Work, and has no relationship to the time within which the obligation to comply with the Contract Documents may be sought to be enforced, nor to the time within which proceedings may be commenced to establish the Contractor's liability with respect to the Contractor's obligations other than specifically to correct the Work.

12.3 ACCEPTANCE OF NONCONFORMING WORK

12.3.1 If the Owner prefers to accept Work which is not in accordance with the requirements of the Contract Documents, the Owner may do so instead of requiring its removal and correction, in which case the Contract Sum will be reduced as appropriate and equitable. Such adjustment shall be effected whether or not final payment has been made.

ARTICLE 13

MISCELLANEOUS PROVISIONS

13.1 GOVERNING LAW

13.1.1 The Contract shall be governed by the law of the place where the Project is located.

13.2 SUCCESSORS AND ASSIGNS

13.2.1 The Owner and Contractor respectively bind themselves, their partners, successors, assigns and legal representatives to the other party hereto and to partners, successors, assigns and legal representatives of such other party in respect to covenants, agreements and obligations contained in the Contract Documents. Neither party to the Contract shall assign the Contract as a whole without written consent of the other. If either party attempts to make such an assignment without such consent, that party shall nevertheless remain legally responsible for all obligations under the Contract.

13.3 WRITTEN NOTICE

13.3.1 Written notice shall be deemed to have been duly served if delivered in person to the individual or a member of the firm or entity or to an officer of the corporation for which it was intended, or if delivered at or sent by registered or certified mail to the last business address known to the party giving notice.

13.4 RIGHTS AND REMEDIES

13.4.1 Duties and obligations imposed by the Contract Documents and rights and remedies available thereunder shall be in addition to and not a limitation of duties, obligations, rights and remedies otherwise imposed or available by law.

13.4.2 No action or failure to act by the Owner, Architect or Contractor shall constitute a waiver of a right or duty afforded them under the Contract, nor shall such action or failure to act constitute approval of or acquiescence in a breach thereunder, except as may be specifically agreed in writing.

13.5 TESTS AND INSPECTIONS

13.5.1 Tests, inspections and approvals of portions of the Work required by the Contract Documents or by laws, ordinances, rules, regulations or orders of public authorities having jurisdiction shall be made at an appropriate time. Unless otherwise provided, the Contractor shall make arrangements for such tests, inspections and approvals with an independent testing laboratory or entity acceptable to the Owner, or with the appropriate public authority, and shall bear all related costs of tests, inspections and approvals. The Contractor shall give the Architect timely notice of when and where tests and inspections are to be made so the Architect may observe such procedures. The Owner shall bear costs of tests, inspections or approvals which do not become requirements until after bids are received or negotiations concluded.

13.5.2 If the Architect, Owner or public authorities having jurisdiction determine that portions of the Work require additional testing, inspection or approval not included under Subparagraph 13.5.1, the Architect will, upon written authorization from the Owner, instruct the Contractor to make arrangements for such additional testing, inspection or approval by an entity acceptable to the Owner, and the Contractor shall give timely notice to the Architect of when and where tests and inspections are to be made so the Architect may observe such procedures.

AIA DOCUMENT A201 • GENERAL CONDITIONS OF THE CONTRACT FOR CONSTRUCTION • FOURTEENTH EDITION
AIA® • ©1987 THE AMERICAN INSTITUTE OF ARCHITECTS, 1735 NEW YORK AVENUE, N.W., WASHINGTON, D.C. 20006

The Owner shall bear such costs except as provided in Subparagraph 13.5.3.

13.5.3 If such procedures for testing, inspection or approval under Subparagraphs 13.5.1 and 13.5.2 reveal failure of the portions of the Work to comply with requirements established by the Contract Documents, the Contractor shall bear all costs made necessary by such failure including those of repeated procedures and compensation for the Architect's services and expenses.

13.5.4 Required certificates of testing, inspection or approval shall, unless otherwise required by the Contract Documents, be secured by the Contractor and promptly delivered to the Architect.

13.5.5 If the Architect is to observe tests, inspections or approvals required by the Contract Documents, the Architect will do so promptly and, where practicable, at the normal place of testing.

13.5.6 Tests or inspections conducted pursuant to the Contract Documents shall be made promptly to avoid unreasonable delay in the Work.

13.6   INTEREST

13.6.1 Payments due and unpaid under the Contract Documents shall bear interest from the date payment is due at such rate as the parties may agree upon in writing or, in the absence thereof, at the legal rate prevailing from time to time at the place where the Project is located.

13.7   COMMENCEMENT OF STATUTORY LIMITATION PERIOD

13.7.1 As between the Owner and Contractor:

> **.1 Before Substantial Completion.** As to acts or failures to act occurring prior to the relevant date of Substantial Completion, any applicable statute of limitations shall commence to run and any alleged cause of action shall be deemed to have accrued in any and all events not later than such date of Substantial Completion;

> **.2 Between Substantial Completion and Final Certificate for Payment.** As to acts or failures to act occurring subsequent to the relevant date of Substantial Completion and prior to issuance of the final Certificate for Payment, any applicable statute of limitations shall commence to run and any alleged cause of action shall be deemed to have accrued in any and all events not later than the date of issuance of the final Certificate for Payment; and

> **.3 After Final Certificate for Payment.** As to acts or failures to act occurring after the relevant date of issuance of the final Certificate for Payment, any applicable statute of limitations shall commence to run and any alleged cause of action shall be deemed to have accrued in any and all events not later than the date of any act or failure to act by the Contractor pursuant to any warranty provided under Paragraph 3.5, the date of any correction of the Work or failure to correct the Work by the Contractor under Paragraph 12.2, or the date of actual commission of any other act or failure to perform any duty or obligation by the Contractor or Owner, whichever occurs last.

ARTICLE 14

TERMINATION OR SUSPENSION OF THE CONTRACT

14.1   TERMINATION BY THE CONTRACTOR

14.1.1 The Contractor may terminate the Contract if the Work is stopped for a period of 30 days through no act or fault of the Contractor or a Subcontractor, Sub-subcontractor or their agents or employees or any other persons performing portions of the Work under contract with the Contractor, for any of the following reasons:

> **.1** issuance of an order of a court or other public authority having jurisdiction;

> **.2** an act of government, such as a declaration of national emergency, making material unavailable;

> **.3** because the Architect has not issued a Certificate for Payment and has not notified the Contractor of the reason for withholding certification as provided in Subparagraph 9.4.1, or because the Owner has not made payment on a Certificate for Payment within the time stated in the Contract Documents;

> **.4** if repeated suspensions, delays or interruptions by the Owner as described in Paragraph 14.3 constitute in the aggregate more than 100 percent of the total number of days scheduled for completion, or 120 days in any 365-day period, whichever is less; or

> **.5** the Owner has failed to furnish to the Contractor promptly, upon the Contractor's request, reasonable evidence as required by Subparagraph 2.2.1.

14.1.2 If one of the above reasons exists, the Contractor may, upon seven additional days' written notice to the Owner and Architect, terminate the Contract and recover from the Owner payment for Work executed and for proven loss with respect to materials, equipment, tools, and construction equipment and machinery, including reasonable overhead, profit and damages.

14.1.3 If the Work is stopped for a period of 60 days through no act or fault of the Contractor or a Subcontractor or their agents or employees or any other persons performing portions of the Work under contract with the Contractor because the Owner has persistently failed to fulfill the Owner's obligations under the Contract Documents with respect to matters important to the progress of the Work, the Contractor may, upon seven additional days' written notice to the Owner and the Architect, terminate the Contract and recover from the Owner as provided in Subparagraph 14.1.2.

14.2   TERMINATION BY THE OWNER FOR CAUSE

14.2.1 The Owner may terminate the Contract if the Contractor:

> **.1** persistently or repeatedly refuses or fails to supply enough properly skilled workers or proper materials;

> **.2** fails to make payment to Subcontractors for materials or labor in accordance with the respective agreements between the Contractor and the Subcontractors;

> **.3** persistently disregards laws, ordinances, or rules, regulations or orders of a public authority having jurisdiction; or

> **.4** otherwise is guilty of substantial breach of a provision of the Contract Documents.

14.2.2 When any of the above reasons exist, the Owner, upon certification by the Architect that sufficient cause exists to justify such action, may without prejudice to any other rights or remedies of the Owner and after giving the Contractor and the Contractor's surety, if any, seven days' written notice, terminate employment of the Contractor and may, subject to any prior rights of the surety:

.1 take possession of the site and of all materials, equipment, tools, and construction equipment and machinery thereon owned by the Contractor;

.2 accept assignment of subcontracts pursuant to Paragraph 5.4; and

.3 finish the Work by whatever reasonable method the Owner may deem expedient.

14.2.3 When the Owner terminates the Contract for one of the reasons stated in Subparagraph 14.2.1, the Contractor shall not be entitled to receive further payment until the Work is finished.

14.2.4 If the unpaid balance of the Contract Sum exceeds costs of finishing the Work, including compensation for the Architect's services and expenses made necessary thereby, such excess shall be paid to the Contractor. If such costs exceed the unpaid balance, the Contractor shall pay the difference to the Owner. The amount to be paid to the Contractor or Owner, as the case may be, shall be certified by the Architect, upon application, and this obligation for payment shall survive termination of the Contract.

14.3 SUSPENSION BY THE OWNER FOR CONVENIENCE

14.3.1 The Owner may, without cause, order the Contractor in writing to suspend, delay or interrupt the Work in whole or in part for such period of time as the Owner may determine.

14.3.2 An adjustment shall be made for increases in the cost of performance of the Contract, including profit on the increased cost of performance, caused by suspension, delay or interruption. No adjustment shall be made to the extent:

.1 that performance is, was or would have been so suspended, delayed or interrupted by another cause for which the Contractor is responsible; or

.2 that an equitable adjustment is made or denied under another provision of this Contract.

14.3.3 Adjustments made in the cost of performance may have a mutually agreed fixed or percentage fee.

STANDARD FORM 254 ARCHITECT-ENGINEER AND RELATED SERVICES QUESTIONNAIRE

Form Approved
OMB No. 9000–0004

STANDARD FORM (SF)

254

Architect-Engineer and Related Services Questionnaire

Purpose:

The policy of the Federal Government in acquiring architectural, engineering, and related professional services is to encourage firms lawfully engaged in the practice of those professions to submit annually a statement of qualifications and performance data. Standard Form 254, "Architect-Engineer and Related Services Questionnaire" is provided for that purpose. Interested A-E firms (including new, small, and/or minority firms) should complete and file SF 254's with the A-E is qualified to perform services. The agency head for each proposed project shall evaluate these qualification resumes, together with any other performance data on file or requested by the agency, in relation to the proposed project. The SF 254 may be used as a basis for selecting firms for discussions, or for screening firms preliminary to inviting submission of additional information.

Definitions:

"**Architect-engineer and related services**" are those professional services associated with research, development, design and construction, alteration, or repair of real property, as well as incidental services that members of these professions and those in their employ may logically or justifiably perform, including studies, investigations, surveys, evaluations, consultations, planning, programming, conceptual designs, plans and specifications, cost estimates, inspections, shop drawing reviews, sample recommendations, preparation of operating and maintenance manuals, and other related services.

"**Parent Company**" is that firm, company, corporation, association or conglomerate which is the major stockholder or highest tier owner of the firm completing this questionnaire, i.e. Firm A is owned by Firm B which is, in turn, a subsidiary of Corporation C. The "parent company" of Firm A is Corporation C.

"**Principals**" are those individuals in a firm who possess legal responsibility for its management. They may be owners, partners, corporate officers, associates, administrators, etc.

"**Discipline**," as used in this questionnaire, refers to the primary technological capability of individuals in the responding firm. Possession of an academic degree, professional registration, certification, or extensive experience in a particular field of practice normally reflects an individual's primary technical discipline.

"**Joint Venture**" is a collaborative undertaking by two or more firms or individuals for which the participants are both jointly and individually responsible.

"**Consultant**," as used in this questionnaire, is a highly specialized individual or firm having significant input and responsibility for certain aspects of a project and possessing unusual or unique capabilities for assuring success of the finished work.

"**Prime**" refers to that firm which may be coordinating the concerted and complementary inputs of several firms, individuals or related services to produce a completed study or facility. The "prime" would normally be

regarded as having full responsibility and liability for quality of performance by itself as well as by subcontractor professionals under its jurisdiction.

"**Branch Office**" is a satellite, or subsidiary extension, of a headquarters office of a company, regardless of any differences in name or legal structure of such a branch due to local or state laws. "Branch offices" are normally subject to the management decisions, bookkeeping, and policies of the main office.

Instructions for Filing (Numbers below correspond to numbers contained in form):

1. Type accurate and complete name of submitting firm, its address, and zip code.

  1a. Indicate whether form is being submitted in behalf of a parent firm or a branch office. (Branch office submissions should list only personnel in, and experience of, that office.)

2. Provide date the firm was established under the name shown in question 1.

3. Show date on which form is prepared. All information submitted shall be current and accurate as of this date.

4. Enter type of ownership, or legal structure, of firm (sole proprietor, partnership, corporation, joint venture, etc.)

  Check appropriate boxes indicating if firm is (a) a small business concern; (b) a small business concern owned and operated by socially and economically disadvantaged individuals; and (c) Women-owned: (See 48 CFR 19.101 and 52.219–9).

5. Branches of subsidiaries of large or parent companies, or conglomerates, should insert name and address of highest-tier owner.

  5a. If present firm is the successor to, or outgrowth of, one or more predecessor firms, show name(s) of former entity(ies) and the year(s) of their original establishment

6. List not more than two principals from submitting firm who may be contacted by the agency receiving this form. (Different principals may be listed on forms going to another agency.) Listed principals must be empowered to speak for the firm on policy and contractual matters.

7. Beginning with the submitting office, list name, location, total number of personnel and telephone numbers for all associated or branch offices, (including any headquarters or foreign offices) which provide A-E and related services.

  7a. Show total personnel in all offices. (Should be sum of all personnel, all branches.)

8. Show total number of employees, by discipline, in submitting office. (If firm is being submitted by main or headquarters office, firm should list total employees, by discipline, in all offices.) While some personnel may be qualified in several disciplines, each person should be counted only once in accord with his or her primary function. Include clerical personnel as "administrative." Write in any additional disciplines—sociologists, biologists, etc.—and number of people in each, in blank spaces.

9. Using chart (below) insert appropriate index number to indicate range of professional services fees received by submitting firm each calendar year for last five years, most recent year first. Fee summaries should be broken down to

NSN 7540-01-152-8073

STANDARD FORM 254 (Rev 10-83)
PRESCRIBED BY GSA, FAR (48 CFR) 53.236-2(b)

Architect-Engineer and Related Services Questionnaire

reflect the fees received each year for (a) work performed directly for the Federal Government (not including grant and loan projects) or as a sub to other professionals performing work directly for the Federal Government; (b) all other domestic work, U.S. and possessions, including Federally-assisted projects, and (c) all other foreign work.

Ranges of Professional Services Fees

INDEX		INDEX	
1.	Less than $100,000	5.	$1 million to $2 million
2.	$100,000 to $250,000	6.	$2 million to $5 million
3.	$250,000 to $500,000	7.	$5 million to $10 million
4.	$500,000 to $1 million	8.	$10 million or greater

10. Select and enter, in numerical sequence, **not more than thirty** (30) "Experience Profile Code" numbers from the listing (next page) which most accurately reflect submitting firm's demonstrated technical capabilities and project experience. **Carefully review list.** (It is recognized some profile codes may be part of other services or projects contained on list; firms are encouraged to select profile codes which best indicate type and scope of services provided on past projects.) For each code number, show total number of projects and gross fees (in thousands) received for profile projects performed by firm during past few years. If firm has one or more capabilities not included on list, insert same in blank spaces at end of list and show numbers in question 10 on the form. In such cases, the filled in listing **must** accompany the complete SF 254 when submitted to the Federal agencies.

11. Using the "Experience Profile Code" numbers in the same sequence as entered in item 10, give details of at least one recent (within last five years) representative project for each code number, up to a **maximum** of thirty (30) separate projects, or portions of projects, for which firm was responsible. (Project examples may be used more than once to illustrate different services rendered on the same job. Example: a dining hall may be part of an auditorium or educational facility.) Firms which select less than thirty "profile codes" may list two or more project examples (to illustrate specialization) for each code number so long as total of all project examples does not exceed thirty (30). After each code number in question 11, show: (a) whether firm was "P," the prime professional, or "C," a consultant, or "JV," part of a joint venture on that particular project (New firms, in existence less than five (5) years may use the symbol "IE" to indicate "Individual Experience" as opposed to firm experience); (b) provide name and location of the specific project which typifies firm's (or individual's) performance under that code category; (c) give name and address of the owner of that project (If government agency indicate responsible office); (d) show the estimated construction cost (or other applicable cost) for that portion of the project for which the firm was primarily responsible. (Where no construction was involved, show approximate cost of firm's work); and (e) state year work on that particular project was, or will be, completed.

12. The completed SF 254 should be signed by a principal of the firm, preferably the chief executive officer.

13. Additional data, brochures, photos, etc. should not accompany this form unless specifically requested.

NEW FIRMS (not reorganized or recently-amalgamated firms) are eligible and encouraged to seek work from the Federal Government in connection with performance of projects for which they are qualified. Such firms are encouraged to complete and submit Standard Form 254 to appropriate agencies. Questions on the form dealing with personnel or experience may be answered by citing experience and capabilities of individuals in the firm, based on performance and responsibility while in the employ of others. In so doing, notation of this fact should be made on the form. In question 9, write in "N/A" to indicate "not applicable" for those years prior to firm's organization.

STANDARD FORM (SF) 254 Architect-Engineer and Related Services Questionnaire	1. Firm Name / Business Address:	2. Year Present Firm Established:	3. Date Prepared:

1a. Submittal is for ☐ Parent Company ☐ Branch or Subsidiary Office

4. Specify type of ownership *and* check below, if applicable.

☐ A. Small Business
☐ B. Small Disadvantaged Business
☐ C. Woman-owned Business

5. Name of Parent Company, if any:

5a. Former Parent Company Name(s), if any, and Year(s) Established:

6. Names of not more than Two Principals to Contact: Title / Telephone

1)
2)

7. Present Offices: City / State / Telephone / No. Personnel Each Office

7a. Total Personnel

8. Personnel by Discipline: *(List each person only once, by primary function.)*

___ Administrative	___ Electrical Engineers	___ Oceanographers
___ Architects	___ Estimators	___ Planners: Urban/Regional
___ Chemical Engineers	___ Geologists	___ Sanitary Engineers
___ Civil Engineers	___ Hydrologists	___ Soils Engineers
___ Construction Inspectors	___ Interior Designers	___ Specification Writers
___ Draftsmen	___ Landscape Architects	___ Structural Engineers
___ Ecologists	___ Mechanical Engineers	___ Surveyors
___ Economists	___ Mining Engineers	___ Transportation Engineers

9. Summary of Professional Services Fees Received: (Insert index number)

Last 5 Years (most recent year first)

Direct Federal contract work, including overseas   19___ 19___ 19___ 19___ 19___
All other domestic work
All other foreign work*

*Firms interested in foreign work, but without such experience, check here: ☐

Ranges of Professional Services Fees

INDEX

1. Less than $100,000
2. $100,000 to $250,000
3. $250,000 to $500,000
4. $500,000 to $1 million
5. $1 million to $2 million
6. $2 million to $5 million
7. $5 million to $10 million
8. $10 million or greater

STANDARD FORM 254 (REV. 10-83)

Experience Profile Code Numbers
for use with questions 10 and 11

001 Acoustics; Noise Abatement
002 Aerial Photogrammetry
003 Agricultural Development; Grain Storage; Farm Mechanization
004 Air Pollution Control
005 Airports; Navaids; Airport Lighting; Aircraft Fueling
006 Airports; Terminals & Hangars; Freight Handling
007 Arctic Facilities
008 Auditoriums & Theatres
009 Automation; Controls; Instrumentation
010 Barracks; Dormitories
011 Bridges
012 Cemeteries (Planning & Relocation)
013 Chemical Processing & Storage
014 Churches; Chapels
015 Codes; Standards; Ordinances
016 Cold Storage; Refrigeration; Fast Freeze
017 Commercial Buildings (low rise); Shopping Centers
018 Communications Systems; TV; Microwave
019 Computer Facilities; Computer Service
020 Conservation and Resource Management
021 Construction Management
022 Corrosion Control; Cathodic Protection; Electrolysis
023 Cost Estimating
024 Dams (Concrete; Arch)
025 Dams (Earth; Rock); Dikes; Levees
026 Desalination (Process & Facilities)
027 Dining Halls; Clubs; Restaurants
028 Ecological & Archeological Investigations
029 Educational Facilities; Classrooms
030 Electronics
031 Elevators; Escalators; People-Movers
032 Energy Conservation; New Energy Sources
033 Environmental Impact Studies, Assessments or Statements
034 Fallout Shelters; Blast-Resistant Design
035 Field Houses; Gyms; Stadiums
036 Fire Protection
037 Fisheries; Fish Ladders
038 Forestry & Forest Products
039 Garages; Vehicle Maintenance Facilities; Parking Decks
040 Gas Systems (Propane; Natural, Etc.)
041 Graphic Design

042 Harbors; Jetties; Piers; Ship Terminal Facilities
043 Heating; Ventilating; Air Conditioning
044 Health Systems Planning
045 Highrise; Air-Rights-Type Buildings
046 Highways; Streets; Airfield Paving; Parking Lots
047 Historical Preservation
048 Hospital & Medical Facilities
049 Hotels; Models
050 Housing (Residential, Multi-Family; Apartments; Condominiums)
051 Hydraulics & Pneumatics
052 Industrial Buildings; Manufacturing Plants
053 Industrial Processes; Quality Control
054 Industrial Waste Treatment
055 Interior Design; Space Planning
056 Irrigation; Drainage
057 Judicial and Courtroom Facilities
058 Laboratories; Medical Research Facilities
059 Landscape Architecture
060 Libraries; Museums; Galleries
061 Lighting (Interiors; Display; Theatre, Etc.)
062 Lighting (Exteriors; Streets; Memorials; Athletic Fields, Etc.)
063 Materials Handling Systems; Conveyors; Sorters
064 Metallurgy
065 Microclimatology; Tropical Engineering
066 Military Design Standards
067 Mining & Mineralogy
068 Missile Facilities (Silos; Fuels; Transport)
069 Modular Systems Design; Pre-Fabricated Structures or Components
070 Naval Architecture; Off-Shore Platforms
071 Nuclear Facilities; Nuclear Shielding
072 Office Buildings; Industrial Parks
073 Oceanographic Engineering
074 Ordnance; Munitions; Special Weapons
075 Petroleum Exploration; Refining
076 Petroleum and Fuel (Storage and Distribution)
077 Pipelines (Cross-Country—Liquid & Gas)
078 Planning (Community, Regional, Areawide and State)
079 Planning (Site, Installation, and Project)
080 Plumbing & Piping Design
081 Pneumatic Structures; Air-Support Buildings
082 Postal Facilities
083 Power Generation, Transmission, Distribution
084 Prisons & Correctional Facilities
085 Product, Machine & Equipment Design

086 Radar; Sonar; Radio & Radar Telescopes
087 Railroad; Rapid Transit
088 Recreation Facilities (Parks, Marinas, Etc.)
089 Rehabilitation (Buildings; Structures; Facilities)
090 Resource Recovery; Recycling
091 Radio Frequency Systems & Shieldings
092 Rivers; Canals; Waterways; Flood Control
093 Safety Engineering; Accident Studies; OSHA Studies
094 Security Systems; Intruder & Smoke Detection
095 Seismic Designs & Studies
096 Sewage Collection, Treatment and Disposal
097 Soils & Geologic Studies; Foundations
098 Solar Energy Utilization
099 Solid Wastes; Incineration; Land Fill
100 Special Environments; Clean Rooms, Etc.
101 Structural Design; Special Structures
102 Surveying; Platting; Mapping; Flood Plain Studies
103 Swimming Pools
104 Storm Water Handling & Facilities
105 Telephone Systems (Rural; Mobile; Intercom, Etc.)
106 Testing & Inspection Services
107 Traffic & Transportation Engineering
108 Towers (Self-Supporting & Guyed Systems)
109 Tunnels & Subways
110 Urban Renewals; Community Development
111 Utilities (Gas & Steam)
112 Value Analysis; Life-Cycle Costing
113 Warehouses & Depots
114 Water Resources; Hydrology; Ground Water
115 Water Supply, Treatment and Distribution
116 Wind Tunnels; Research/Testing Facilities Design
117 Zoning; Land Use Studies
201 _____
202 _____
203 _____
204 _____
205 _____

STANDARD FORM 254 (REV 10-83)

3

10. Profile of Firm's Project Experience, Last 5 Years

Profile Code	Number of Projects	Total Gross Fees (in thousands)	Profile Code	Number of Projects	Total Gross Fees (in thousands)	Profile Code	Number of Projects	Total Gross Fees (in thousands)
1)			11)			21)		
2)			12)			22)		
3)			13)			23)		
4)			14)			24)		
5)			15)			25)		
6)			16)			26)		
7)			17)			27)		
8)			18)			28)		
9)			19)			29)		
10)			20)			30)		

11. Project Examples, Last 5 Years

Profile Code	"P", "C", "JV", or "IE"	Project Name and Location	Owner Name and Address	Cost of Work (in thousands)	Completion Date (Actual or Estimated)
		1			
		2			
		3			
		4			

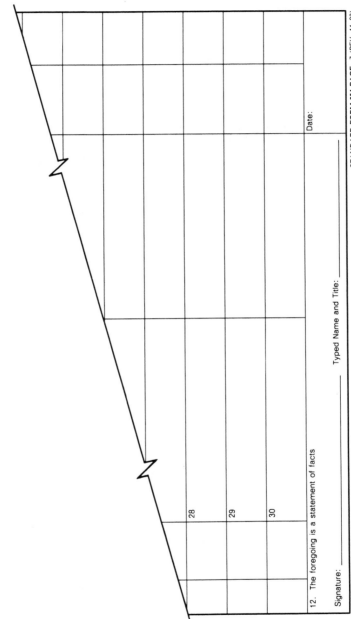

28

29

30

12. The foregoing is a statement of facts

Signature: _____ Typed Name and Title: _____

Date:

APPENDIX D

STANDARD FORM 255 ARCHITECT-ENGINEER AND RELATED SERVICES QUESTIONNAIRE FOR SPECIFIC PROJECT

STANDARD
FORM (SF)

255

Architect-Engineer and Related Services Questionnaire for Specific Project

Purpose:

This form is a supplement to the "Architect-Engineer and Related Services Questionnaire" (SF 254). Its purpose is to provide additional information regarding the qualifications of interested firms to undertake a specific Federal A-E project. Firms, or branch offices of firms, submitting this form should enclose (or already have on file with the appropriate office of the agency) a current (within the past year) and accurate copy of the SF 254 for that office.

The procurement official responsible for each proposed project may request submission of the SF 255 "Architect-Engineer and Related Services Questionnaire for Specific Project" in accord with applicable civilian and military procurement regulations and shall evaluate such submissions, as well as other related information contained on the Standard Form 254, and any other performance data on file with the agency, and shall select firms for subsequent discussions leading to contract award in conformance with Public Law 92-582. This form should only be filed by an architect engineer or related services firm when requested to do so by the agency or by a public announcement. Responses should be as complete and accurate as possible, contain data relative to the specific project for which you wish to be considered, and should be provided, by the required due date, to the office specified in the request or public announcement.

This form will be used only for the specified project. Do not refer to this submittal in response to other requests or public announcements.

Definitions:

"Architect-engineer and related services" are those professional services associated with research, development, design and construction, alteration, or repair of real property, as well as incidental services that members of these professions and those in their employ may logically or justifiably perform, including studies, investigations, surveys, evaluations, consultations, planning, programming, conceptual designs, plans and specifications, cost estimates, inspections, shop drawing reviews, sample recommendations, preparation of operating and maintenance manuals, and other related services.

"Principals" are those individuals in a firm who possess legal responsibility for its management. They may be owners, partners, corporate officers, associates, administrators, etc.

"Discipline", as used in this questionnaire, refers to the primary technological capability of individuals in the responding firm. Possession of an academic degree, professional registration, certification, or extensive experience in a particular field of practice normally reflects an individual's primary technical discipline.

"Joint Venture", is a collaborative undertaking of two or more firms or individuals for which the participants are both jointly and individually responsible.

"Key Persons, Specialists, and Individual Consultants", as used in this questionnaire, refer to individuals who will have major project responsibility or will provide unusual or unique capabilities for the project under consideration.

Instructions for Filing (Numbers below correspond to numbers contained in form):

1. Give name and location of the project for which this form is being submitted.

2. Provide appropriate data from the Commerce Business Daily (CBD) identifying the particular project for which this form is being filed.

   2a. Give the date of the Commerce Business Daily in which the project announcement appeared, or indicate "not applicable" (N/A) if the source of the announcement is other than the CBD.

   2b. Indicate Agency identification or contract number as provided in the CBD announcement.

3. Show name and address of the individual or firm (or joint venture) which is submitting this form for the project.

   3a. List the name, title, and telephone number of that principal who will serve as the point of contact. Such an individual must be empowered to speak for the firm on policy and contractual matters and should be familiar with the programs and procedures of the agency to which this form is directed.

   3b. Give the address of the specific office which will have responsibility for performing the announced work.

4. Insert the number of personnel by discipline presently employed (on date of this form) at office specified in block 3b. While some personnel may be qualified in several disciplines, each person should be counted only once in accord with his or her primary function. Include clerical personnel as "administrative." Write in any additional disciplines—sociologists, biologists, etc.—and number of people in each, in blank spaces.

5. Answer only if this form is being submitted by a joint venture of two or more collaborating firms. Show the names and addresses of all individuals or organizations expected to be included as part of the joint venture and describe their particular areas of anticipated responsibility, (i.e. technical disciplines, administration, financial, sociological, environmental, etc.)

   5a. Indicate, by checking the appropriate box, whether this particular joint venture has worked together on other projects.

STANDARD FORM 255 (Rev. 10-83)
PRESCRIBED BY GSA, FAR (48 CFR) 53.236-2(c)

NSN 7540-01-152-8074

1

483

Architect-Engineer and Related Services Questionnaire for Specific Project

Standard Form 255
General Services Administration,
Washington, D. C. 20405
Fed. Proc. Reg (41 CFR) 1-16 . 803
Armed Svc. Proc. Reg. 18-403

Each firm participating in the joint venture should have a Standard Form 254 on file with the contracting office receiving this form. Firms which do not have such forms on file should provide same immediately along with a notation at the top of page 1 of the form regarding their association with this joint venture submittal.

6. If respondent is not a joint venture, but intends to use outside (as opposed to in-house or permanently and formally affiliated) consultants or associates, he should provide names and addresses of all such individuals or firms, as well as their particular areas of technical/professional expertise, as it relates to this project. Existence of previous working relationships should be noted. If more than eight outside consultants or associates are anticipated, attach an additional sheet containing requested information.

7. Regardless of whether respondent is a joint venture or an independent firm, provide brief resumes of key personnel expected to participate on this project. Care should be taken to limit resumes to only those personnel and specialists who will have major project responsibilities. Each resume must include: (a) name of each key person and specialist and his or her title, (b) the project assignment or role which that person will be expected to fulfill in connection with this project, (c) the name of the firm or organization, if any, with whom that individual is presently associated, (d) years of relevant experience with present firm and other firms, (e) the highest academic degree achieved and the discipline covered (if more than one highest degree, such as two Ph.D.'s, list both), the year received and the particular technical/professional discipline which that individual will bring to the project, (f) if registered as an architect, engineer, surveyor, etc., show only the field of registration and the year that such registration was first acquired. If registered in several states, do not list states, and (g) a synopsis of experience, training, or other qualities which reflect individual's potential contribution to this project. Include such data as: familiarity with Government or agency procedures, similar type of work performed in the past, management abilities, familiarity with the geographic area, relevant foreign language capabilities, etc. Please lim... synopsis of experience to directly relevant information.

8. List up to ten projects which demonstrate the firm's or joint venture's competence to perform work similar to that likely to be required on this project. The more recent such projects, the better. Prime consideration will be given to

projects which illustrate respondent's capability for performing work similar to that being sought. Required information must include: (a) name and location of project, (b) brief description of type and extent of services provided for each project (submissions by joint ventures should indicate which member of the joint venture was the prime on that particular project and what role it played), (c) name and address of the owner of that project (if Government agency, indicate responsible office), (d) completion date (actual when available, otherwise estimated), (e) total construction cost of completed project (or where no construction was involved, the approximate cost of your work) and that portion of the cost of the project for which the named firm was/is responsible.

9. List only those projects which the A·E firm or joint venture, or members of the joint venture, are currently performing under direct contract with an agency or department of the Federal Government. Exclude any grant or loan projects being financed by the Federal Government but being performed under contract to other non Federal governmental entities. Information provided under each heading is similar to that requested in the preceding Item 8, except for (d) "Percent Complete." Indicate in this item the percentage of A·E work completed upon filing this form.

10. Through narrative discussion, show reason why the firm or joint venture submitting this questionnaire believes it is especially qualified to undertake the project. Information provided should include, but not be limited to, such data as: specialized equipment available for this work, any awards or recognition received by a firm or individuals for similar work, required security clearances, special approaches or concepts developed by the firm relevant to this project, etc. Respondents may say anything they wish in support of their qualifications. When appropriate, respondents may supplement this proposal with graphic material and photographs which best demonstrate design capabilities of the team proposed for this project.

11. Completed forms should be signed by the chief executive officer of the joint venture (thereby attesting to the concurrence and commitment of all members of the joint venture), or by the architect-engineer principal responsible for the conduct of the work in the event it is awarded to the organization submitting this form. Joint ventures selected for subsequent discussions regarding this project must make available a statement of participation signed by a principal of each member of the joint venture. ALL INFORMATION CONTAINED IN THE FORM SHOULD BE CURRENT AND FACTUAL.

OMB Approval No. **3090-0029**

| STANDARD FORM (SF) **255** Architect-Engineer Related Services for Specific Project | 1. Project Name / Location for which Firm is Filing: | 2a. *Commerce Business Daily* Announcement Date, if any: | 2b. Agency Identification Number, if any: |

3. Firm (or Joint-Venture) Name & Address

3a. Name, Title & Telephone Number of Principal to Contact

3b. Address of office to perform work, if different from Item 3

4. Personnel by Discipline: (List each person only once, by primary function.)

____ Administrative
____ Architects
____ Chemical Engineers
____ Civil Engineers
____ Construction Inspectors
____ Draftsmen
____ Ecologists
____ Economists
____ Electrical Engineers
____ Estimators
____ Geologists
____ Hydrologists
____ Interior Designers
____ Landscape Architects
____ Mechanical Engineers
____ Mining Engineers
____ Oceanographers
____ Planners: Urban/Regional
____ Sanitary Engineers
____ Soils Engineers
____ Specification Writers
____ Structural Engineers
____ Surveyors
____ Transportation Engineers
____ Total Personnel

5. If submittal is by JOINT-VENTURE list participating firms and outline specific areas of responsibility (including administrative, technical and financial) for each firm: (Attach SF 254 for each if not on file with Procuring Office.)

5a. Has this Joint-Venture previously worked together?   ☐ yes   ☐ no

STANDARD FORM 255 (Rev. 10-83)

3

485

6. If respondent is not a joint-venture, list outside key Consultants/Associates anticipated for this project (Attach SF 254 for Consultants/Associates listed, if not already on file with the Contracting Office).

Name & Address	Specialty	Worked with Prime before (Yes or No)
1)		
2)		
3)		
4)		
5)		
6)		
7)		
8)		

STANDARD FORM 255 (Rev. 10-83)

4

7. Brief resume of key persons, specialists, and individual consultants anticipated for this project.

a. Name & Title:	a. Name & Title:
b. Project Assignment:	b. Project Assignment:
c. Name of Firm with which associated:	c. Name of Firm with which associated:
d. Years experience: With This Firm ____ With Other Firms ____	d. Years experience: With This Firm ____ With Other Firms ____
e. Education: Degree(s) / Year / Specialization	e. Education: Degree(s) / Years / Specialization
f. Active Registration: Year First Registered/Discipline	f. Active Registration: Year First Registered/Discipline
g. Other Experience and Qualifications relevant to the proposed project:	g. Other Experience and Qualifications relevant to the proposed project:

STANDARD FORM 255 (Rev. 10-83)

REPEAT PAGE AS NEEDED

7. Brief resume of key persons, specialists, and individual consultants anticipated for this project.

a. Name & Title:	a. Name & Title:
b. Project Assignment:	b. Project Assignment:
c. Name of Firm with which associated:	c. Name of Firm with which associated:
d. Years experience: With This Firm ____ With Other Firms ____	d. Years experience: With This Firm ____ With Other Firms ____
e. Education: Degree(s) / Year / Specialization	e. Education: Degree(s) / Years / Specialization
f. Active Registration: Year First Registered/Discipline	f. Active Registration: Year First Registered/Discipline
g. Other Experience and Qualifications relevant to the proposed project:	g. Other Experience and Qualifications relevant to the proposed project:

STANDARD FORM 255 (Rev. 10-83)

6

8. Work by firm or joint-venture members which best illustrates current qualifications relevant to this project (list not more than 10 projects).

a. Project Name & Location	b. Nature of Firm's Responsibility	c. Project Owner's Name & Address	d. Completion Date (actual or estimated)	e. Estimated Cost (in thousands)	
				Entire Project	Work for which Firm was/is responsible
(1)					
(2)					
(3)					
(4)					
(5)					
(6)					
(7)					
(8)					
(9)					
(10)					

STANDARD FORM 255 (Rev. 10-83)

9

9. All work by firms or joint-venture members currently being performed directly for Federal agencies.

a. Project Name & Location	b. Nature of Firm's Responsibility	c. Agency (Responsible Office) Name & Address	d. Percent complete	e. Estimated Cost (In Thousands)	
				Entire Project	Work for which firm is responsible

10. Use this space to provide any additional information or description of resources (including any computer design capabilities) supporting your firm's qualifications for the proposed project.

11. The foregoing is a statement of facts.

Signature: _____ Typed Name and Title: _____ Date:

★ U.S. GOVERNMENT PRINTING OFFICE: 1988-211-351

STANDARD FORM 255 (Rev. 10-83)

SELECTED REFERENCES

Bes, J. *Chartering and Shipping Terms*. Vol. 1, 10th ed. Baker and Howard Ltd., London, 1977.

Brimson, James A. *Activity Accounting: An Activity-Based Costing Approach*. Wiley, New York, 1991.

Burr, Irving W. *Elementary Statistical Quality Control*. Marcel Dekker, New York, 1979.

Clough, Richard H. *Construction Contracting*. 5th ed. Wiley, New York, 1986.

Corley, R. N., and W. J. Robert. *Principles of Business Law*. Prentice-Hall, New Jersey, 1988.

Deming, W. Edwards. *Out of the Crisis*. Center for Advanced Engineering Study, Cambridge, Mass. 1986.

Drucker, Peter F. *Management, Tasks, Responsibilities, Practices*. Harper, New York, 1985.

Granholm, Axel R. *Human Resource Director's Portfolio of Personnel Forms, Records, and Reports*. Prentice-Hall, New Jersey, 1988.

Heisler, Sanford I. *The Wiley Engineer's Desk Reference*. Wiley, New York, 1984.

Henley, Ernest, J., and Hiromitsu Kumamoto. *Reliability Engineering and Risk Assessment*. Prentice-Hall, New Jersey, 1981.

Hodges, John C., and Mary E. Whitten. *Harbrace College Handbook*. 11th ed. Harcourt, New York, 1990.

Kaye, Harvey. *Inside the Technical Consulting Business*. Wiley, New York, 1986.

Lindert, P. H. *Prices, Jobs, and Growth: An Introduction to Macroeconomics*. Little Brown, Boston, 1976.*

* May not be readily available but worth the search.

Mahoney, William P. *Means Construction Data*. R. S. Means Co. Inc., 100 Construction Plaza, PO Box 800, Kingston, MA 02364. (Updated annually)

McGraw-Hill Information Systems. *The Dodge Construction Cost Data*. McGraw-Hill, Princeton, NJ. (Updated annually)

Messina, William S. *Statistical Quality Control for Manufacturing Managers*. Wiley, New York, 1987.

Owen, Robert R., Daniel R. Garner, and Dennis S. Bunder. *The Arthur Young Guide to Financing for Growth*. Wiley, New York, 1986

Rase, Howard F., and M. H. Barrow. *Project Engineering of Process Plants*. Wiley, New York, 1957

Richardson Engineering Services Inc. *The Richardson Rapid System*, 1742 S. Fraser Dr., PO Box 9103, Mesa, AZ 85214. (Updated annually)

Sayle, Hans, M.D. *Stress without Distress*. NAL-Dutton, New York, 1975.

Smith, Preston, G., and Donald G. Reinertsen. *Developing Products in Half the Time*. Van Nostrand Reinhold, New York, 1991.

Stalk, George, Jr., and Thomas M. Hout. *Competing against Time*. Free Press, New York, 1990.

Standard & Poors. *Industry Surveys*. Standard & Poors, New York. (Updated annually)

Streeter, Harrison. *Professional Liability of Architects and Engineers*. Wiley, New York, 1988.

Stewart, Rodney D. *Cost Estimating*. 2d ed. Wiley, New York, 1991.

Strunk, William, Jr., and E. B. White. *The Elements of Style*. 3d ed. Macmillan, New York, 1979.

Sweets Group. *Sweets Engineering & Retrofit Catalog File*. McGraw-Hill, New York. (Updated annually)

Walker, F. R. *Walker's Building Estimator's Reference Book*. 22d. ed. Frank R. Walker Company, 5030 N. Harlem Ave., Chicago, IL 60656

INDEX